AP

Advanced Placement Testing

Calculus

Thomas Michael Mattson
XAMonline, Inc.

To obtain permission(s) to use the material from this work for any purpose including workshops or seminars, please submit a written request to:

XAMonline, Inc.
21 Orient Avenue
Melrose, MA 02176
Toll Free: 1-800-301-4647
Email: info@xamonline.com
Web: www.xamonline.com
Fax: 1-617-583-5552

Library of Congress Cataloging-in-Publication Data
Mattson, Thomas M.

AP Calculus / Thomas M. Mattson
   ISBN: 978-1-60787-638-0

1. AP   2. Study Guides   3. Mathematics

Disclaimer:

The opinions expressed in this publication are the sole works of XAMonline and were created independently from The College Board, or other testing affiliates. Between the time of publication and printing, specific test standards as well as testing formats and website information may change that are not included in part or in whole within this product. XAMonline develops sample test questions, and they reflect similar content as on real tests; however, they are not former tests. XAMonline assembles content that aligns with test standards but makes no claims nor guarantees candidates a passing score.

Cover photos provided by © CanStockPhoto/sdmix/5783896, CanStockPhoto/airdone/26886042, CanStockPhoto/kmitu/0405102, CanStockPhoto/vtorous/7098607, CanStockPhoto/yurolaitsalbert/16472806

Printed in the United States of America

AP Calculus
ISBN: 978-1-60787-638-0

Contents

PART I:
Introduction

About the Advanced Placement Calculus Program

The Advanced Placement Program® is designed to offer students college credit while still in high school. The more than 30 AP courses culminate in an intensive final exam given every year in May.

AP Calculus AB covers the material that would be offered in the first semester of a freshman-level calculus course, and AP Calculus BC covers the material that would be offered in the first two semesters of such a course. The curriculum requires students to understand calculus via a variety of representations: analytical, numerical, graphical, and verbal. Furthermore, students are required to make connections between these representations. Use of technology in solving calculus problems is also included in the curriculum via graphing calculators.

Successful completion of the course and a passing score on the Exam not only provides students with a deep sense of accomplishment, but also gives them a jumpstart on their college careers. AP credit is almost universally accepted by post-secondary schools, however, each school has different guidelines as to what scores they will accept.

According to College Board, the objectives of the AP Calculus AB/BC course are as follows (from http://media.collegeboard.com/digitalServices/pdf/ap/ap-course-overviews/ap-calculus-ab-course-overview.pdf):

- Work with functions represented in multiple ways: graphical, numerical, analytical, or verbal. They should understand the connections among these representations.
- Understand the meaning of the derivative in terms of a rate of change and local linear approximation and use derivatives to solve problems.
- Understand the meaning of the definite integral as a limit of Riemann sums and as the net accumulation of change and use integrals to solve problems.
- Understand the relationship between the derivative and the definite integral as expressed in both parts of the Fundamental Theorem of Calculus.
- Communicate mathematics and explain solutions to problems verbally and in writing.
- Model a written description of a physical situation with a function, a differential equation, or an integral.
- Use technology to solve problems, experiment, interpret results, and support conclusions.

- Determine the reasonableness of solutions, including sign, size, relative accuracy, and units of measurement.
- Develop an appreciation of calculus as a coherent body of knowledge and as a human accomplishment.

About the Exam

The AP Calculus AB Exam is designed to assess students' grasp of the concepts of calculus, their ability to use their understanding of these concepts to solve problems, and their ability to make connections among the various representations of calculus mentioned above. The AP Calculus BC Exam is designed to accomplish the same ends, but while covering more topics. Students who take the BC Exam are given an AB Exam subscore to notify them of their performance on AB topics. Note: Students are not allowed to take both the AB and BC Exams in the same year.

While the AP Calculus course does cover all of the topics that are tested on both the AB and BC Exams, it should be noted that a student should not be considered prepared unless he or she has also mastered the topics covered in the prerequisite courses. These topics include algebra, geometry, trigonometry, and elementary functions.

The Curriculum

The topics tested on the AP Calculus Exams are as follows (from http://media. collegeboard.com/digitalServices/pdf/ap/ap-course-overviews/ap-calculus-ab-course-overview.pdf):

I. Functions, Graphs, and Limits
- Analysis of Graphs
- Limits of Functions
- Asymptotic and Unbounded Behavior
- Continuity as a Property of Functions
- Parametric, Polar, and Vector Functions (BC Only)

II. Differential Calculus
- Concept of the Derivative
- Derivative at a Point
- Derivative as a Function
- Second Derivatives
- Applications and Computation of Derivatives

III. Integral Calculus
- Interpretations and Properties of Definite Integrals
- Applications of Integrals
- Fundamental Theorem of Calculus
- Techniques and Applications of Antidifferentiation
- Numerical Approximations to Definite Integrals

IV. Polynomial Approximations and Series (BC Only)
- Concept of Series
- Series of Constants
- Taylor Series

The Make Up of the Exam

The AP Calculus Exams are administered every May. They are timed, and they are each 3 hours 15 minutes long. The structure of each exam is as follows.

Section I: Multiple-Choice Questions – 50% of Exam Score
Part A: 28 Questions, 55 Minutes, No Calculator Permitted
Part B: 17 Questions, 50 Minutes, Graphing Calculator Permitted
Section II: Free-Response Questions – 50% of Exam Score
Part A: 2 Problems, 30 Minutes, Graphing Calculator Permitted
Part B: 4 Problems, 60 Minutes, No Calculator Permitted

An example of a Multiple-Choice Question that assesses student's computational skills is given below.

1. Which of the following is equal to $\dfrac{d}{dx}\left[x^2 \cos(3x)\right]$?
   A. $2x \sin(3x) + 3x^2 \cos(3x)$
   B. $2x \sin(3x) - 3x^2 \cos(3x)$
   C. $2x \cos(3x) + 3x^2 \sin(3x)$
   D. $2x \cos(3x) - 3x^2 \sin(3x)$
   E. $2x \cos(3x) - x^2 \sin(3x)$

1. **Answer – D**

   This question requires the student to be able to apply both the Product Rule and the Chain Rule to compute the derivative.

   An example of a Multiple-Choice Question that assesses the student's graphical reasoning skills is given below.

1. Which of the following is a slope field for the differential equation $\dfrac{dy}{dx} = x + y$?

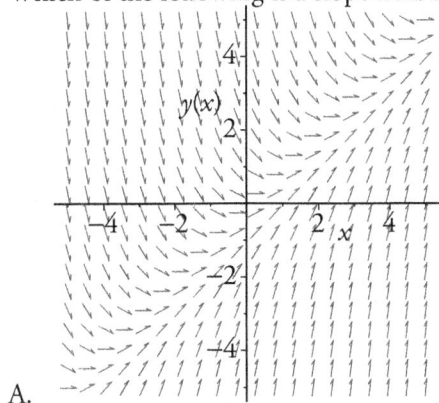

A.

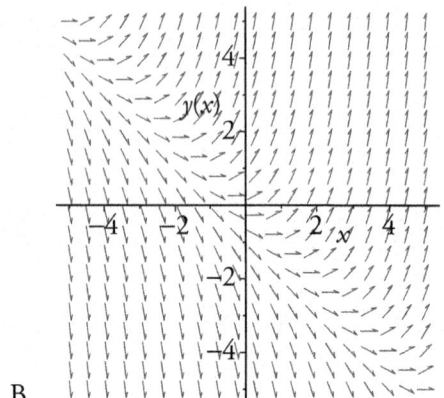

B.

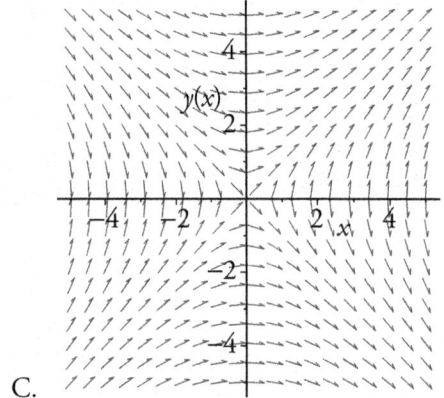

C.

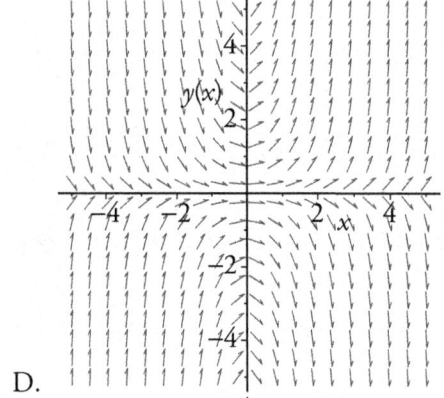

D.

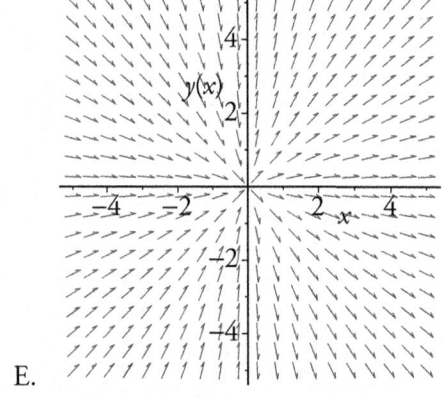

E.

2. **Answer – B**

This question requires the student to be able determine the correct answer by understanding the connection between the differential equation and a visual representation of its solutions. The student is certainly *not* expected to compute the solutions and then plot them. In this way, the second example is of a very different nature than the first.

Permissible Calculators

For two parts of the exam (Part B of Section I and Part A of Section II), you are permitted to use a graphing calculator. Some questions actually *require* one. On the exam, you are encouraged to use a calculator that can perform the following functions.

- Graph a function.
- Find the zeros of a function.
- Numerically calculate derivatives.
- Numerically calculate definite integrals.

You *may not* use a nongraphing scientific calculator or a portable or handheld computer on the exam. In addition to these restrictions, you also may not use any device on the exam that:

- has a QWERTY keyboard,
- uses a stylus or a touch screen for input,
- has wireless or Bluetooth capabilities,
- makes noise,
- requires an electrical outlet,
- has cell phone, audio, or video recording capability,
- is capable of accessing the internet, or
- has camera or scanning capability.

College Board's detailed policies for the AB and BC Exams are at the following web addresses.

https://apstudent.collegeboard.org/apcourse/ap-calculus-ab/calculator-policy

https://apstudent.collegeboard.org/apcourse/ap-calculus-bc/calculator-policy

On these pages you will find a long list of approved calculators.

How the Exam Is Scored

The Multiple-Choice part of the test is scored by machine and the Free-Response portion is scored by hand. Once both scores have been tallied, they are combined and then scaled. This raw score is then changed into a composite score ranging from 1–5.

The College Board proposes the following qualifications for each of the potential score:

Exam Grade	Recommendation
5	Extremely Well Qualified
4	Well Qualified
3	Qualified
2	Possibly Qualified
1	No recommendation

The minimum score required for college credit to be granted is a 3. As mentioned above, many schools require scores of 4 or 5 in order to grant credit.

For comparison, the College Board makes the equivalents of the AP Exam scores at follows:

AP Exam Grade	Letter Grade Equivalent
5	A
4	A–, B+, B
3	B–, C+, C
2	None
1	None

For reference, the 2015 administration of the AP Calculus Exams had the following distributions (from https://apscore.collegeboard.org/scores/about-ap-scores/score-distributions):

Exam Score	Percentage of Students AB Exam	Percentage of Students BC Exam
5	56.1	45.4
4	16.3	16.4
3	12.8	18.0
2	5.5	5.5
1	9.4	14.8

As is evident from the above distributions, students had a high rate of success, with 72.4% earning a score of at least 4 on the AB Exam, and 61.8% earning at least the same mark on the BC Exam.

What to Expect in This Book

As you move forward through these next pages, you will see a variety of information. The first section is a review of major calculus concepts that you should know, or at least with which you should be familiar. You will see they are broken down by main topic (Limits, Derivatives, etc). Additionally, for each topic you will find many fully worked examples.

You will also find some sample tests at the end of this book. These are designed to give you hands-on experience that simulates the actual exam you will be taking. Each question on these tests has a detailed answer as to why it is correct and why the incorrect answers are wrong. Use this information to help guide your learning.

PART II:
Strategies

Strategies for the AP Calculus Exam

Strategy 1: Know Your Stuff

This may seem obvious, but it needs to be mentioned. No amount of coaching will help you on the exam if you don't know the material. Here is a list of things that you should have a working knowledge of before you enter the exam. When we say "working knowledge," we mean that you should know these things like you know your own name. Please be advised that this list is not exhaustive. However, it is a list of things that you cannot do without!

Pre-calculus

Unit circle values

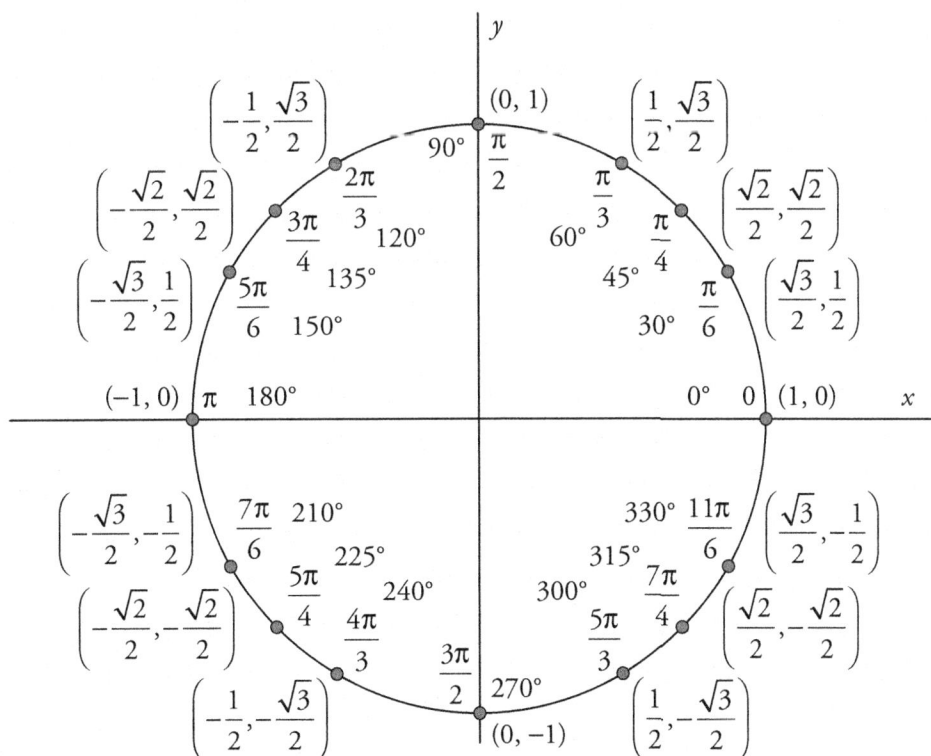

Trigonometric identities

Reciprocal Identities:

$$\csc(\theta) = \frac{1}{\sin(\theta)}$$

$$\sec(\theta) = \frac{1}{\cos(\theta)}$$

$$\cot(\theta) = \frac{1}{\tan(\theta)}$$

Pythagorean Identities:

$$\sin^2(\theta) + \cos^2(\theta) = 1$$

$$\tan^2(\theta) + 1 = \sec^2(\theta)$$

$$\cot^2(\theta) + 1 = \csc^2(\theta)$$

Double Angle Identities

$$\sin(2\theta) = 2\sin(\theta)\cos(\theta)$$

$$\cos(2\theta) = 2\cos^2(\theta) - 1 = 1 - 2\sin^2(\theta)$$

Power Reduction Identities

$$\sin^2(\theta) = \frac{1 - \cos(2\theta)}{2}$$

$$\cos^2(\theta) = \frac{1 + \cos(2\theta)}{2}$$

Limits and Continuity

Continuity at a Point

$f(x)$ is *continuous* at $x = c$ if the following three conditions are satisfied.

1. $f(x)$ is defined at $x = c$.

2. $\lim\limits_{x \to c} f(x)$ exists.

3. $\lim\limits_{x \to c} f(x) = f(c)$

L'Hôpital's Rule

If $f(x)$ and $g(x)$ are differentiable, and $\lim\limits_{x \to c} \dfrac{f(x)}{g(x)}$ yields any of the indeterminate forms $\dfrac{0}{0}, \dfrac{\infty}{\infty}, \dfrac{-\infty}{\infty}, \dfrac{\infty}{-\infty},$ or $\dfrac{-\infty}{-\infty}$, then $\lim\limits_{x \to c} \dfrac{f(x)}{g(x)} = \lim\limits_{x \to c} \dfrac{f'(x)}{g'(x)}$.

Derivatives

Definitions of the Derivative

Derivative as a Function: $\quad f'(x) = \lim\limits_{h \to 0} \dfrac{f(x+h) - f(x)}{h}$

Derivative at a point: $\quad f'(c) = \lim\limits_{x \to c} \dfrac{f(x) - f(c)}{x - c}$

Parametric Form of the Derivative (BC)

If $x = f(t)$ and $y = g(t)$, then $\dfrac{dy}{dx} = \dfrac{dy/dt}{dx/dt} = \dfrac{g'(t)}{f'(t)}$.

The graph of y has a *horizontal tangent* whenever $g'(t) = 0$ and $f'(t) \neq 0$, and a *vertical tangent* whenever $f'(t) = 0$ and $g'(t) \neq 0$

Polar Form of the Derivative (BC)

If $r = f(\theta)$, $x = r\cos(\theta)$, and $y = r\sin(\theta)$, then

$$\frac{dy}{dx} = \frac{dy/d\theta}{dx/d\theta} = \frac{f(\theta)\cos(\theta) + f'(\theta)\sin(\theta)}{-f(\theta)\sin(\theta) + f'(\theta)\cos(\theta)}.$$

The graph of $r = f(\theta)$ has a *horizontal tangent* whenever $\dfrac{dy}{d\theta} = 0$ and $\dfrac{dx}{d\theta} \neq 0$, and a *vertical tangent* whenever $\dfrac{dx}{d\theta} = 0$ and $\dfrac{dy}{d\theta} \neq 0$. $\theta = \alpha$ is a *tangent at the pole* of the graph of $r = f(\theta)$ if $f(\alpha) = 0$ and $f'(\alpha) \neq 0$.

Derivatives of Elementary Functions

1. $\dfrac{d}{dx}(C) = 0$

2. $\dfrac{d}{dx}(x^n) = nx^{n-1}$

3. $\dfrac{d}{dx}(\sin(x)) = \cos(x)$

4. $\dfrac{d}{dx}(\cos(x)) = -\sin(x)$

5. $\dfrac{d}{dx}(\tan(x)) = \sec^2(x)$

6. $\dfrac{d}{dx}(\sec(x)) = \sec(x)\tan(x)$

7. $\dfrac{d}{dx}(\csc(x)) = -\csc(x)\cot(x)$

8. $\dfrac{d}{dx}(\cot(x)) = -\csc^2(x)$

9. $\dfrac{d}{dx}(e^x) = e^x$

10. $\dfrac{d}{dx}(a^x) = \ln(a)a^x$

11. $\dfrac{d}{dx}(\ln(x)) = \dfrac{1}{x}$

12. $\dfrac{d}{dx}(\log_a(x)) = \dfrac{1}{\ln(a)x}$

13. $\dfrac{d}{dx}(\sin^{-1}(x)) = \dfrac{1}{\sqrt{1-x^2}}$

14. $\dfrac{d}{dx}(\cos^{-1}(x)) = -\dfrac{1}{\sqrt{1-x^2}}$

15. $\dfrac{d}{dx}\left(\sec^{-1}(x)\right) = \dfrac{1}{|x|\sqrt{x^2-1}}$

16. $\dfrac{d}{dx}\left(\csc^{-1}(x)\right) = -\dfrac{1}{|x|\sqrt{x^2-1}}$

17. $\dfrac{d}{dx}\left(\tan^{-1}(x)\right) = \dfrac{1}{x^2+1}$

18. $\dfrac{d}{dx}\left(\cot^{-1}(x)\right) = -\dfrac{1}{x^2+1}$

Differentiation Theorems

Product Rule:	$\dfrac{d}{dx}\left(f(x)g(x)\right) = f'(x)g(x) + f(x)g'(x)$
Quotient Rule	$\dfrac{d}{dx}\left(\dfrac{f(x)}{g(x)}\right) = \dfrac{f'(x)g(x) - f(x)g'(x)}{\left(g(x)\right)^2}$
Chain Rule:	$\dfrac{d}{dx}\left(f(g(x))\right) = f'(g(x))g'(x)$

Increasing and Decreasing Functions, and Relative Extrema

$f(x)$ is *increasing* if $f'(x) > 0$ and *decreasing* if $f'(x) < 0$.

Let $f(c)$ be defined.

$x = c$ is a *critical number* of $f(x)$ if $f'(c) = 0$ or if $f'(c)$ is undefined.

First Derivative Test:	If $f(x)$ changes from *decreasing* to *increasing* at $x = c$, then $f(x)$ has a *relative minimum* at the point $(c, f(c))$.
	If $f(x)$ changes from *increasing* to *decreasing* at $x = c$, then $f(x)$ has a *relative maximum* at the point $(c, f(c))$.
Second Derivative Test:	If $f''(c) > 0$, then $f(x)$ has a *relative minimum* at the point $(c, f(c))$.
	If $f''(c) < 0$, then $f(x)$ has a *relative maximum* at the point $(c, f(c))$.
	If $f''(c) = 0$, then the Second Derivative Test is inconclusive. Use the First Derivative Test.

Concavity and Points of Inflection

The graph of $f(x)$ is *concave up* if $f''(x) > 0$ and *concave down* if $f''(x) < 0$.

Let $f(c)$ be defined.

The graph of $f(x)$ *may* have a *point of inflection* $(c, f(c))$ if $f''(c) = 0$ or if $f''(c)$ is undefined. The graph of $f(x)$ *does* have a point of inflection at $(c, f(c))$ if it *changes concavity* at $x = c$.

Intermediate Value Theorem

If $f(x)$ is continuous on $[a,b]$, and k is between $f(a)$ and $f(b)$, then there exists a number c in $[a,b]$ such that $f(c) = k$.

Rolle's Theorem

If f is continuous on $[a,b]$ and differentiable on (a,b), and if $f(a) = f(b)$, then there exists a number c in (a,b) such that $f'(c) = 0$.

Mean Value Theorem for Derivatives

If f is continuous on $[a,b]$ and differentiable on (a,b), then there exists a number c in (a,b) such that $f'(c) = \dfrac{f(b) - f(a)}{b - a}$.

Integrals

Antiderivatives of Elementary Functions

1. $\int kf(x)\,dx = k\int f(x)\,dx$

2. $\int \big(f(x) \pm g(x)\big)\,dx = \int f(x)\,dx \pm \int g(x)\,dx$

3. $\int 0\,dx = C$

4. $\int k\,dx = kx + C$

5. $\int x^n dx = \dfrac{1}{n+1}x^n + C\,(n \neq -1)$

6. $\int \dfrac{1}{x}\,dx = \ln|x| + C$

7. $\int e^x dx = e^x + C$

8. $\int a^x dx = \dfrac{1}{\ln(a)}a^x + C\,(a > 0, a \neq 1)$

9. $\int \sin(x)\,dx = -\cos(x) + C$

10. $\int \cos(x)\,dx = \sin(x) + C$

11. $\int \sec^2(x)\,dx = \tan(x) + C$

12. $\int \csc^2(x)\,dx = -\cot(x)+C$

13. $\int \sec(x)\tan(x)\,dx = \sec(x)+C$

14. $\int \csc(x)\cot(x)\,dx = -\csc(x)+C$

15. $\int \dfrac{1}{\sqrt{a^2-x^2}}\,dx = \arcsin\left(\dfrac{x}{a}\right)+C\,(a>0)$

16. $\int \dfrac{1}{x^2+a^2}\,dx = \dfrac{1}{a}\arctan\left(\dfrac{x}{a}\right)+C\,(a>0)$

17. $\int \dfrac{1}{x\sqrt{x^2-a^2}}\,dx = \dfrac{1}{a}\operatorname{arcsec}\left(\dfrac{|x|}{a}\right)+C$

18. $\int \tan(x)\,dx = -\ln\left|\cos(x)\right|+C$

19. $\int \cot(x)\,dx = \ln\left|\sin(x)\right|+C$

20. $\int \sec(x)\,dx = \ln\left|\sec(x)+\tan(x)\right|+C$

21. $\int \csc(x)\,dx = -\ln\left|\csc(x)+\cot(x)\right|+C$

22. $\int f\left(g(x)\right)g'(x)\,dx = \int f\left(u\right)du$

Integration by Parts (BC)

$$\int u\,dv = uv - \int v\,du$$

Definition of Definite Integral

$$\int_a^b f(x)\,dx = \lim_{\max \Delta x_i \to 0} \sum_{i=1}^{n} f\left(x_i^*\right)\Delta x_i = \lim_{n\to\infty} \sum_{i=1}^{n} f\left(x_i^*\right)\Delta x_i$$

Approximate Definite Integration

Left Riemann Sum:	$\displaystyle\int_a^b f(x)\,dx \approx \lim_{n\to\infty}\sum_{i=1}^{n} f(x_{i-1})\,\Delta x_i$
Right Riemann Sum:	$\displaystyle\int_a^b f(x)\,dx \approx \lim_{n\to\infty}\sum_{i=1}^{n} f(x_i)\,\Delta x_i$
Midpoint Riemann Sum:	$\displaystyle\int_a^b f(x)\,dx \approx \lim_{n\to\infty}\sum_{i=1}^{n} f\left(\frac{x_{i-1}+x_i}{2}\right)\Delta x_i$
Trapezoidal Sum:	$\displaystyle\int_a^b f(x)\,dx \approx \frac{b-a}{2n}\left[f(x_0)+2f(x_1)+\cdots+2f(x_{n-1})+f(x_n)\right]$

The Fundamental Theorems of Calculus

If f is continuous on $[a,b]$ and F is an antiderivative of f on $[a,b]$, then

$$\int_a^b f(x)\,dx = F(b)-F(a).$$

If f is continuous on an open interval containing a, then for all x in the interval

$$\frac{d}{dx}\left(\int_a^x f(t)\,dt\right) = f(x).$$

Furthermore, if $\alpha(x)$ and $\beta(x)$ are differentiable,

$$\frac{d}{dx}\left(\int_{\alpha(x)}^{\beta(x)} f(t)\,dt\right) = f(\beta(x))\beta'(x) - f(\alpha(x))\alpha'(x).$$

Mean Value Theorem for Integrals

If f is continuous on $[a,b]$, then there exists a number c in $[a,b]$ such that $\displaystyle\int_a^b f(x)\,dx = f(c)(b-a)$.

The *mean value* of f on $[a,b]$ is $\displaystyle \overline{f} = \frac{1}{b-a}\int_a^b f(x)\,dx$.

Applications of Integration

Arc Length:	$s = \int\limits_a^b \sqrt{1 + \left(f'(x)\right)^2}\, dx$
Disk Method:	$V = \pi \int\limits_a^b \left(R(x)\right)^2 dx$
Washer Method:	$V = \pi \int\limits_a^b \left(\left(R(x)\right)^2 - \left(r(x)\right)^2\right) dx$
Shell Method:	$V = 2\pi \int\limits_a^b r(x) h(x)\, dx$
Parametric Arc Length (BC):	$s = \int\limits_a^b \sqrt{\left(f'(t)\right)^2 + \left(g'(t)\right)^2}\, dt$
Polar Area (BC):	$A = \dfrac{1}{2} \int\limits_\alpha^\beta \left(f(\theta)\right)^2 d\theta$

The Shell Method no longer explicitly appears on the AP Calculus Exam, but it couldn't hurt to know it. Sometimes it is easier to use than either the Disk or Washer Methods.

Motion

1D:	$v(t) = x'(t)$ $a(t) = v'(t) = x''(t)$ $Avg\ Velocity = \dfrac{\Delta x}{\Delta t}$	$Distance = \int\limits_a^b \lvert v(t) \rvert\, dt$ $Displacement = \int\limits_a^b v(t)\, dt$
2D:	$\vec{r}(t) = x(t), y(t)$ $\vec{v}(t) = x'(t), y'(t)$ $\vec{a}(t) = x''(t), y''(t)$ $Avg\ Velocity = \dfrac{\Delta \vec{r}}{\Delta t}$	$Distance = \int\limits_a^b \sqrt{\left(x'(t)\right)^2 + \left(y'(t)\right)^2}\, dt$ $Displacement = \int\limits_a^b x'(t), y'(t)\, dt$

Euler's Method

$$\frac{dy}{dx} = f(x, y) \text{ and } y(x_0) = y_0.$$

$$x_n = x_{n-1} + h$$

$$y_n = y_{n-1} + f(x_{n-1}, y_{n-1}) \cdot h$$

Series

Tests of Infinite Series

Test	Series	Convergence Condition	Divergence Condition
nth-Term	$\sum_{n=1}^{\infty} a_n$	None	$\lim_{n \to \infty} a_n \neq 0$
Geometric Series	$\sum_{n=0}^{\infty} ar^n$	$\|r\| < 1$ $S = \dfrac{a}{1-r}$	$\|r\| \geq 1$
Telescoping Series	$\sum_{n=1}^{\infty} (a_n - a_{n+1})$	$\lim_{n \to \infty} a_n = L$ $S = a_1 - L$	None
p-Series	$\sum_{n=1}^{\infty} \dfrac{1}{n^p}$	$p > 1$	$p \leq 1$
Alternating Series	$\sum_{n=1}^{\infty} (-1)^n a_n,\ a_n \geq 0$	$a_{n+1} \leq a_n$ and $\lim_{n \to \infty} a_n = 0$	None
Integral	$\sum_{n=1}^{\infty} a_n,\ a_n = f(n) \geq 0$ $f(x)$ continuous, positive, decreasing	$\int_1^{\infty} f(x)\,dx$ converges	$\int_1^{\infty} f(x)\,dx$ diverges
Root	$\sum_{n=1}^{\infty} a_n$	$\lim_{n \to \infty} \sqrt[n]{\|a_n\|} < 1$	$\lim_{n \to \infty} \sqrt[n]{\|a_n\|} > 1$
Ratio	$\sum_{n=1}^{\infty} a_n$	$\lim_{n \to \infty} \left\|\dfrac{a_{n+1}}{a_n}\right\| < 1$	$\lim_{n \to \infty} \left\|\dfrac{a_{n+1}}{a_n}\right\| > 1$

Test	Series	Convergence Condition	Divergence Condition
Direct Comparison	$\sum\limits_{n=1}^{\infty} a_n$	$0 < a_n \leq b_n$ and $\sum\limits_{n=1}^{\infty} b_n$ converges	$0 < b_n \leq a_n$ and $\sum\limits_{n=1}^{\infty} b_n$ diverges
Limit Comparison	$\sum\limits_{n=1}^{\infty} a_n$	$\lim\limits_{n\to\infty} \dfrac{a_n}{b_n} = L > 0$ and $\sum\limits_{n=1}^{\infty} b_n$ converges	$\lim\limits_{n\to\infty} \dfrac{a_n}{b_n} = L > 0$ and $\sum\limits_{n=1}^{\infty} b_n$ diverges

Alternating Series Error Bound

If the alternating series $\sum\limits_{n=1}^{\infty} (-1)^n a_n$ ($a_n \geq 0$) converges to a sum S, and $S_N = \sum\limits_{n=1}^{N} (-1)^n a_n$ is the Nth partial sum of the series, then $|S - S_N| \leq a_{N+1}$.

Maclaurin Series for Elementary Functions

1. $e^x = \sum\limits_{n=0}^{\infty} \dfrac{x^n}{n!} = 1 + x + \dfrac{x^2}{2!} + \dfrac{x^3}{3!} + \cdots$

2. $\sin(x) = \sum\limits_{n=0}^{\infty} \dfrac{(-1)^n x^{2n+1}}{(2n+1)!} = x - \dfrac{x^3}{3!} + \dfrac{x^5}{5!} - \dfrac{x^7}{7!} + \cdots$

3. $\cos(x) = \sum\limits_{n=0}^{\infty} \dfrac{(-1)^n x^{2n}}{(2n)!} = 1 - \dfrac{x^2}{2!} + \dfrac{x^4}{4!} - \dfrac{x^6}{6!} + \cdots$

4. $\dfrac{1}{1-x} = \sum\limits_{n=0}^{\infty} x^n = 1 + x + x^2 + x^3 + \cdots$

5. $\ln(x+1) = \sum\limits_{n=1}^{\infty} \dfrac{(-1)^{n+1} x^n}{n} = x - \dfrac{x^2}{2} + \dfrac{x^3}{3} - \dfrac{x^4}{4} + \cdots$

Taylor Series

If f is smooth at $x = c$, then $f(x) = \sum\limits_{n=0}^{\infty} \dfrac{f^{(n)}(c)}{n!}(x-c)^n$ is the Taylor series for f centered at c, and $P_N(x) = \sum\limits_{n=0}^{N} \dfrac{f^{(n)}(c)}{n!}(x-c)^n$ is the Nth degree polynomial for f centered at c.

Lagrange Error Bound

If $P_N(x)$ is the Nth degree polynomial for f centered at c, and $\left|f^{(n+1)}(t)\right| \leq M$ for all t between x and c, then $\left|f(x) - P_N(x)\right| \leq \dfrac{M}{(n+1)!}|x-c|^{n+1}$.

Strategy 2: Time Management

In this part of the book, we go over some test-taking tips to help you maximize your score. The exam is divided into two sections. Section I contains Multiple-Choice Questions (MCQs), and Section II contains Free-Response Questions (FRQs). We will give some helpful tips that are specific to those sections in the following two sections of this guide, but for now, we will give some advice on time management that is useful on the exam as a whole.

Use your time wisely.

The most important thing you can do (aside from knowing your calculus!) to ensure your success on the AP Calculus Exam is to learn how much time to allocate to an individual question. In Section I, you have just under 2 minutes per question in Part A and just under 3 minutes per question in Part B. In Section II, you have 15 minutes per question in both parts. If you're spending more than the average allotted time on any question, *move on*; that time would be better spent elsewhere. Remember: *All MCQs are weighted the same, and all FRQs are weighted the same.* Your objective is to use your time to score as many points as you can. This leads us to the next tip.

Learn the directions *before* you get to the exam.

In this book you will find four full-length practice AP Calculus exams (2 AB and 2 BC). The directions on those practice exams are exactly as they appear on the real thing. If you go take a look (go ahead, we'll wait), you'll find that the directions are long. If you learn these directions before you get to the exam, then while everyone else in the room is reading them, you'll be solving problems.

Strategy 3: Multiple Choice Questions _____

Answer the easy questions first.

You don't have to answer the questions in the order in which they are presented to you. Plan to make two passes through each Part of Section I. On your first pass, answer only those questions that you can quickly and accurately work out. Chances are that you will spend less than the average allotted time on these. On your second pass, use the rest of your time to tackle the tougher questions. Not only will you have more time per question on your second pass, but also you'll have the peace of mind of knowing that you already have the points from the easier ones.

So, what are you to do about those difficult questions? We have some suggestions for those.

Use the process of elimination.

There may be some multiple choice questions that you just can't get. This does not mean that you're dead in the water. Even if you can't positively identify the *right* answer, it is possible that you can identify one or more of the *wrong* ones. Consider the following example from Part A of Section I.

1. Which of the following is a slope field for the differential equation $\frac{dy}{dx} = x + y$?

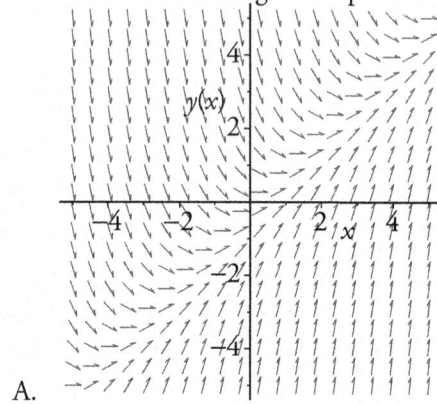

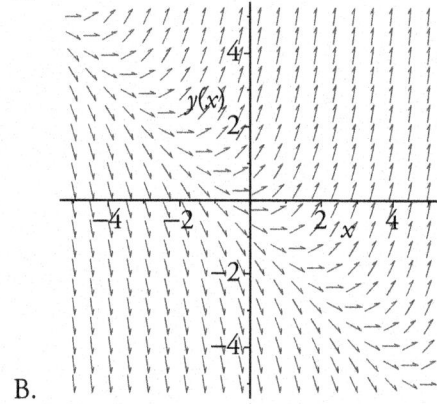

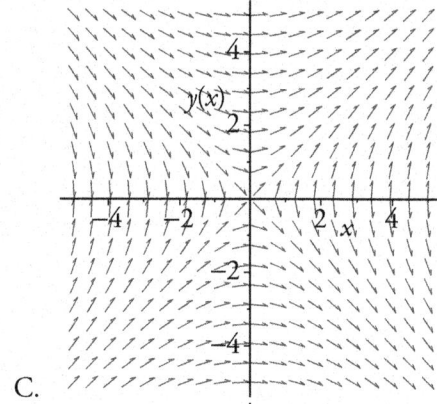

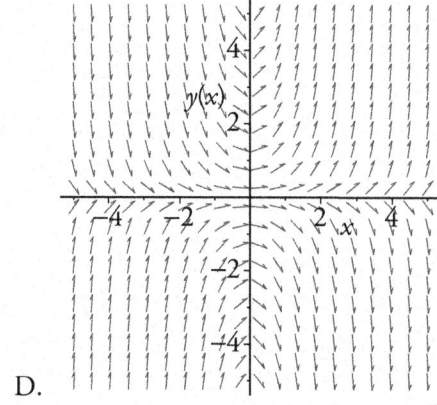

A.

B.

C.

D.

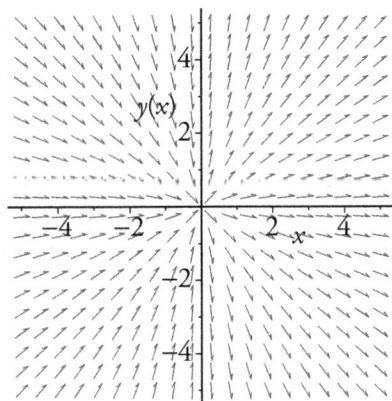

E.

If you don't know how to solve this problem, then you *could* just take a guess. You have 1 shot in 5 of getting it right. A better way would be to use what you *do* know to stack the deck in your favor. Recall that $\dfrac{dy}{dx}$ gives the *slope* of y as a function of x. So what is the slope of the solution at, say, the origin? Use the differential equation to find out.

$$\left.\frac{dy}{dx}\right|_{(0,0)} = \left.(x+y)\right|_{(0,0)} = 0+0 = 0$$

So the slope of the solution curve y at the origin is 0. Looking at the answers, we can see that choices C and E most definitely do *not* have a zero slope at the origin. Eliminate them. Now you've got a 1 in 3 chance of guessing the correct answer, which is an improvement. (The correct answer is B, by the way.)

Work backwards.

Two of the "Big Ideas" of the AP Calculus curriculum are derivatives and integrals, which are (sort of) inverse operations. We say "sort of" because of the non-uniqueness of antiderivatives. Consider the following problem from Part A of Section I.

1. $\int \sin(2x)\cos(x)\,dx =$

    A. $-\dfrac{1}{2}\cos(2x)\sin(x)+C$

    B. $-\dfrac{1}{2}\cos(2x)\cos(x)+\sin(2x)\sin(x)+C$

    C. $\dfrac{2}{3}\cos^3(x)+C$

    D. $-\dfrac{2}{3}\cos^3(x)+C$

    E. $\dfrac{2}{3}\sin^3(x)+C$

First, we'll apply the process of elimination. Even if you don't know how to do this integral, you know that the integral of a product of two functions is **not** equal to the product of the integrals. That is, $\int f(x)\cdot g(x)\,dx \neq \int f(x)\,dx \cdot \int g(x)\,dx$. That's what was done in answer choice A. Eliminate it. Also, you know that there is no Product Rule for integrals. That is, $\int f(x) \cdot g(x)\,dx \neq \int f(x)\,dx \cdot g(x) + f(x) \cdot \int g(x)\,dx$. That's what was done in answer choice B. Eliminate it.

So we're down to three choices. Rather than guess, we'll *work backwards*. If we take the derivative of the correct answer choice, it must simplify to the integrand. Let's try it with answer choice C. We will need the Chain Rule here.

$$\frac{d}{dx}\left(\frac{2}{3}\cos^3(x) + C\right) = \frac{2}{3} \cdot 3\cos^2(x) \cdot \frac{d}{dx}(\cos(x)) = -2\cos^2(x)\sin(x)$$

This doesn't look like the integrand. However, the presence of $\sin(2x)$ in the integrand should remind you that you can use the double-angle trig identity $\sin(2x) = 2\sin(x)\cos(x)$.

$$\frac{d}{dx}\left(\frac{2}{3}\cos^3(x) + C\right) = -2\sin(x)\cos(x)\cos(x) = -2\sin(2x)\cos(x)$$

This *almost* looks like the integrand. We're off by a negative sign. So the correct answer is not C, but rather D.

On Part B you are allowed a calculator. Use it.

There is no partial credit for any of the multiple choice questions. A machine grades them, and it doesn't care how you arrive at any of your answers. So any time your calculator can save you some time, you should take advantage of it. Consider the following problem from Part B of Section I.

1. $\displaystyle\int_0^1 xe^{-x^2}\,dx =$

   A. 0
   B. 0.316
   C. 0.632
   D. 0.684
   E. 1.378

Of course, you *could* integrate this with a *u*-substitution, but there's no need. Just enter the integral into your graphing calculator, choose answer B, and be done with it.

Strategy 4: Attacking the Free Response Questions _____

All FRQs are scored out of 9 points, and they are all equally weighted. However, each FRQ is divided into 2–4 parts, which are *not* necessarily equally weighted. For each part, the answer is always worth 1 point, while the rest of the points are allocated to the steps taken in obtaining the answer. Unlike the MCQs, the FRQs are graded by a living, breathing human being who will give you partial credit for partially correct solutions. The grader is not just interested in seeing you get the correct answer; he or she is also interested in seeing whether you know how to solve the problem.

Answer the easy parts first.

Just as with Section I, you do not have to answer the questions (or even the parts of the questions) of Section II in the order in which they are presented. Sometimes the answer to one part of a FRQ is needed to answer a subsequent part. If this happens, and you don't know how to find the answer to one part, but you *could* find the answer to the next part *if* you knew the answer to the previous part, then do what you can to show the grader that you know what you're doing on the subsequent part.

Show *all* work.

As noted previously, the answer is only worth 1 point out of the total score for a part of a question. Simply writing down an answer for each part isn't going to get you a passing score on the exam. You need to show all of your work to get full credit. Even if you use your calculator, you should write down clearly what you're doing. For instance, if you need to calculate $\int_0^{\pi} e^x \sin(x)\,dx$ to answer part of a question, and your calculator tells you that the result is 12.07 (to the nearest hundredth), don't just write "12.07". Explicitly set this number equal to the integral, like this:

$$\int_0^{\pi} e^x \sin(x)\,dx = 12.07$$

If you find that you made a wrong turn in solving a problem and need to start over, don't bother erasing your incorrect solution. That just wastes time that you could be spending on the new solution. Instead, clearly cross out the incorrect solution. The grader will know that you don't want that solution graded.

On Part B you are allowed a calculator. Use it…sparingly!

A grader will give you full credit for clear, correct solutions if you limit your calculator use to the following tasks:

- Do numerical calculations.
- Evaluate derivatives numerically.
- Evaluate definite integrals.

You will **not** get full credit if you use your calculator to do the following tasks.

- Find intervals on which a function is increasing/decreasing.
- Locate relative or absolute extrema.
- Find intervals on which a function is concave up/down.
- Locate inflection points.

You are certainly encouraged to use your calculator to *investigate* these things, but you must explicitly write the justification for your answers in order to get full credit for these tasks.

PART III:
Content Review

Prelude:
The Functions of Precalculus

Introduction

A thorough knowledge of the functions of precalculus is essential to doing well on the AP Calculus Exam. You should know their properties and essential features of their graphs. The functions that you should know inside and out will be reviewed in the subsequent sections. But first, let's go over what exactly a function is. Suppose $y = f(x)$, where $f(x)$ is an expression in x. The set of all real numbers x for which $f(x)$ is defined is called the **domain** of f. As x varies over the domain, the expression $y = f(x)$ also takes on a set of values, and this set is called the **range** of f. Here's the really important part: f is a **function** if it assigns every x in its domain to *exactly one* y in its range.

Vertical Line Test

A quick graphical test to determine if y is a function of x is the **vertical line test**. If a vertical line intersects the graph of y vs x at more than one point, then y is *not* a function of x. See the figures below.

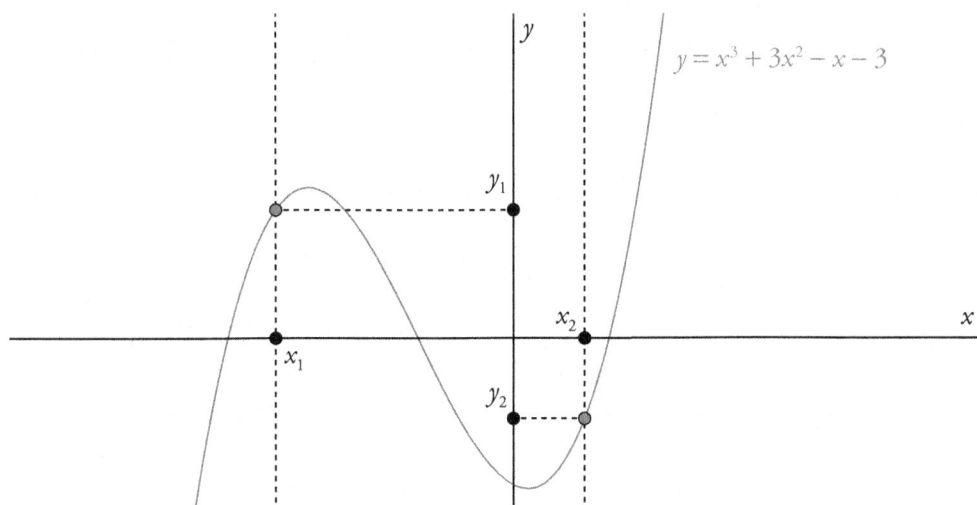

$$y = x^3 + 3x^2 - x - 3$$

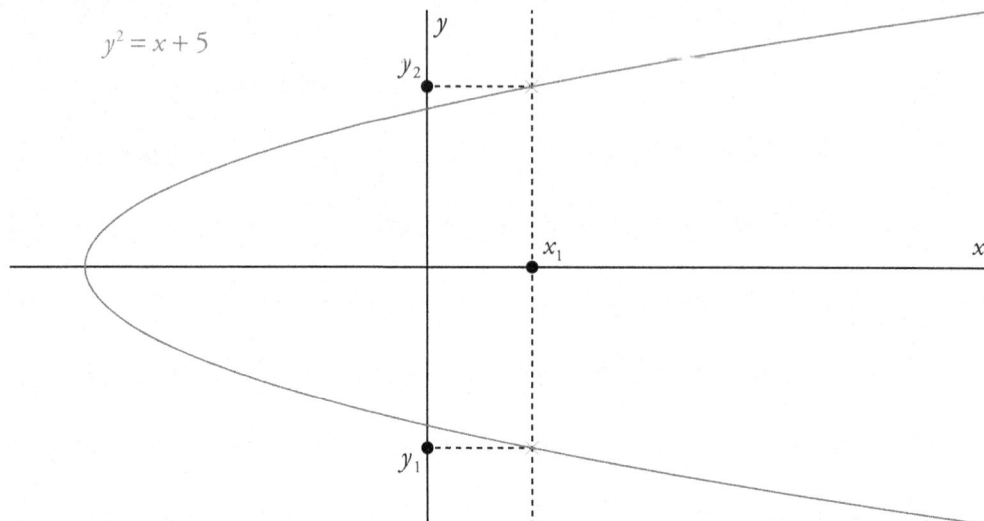

$y^2 = x + 5$

In the first figure, no vertical line intersects the graph of $y = x^3 + 3x^2 - x - 3$ more than once. We have sketched in two vertical lines as examples: $x = x_1$ and $x = x_2$. We can see that x_1 is assigned to y_1, and x_2 is assigned to y_2. The formula $y = x^3 + 3x^2 - x - 3$ does indeed give y as a function of x.

In the second figure, the vertical line at $x = x_1$ intersects the graph at *two* points. x_1 is assigned to y_1 *and* to y_2. Since a function assigns each x in its domain to exactly one y in its range, the formula $y^2 = x + 5$ *does not* give y as a function of x.

Zeros of a Function

In order for you to do well on the exam, it will be essential for you to be able to find the **zeros** of a function. A zero of a function f is a number a such that $f(a) = 0$. The zeros of f tell you where the x-intercepts of the graph of f are. The importance of zeros does not stop there, however. The zeros of f' are critical numbers of f, and the zeros of f'' are the possible points of inflection of f. You will have to be able to find **all** of these things on the exam. Hopefully this makes it clear that finding zeros is pretty important!

Polynomial Functions

A **polynomial function of degree n** is a function of the form $y = f(x)$, where

$$f(x) = a_n x^n + a_{n-1} x^{n-1} + \cdots + a_1 x + a_0.$$

Here, $a_n, a_{n-1}, \ldots, a_1, a_0$ are real numbers, and $a_n \neq 0$. a_n is called the **leading coefficient**. The domain of any polynomial function is the set of all real numbers, since the expression $f(x)$ is defined for any real number x. Also, polynomial functions are **continuous** for all real numbers x. That means that you can sketch the graph of a polynomial function without picking your pencil up off the paper. We will look at polynomial functions of degree 0, 1, and 2 before reviewing polynomial functions in general.

Polynomial Functions of Degree 0: $f(x) = a_0$

Here, y is a *constant*, and the graph of f is a horizontal line whose y-intercept is $(0, a_0)$.

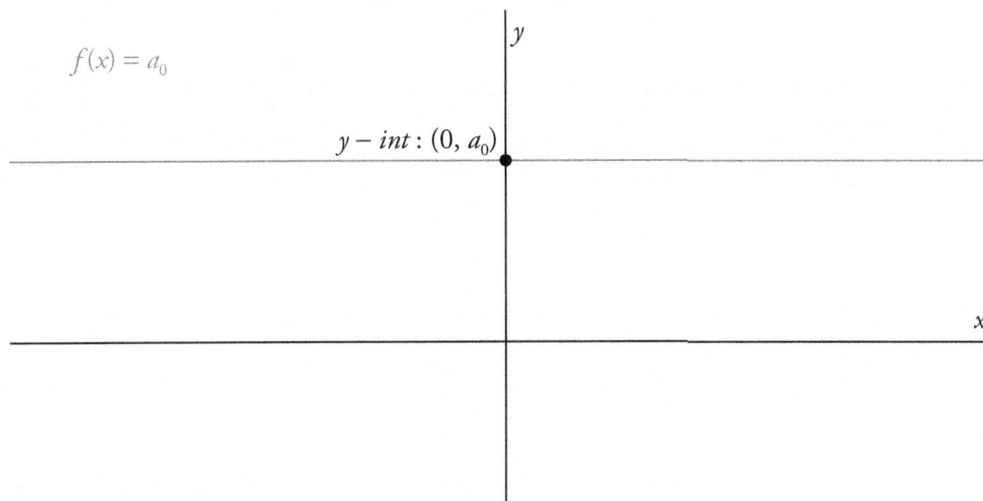

$f(x) = a_0$

$y - int : (0, a_0)$

Polynomial Functions of Degree 1: $f(x) = a_1x + a_0$

These functions are sometimes written in the form $y = mx + b$, which you should recognize this as a *linear* function. $f(x) = a_1x + a_0$ is called linear because its graph is a line with slope a_1 and y-intercept $(0, a_0)$.

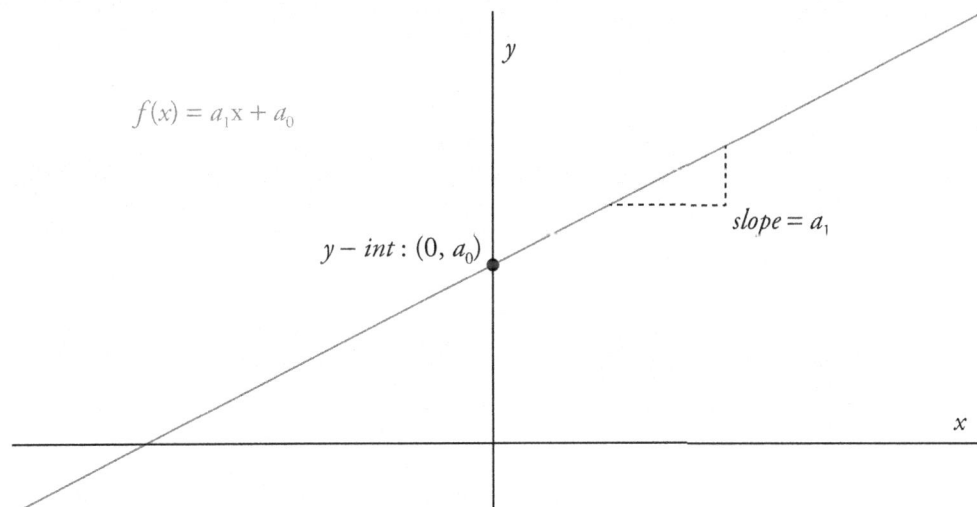

$f(x) = a_1 x + a_0$

$y - int : (0, a_0)$

$slope = a_1$

Polynomial Functions of Degree 2: $f(x) = a_2x^2 + a_1x + a_0$

These functions are called *quadratic*, and the graph of a quadratic function is a *parabola*.

The **standard form** of a quadratic function is $f(x) = a_2(x - h)^2 + k$, where $h = -\dfrac{a_1}{2a_2}$ and $k = f\left(-\dfrac{a_1}{2a_2}\right)$. The point $(0, a_0)$ is the y-intercept, the point (h, k) is the **vertex** (or **turning point**), and the line $x = h$ is the **axis of symmetry**.

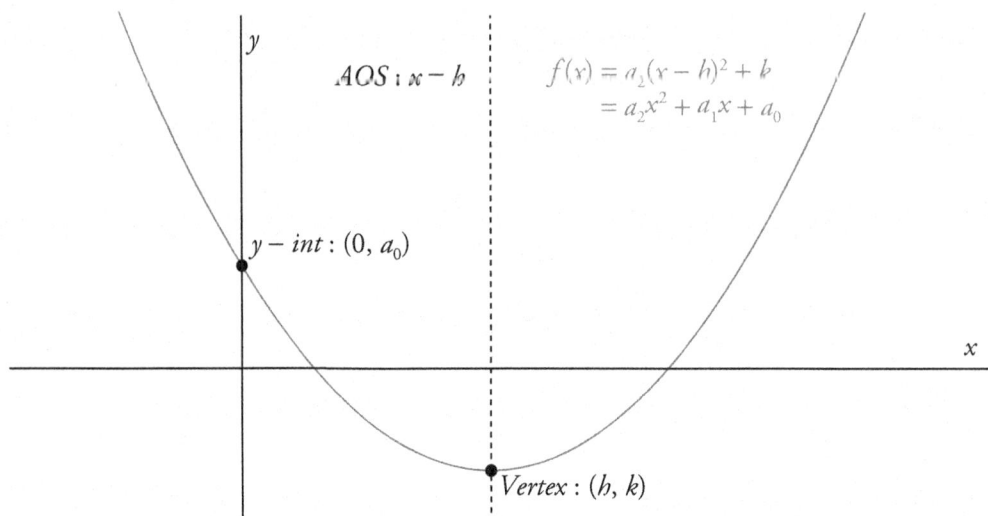

$AOS : x = h$

$f(x) = a_2(x - h)^2 + k$
$= a_2x^2 + a_1x + a_0$

$y - int : (0, a_0)$

$Vertex : (h, k)$

The parabola in the above figure is concave up, but a parabola can be either concave up or concave down. The concavity is determined by the sign of a_2. If $a_2 > 0$ then the parabola is concave up, and if $a_2 < 0$ then the parabola is concave down.

Polynomial Functions of Any Degree

We will review two important concepts here: **zeros** and **end behavior** of polynomial functions.

Zeros of a Polynomial Function

As we mentioned before, $x = a$ is a zero of f if $f(a) = 0$. If $f(a) = 0$, then the graph of f has an x-intercept at the point $(a, 0)$, and vice versa. If f is a polynomial function, then we can say a little more. If $f(a) = 0$, then $(x - a)$ is a *factor* of $f(x)$, and vice versa. We summarize these facts below.

The following statements are equivalent for a polynomial function f:

1. $x = a$ is a *zero* of f.
2. $x = a$ is a *solution* of the equation $f(x) = 0$.
3. $(x - a)$ is a *factor* of $f(x)$.
4. $(a, 0)$ is an *x-intercept* of the graph of f.

Since the zeros of a polynomial are connected to the solutions of $f(x) = 0$ and to the factors of $f(x)$, it follows that you should be comfortable using both the quadratic formula and the factoring rules from algebra. However, you should be aware that those may not be enough to get you through the exam. For instance, how would you find the zeros of $f(x) = x^3 - 7x + 6$ without a calculator? For this you would need the following result:

Let $f(x) = a_nx^n + a_{n-1}x^{n-1} + \cdots + a_1x + a_0$ with $a_n, a_0 \neq 0$. The **Rational Zero Theorem** states that all rational zeros of f are of the form $\frac{p}{q}$, where p is an integer factor of a_0, and q is an integer factor of a_n.

To apply this theorem, take the following steps:

1. Generate a list of all possible rational zeros of f.
2. Plug numbers from the list into f until you find a zero.
3. If $x = a$ is a zero of f, then divide $f(x)$ by $(x - a)$ to obtain a quotient $g(x)$, so that $f(x) = (x - a)g(x)$.
4. Repeat with $g(x)$ as necessary.

We'll illustrate these steps with our example $f(x) = x^3 - 7x + 6$.

1. $a_3 = 1$ and $a_0 = 6$. The integer divisors of a_3 are ± 1, and the integer divisors of a_0 are ± 1, ± 2, ± 3, and ± 6. The possible rational zeros of f are all of the rational numbers that can be formed with an integer divisor of a_0 in the numerator and an integer divisor of a_3 in the denominator. These are ± 1, ± 2, ± 3, and ± 6.
2. If we try plugging these into $f(x)$, we find that $f(1) = 0$, so $x = 1$ is a zero of f. That means that $x - 1$ is a factor of $f(x)$.
3. If we divide $x^3 - 7x + 6$ by $x - 1$ (using either polynomial or synthetic division), we find that $(x^3 - 7x + 6) \div (x - 1) = x^2 + x - 6$, or $f(x) = (x - 1)(x^2 + x - 6)$.
4. It is not necessary to apply the Rational Zero Test to $x^2 + x - 6$, since this polynomial is easily factored into $(x - 2)(x + 3)$. The complete factorization of $f(x)$ is then $f(x) = (x - 1)(x - 2)(x + 3)$, and the zeros of f are $x = 1$, 2, and -3.

End Behavior of a Polynomial Function

End behavior refers to how the graph of the function behaves as we look far away from the origin on the x-axis (that is, as $x \to \pm\infty$). To determine a polynomial function's end behavior, use the **Leading Coefficient Test**, which states that the end behavior can be determined by both the *sign* of a_n and the *evenness or oddness* of n. The results are summarized in the following table:

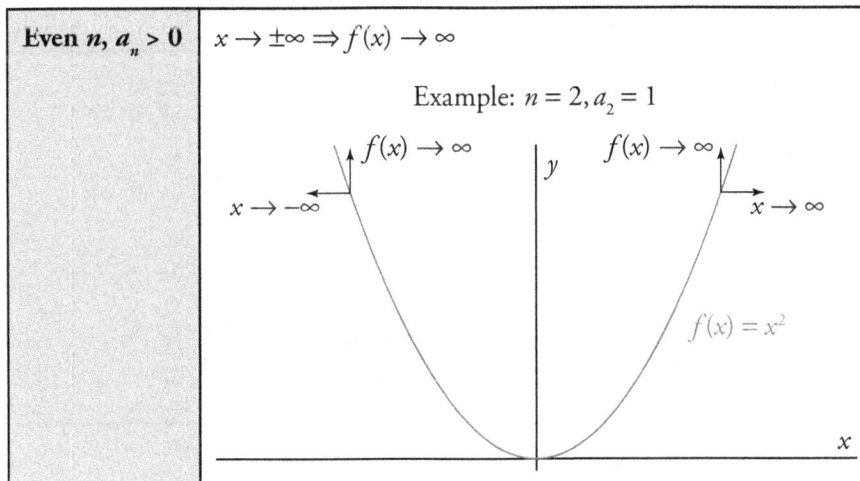

Even n, $a_n > 0$	$x \to \pm\infty \Rightarrow f(x) \to \infty$

Example: $n = 2$, $a_2 = 1$

Even n, $a_n < 0$	$x \to \pm\infty \Rightarrow f(x) \to -\infty$ Example: $n = 2, a_2 = -1$
Odd n, $a_n > 0$	$x \to -\infty \Rightarrow f(x) \to -\infty$ $x \to \infty \Rightarrow f(x) \to \infty$ Example: $n = 1, a_1 = 1$
Odd n, $a_n < 0$	$x \to -\infty \Rightarrow f(x) \to \infty$ $x \to \infty \Rightarrow f(x) \to -\infty$ Example: $n = 1, a_1 = -1$

To help you remember the Leading Coefficient Test, remember the end behaviors of these simple functions.

Rational Functions

A **rational function** is a function of the form $f(x) = \dfrac{p(x)}{q(x)}$, where $p(x)$ is a polynomial function of degree m, and $q(x)$ is a polynomial function of degree n. A rational function is called **proper** if $m < n$, and it is called **improper** if $m \geq n$. An improper rational function can be written as the sum of a polynomial and a proper rational function by using polynomial division. Two examples are given below.

Proper Rational Function

$$f(x) = \frac{-2x}{x^2 - 1}$$

Here, $m = 1$ and $n = 2$, so $m < n$.

Improper Rational Function

$$g(x) = \frac{2x^3 + 5x^2 - 3x + 4}{x^2 + 2x + 3}$$

Here, $m = 3$ and $n = 2$, so $m \geq n$.

Using polynomial division, the second example above can be written as:

$$g(x) = 2x + 1 + \frac{-11x + 1}{x^2 + 2x + 3}.$$

Zeros and Domain of a Rational Function

The zeros of a rational function $f(x) = \dfrac{p(x)}{q(x)}$ is all real numbers $x = a$ for which $p(a) = 0$ *and* $q(a) \neq 0$. The reason for the second requirement is that $f(a)$ is undefined when $q(a) = 0$. In fact, the domain of f is the set of all real numbers x for which $q(x) \neq 0$.

Continuity (and Discontinuity) of a Rational Function

A rational function f is continuous for all real numbers x in its domain, and it is discontinuous for all real numbers x not in its domain. Discontinuities of rational functions come in two varieties: **removable discontinuities** and **vertical asymptotes**.

Removable Discontinuities

If f is discontinuous at $x = a$, then the discontinuity is called *removable* if f could be made continuous simply by defining f at $x = a$. An example is shown below.

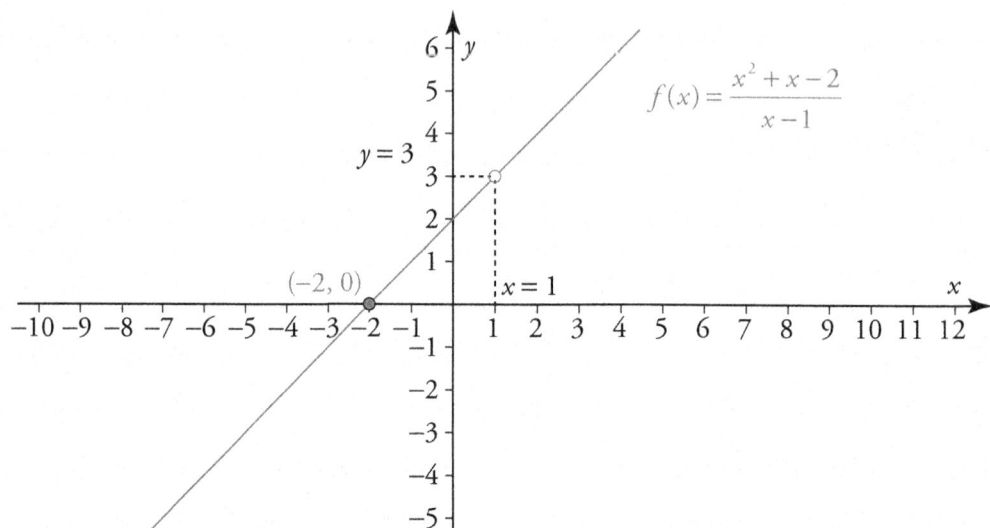

$$f(x) = \frac{x^2 + x - 2}{x - 1}$$

The zeros of $x^2 + x - 2$ are $x = -2$ and $x = 1$, and the zero of the denominator is $x = 1$. Thus, the x-intercept of f is $(-2, 0)$, and the domain of f is all real numbers *except* $x = 1$. The discontinuity of f at $x = 1$ is removable, because if we were to make the definition $f(1) = 3$, f would be continuous there.

Vertical Asymptotes

If f is discontinuous at $x = a$, then the discontinuity is called a *vertical asymptote* if $f(x) \to \pm\infty$ as $x \to a$ from either the left or the right. An example is shown below.

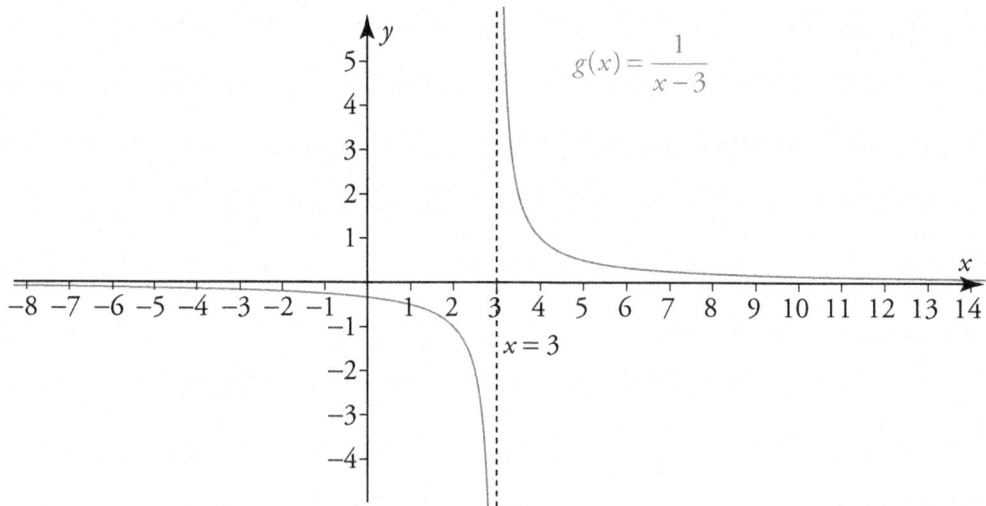

$$g(x) = \frac{1}{x - 3}$$

The numerator of $g(x)$ has no zeros, and so the graph of g has no x-intercepts. The zero of the denominator is $x = 3$, and so the domain of g is all real numbers *except* $x = 3$. The discontinuity of g at $x = 3$ is a vertical asymptote, because $f(x) \to -\infty$ as x approaches 3 from the left, and $f(x) \to \infty$ as x approaches 3 from right.

End Behavior of a Rational Function

The end behavior of a rational function f can be characterized in one of three ways.

First, if $m < n$, then $f(x) \to 0$ as $x \to \pm\infty$. In this case, the line $y = 0$ is called a **horizontal asymptote** of the graph of f. An example is shown below.

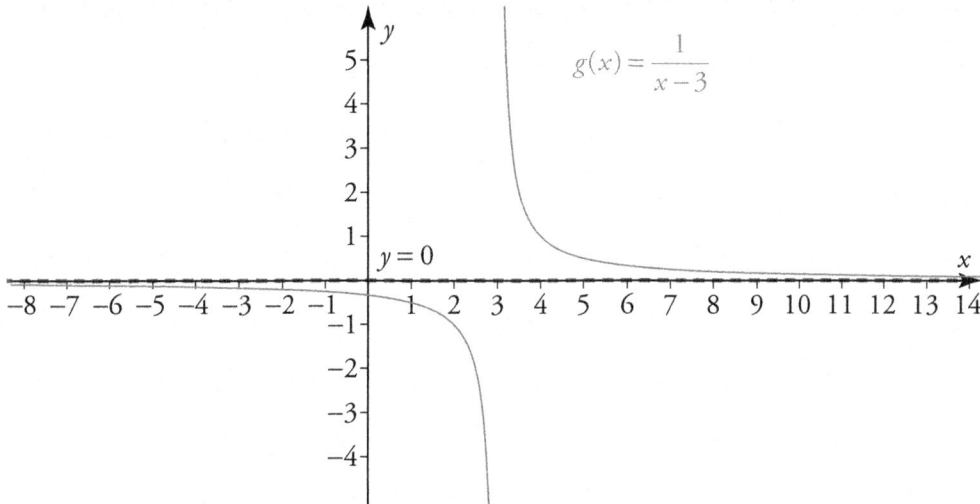

$$g(x) = \frac{1}{x-3}$$

Second, if $m = n$ then f takes the form

$$f(x) = \frac{p(x)}{q(x)} = \frac{a_n x^n + a_{n-1} x^{n-1} + \cdots + a_1 x + a_0}{b_n x^n + b_{n-1} x^{n-1} + \cdots + b_1 x + b_0},$$

where $a_n, b_n \neq 0$. As $x \to \pm\infty$ the *leading term* of each polynomial dominates, and so $f(x) \to \frac{a_n}{b_n}$, and the line $y = \frac{a_n}{b_n}$ is a horizontal asymptote of the graph of f. An example is shown below.

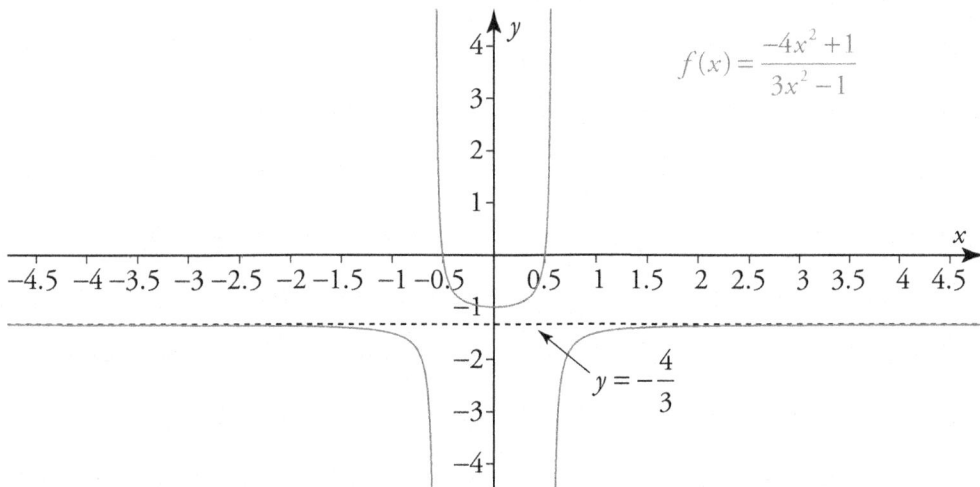

$$f(x) = \frac{-4x^2 + 1}{3x^2 - 1}$$

$$y = -\frac{4}{3}$$

In the above example $a_2 = -4$, and $b_2 = 3$, so the horizontal asymptote is $y = -\frac{4}{3}$.

And third, if $m > n$ then the graph of f has no horizontal asymptote. Instead, $f(x)$ is unbounded as $x \to \pm\infty$. An example is shown below.

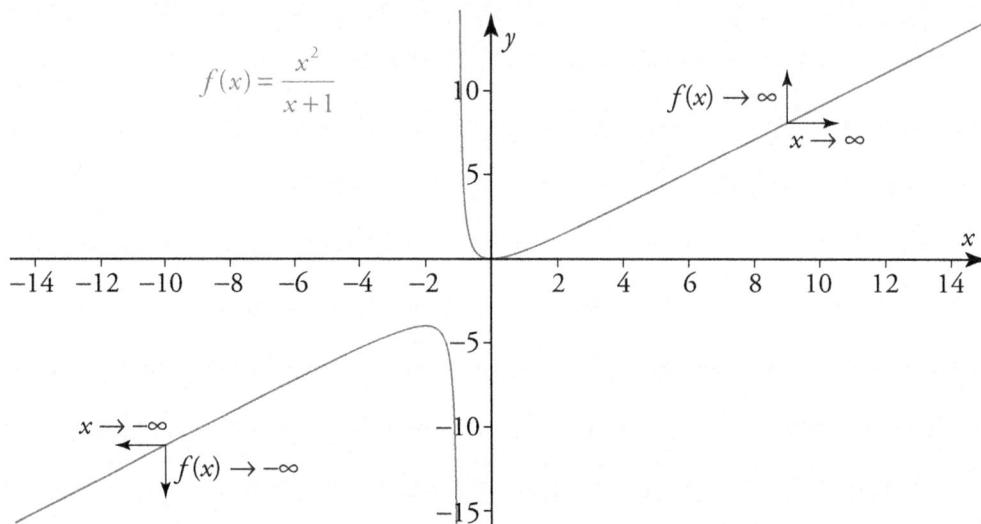

$$f(x) = \frac{x^2}{x+1}$$

In the above example $m = 2$ and $n = 1$, and f is unbounded as $x \to \pm\infty$.

Radical Functions

A simple radical function is any function of the form $f(x) = \sqrt[n]{x}$. n is called the **index** of the radical, and x is called the **radicand**. If n is odd then the domain of f is all real numbers. If n is even, then the domain of f is the interval $[0, \infty)$. In either case f has a zero at $x = 0$. Two examples are graphed below.

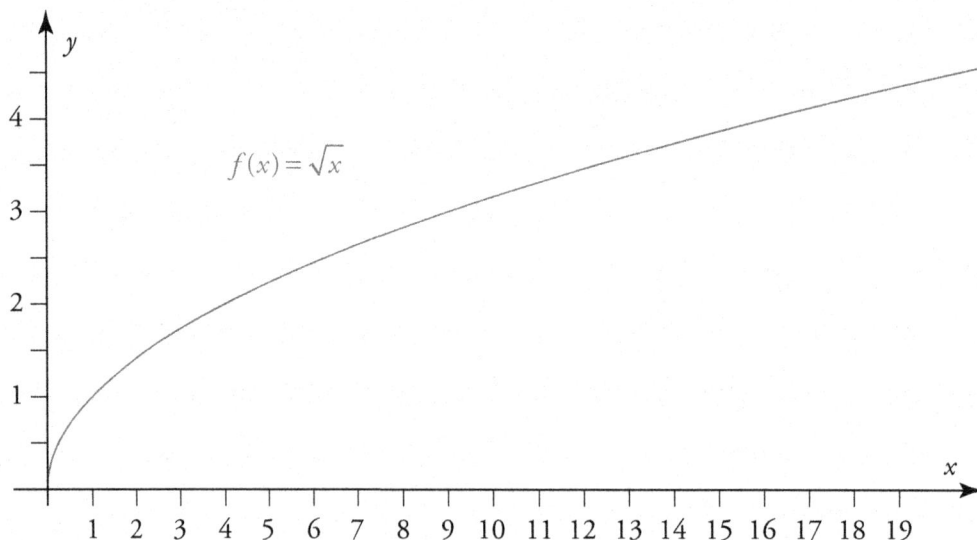

$$f(x) = \sqrt{x}$$

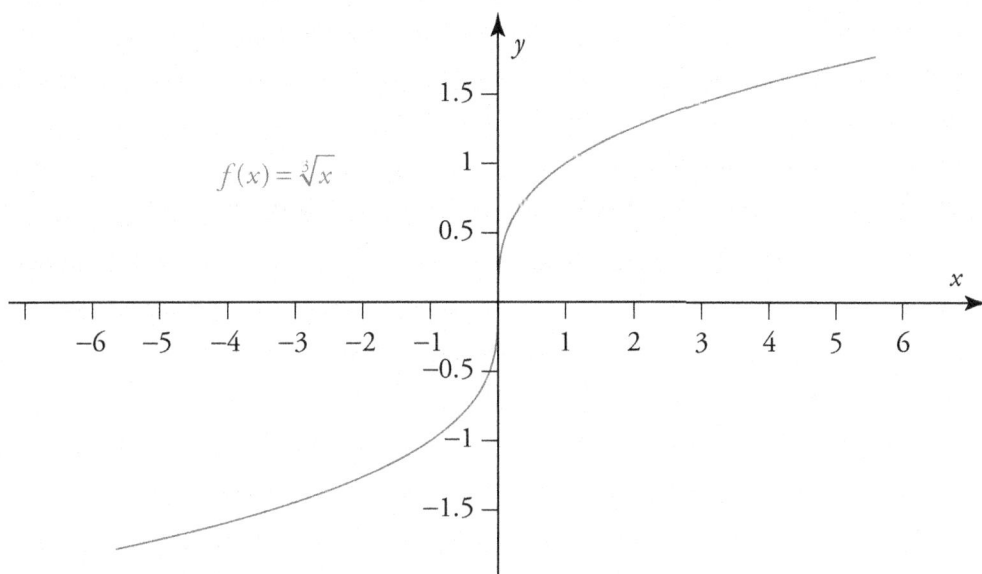

$f(x) = \sqrt[3]{x}$

Composite Functions

Let $f(x)$ and $g(x)$ be functions. The **composition** of f and g, denoted $f \circ g$, is given by $(f \circ g)(x) = f(g(x))$. The domain of $f \circ g$ is the set of all x in the domain of g such that $g(x)$ is in the domain of $f(x)$.

For example, let $f(x) = x^2$, and let $g(x) = \sqrt{x}$. Then $f \circ g(x) = f(g(x)) = f(\sqrt{x}) = (\sqrt{x})^2 = x$, and the domain of $f \circ g$ is the interval $[0, \infty)$. Note that the domain of $f \circ g$ is *not* all real numbers, despite the fact that $f(g(x))$ is defined for all real numbers.

As another example let $f(x) = x^2 + 1$, and let $g(x) = 2x + 1$. We compute $(f \circ g)(x)$ and $(g \circ f)(x)$ below.

$$(f \circ g)(x) = f\big(g(x)\big) = f(2x+1) = (2x+1)^2 + 1 = 4x^2 + 4x + 2$$

$$(g \circ f)(x) = g\big(f(x)\big) = g(x^2+1) = 2(x^2+1) + 1 = 2x^2 + 3$$

Note that $(f \circ g)(x) \neq (g \circ f)(x)$, which means that composition of functions is not a commutative operation.

Inverse Functions

A function $y = f(x)$ is **invertible** if there exists a function $x = f^{-1}(y)$ that satisfies the following conditions:

1. $f^{-1}(f(x)) = x$ for all x in the domain of f.
2. $f(f^{-1}(y)) = y$ for all y in the domain of f^{-1}.

f^{-1} is called the **inverse function** of f. Furthermore, the domain and range of f are, respectively, the range and domain of f^{-1}, and vice versa. Note that this implies if the point (a, b) is on the graph of f, then the point (b, a) is on the graph of f^{-1}. This is illustrated in the figure below.

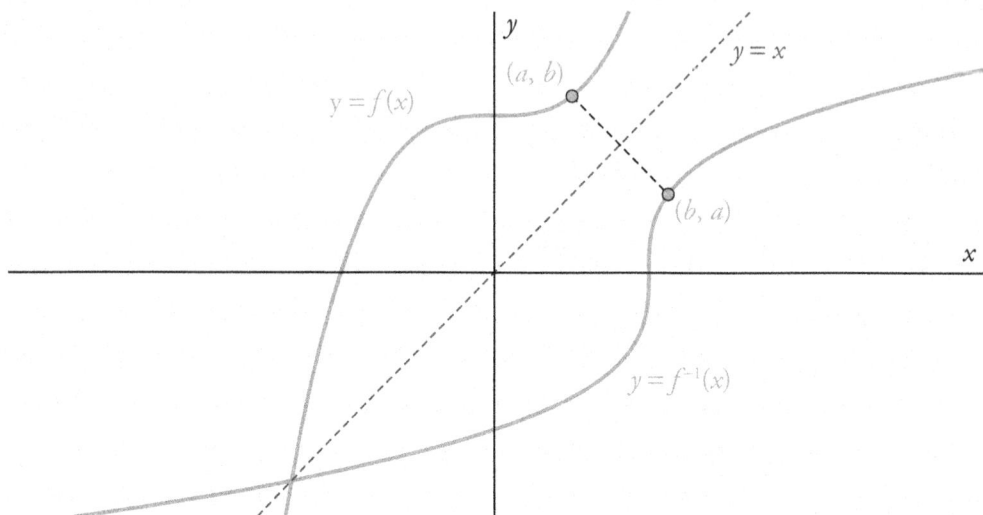

The graphs of f and f^{-1} are mirror images of each other, if the mirror is along the line $y = x$.

One-to-One Functions and the Horizontal Line Test

Not every function f has an inverse function f^{-1}. In order for an inverse to exist, each y in the range of f must have *exactly one x* in the domain assigned to it. For example, consider the function $f(x) = x^2$. For this function $f(-1) = 1$ and $f(1) = 1$, so there's no way to define a function that assigns $y = 1$ back to a unique value of x.

In order for f to have an inverse, it must be **one-to-one**, which means that for all x_1, x_2 in the domain of f, $x_1 \neq x_2 \Rightarrow f(x_1) \neq f(x_2)$. This is equivalent to saying that x_1, x_2 in the domain of f, $f(x_1) = f(x_2) \Rightarrow x_1 = x_2$. There is a graphical test to determine if $f(x)$ is one-to-one, and it is called **horizontal line test**. If a horizontal line intersects the graph of y vs x at more than one point, then y is *not* a function of x. See the figures below.

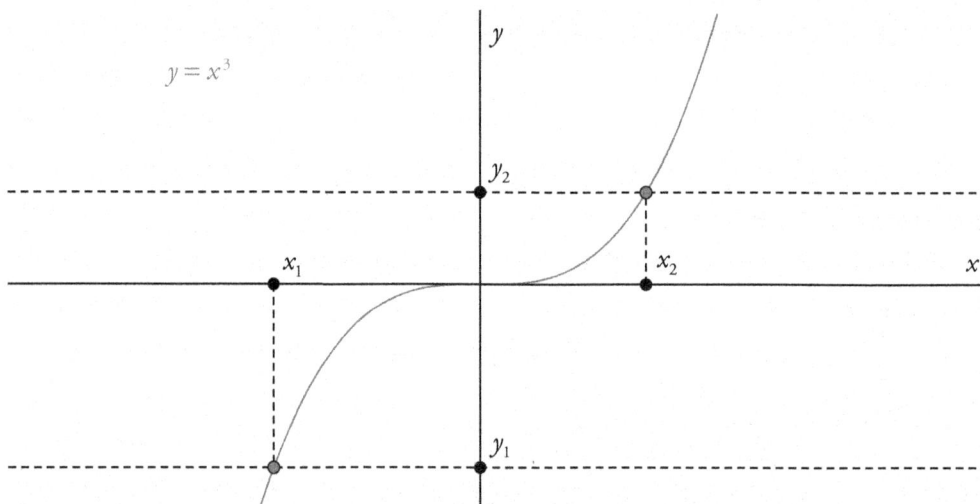

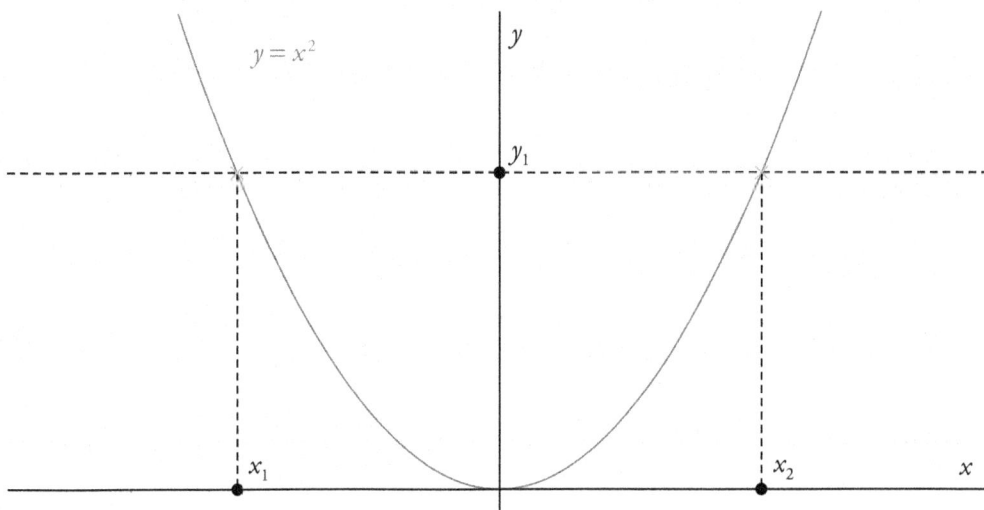

In the first figure, no horizontal line intersects the graph of $y = x^3$ more than once. Two horizontal lines have been sketched in as examples: $y = y_1$ and $y = y_2$. We can see that x_1 is assigned to y_1, and x_2 is assigned to y_2. The formula $y = x^3$ indeed give y as a one-to-one function of x.

In the second figure, the horizontal line at $y = y_1$ intersects the graph at *two* points. x_1 and x_2 are *both* assigned to y_1. Since each y in the range of a one-to-one function has exactly one x in its domain assigned to it, the formula $y = x^2$ *does not* give y as a one-to-one function of x.

Exponential Functions

Let $a > 0$ and $a \neq 1$. Then $f(x) = a^x$ is an **exponential function with base a**. Two examples, $f(x) = 2^x$ and $g(x) = \left(\dfrac{1}{2}\right)^x$, are graphed below.

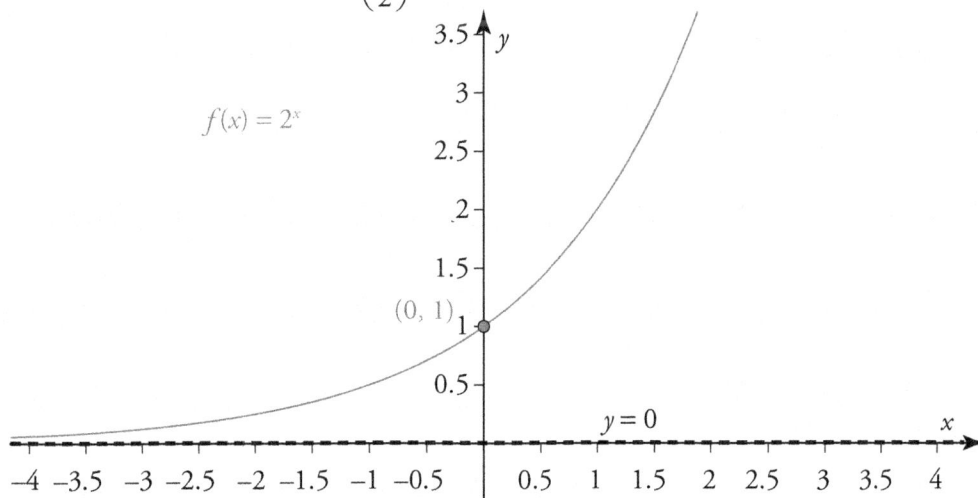

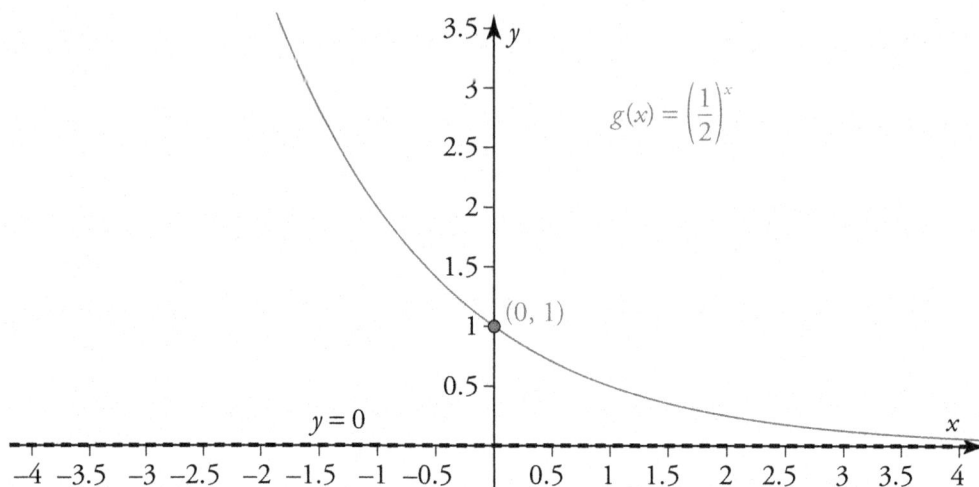

We now make some observations from the graphs. f is increasing for all x, and g is decreasing for all x. This is because the base of f is in the interval $(1, \infty)$, and the base of g is in the interval $(0, 1)$. Some additional properties common to the two functions (and indeed, *all* exponential functions) are listed below.

- y-Intercept: $(0, 1)$
- Horizontal asymptote: $y = 0$
- Domain: All Reals
- Range: $(0, \infty)$
- Both functions are one-to-one.

The fact that exponential functions are one-to-one means that they have inverse functions. We will have more to say about this in the next section, which is on logarithmic functions.

Two Special Bases

Two special bases are 10 and e, where $e \approx 2.71828$ is the value of the expression $\left(1 + \dfrac{1}{n}\right)^n$ as $n \to \infty$. 10 is called the **common base**, and e is called the **natural base**.

Other Properties of Exponential Functions

Let $a > 0$ and $a \neq 1$, and let n, x_1, x_2 be any real numbers. Then the following properties hold:

- $a^0 = 1$

- $a^{x_1} a^{x_2} = a^{x_1 + x_2}$

- $\dfrac{a^{x_1}}{a^{x_2}} = a^{x_1 - x_2}$

- $\left(a^{x_1}\right)^n = a^{nx_1}$

- $a^{x_1} = a^{x_2} \Rightarrow x_1 = x_2$

The last property is called the **one-to-one property of exponential functions**.

Logarithmic Functions

The inverse functions of exponential functions are logarithmic functions. Once again, let $a > 0$ and $a \neq 1$. If $f(x) = a^x$, then $f^{-1}(x) = \log_a(x)$. $\log_a(x)$ is the **logarithmic function with base a**. If no base is written, then the base is understood to be 10. We call $\log(x)$ the **common logarithm**. We show the graphs of two examples of exponential functions and their corresponding logarithmic functions below.

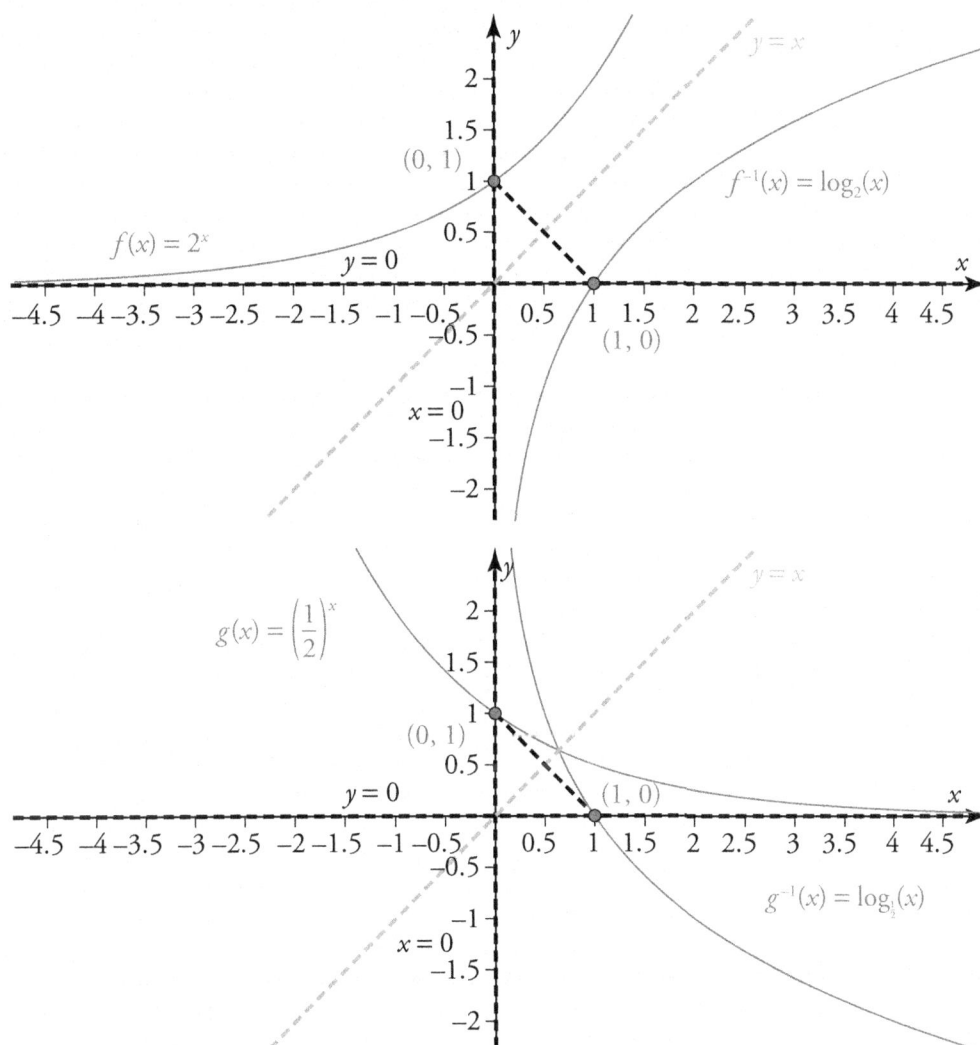

As we can see from the graphs, f^{-1} is increasing for all x, and g^{-1} is decreasing for all x. As it happens, whenever an invertible function is increasing, its inverse function is increasing as well. Likewise, when an invertible function is decreasing, its inverse function is also decreasing. Some additional properties common to the two functions (and all logarithmic functions) are listed below.

- x-Intercept: $(1, 0)$
- Vertical Asymptote: $x = 0$

- Domain: $(0, \infty)$
- Range: All Reals
- Both functions are one-to-one.

Two Special Bases

Just as with exponential functions, 10 and e are special bases of logarithmic functions. We denote $\log_{10}(x)$ as simply $\log(x)$, and we call this logarithm the **common logarithm**. We denote $\log_e(x)$ as $\ln(x)$, and we call it the **natural logarithm**.

Other Properties of Logarithmic Functions

Let $a, b > 0$ and $a, b \neq 1$, let $x_1, x_2 > 0$, and let n be any real number. Then the following properties hold:

- $\log_a(1) = 0$

- $\log_a(x_1 x_2) = \log_a(x_1) + \log_a(x_2)$

- $\log_a\left(\dfrac{x_1}{x_2}\right) = \log_a(x_1) - \log_a(x_2)$

- $\log_a(x_1^n) = n\log_a(x_1)$

- $\log_b(x_1) = \dfrac{\log_a(x_1)}{\log_a(b)}$

- $\log_a(x_1) = \log_a(x_2) \Rightarrow x_1 = x_2$

The last two properties in the list are called the **change of base property** and the **one-to-one property of logarithms**, respectively.

One last useful property for dealing with exponential and logarithmic equations is the following equivalence:

$$a^{x_1} = y_1 \Leftrightarrow \log_a(y_1) = x_1$$

This property is a direct consequence of the fact that a^x and $\log_a(x)$ are inverses. It says that an exponential equation can be rewritten in logarithmic form, and that a logarithmic equation can be rewritten in exponential form.

Trigonometric Functions

There are six trigonometric functions, and you should be familiar with all of them. We will cover them in pairs.

Sine and Cosine

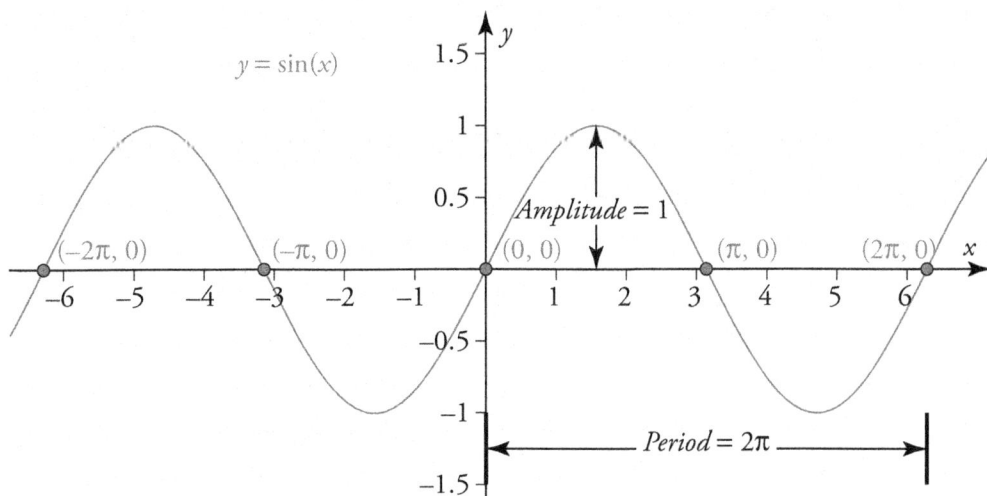

$y = \sin(x)$

$Amplitude = 1$

$(-2\pi, 0)$ $(-\pi, 0)$ $(0, 0)$ $(\pi, 0)$ $(2\pi, 0)$

$Period = 2\pi$

Domain: All Reals

Range: $[-1,1]$

Zeros: $x = n\pi$, $n = 0, \pm 1, \pm 2, \ldots$

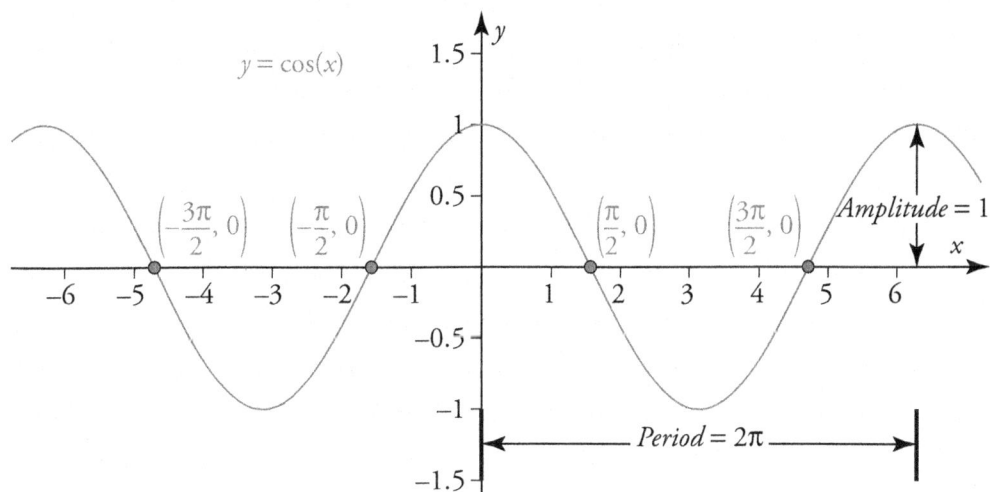

$y = \cos(x)$

$\left(-\dfrac{3\pi}{2}, 0\right)$ $\left(-\dfrac{\pi}{2}, 0\right)$ $\left(\dfrac{\pi}{2}, 0\right)$ $\left(\dfrac{3\pi}{2}, 0\right)$ $Amplitude = 1$

$Period = 2\pi$

Domain: All Reals

Range: $[-1,1]$

Zeros: $x = (2n+1)\dfrac{\pi}{2}$, $n = 0, \pm 1, \pm 2, \ldots$

Both $\sin(x)$ and $\cos(x)$ are continuous for all x.

Cosecant and Secant

Since $csc(x) = \dfrac{1}{sin(x)}$ and $sec(x) = \dfrac{1}{cos(x)}$, we expect that $csc(x)$ and $sec(x)$ are undefined at the zeros of $sin(x)$ and $cos(x)$, respectively.

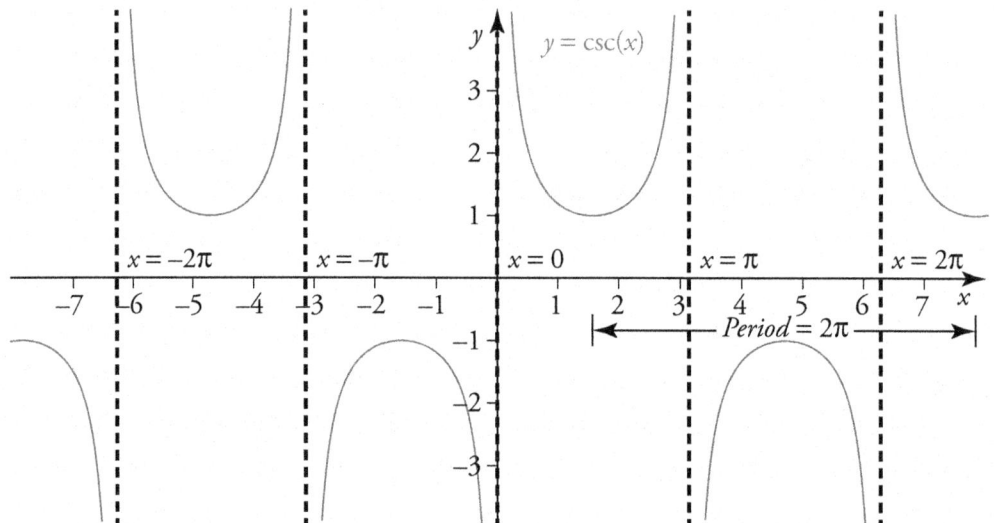

Domain: $x \neq n\pi$, $n = 0, \pm1, \pm2, \ldots$
Range: $(-\infty, -1] \cup [1, \infty)$

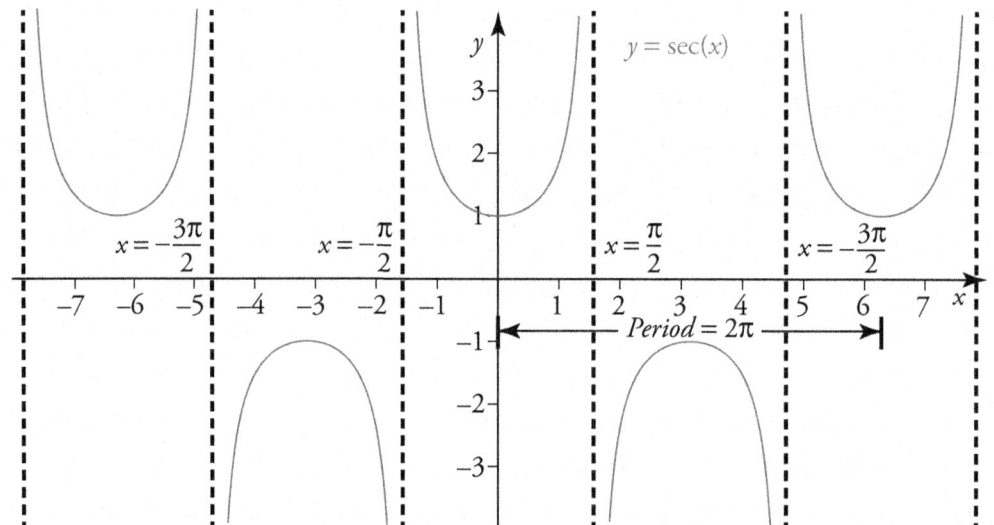

Domain: $x \neq (2n+1)\dfrac{\pi}{2}$, $n = 0, \pm1, \pm2, \ldots$
Range: $(-\infty, -1] \cup [1, \infty)$

Both $csc(x)$ and $sec(x)$ are continuous for all x in their respective domains. The graph of each function has a vertical asymptote for each value of x at which the function is undefined.

Cotangent and Tangent

Since $\cot(x) = \dfrac{\cos(x)}{\sin(x)}$ and $\tan(x) = \dfrac{\sin(x)}{\cos(x)}$, we expect that $\cot(x)$ and $\tan(x)$ are undefined at the zeros of $\sin(x)$ and $\cos(x)$, respectively.

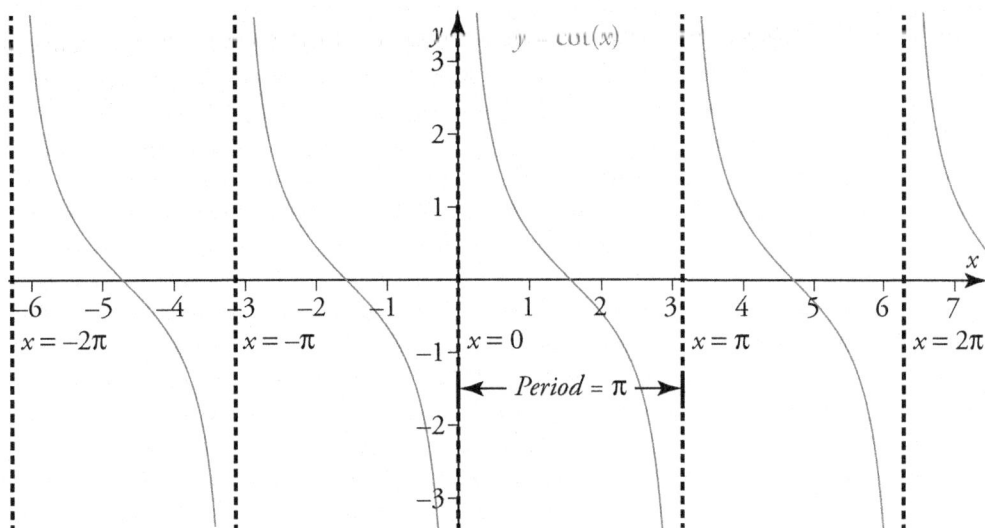

Domain: $x \neq n\pi$, $n = 0, \pm 1, \pm 2, \ldots$
Range: All Reals

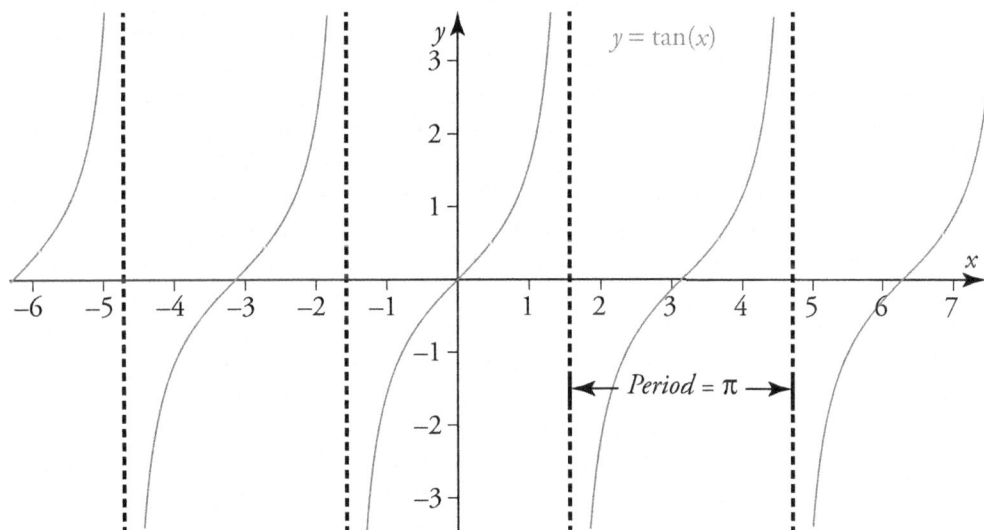

Domain: $x \neq (2n+1)\dfrac{\pi}{2}$, $n = 0, \pm 1, \pm 2, \ldots$
Range: All Reals

Both $\cot(x)$ and $\tan(x)$ are continuous for all x in their respective domains. The graph of each function has a vertical asymptote for each value of x at which the function is undefined.

Inverse Trigonometric Functions

You'll notice that none of the trigonometric functions pass the Horizontal Line Test, which means that none of them are one-to-one. This in turn means that *none of the trigonometric functions has an inverse function*. So what do we mean when we say "inverse trigonometric functions"? In order define inverse trigonometric functions, we must *restrict the domain* of each of the trigonometric functions such that each one passes the Horizontal Line Test (and hence, is one-to-one) on its restricted domain. We do this as follows:

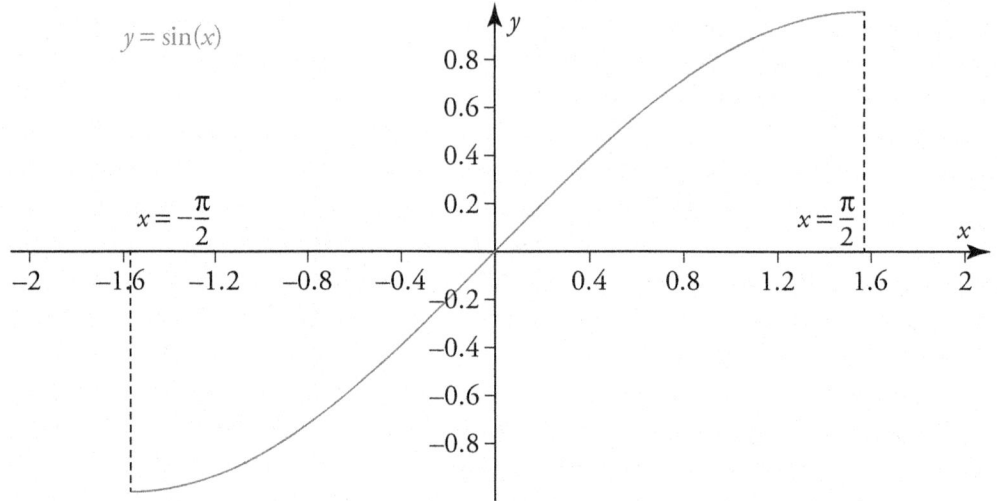

$y = \sin(x)$

Domain: $\left[-\dfrac{\pi}{2}, \dfrac{\pi}{2}\right]$

Range: $[-1, 1]$

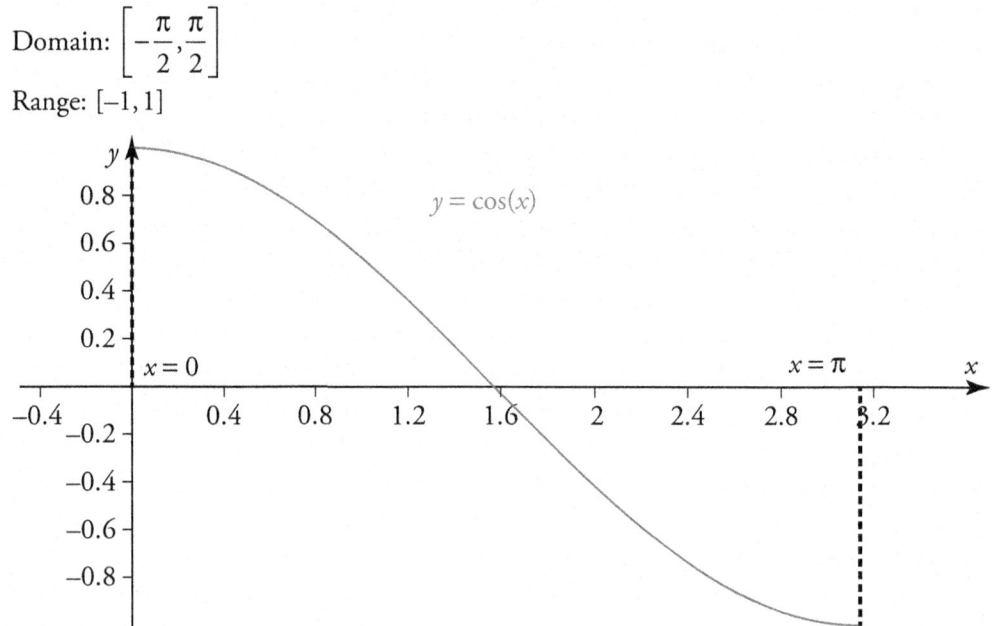

$y = \cos(x)$

Domain: $[0, \pi]$

Range: $[-1, 1]$

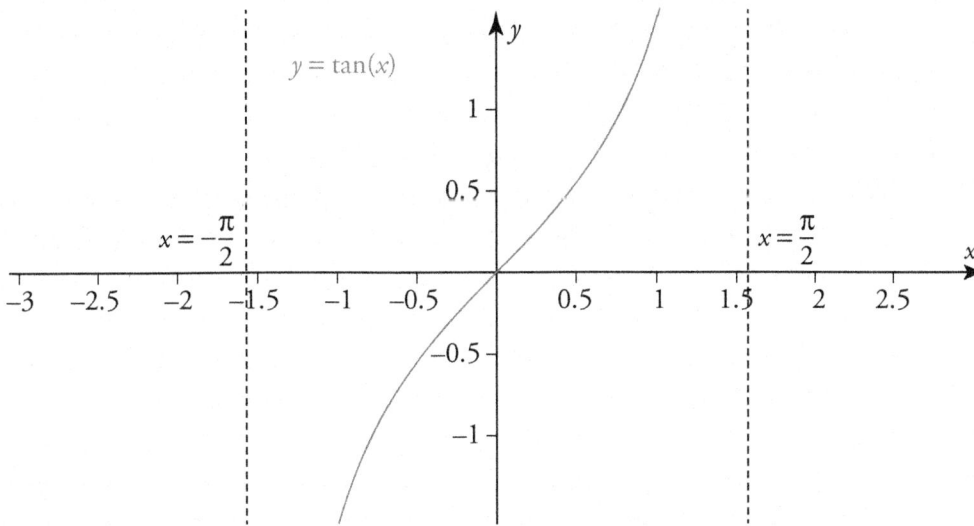

Domain: $\left(-\dfrac{\pi}{2}, \dfrac{\pi}{2}\right)$

Range: All Reals

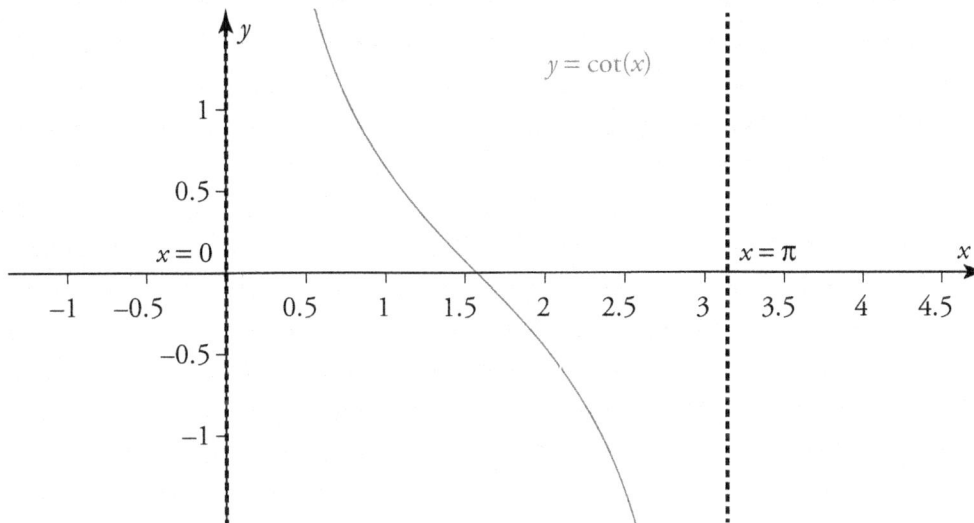

Domain: $(0, \pi)$

Range: All Reals

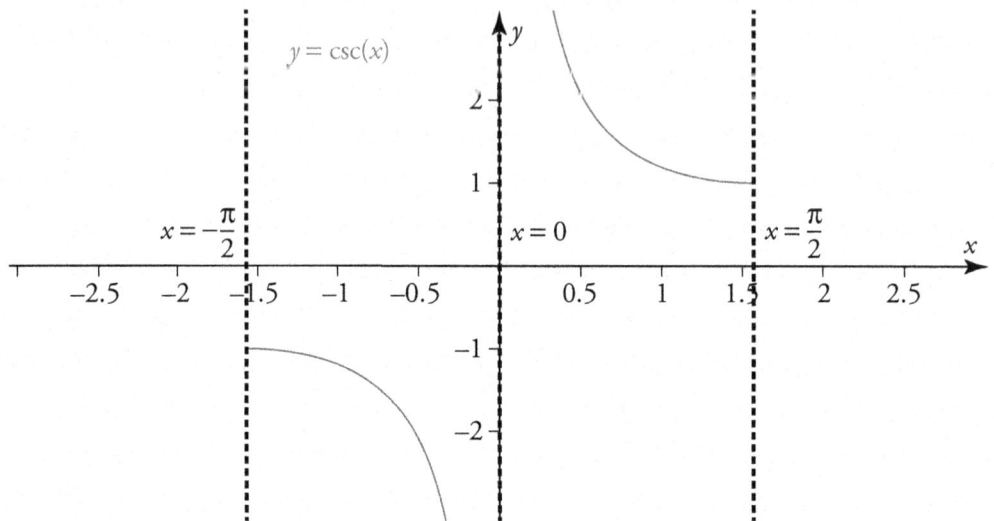

$y = \csc(x)$

$x = -\dfrac{\pi}{2}$ $x = 0$ $x = \dfrac{\pi}{2}$

Domain: $\left[-\dfrac{\pi}{2},0\right) \cup \left(0,\dfrac{\pi}{2}\right]$

Range: $(-\infty,-1] \cup [1,\infty)$

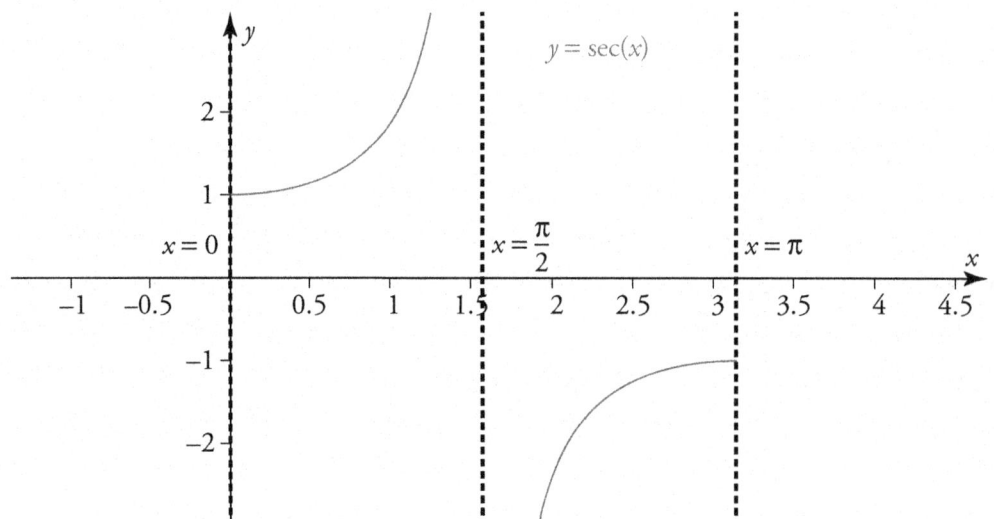

$y = \sec(x)$

$x = 0$ $x = \dfrac{\pi}{2}$ $x = \pi$

Domain: $\left[0,\dfrac{\pi}{2}\right) \cup \left(\dfrac{\pi}{2},\pi\right]$

Range: $(-\infty,-1] \cup [1,\infty)$

On these restricted domains, each trigonometric function *does* have an inverse function. We review them below.

Inverse Sine and Cosine Functions

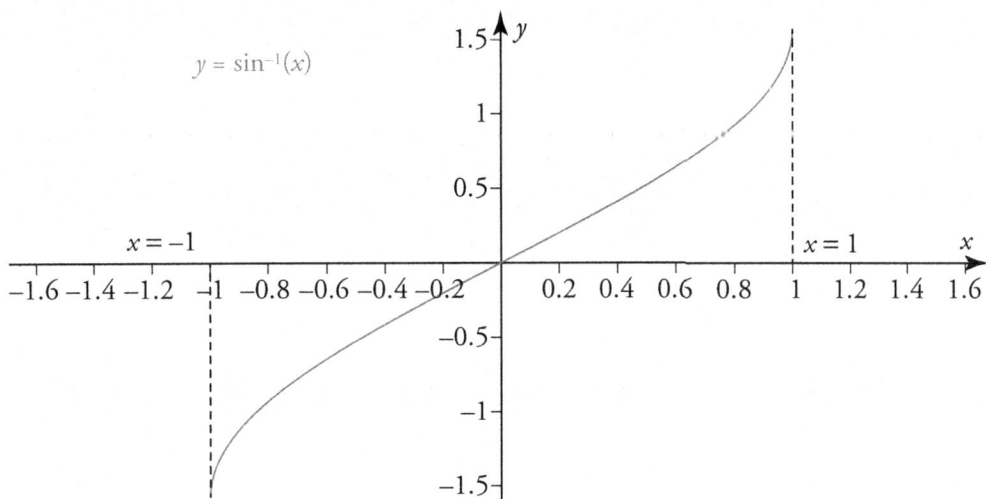

$y = \sin^{-1}(x)$

$x = -1$

$x = 1$

Domain: $[-1, 1]$

Range: $\left[-\dfrac{\pi}{2}, \dfrac{\pi}{2} \right]$

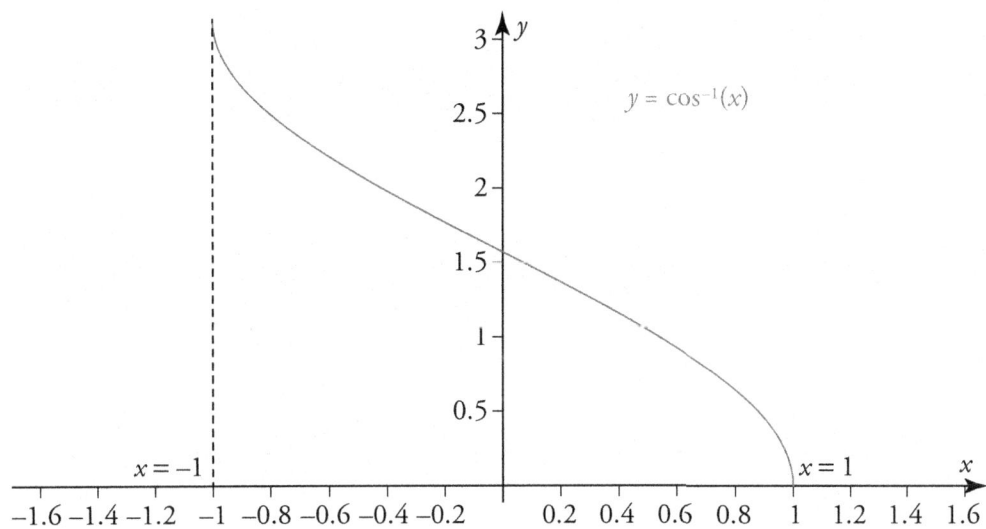

$y = \cos^{-1}(x)$

$x = -1$

$x = 1$

Domain: $[-1, 1]$

Range: $[0, \pi]$

Inverse Tangent and Cotangent Functions

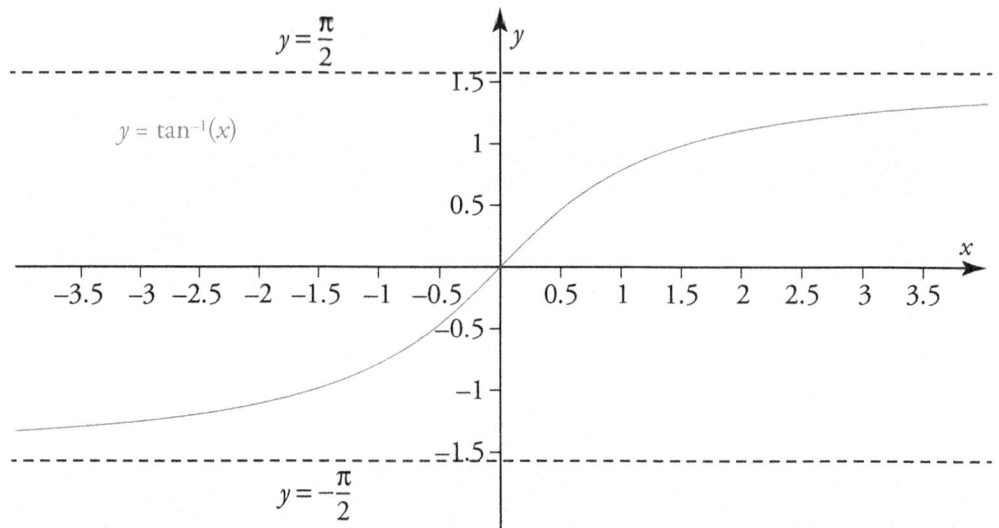

Domain: All Reals

Range: $\left(-\dfrac{\pi}{2},\dfrac{\pi}{2}\right)$

Horizontal Asymptotes: $y=\pm\dfrac{\pi}{2}$

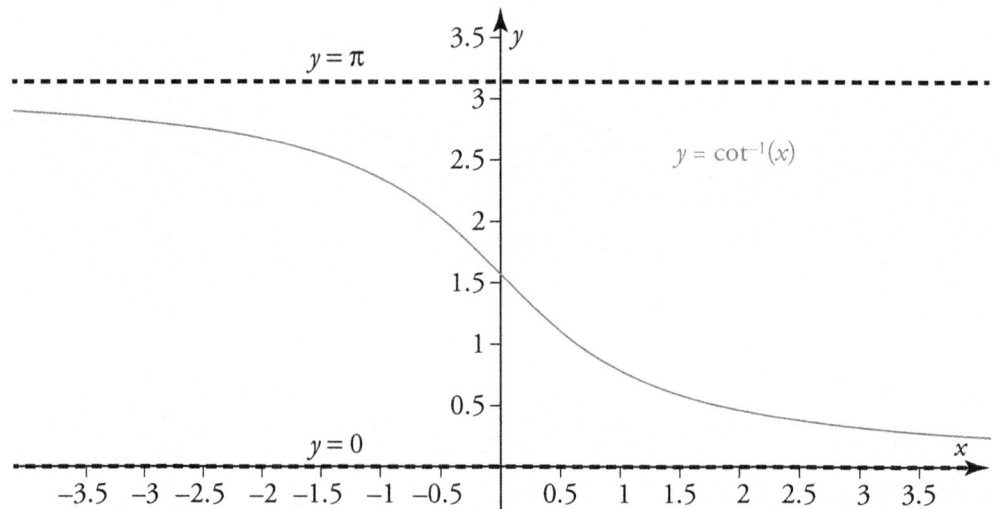

Domain: All Reals

Range: $(0,\pi)$

Horizontal Asymptotes: $y=0$ and $y=\pi$

Inverse Cosecant and Secant Functions

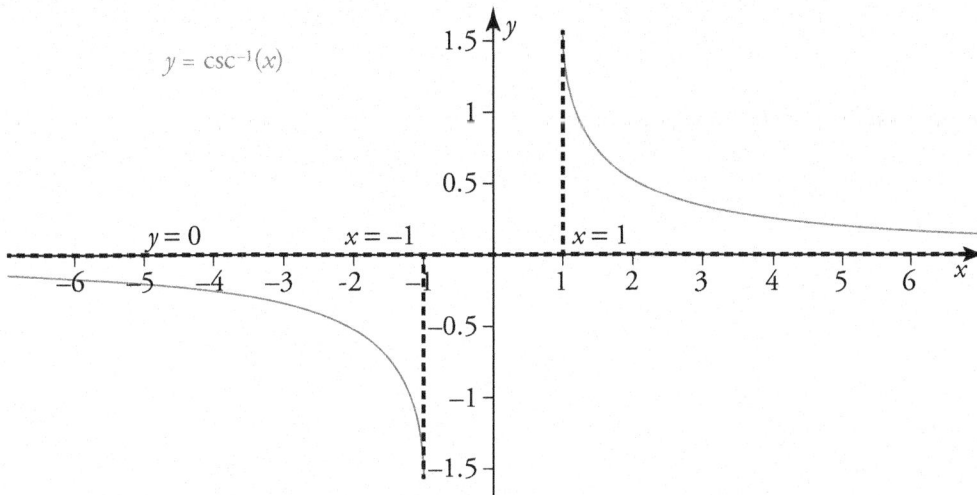

$y = \csc^{-1}(x)$

$y = 0$

$x = -1$

$x = 1$

Domain: $(-\infty, -1] \cup [1, \infty)$

Range: $\left[-\dfrac{\pi}{2}, 0\right) \cup \left(0, \dfrac{\pi}{2}\right]$

Horizontal Asymptote: $y = 0$

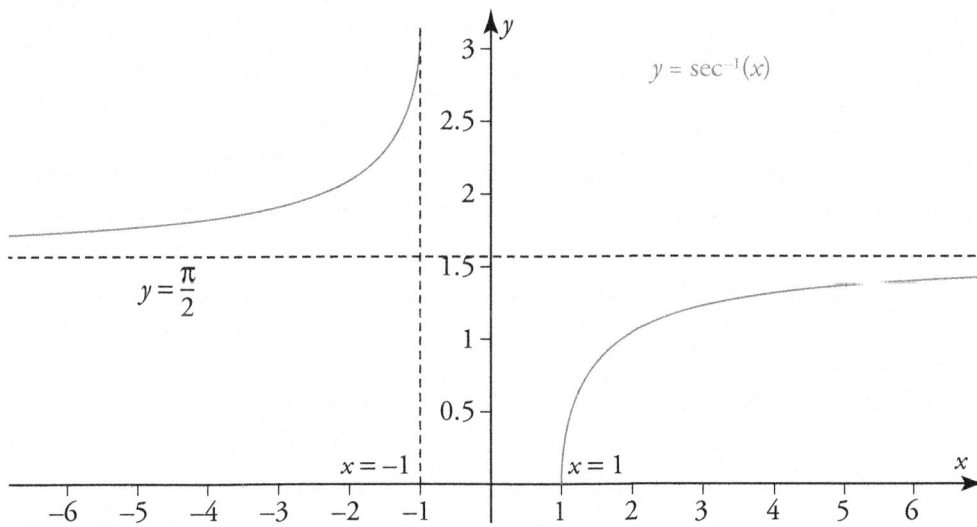

$y = \sec^{-1}(x)$

$y = \dfrac{\pi}{2}$

$x = -1$

$x = 1$

Domain: $(-\infty, -1] \cup [1, \infty)$

Range: $\left[0, \dfrac{\pi}{2}\right) \cup \left(\dfrac{\pi}{2}, \pi\right]$

Horizontal Asymptote: $y = \dfrac{\pi}{2}$

Miscellaneous Functions _____

Two functions that are covered in precalculus that do not fit into any of the above categories are the **absolute value function** and the **greatest integer function**.

The Absolute Value Function

This function is given by $f(x) = |x|$, and it is defined as follows:

$$f(x) = \begin{cases} -x & x < 0 \\ x & x \geq 0 \end{cases}$$

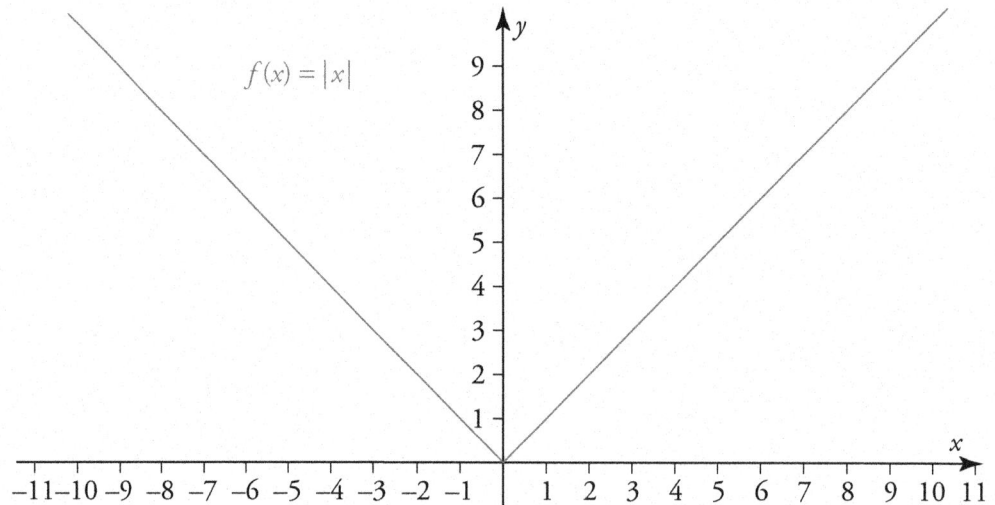

The domain and range of this function are all reals and $[0, \infty)$, respectively. On $(-\infty, 0)$ the function is linear with slope -1, and on $(0, \infty)$ it is linear with slope 1. The function is continuous for all x, but at $x = 0$ the graph comes to a sharp point. We say that the function is **nonsmooth** at that point, and it is not possible to define a slope there.

The Greatest Integer Function

The greatest integer function is denoted by $f(x) = x$, and for any real number x the function returns the greatest integer that is less than or equal to x. The graph of the greatest integer function is shown below.

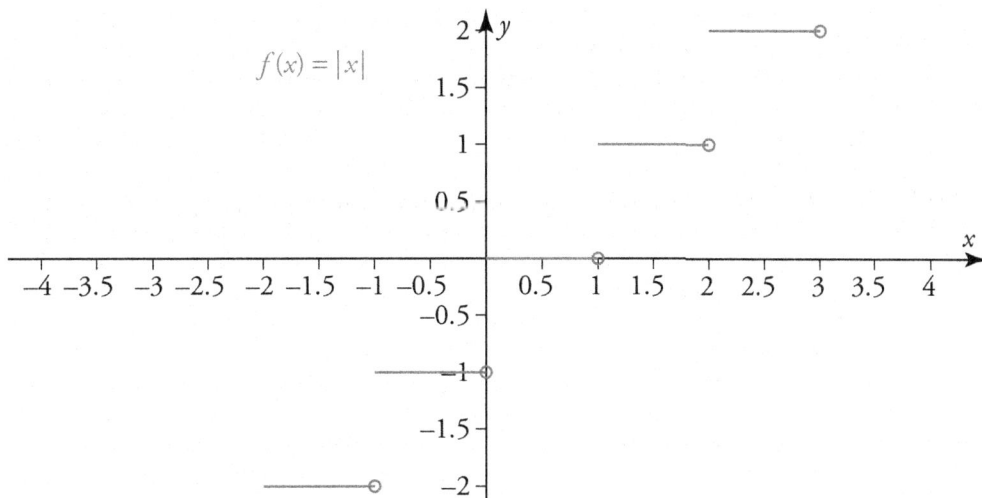

$f(x) = |x|$

The domain of the greatest integer function is all reals, and its range is all integers. It has a **jump discontinuity** at every integer value of x. A jump discontinuity is a type of *non*removable discontinuity.

Symmetry

On the exam, you will need to know about two different types of symmetry: **even symmetry** and **odd symmetry**.

Even Symmetry

A function $y = f(x)$ has even symmetry if $f(-x) = f(x)$ for all x in the domain of f. What this implies is that, for every point (x, y) on the graph of f, the point $(-x, y)$ is also on the graph. A function with even symmetry is illustrated below.

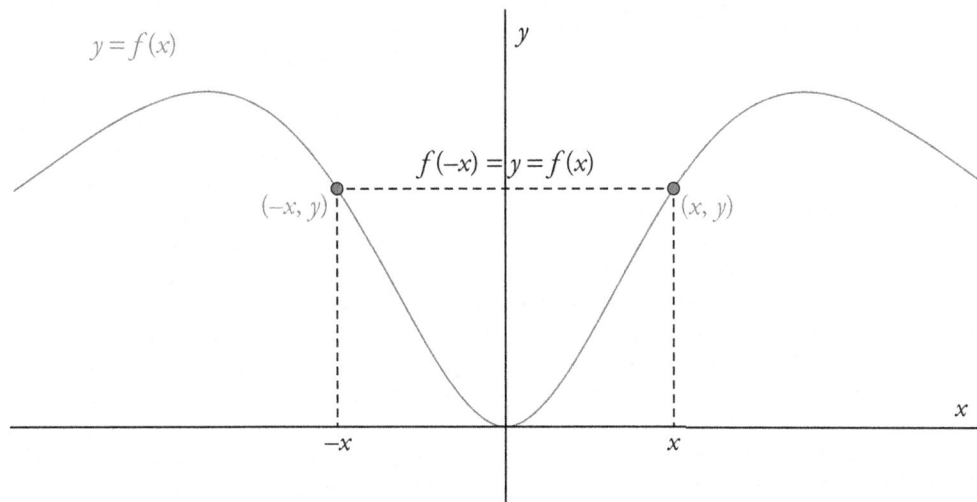

Even symmetry is also called *y*-**axis symmetry**. If you fold the graph along the y-axis, the right branch of the graph is coincident with the left branch.

Odd Symmetry

A function $y = f(x)$ has even symmetry if $f(-x) = -f(x)$ for all x in the domain of f. Consequently, for every point (x, y) on the graph of f, the point $(-x, -y)$ is also on the graph. A function with odd symmetry is illustrated below.

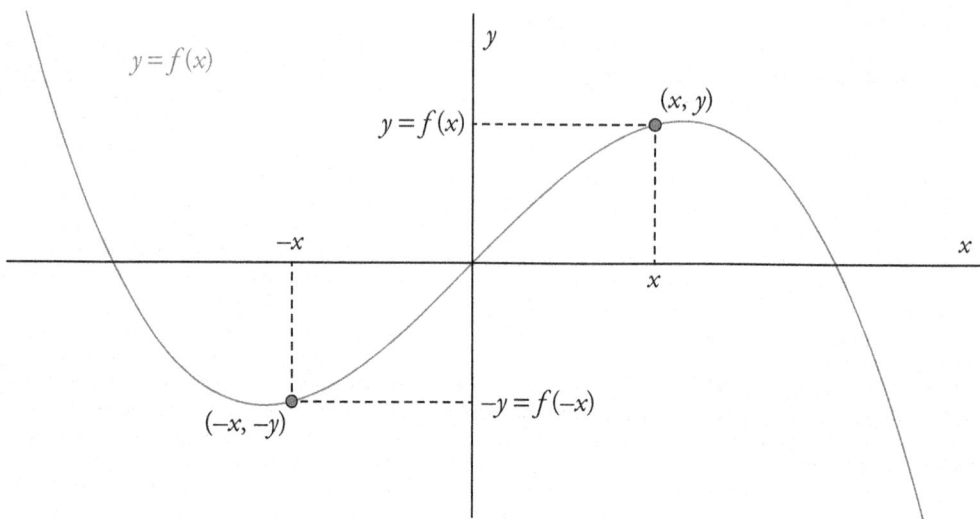

Odd symmetry is also called **origin symmetry**. If you rotate the branch of the graph on the right side of the origin and rotate it 180° about the origin (in either direction), it will be coincident with the branch of the graph on the left side.

Transformations of Functions

And now, we come to the last topic of this chapter. Transformations of functions come in three varieties: translations, reflections, and dilations.

Translations

A translation is a rigid transformation in which every point on a graph is shifted by the same distance and in the same direction. Let $c > 0$.

Type	Example
Vertical Translation: $f(x) \pm c$ • $f(x) + c$ shifts the graph of $f(x)$ up by c units. • $f(x) - c$ shifts the graph of $f(x)$ down by c units.	

Horizontal Translation: $f(x \pm c)$	
• $f(x - c)$ shifts the graph of $f(x)$ to the right by c units. • $f(x + c)$ shifts the graph of $f(x)$ to the left by c units.	

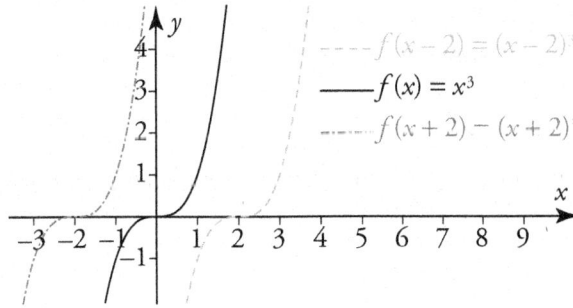

Reflections

A reflection is a rigid transformation in which every point on a graph is reflected in the same line.

Type	Example
Reflection in the x-Axis: $-f(x)$	
Reflection in the y-Axis: $f(-x)$	

Dilations

A dilation is a nonrigid transformation which either stretches or shrinks the graph in the direction of a given line.

Type	Example
Vertical Dilation: $cf(x)$ • $cf(x)$ stretches the graph of $f(x)$ vertically if $c > 1$. • $cf(x)$ shrinks the graph of $f(x)$ vertically if $0 < c < 1$.	$2f(x) = 2\sin(x)$ $f(x) = \sin(x)$ $\tfrac{1}{2}f(x) = \tfrac{1}{2}\sin(x)$
Horizontal Dilation: $f(cx)$ • $f(cx)$ shrinks the graph of $f(x)$ horizontally if $c > 1$. • $f(cx)$ stretches the graph of $f(x)$ horizontally if $0 < c < 1$.	$f\left(\tfrac{1}{2}x\right) = \cos\left(\tfrac{1}{2}x\right)$ $f(x) = \cos(x)$ $f(2x) = \cos(2x)$

Chapter 1: Limits and Continuity

Chapter 1 Overview

Chapter 1 is divided into the following two Enduring Understandings:

- **EU 1.1:** The concept of a limit can be used to understand the behavior of functions.
- **EU 1.2:** Continuity is a key property of functions that is defined using limits.

EU 1.1 – Limits

LO 1.1A(a,b) – Limits Expressed Symbolically

Since the two parts of the first Learning Objective are intertwined, we will review them together. The full statements of the LO's are given below.

LO 1.1A(a): Express limits symbolically using correct notation.

LO 1.1A(b): Interpret limits expressed symbolically.

The first item of Essential Knowledge under **LO 1.1A** is a basic description of a limit.

EK 1.1A1: Given a function f, the limit of $f(x)$ as x approaches c is a finite real number R if $f(x)$ can be made arbitrarily close to R by taking x sufficiently close to c (but not equal to c). If the limit exists and is a real number, then the common notation is $\lim_{x \to c} f(x) = R$.

The above description of limit may seem confusing. Let's try to understand what the limit means with help of an example.

Example 1.1.1: Consider the function $f(x) = \dfrac{9}{x} + 1$ and its values around $x = 3$ at $x = 2.990, 2.995, 2.999, 3.001, 3005,$ and 3.010. These values are listed in the table below.

x	2.990	2.995	2.999	3.001	3.005	3.010
$f(x)$	4.010	4.005	4.001	3.999	3.995	3.990

In the table, as we move inwards from the outer cells to the center, x is approaching 3 from each side. The statement "x approaches 3" is denoted by $x \to 3$. We also see that the value of $f(x)$ approaches 4 from each side. This is the *limit* of $f(x)$ as $x \to 3$. The statement "the limit of $f(x)$ as x approaches 3 is 4" is written in the notation of calculus as follows:

$$\lim_{x \to 3} f(x) = 4$$

Example 1.1.2: It is given that the value of a function $f(x)$ approaches 20 as x approaches 4. Express this using limit notation.

Even though we don't know the function, we have everything we need to translate this into limit notation.

$$\lim_{x \to 4} f(x) = 20$$

Example 1.1.3: Express the following limit notation in words:

$$\lim_{y \to -4} h(y) = -25$$

Here, the independent variable of the function $h(y)$ is y, so the limit notation can be expressed in words in this way: as y approaches -4, $h(y)$ approaches -25.

With our final example for this EK, we wish to point out that it is possible for $\lim_{x \to c} f(x) = R$ even if $f(c) \neq R$, or even if $f(c)$ is *undefined*.

Example 1.1.4: Consider the function $f(x) = \dfrac{x^2 - 1}{x - 1}$, which is graphed below.

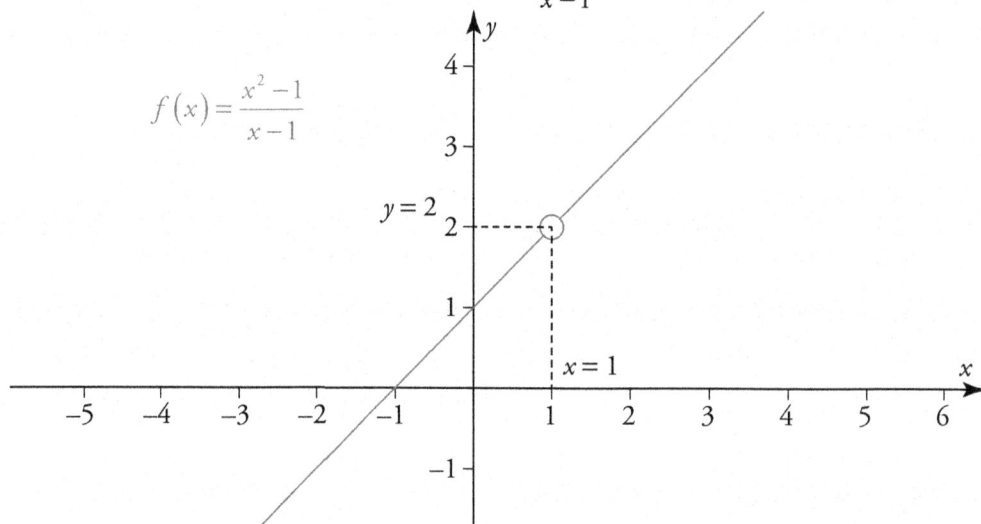

The graph of this function looks just like that of $g(x) = x + 1$, except at $x = 1$, where f is not defined. However, it is still the case that $\lim_{x \to 1} f(x) = 2$, because the value of $f(x)$ *approaches* 2 as x approaches 1.

Next, we will look at some special types of limits.

EK 1.1A2: The concept of a limit can be extended to include one-sided limits, limits at infinity, and infinite limits.

One-Sided Limits

We hinted at one-sided limits when we considered $\lim\limits_{x\to 3}\left(\dfrac{9}{x}+1\right)$ in Example 1.1.1, when we considered what happens when we approach $x = 3$ from *each side*. More generally, the limit of a function $f(x)$ as x approaches c *from the left* is denoted $\lim\limits_{x\to c^-} f(x)$, and the limit of $f(x)$ as x approaches c *from the right* is denoted $\lim\limits_{x\to c^+} f(x)$. These are referred to as *left-* and *right-handed limits*, respectively, and they are collectively referred to as *one-sided limits*. We'll illustrate with the function from Example 1.1.1.

Example 1.1.5: Consider the function $f(x)=\dfrac{9}{x}+1$ near $x = 3$, which was the topic of discussion in **EK 1.1A1**. On the number line, the right side of $x = 3$ implies the region where $x > 3$ and the left side of $x = 3$ implies the region where $x < 3$. So when x increases from any value less than 3, and approaches $x = 3$, it is said that x *approaches 3 from the left* and is mathematically expressed as $x \to 3^-$. Similarly, when x decreases from any value greater than 3, and approaches $x = 3$ it is said that x *approaches 3 from the right* and is mathematically expressed as $x \to 3^+$. This is also explained with the graph of $f(x)$ below.

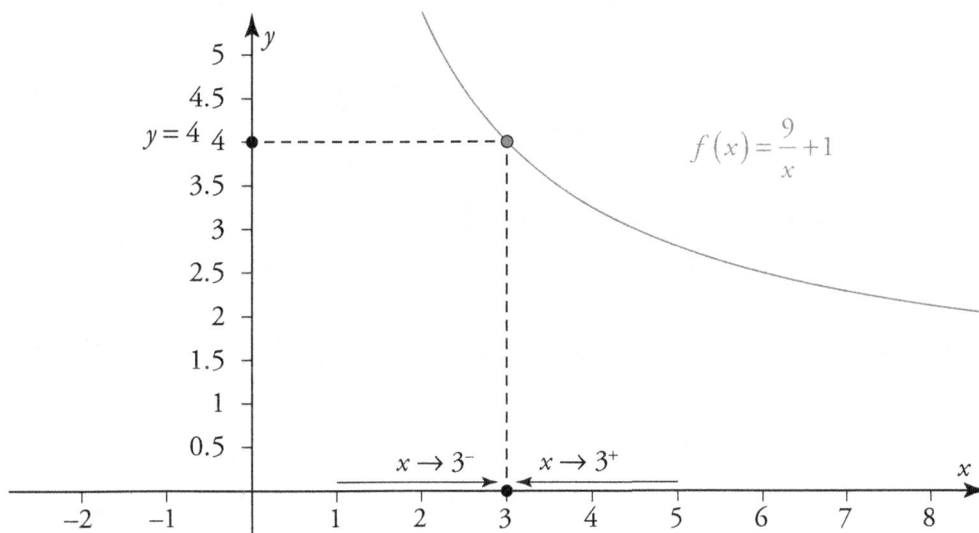

As $x \to 3^-$, $f(x) \to 4$. We express this by saying that the *left-handed limit* of $f(x)$ as x approaches 3 is 4. In mathematical notation, we express it as follows:

$$\lim_{x\to 3^-} f(x) = 4$$

As $x \to 3^+$, $f(x) \to 4$. We say that the *right-handed limit* of $f(x)$ as x approaches 3 is 4, and we express this statement in symbols as follows:

$$\lim_{x\to 3^+} f(x) = 4$$

Limits at Infinity

Sometimes the value of a function $f(x)$ approaches a finite value R as x approaches either $\pm\infty$. We express these as $\lim\limits_{x\to-\infty} f(x) = R$ and $\lim\limits_{x\to\infty} f(x) = R$, respectively. We will illustrate these by revisiting two examples that we discussed in **Prelude: The Functions of Precalculus**.

Example 1.1.6: Consider the function $f(x) = \dfrac{-4x^2+1}{3x^2-1}$, which is graphed below.

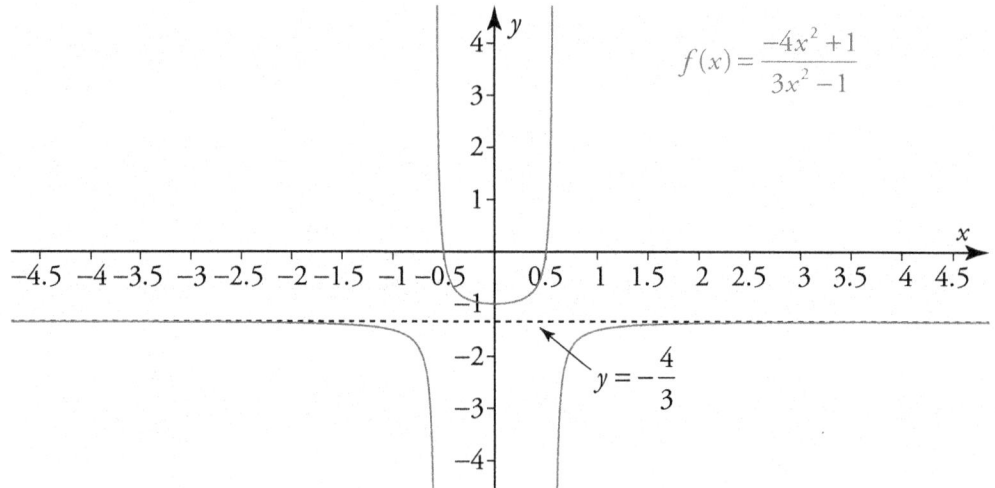

If we were studying this function in a precalculus course, we would describe the end behavior of this function by saying $f(x) \to -\dfrac{4}{3}$ as $x \to \pm\infty$. What we really mean by that is $\lim\limits_{x\to-\infty} \dfrac{-4x^2+1}{3x^2-1} = -\dfrac{4}{3}$ and $\lim\limits_{x\to\infty} \dfrac{-4x^2+1}{3x^2-1} = -\dfrac{4}{3}$.

Example 1.1.7: Consider the function $f(x) = \tan^{-1}(x)$, which is graphed below.

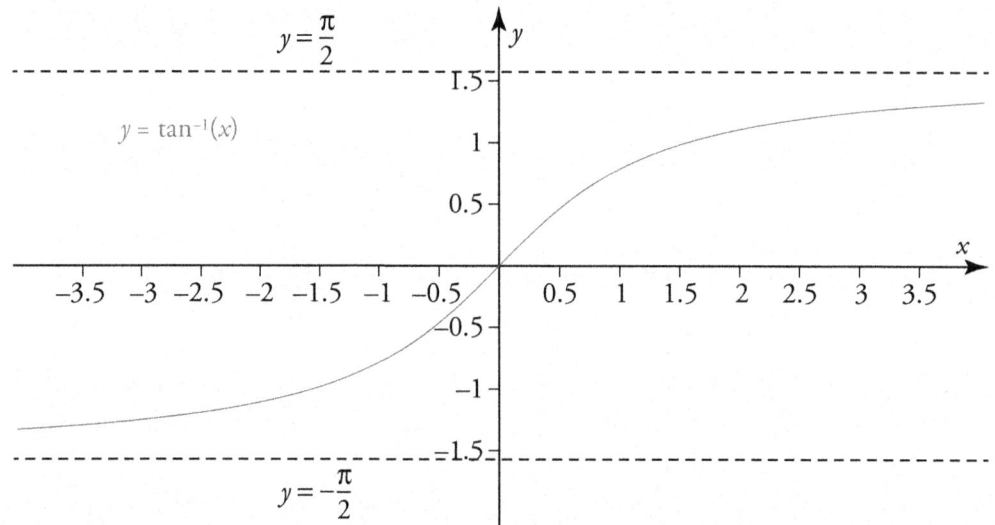

We can describe the end behavior of this function with the limits $\lim\limits_{x\to-\infty} \tan^{-1}(x) = -\dfrac{\pi}{2}$ and $\lim\limits_{x\to\infty} \tan^{-1}(x) = \dfrac{\pi}{2}$.

Infinite Limits

If a function $f(x)$ is *unbounded* as $x \to c$, then we express this mathematically using an *infinite limit*. If $f(x) \to -\infty$ as $x \to c$, then we say $\lim_{x \to c} f(x) = -\infty$, and if $f(x) \to -\infty$ as $x \to c$, then we say $\lim_{x \to c} f(x) = \infty$. Just as with limits at infinity, we'll illustrate this with an example from **Prelude: The Functions of Precalculus**.

Example 1.1.8: Consider the function $g(x) = \dfrac{1}{x-3}$, which is graphed below.

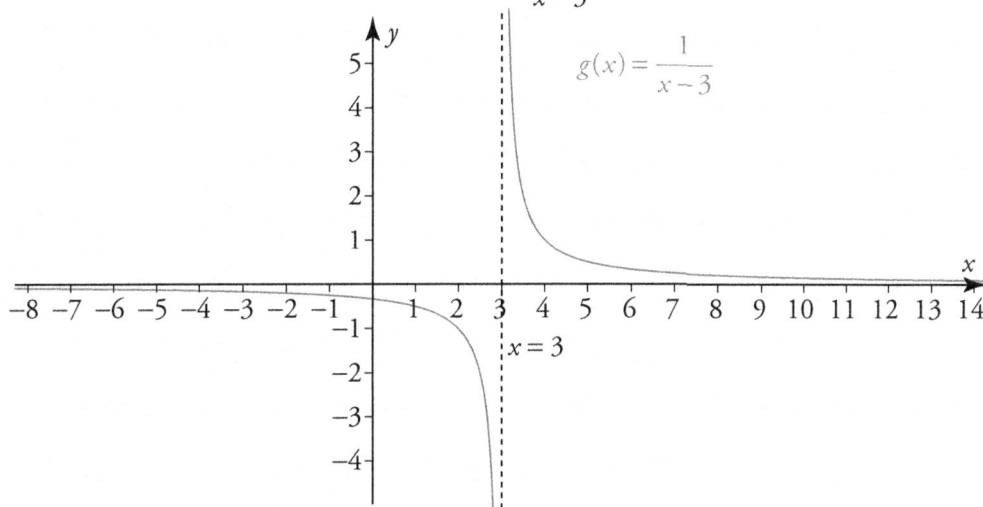

As $x \to 3^-$, $g(x) \to -\infty$, and as $x \to 3^+$, $g(x) \to \infty$. We express these using the infinite limits as $\lim_{x \to 3^-} \dfrac{1}{x-3} = -\infty$ and $\lim_{x \to 3^+} \dfrac{1}{x-3} = \infty$.

Suppose that we did not have the graph of g. We could still tell that the one-sided limits are infinite by trying to evaluate the limit by direct substitution:

$$\lim_{x \to 3} \frac{1}{x-3} \to \frac{1}{3-3} \to \frac{1}{0}$$

If evaluating a limit by direct substitution yields the form $\dfrac{a}{0}$, *where a is a nonzero number, then the corresponding one-sided limits will be infinite.* Every time. You just have to figure out if they're positively or negatively infinite.

For this example, when considering $\lim_{x \to 3^-} \dfrac{1}{x-3}$, imagine inserting a number that is slightly *less* than 3 in for x. Then you'll have a *positive* numerator over a *negative* denominator, which gives you a *negative* quotient, and so $\lim_{x \to 3^-} \dfrac{1}{x-3} = -\infty$. When considering $\lim_{x \to 3^+} \dfrac{1}{x-3}$, imagine plugging in a number that is slightly greater than 3 in for x. Then both the numerator and denominator are *positive*, and so the quotient is *positive*, hence $\lim_{x \to 3^+} \dfrac{1}{x-3} = \infty$.

Infinite Limits at Infinity

Here we combine the last two concepts. A function $f(x)$ has an *infinite limit at infinity* if either $\lim\limits_{x \to -\infty} f(x) = \pm\infty$ or $\lim\limits_{x \to \infty} f(x) = \pm\infty$.

Example 1.1.9: Consider the function $f(x) = x^2$, whose graph is shown below.

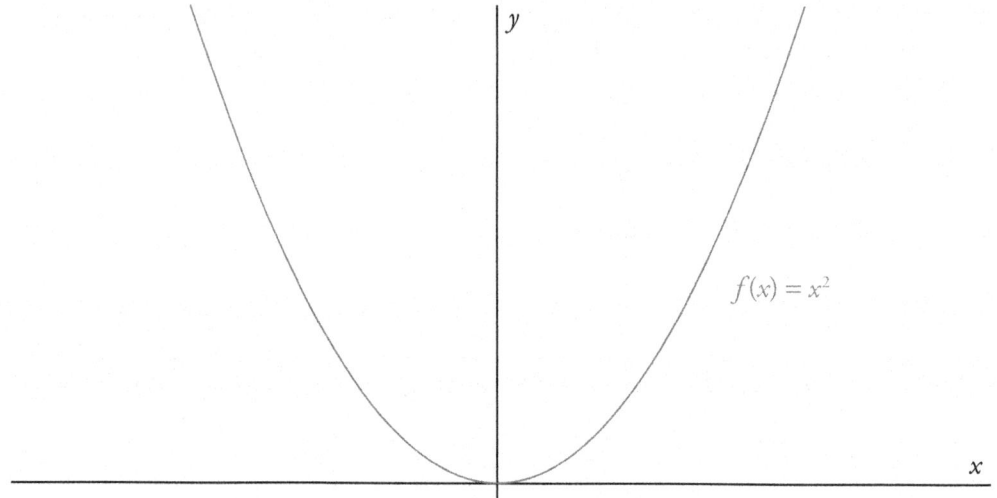

As $x \to \pm\infty$, $f(x) \to \infty$. We express this in the language of calculus with the limits $\lim\limits_{x \to -\infty} x^2 = \infty$ and $\lim\limits_{x \to \infty} x^2 = \infty$.

It should be noted that, as far as the AP Calculus Exam is concerned, infinite limits are examples of limits that *do not exist*. We'll spend more time on limits that don't exist while reviewing the next EK.

EK 1.1A3: A limit might not exist for some functions at particular values of x. Some ways wherein the limit might not exist are if the function is unbounded, if the function is oscillating near this value, or if the limit from the left does not equal the limit from the right.

Before we get to the conditions under which a limit *doesn't* exist, let us be clear about the condition under which a limit *does* exist: $\lim\limits_{x \to c} f(x) = R$ if and only if $\lim\limits_{x \to c^-} f(x) = R$ and $\lim\limits_{x \to c^+} f(x) = R$.

Unbounded Functions / Infinite Limits

We'll illustrate this with a simple example.

Example 1.1.10: Consider the function $f(x) = \dfrac{1}{x^2}$, which is graphed below.

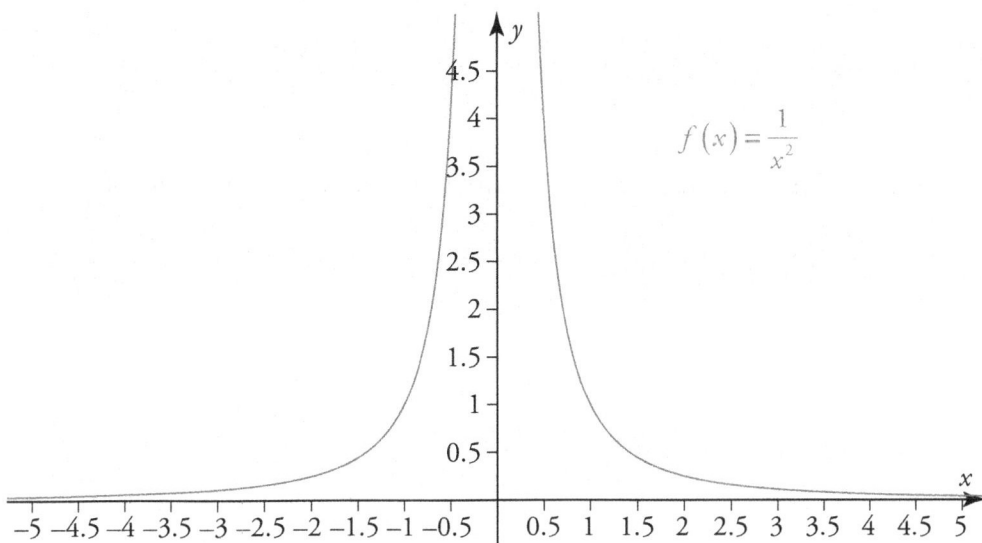

$f(x) = \dfrac{1}{x^2}$

As you can see from the graph, $\lim\limits_{x \to 0} \dfrac{1}{x^2} = \infty$. $f(x) = \dfrac{1}{x^2}$ is unbounded at $x = 0$. Since the limit is not equal to a real number R, it does not exist.

Oscillating Functions

$\lim\limits_{x \to c} f(x)$ does not exist if $f(x)$ oscillates between two fixed values as $x \to c$.

Example 1.1.11: Consider the function $f(x) = \sin\left(\dfrac{1}{x}\right)$ and its limit at $x = 0$. To examine existence of the limit, first evaluate $f(x)$ around $x = 0$, i.e. on both left and right of $x = 0$.

x:	$-\dfrac{2}{\pi}$	$-\dfrac{2}{3\pi}$	$-\dfrac{2}{5\pi}$	$-\dfrac{2}{7\pi}$	0
$f(x)$:	-1	1	-1	1	???

x:	$\dfrac{2}{\pi}$	$\dfrac{2}{3\pi}$	$\dfrac{2}{5\pi}$	$\dfrac{2}{7\pi}$	0
$f(x)$:	1	-1	1	-1	???

As you can see from the table, as $x \to 0$ from either the left or the right, $f(x)$ does not approach a unique value. The graph of this function is shown below. Note the rapid oscillation of $f(x)$ between -1 and 1 near $x = 0$.

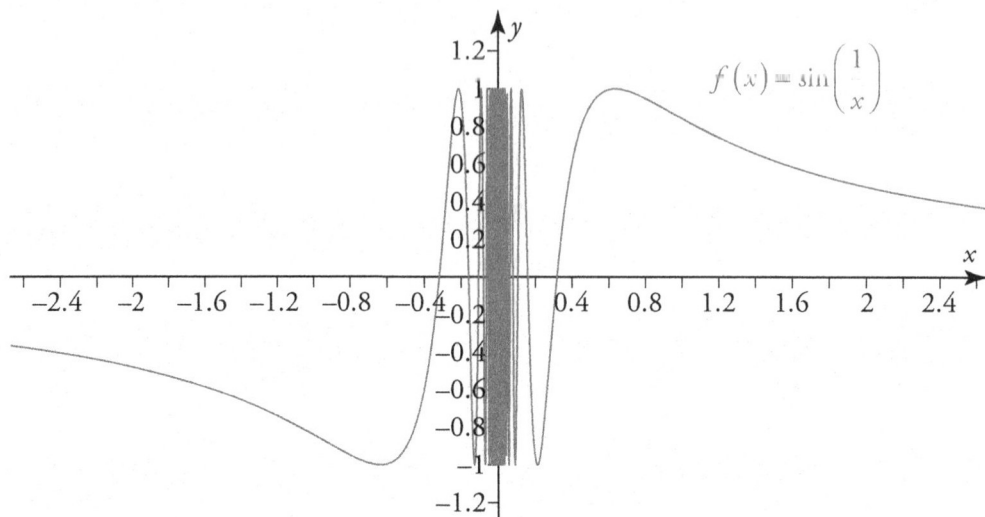

$f(x) = \sin\left(\dfrac{1}{x}\right)$

Piecewise Functions

Recall that piecewise functions are those which are defined differently over different intervals of the independent variable. The limit of piecewise functions defined at the boundary points exist if the left- and right-hand limit at the boundary point of adjoining "pieces" (domains of definition) are equal.

In other words, if a function $f(x)$ is defined as

$$f(x) = \begin{cases} g(x) & x < c \\ h(x) & x > c \end{cases},$$

then $\lim\limits_{x \to c} f(x) = R$ exists if and only if $\lim\limits_{x \to c^-} g(x) = R = \lim\limits_{x \to c^+} h(x)$.

Example 1.1.12: Consider the function $f(x) = \dfrac{|x|}{x}$. Evaluate $\lim\limits_{x \to 0^-} f(x)$ and $\lim\limits_{x \to 0^+} f(x)$, and hence show that $\lim\limits_{x \to 0} f(x)$ does not exist.

This function is defined piecewise as follows:

$$f(x) = \frac{|x|}{x} = \begin{cases} -1 & x < 0 \\ 1 & x > 0 \end{cases}$$

The graph is shown below.

$$f(x) = \frac{|x|}{x}$$

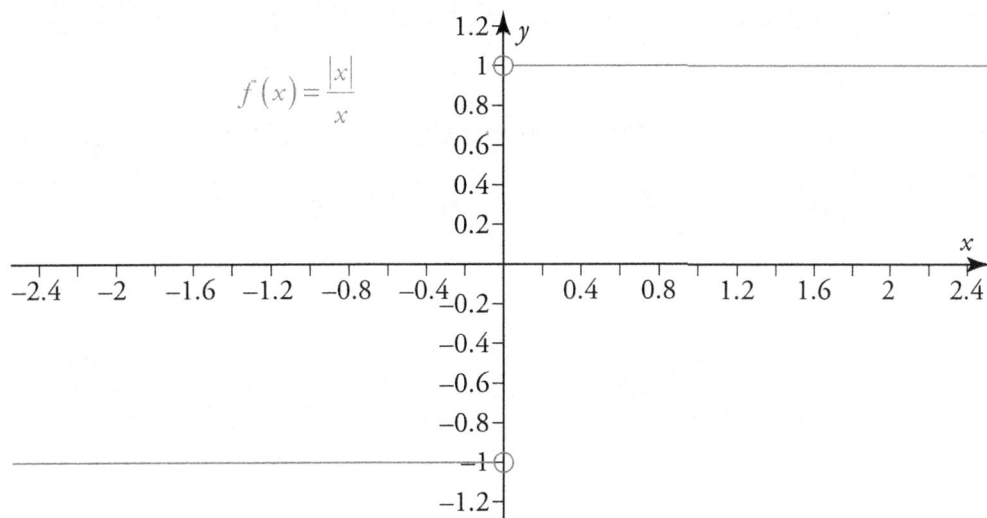

From the graph we can see that $\lim\limits_{x \to 0^-} \dfrac{|x|}{x} = -1$ and $\lim\limits_{x \to 0^+} \dfrac{|x|}{x} = 1$. Since $\lim\limits_{x \to 0^-} \dfrac{|x|}{x} \neq \lim\limits_{x \to 0^+} \dfrac{|x|}{x}$, it follows that $\lim\limits_{x \to 0} \dfrac{|x|}{x}$ does not exist.

Example 1.1.13: Suppose $f(x) = \dfrac{1}{x}$. Evaluate $\lim\limits_{x \to 0^-} f(x)$ and $\lim\limits_{x \to 0^+} f(x)$, and hence show that $\lim\limits_{x \to 0} f(x)$ does not exist.

The graph of the function is shown below.

$$f(x) = \frac{1}{x}$$

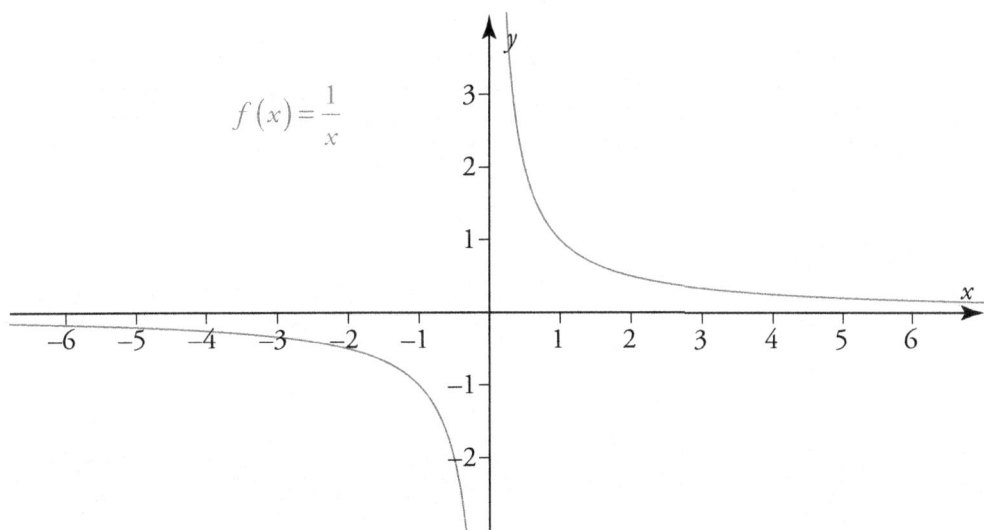

As we can see from the graph, $\lim\limits_{x \to 0^-} \dfrac{1}{x} = -\infty$, and $\lim\limits_{x \to 0^+} \dfrac{1}{x} = \infty$. Since both limits are infinite limits, $\lim\limits_{x \to 0} \dfrac{1}{x}$ does not exist. Indeed, we cannot even say that $\lim\limits_{x \to 0} \dfrac{1}{x} = -\infty$ or $\lim\limits_{x \to 0} \dfrac{1}{x} = \infty$, because $\lim\limits_{x \to 0^-} \dfrac{1}{x} \neq \lim\limits_{x \to 0^+} \dfrac{1}{x}$.

LO 1.1B – Evaluation of limits of functions

The full Learning Objective of this section is as follows.

LO 1.1B: Estimate limits of functions.

The first and only item of Essential Knowledge under LO 1.1B discusses how to use numerical and graphical methods to evaluate limits of functions at given points.

EK 1.1B1: Numerical and graphical information can be used to estimate limits.

Numerical Method

We have already reviewed this method in Examples 1.1.1 and 1.1.11. In this method, you estimate $\lim_{x \to c} f(x)$ by calculating $f(x)$ for values of x that approach c from each side. We will show one more example.

Example 1.1.14: Estimate $\lim_{x \to 3} \dfrac{2x^3 - 3}{5x + 2}$.

We need to find the value of $\dfrac{2x^3 - 3}{5x + 2}$ as $x \to 3$. We first build a table, with help of a calculator, with values of x and the function where x is sufficiently close to 3 from both the right and left side. Then, based on the trend of the values of the function, we conclude about the limit.

	$x \to 3^-$			$x \to 3^+$		
x	2.990	2.995	2.999	3.001	3.005	3.009
$f(x) = \dfrac{2x^3 - 3}{5x + 2}$	2.997	2.989	2.998	3.002	3.011	3.021

It is evident that $\lim_{x \to 3^-} \dfrac{2x^3 - 3}{5x + 2} \approx 3$ and $\lim_{x \to 3^+} \dfrac{2x^3 - 3}{5x + 2} \approx 3$. When we review **EK 1.1C1**, we will see that both of these limits are *exactly* equal to 3.

Graphical Method

We reviewed this method in Examples 1.1.4 through 1.1.10, and Examples 1.1.12-1.1.13. In this method, you need to plot the graph of the function around the point at which the limit is considered. From that graph, you need to make observations about the function around the points and then conclude about the limit. This method may not be accurate if the limit is not a whole number and if the graph is not drawn to a reasonably high scale for both x- and y-axes. However, this method is helpful for piecewise-defined functions, as you will see in our next example.

Example 1.1.15: A piecewise function is graphed below. Each square corresponds to 0.5 units.

$y = f(x)$

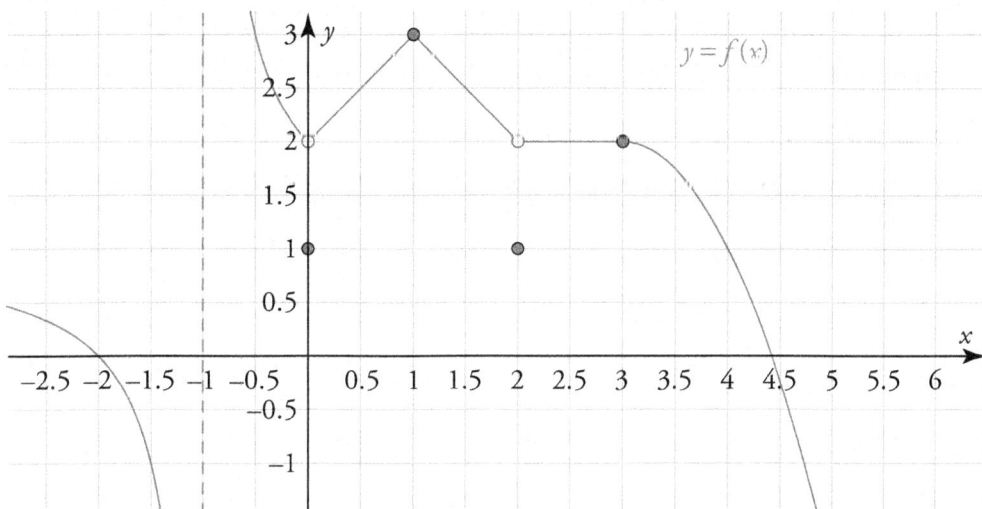

Use the graph to estimate $\lim_{x \to c} f(x)$ for $c = -1, 0, 1, 2, 3$, if the limit exists. If the limit does not exist, state the values of the one-sided limits.

It turns out that we can get exact values from this graph.

- $\lim_{x \to -1^-} f(x) = -\infty$ and $\lim_{x \to -1^+} f(x) = \infty$, so $\lim_{x \to -1} f(x)$ does not exist.
- $\lim_{x \to 0} f(x) = 2$. This is despite the fact that $f(0) = 1$. The value of $f(x)$ approaches 2 on both sides of $x = 0$.

- $\lim_{x \to 1} f(x) = 3$

- $\lim_{x \to 2} f(x) = 2$

- $\lim_{x \to 3} f(x) = 2$

LO 1.1C – Determine Limits of Functions

The Learning Objective states exactly the same thing as the section title.

LO 1.1C: Determine the limits of functions.

We're going to abandon the numerical and graphical approaches to limits in this EK. Here we will review the limit theorems that are used to evaluate limits *analytically*. Before we get to the EK, we'll get warmed up with some simple limit theorems.

Basic Limit Theorems

For real numbers c and k and positive integers n, the following limit rules hold.

$$\lim_{x \to c} k = k$$

$$\lim_{x \to c} x = c$$

$$\lim_{x \to c} x^n = c^n$$

What can we do with these? Not a whole lot, but we'll illustrate what we can do with an example.

Example 1.1.16: Evaluate the following limits:

$$\lim_{x \to -3} 6$$

$$\lim_{x \to 5} x$$

$$\lim_{x \to -2} x^3$$

Straightforwardly applying the three rules given above, we obtain the following:

$$\lim_{x \to -3} 6 = 6$$

$$\lim_{x \to 5} x = 5$$

$$\lim_{x \to -2} x^3 = (-2)^3 = -8$$

The first item of Essential Knowledge under LO 1.1C expands this list of limit theorems so that we can do more.

EK 1.1C1: The limits of sums, differences, products, quotients, and composite functions can be found using the basic theorems of limits and algebraic rules.

More Limit Theorems

Here are the limit theorems that pertain to this EK. Let c and k be real numbers, let n be a positive integer, and let f and g be functions with $\lim_{x \to c} f(x) = R$ and $\lim_{x \to c} g(x) = S$. Then the following rules hold.

Scalar Multiple Rule: $\lim_{x \to c} \left(kf(x) \right) = k \lim_{x \to c} f(x) = kR$

Sum and Difference Rule: $\lim_{x \to c} \left(f(x) \pm g(x) \right) = \lim_{x \to c} f(x) \pm \lim_{x \to c} g(x) = R \pm S$

Product Rule: $\lim_{x \to c} \left(f(x) g(x) \right) = \left(\lim_{x \to c} f(x) \right)\left(\lim_{x \to c} g(x) \right) = RS$

Quotient Rule: $\lim_{x \to c} \dfrac{f(x)}{g(x)} = \dfrac{\lim_{x \to c} f(x)}{\lim_{x \to c} g(x)} = \dfrac{R}{S}, \ S \neq 0$

Power Rule: $\lim_{x \to c} \left(f(x) \right)^n = \left(\lim_{x \to c} f(x) \right)^n = R^n$

Limits of Polynomial Functions

Now we're getting somewhere. With these rules, we can now take the limit of any polynomial function, as in the following example:

Example 1.1.17: Evaluate $\lim_{x \to 3} \left(2x^2 - 5x + 3 \right)$.

We first apply the Sum and Difference Rule.

$$\lim_{x \to 3}\left(2x^2 - 5x + 3\right) = \lim_{x \to 3} 2x^2 - \lim_{x \to 3} 5x + \lim_{x \to 3} 3$$

Next, we apply the Scalar Multiple Rule.

$$\lim_{x \to 3}\left(2x^2 - 5x + 3\right) = 2\lim_{x \to 3} x^2 - 5\lim_{x \to 3} x + \lim_{x \to 3} 3$$

Finally, we apply the Basic Limit Theorems that we reviewed right before Example 1.1.16.

$$\lim_{x \to 3}\left(2x^2 - 5x + 3\right) = 2 \cdot 3^2 - 5 \cdot 3 + 3 = 6$$

You may have noticed that we could have calculated this limit simply by plugging $x = 3$ into $2x^2 - 5x + 3$. We can generalize this. If $f(x)$ is a *polynomial* function, then $\lim_{x \to c} f(x) = f(c)$. In other words, limits of polynomial functions can always be evaluated by *direct substitution*.

Limits of Rational Functions

We can also use the limit rules to calculate the limit of any rational function, as in the following example:

Example 1.1.18: Evaluate $\lim_{x \to 1}\dfrac{x^2 - 2x + 7}{x^2 + 6x + 3}$.

First we apply the Quotient Rule.

$$\lim_{x \to 1}\frac{x^2 - 2x + 7}{x^2 + 6x + 3} = \frac{\lim_{x \to 1}\left(x^2 - 2x + 7\right)}{\lim_{x \to 1}\left(x^2 + 6x + 3\right)}$$

Now we have a quotient of limits of polynomials, and as we just reviewed, limits of polynomials can be evaluated by direct substitution.

$$\lim_{x \to 1}\frac{x^2 - 2x + 7}{x^2 + 6x + 3} = \frac{1^2 - 2(1) + 7}{1^2 + 6(1) + 3} = \frac{3}{5}$$

This limit could also have been evaluated by direct substitution, and we can generalize this just like we did for polynomials. If $f(x)$ is a *rational* function, then $\lim_{x \to c} f(x) = f(c)$, provided that $f(c)$ is defined.

This EK does not specifically mention radical functions, but we're going to review them anyway because they will show up on the exam.

Limits of Radical Functions

Let n be a positive integer. Then the limit rule $\lim_{x \to c}\sqrt[n]{x} = \sqrt[n]{c}$ holds for *all* real numbers c if n is odd, and for all *positive* real numbers c if n is even.

Example 1.1.19: Evaluate the following limits:

$$\lim_{x \to -8} \sqrt[3]{x}$$

$$\lim_{x \to 12} \sqrt{x}$$

Applying the above rule, we obtain the following:

$$\lim_{x \to -8} \sqrt[3]{x} = \sqrt[3]{-8} = -2$$

$$\lim_{x \to 12} \sqrt{x} = \sqrt{12} = 2\sqrt{3}$$

There are other function types whose limits you should know how to compute. We review them below.

Limits of Other Functions

By "other functions", we mean any of the other function types that you saw in **Prelude: The Functions of Precalculus**, which are listed below.

- Exponential Functions
- Logarithmic Functions
- Trigonometric Functions
- Inverse Trigonometric Functions
- Absolute Value Functions

If $f(x)$ is a function of *any* of the types listed above, then $\lim_{x \to c} f(x) = f(c)$, provided that $f(c)$ is defined. We'll illustrate with a multi-part example.

Example 1.1.20: Evaluate each of the following limits, if it exists.

$$\lim_{x \to \frac{\pi}{3}} \sin(x)$$

$$\lim_{x \to \frac{\pi}{2}} \tan(x)$$

$$\lim_{x \to \ln(2)} e^x$$

$$\lim_{x \to -1} \ln(x)$$

$$\lim_{x \to -\frac{1}{2}} \cos^{-1}(x)$$

$$\lim_{x \to 0} |x|$$

A thorough knowledge of the functions of precalculus is needed here.

The sine function is defined for all real x, so this one is easy.

$$\lim_{x \to \frac{\pi}{3}} \sin(x) = \sin\left(\frac{\pi}{3}\right) = \frac{\sqrt{3}}{2}$$

The tangent function is not defined at $x = \dfrac{\pi}{2}$. This limit does not exist, but that doesn't mean that we can't say anything about it. Go back and look at the graph of the tangent function. It has a vertical asymptote at $x = \dfrac{\pi}{2}$. From the graph, we can determine the one sided limits: $\lim\limits_{x \to \frac{\pi}{2}^{-}} \tan(x) = \infty$ and $\lim\limits_{x \to \frac{\pi}{2}^{+}} \tan(x) = -\infty$.

The exponential function is defined everywhere. Evaluate by direct substitution.

$$\lim_{x \to \ln(2)} e^x = e^{\ln(2)} = 2$$

The natural logarithmic function is not defined at or even around $x = -1$. This limit does not exist.

$x = -\dfrac{1}{2}$ is in the domain of the inverse cosine function. This one can be done by direct substitution.

$$\lim_{x \to -\frac{1}{2}} \cos^{-1}(x) = \cos^{-1}\left(-\frac{1}{2}\right) = \frac{2\pi}{3}$$

The absolute value function is defined everywhere. Plug in to find the limit.

$$\lim_{x \to 0} |x| = |0| = 0$$

We'll close our review of this EK by discussing composite functions. These other function types that we just reviewed can sneak into the exam via composite functions, so you'll want to make sure that you know how to deal with them.

Limits of Composite Functions

If f and g are functions such that $\lim\limits_{x \to c} g(x) = S$ and $\lim\limits_{x \to S} f(x) = f(S)$, then

$$\lim_{x \to c} f\big(g(x)\big) = f\left(\lim_{x \to c} g(x)\right) = f(L)$$

This *really* expands the class of functions whose limits we can compute, because we can combine any of the previous function types together.

Example 1.1.21: Evaluate each of the following limits.

$$\lim_{x \to 3} \sqrt{2x^2 + 1}$$

$$\lim_{x \to \sqrt{\frac{\pi}{4}}} \cos\left(x^2\right)$$

$$\lim_{x \to 4} e^{1/\sqrt{x}}$$

As you can see, each of these is a limit of a composite function.

$$\lim_{x \to 3} \sqrt{2x^2 + 1} = \sqrt{\lim_{x \to 3}\left(2x^2 + 1\right)} = \sqrt{19}$$

$$\lim_{x \to \sqrt{\frac{\pi}{4}}} \cos\left(x^2\right) = \cos\left(\lim_{x \to \sqrt{\frac{\pi}{4}}} x^2\right) = \cos\left(\frac{\pi}{4}\right) = \frac{\sqrt{2}}{2}$$

$$\lim_{x \to 4} e^{\sqrt{x}} = e^{\lim_{x \to 4}\left(\sqrt{x}\right)} = e^2$$

In the next EK, we'll expand our bag of tricks even further.

EK 1.1C2: The limit of a function may be found by using algebraic manipulation, alternate forms of trigonometric functions, or the Squeeze Theorem.

Algebraic Manipulation

When computing $\lim_{x \to c} f\left(x\right)$, sometimes the result sometimes the result cannot be obtained by direct substitution, even if the limit exists. In such cases, simplification of the expression and bringing it to a form which can be evaluated is necessary using algebraic manipulation techniques. We begin by reviewing an important theorem.

If $f(x) = g(x)$ for all x in an open interval containing c, but *not* at $x = c$ itself, and if $\lim_{x \to c} g\left(x\right)$ exists, then $\lim_{x \to c} f\left(x\right)$ also exists, and furthermore $\lim_{x \to c} f\left(x\right) = \lim_{x \to c} g\left(x\right)$.
We will use this theorem in each of the next two examples.

Example 1.1.22: Evaluate $\lim_{x \to 2} \dfrac{x^2 - 4}{x - 2}$.

Direct substitution here yields $\dfrac{0}{0}$, which is not a number. We call this an **indeterminate form**. We'll begin our algebraic manipulation by *factoring*.

$$\lim_{x \to 2} \frac{x^2 - 4}{x - 2} = \lim_{x \to 2} \frac{\left(x - 2\right)\left(x + 2\right)}{x - 2}$$

The function $f\left(x\right) = \dfrac{x^2 - 4}{x - 2}$ is equal to the function $g(x) = x + 2$ for all $x \neq 2$. The theorem that we just reviewed tells us that $\lim_{x \to 2} f\left(x\right) = \lim_{x \to 2} g\left(x\right)$, and $\lim_{x \to 2} g\left(x\right)$ *can* be evaluated by direct substitution. Pressing on, we obtain the following:

$$\lim_{x \to 2} \frac{x^2 - 4}{x - 2} = \lim_{x \to 2} (x + 2) = 4$$

Example 1.1.23: Evaluate $\displaystyle\lim_{x \to 4} \frac{\sqrt{x} - 2}{x - 4}$.

Again, direct substitution produces the indeterminate form $\dfrac{0}{0}$, so we'll need to use an algebraic manipulation if we're going to evaluate this. The manipulation that we'll use is *rationalizing the numerator*. We'll do this by multiplying the numerator and denominator by the *conjugate* of the numerator: $\sqrt{x} + 2$.

$$\lim_{x \to 4} \frac{\sqrt{x} - 2}{x - 4} = \lim_{x \to 4} \left(\frac{\sqrt{x} - 2}{x - 4} \cdot \frac{\sqrt{x} + 2}{\sqrt{x} + 2} \right) = \lim_{x \to 4} \frac{x - 4}{(x - 4)(\sqrt{x} + 2)}$$

For all $x \neq 4$, the function $f(x) = \dfrac{x - 4}{(x - 4)(\sqrt{x} + 2)}$ is equal to the function $g(x) = \dfrac{1}{\sqrt{x} + 2}$, and so according to the previous theorem, their limits at $x = 4$ are equal, so $\displaystyle\lim_{x \to 4} f(x) = \lim_{x \to 4} g(x)$. Furthermore, thanks to the limit rules of **EK 1.1C1**, $\displaystyle\lim_{x \to 4} g(x)$ can be evaluated by direct substitution. We'll use this information to finish this off.

$$\lim_{x \to 4} \frac{\sqrt{x} - 2}{x - 4} = \lim_{x \to 4} \frac{1}{\sqrt{x} + 2} = \frac{1}{4}$$

Before we get to our next example, we will review another important theorem.

If c is a real number and n is a positive rational number, then the following two limit rules hold.

- $\displaystyle\lim_{x \to \infty} \frac{c}{x^n} = 0$

- $\displaystyle\lim_{x \to -\infty} \frac{c}{x^n} = 0$, provided x^n is defined when $x < 0$.

We'll use this result in our next example.

Example 1.1.24: Evaluate $\displaystyle\lim_{x \to \infty} \frac{2x^2 - 5x + 1}{-3x^2 + 7x + 4}$.

As $x \to \infty$, the numerator tends to ∞, and the denominator tends to $-\infty$, so this limit produces the form $\dfrac{\infty}{-\infty}$, which, like $\dfrac{0}{0}$, is indeterminate. The algebraic manipulation that we'll use here is *dividing* the numerator and denominator by x^2.

$$\lim_{x \to \infty} \frac{2x^2 - 5x + 1}{-3x^2 + 7x + 4} = \lim_{x \to \infty} \frac{\dfrac{2x^2 - 5x + 1}{x^2}}{\dfrac{-3x^2 + 7x + 4}{x^2}} = \lim_{x \to \infty} \frac{\dfrac{2x^2}{x^2} - \dfrac{5x}{x^2} + \dfrac{1}{x^2}}{-\dfrac{3x^2}{x^2} + \dfrac{7x}{x^2} + \dfrac{4}{x^2}} = \lim_{x \to \infty} \frac{2 - \dfrac{5}{x} + \dfrac{1}{x^2}}{-3 + \dfrac{7}{x} + \dfrac{4}{x^2}}$$

Taking the limit of each term in the numerator and denominator separately, and applying the previous theorem, we get our final result.

$$\lim_{x \to \infty} \frac{2x^2 - 5x + 1}{-3x^2 + 7x + 4} = \frac{2 - 0 + 0}{-3 + 0 + 0} = -\frac{2}{3}$$

Alternate Forms of Trigonometric Functions

This technique will require you to know your trigonometric identities. We'll get right to an example.

Example 1.1.25: Evaluate $\lim\limits_{x \to 0} \dfrac{1 - \cos(x)}{\sin(x)}$.

Direct substitution yields the indeterminate form $\dfrac{0}{0}$. We can get around this by multiplying the numerator and denominator by the *Pythagorean conjugate* of the numerator: $1 + \cos(x)$.

$$\lim_{x \to 0} \frac{1 - \cos(x)}{\sin(x)} = \lim_{x \to 0} \left(\frac{1 - \cos(x)}{\sin(x)} \cdot \frac{1 + \cos(x)}{1 + \cos(x)} \right) = \lim_{x \to 0} \frac{1 - \cos^2(x)}{\sin(x)(1 + \cos(x))}$$

Here's where the trig identity comes in. Recognize that $1 - \cos^2(x) = \sin^2(x)$.

$$\lim_{x \to 0} \frac{1 - \cos(x)}{\sin(x)} = \lim_{x \to 0} \frac{\sin^2(x)}{\sin(x)(1 + \cos(x))}$$

Notice that $f(x) = \dfrac{\sin^2(x)}{\sin(x)(1 + \cos(x))}$ is equal to $g(x) = \dfrac{\sin(x)}{1 + \cos(x)}$ everywhere in the open interval $(-\pi, \pi)$ except at $x = 0$, so $\lim\limits_{x \to 0} f(x) = \lim\limits_{x \to 0} g(x)$. Furthermore, $\lim\limits_{x \to 0} g(x)$ can be evaluated by direct substitution. This lets us finish as follows:

$$\lim_{x \to 0} \frac{1 - \cos(x)}{\sin(x)} = \lim_{x \to 0} \frac{\sin(x)}{1 + \cos(x)} = \frac{\sin(0)}{1 + \cos(0)} = 0$$

The Squeeze Theorem

The Squeeze Theorem, or Sandwich Theorem, states that if there exist functions f, g, and h such that $h(x) \leq f(x) \leq g(x)$ for all x in an open interval containing c, except possibly at c, and if $\lim\limits_{x \to c} h(x) = R = \lim\limits_{x \to c} f(x)$, then $\lim\limits_{x \to c} g(x) = R$ also. We illustrate this with an example.

Example 1.1.26: Evaluate $\lim\limits_{x \to 0} x^2 \sin\left(\dfrac{1}{x} \right)$.

Direct substitution yields $0 \cdot \sin\left(\dfrac{1}{0} \right)$, which is mathematical gibberish. We'll try to *squeeze* the function $g(x) = x^2 \sin\left(\dfrac{1}{x} \right)$ in between two functions $f(x)$ and $h(x)$ near $x = 0$. First we note that, for all $x \neq 0$,

$$-1 \le \sin\left(\frac{1}{x}\right) \le 1$$

Multiplying all three members of the inequality by x^2, we get the following.

$$-x^2 \le x^2 \sin\left(\frac{1}{x}\right) \le x^2$$

Note that $g(x)$ is in the center of the inequality. Let $h(x) = -x^2$ and $f(x) = x^2$, and recognize that $\lim_{x \to c} h(x) = 0 = \lim_{x \to c} f(x)$. Therefore, by the Squeeze Theorem, $\lim_{x \to 0} g(x) = \lim_{x \to 0} x^2 \sin\left(\frac{1}{x}\right) = 0$. If all that is a lot to process, then perhaps the following illustration will help.

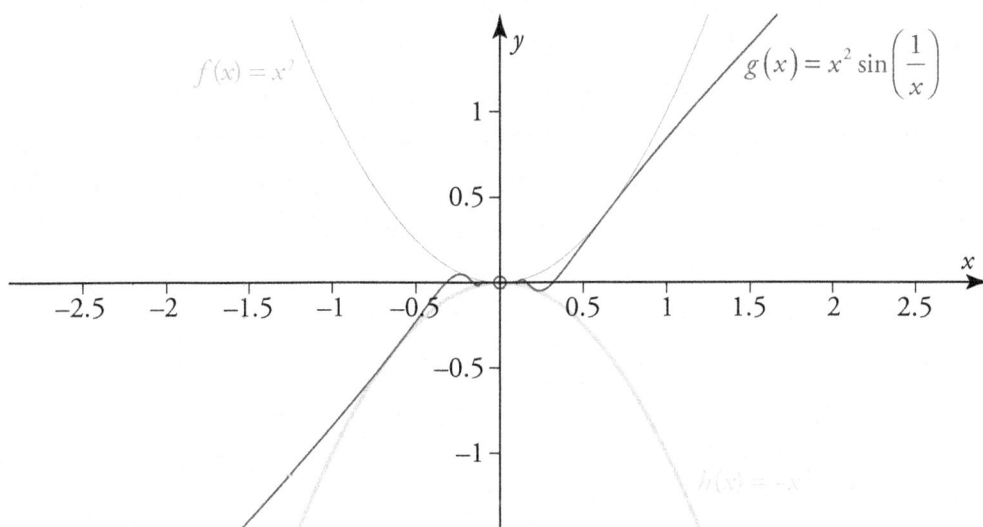

As you can see from the graph, since $g(x)$ is trapped between $h(x)$ and $f(x)$ near $x = 0$, it can't help but have the same limit as those two functions at $x = 0$.

We've mentioned *indeterminate forms* in Examples 1.1.22 through 1.1.25. In the next EK we present a powerful method for dealing with such forms.

EK 1.1C3: Limits of the indeterminate forms $\frac{0}{0}$ and $\frac{\infty}{\infty}$ may be evaluated using L'Hôpital's Rule.

Note: Although L'Hôpital's Rule appears in this chapter, it does depend on differential calculus as a prerequisite. You may wish to postpone this EK until after you have reviewed **LO 2.1C**.

L'Hôpital's Rule

Let (a, b) be an open interval containing c.

If:

- f and g are differentiable on (a, b) (except possibly at c),

- $g'(x) \neq 0$ on (a, b) (except possibly at c), and

- $\lim\limits_{x \to c} \dfrac{f(x)}{g(x)}$ produces either of the indeterminate forms $\dfrac{0}{0}$ or $\dfrac{\infty}{\infty}$,

then $\lim\limits_{x \to c} \dfrac{f(x)}{g(x)} = \lim\limits_{x \to c} \dfrac{f'(x)}{g'(x)}$, provided the limit on the right side of the equation exists.

In fact, L'Hôpital's Rule also applies if the indeterminate forms $\dfrac{-\infty}{\infty}, \dfrac{\infty}{-\infty}$, or $\dfrac{-\infty}{-\infty}$ are produced.

Example 1.1.27: Evaluate $\lim\limits_{x \to 0} \dfrac{e^{3x} - 1}{2x}$.

Direct substitution produces $\dfrac{0}{0}$, so let's try L'Hôpital's Rule.

$$\lim_{x \to 0} \frac{e^{3x} - 1}{2x} = \lim_{x \to 0} \frac{\dfrac{d}{dx}\left(e^{3x} - 1\right)}{\dfrac{d}{dx}(2x)} = \lim_{x \to 0} \frac{3e^{3x}}{2} = \lim_{x \to \infty} \frac{3}{2}e^{3x}$$

The limit on the right side can be done by direct substitution.

$$\lim_{x \to 0} \frac{e^{3x} - 1}{2x} = \frac{3}{2}e^{3(0)} = \frac{3}{2}$$

Example 1.1.28: Evaluate $\lim\limits_{x \to \infty} \dfrac{\ln(x)}{x^2}$.

This limit produces the form $\dfrac{\infty}{\infty}$, so again, let's try L'Hôpital's Rule.

$$\lim_{x \to \infty} \frac{\ln(x)}{x^2} = \lim_{x \to \infty} \frac{\dfrac{d}{dx}\left(\ln(x)\right)}{\dfrac{d}{dx}\left(x^2\right)} = \lim_{x \to \infty} \frac{\dfrac{1}{x}}{2x} = \lim_{x \to \infty} \frac{1}{2x^2}$$

This limit can be evaluated by the theorem that we reviewed immediately before Example 1.1.24.

$$\lim_{x \to \infty} \frac{\ln(x)}{x^2} = 0$$

Sometimes L'Hôpital's Rule must be applied more than once in order to evaluate a limit. Our next example illustrates this.

Example 1.1.29: Evaluate $\displaystyle\lim_{x\to\infty}\frac{e^{2x}}{x^2}$.

This limit produces the form $\frac{\infty}{\infty}$. Let's turn to Professor L'Hôpital again.

$$\lim_{x\to\infty}\frac{e^{2x}}{x^2}=\lim_{x\to\infty}\frac{\dfrac{d}{dx}\left(e^{2x}\right)}{\dfrac{d}{dx}\left(x^2\right)}=\lim_{x\to\infty}\frac{2e^{2x}}{2x}=\lim_{x\to\infty}\frac{e^{2x}}{x}$$

The limit on the right *again* produces the form $\frac{\infty}{\infty}$, so we'll try L'Hôpital's Rule a second time.

$$\lim_{x\to\infty}\frac{e^{2x}}{x^2}=\lim_{x\to\infty}\frac{\dfrac{d}{dx}\left(e^{2x}\right)}{\dfrac{d}{dx}\left(x\right)}=\lim_{x\to\infty}\frac{2e^{2x}}{1}=\lim_{x\to\infty}2e^{2x}$$

This time, the limit on the right does *not* produce an indeterminate form, and so we may not use L'Hôpital's Rule. To finish this off, we need to know that the exponential function is unbounded as $x\to\infty$.

$$\lim_{x\to\infty}\frac{e^{2x}}{x^2}=\infty$$

Other Indeterminate Forms

There do exist indeterminate forms that aren't named in the statement of L'Hôpital's Rule, but could nevertheless show up on the exam. Those forms are $0\cdot(\pm\infty)$, 1^∞, 0^0, and $\infty-\infty$. The objective in dealing with these forms is to use algebraic manipulation to try to work them into one of the forms that is covered by the hypotheses of L'Hôpital's Rule. We will show some examples to illustrate.

Example 1.1.30: Evaluate $\displaystyle\lim_{x\to0^+}x^3\ln\left(x\right)$.

As $x\to0^+$, $x^3\to0$ and $\ln(x)\to-\infty$, so this limit produces the indeterminate form $0\cdot(-\infty)$. We'll manipulate it by writing x^3 as $\dfrac{1}{x^{-3}}$ and continuing as follows:

$$\lim_{x\to0^+}x^3\ln\left(x\right)=\lim_{x\to0^+}\frac{\ln\left(x\right)}{x^{-3}}$$

The limit on the right produces the indeterminate form $\frac{-\infty}{\infty}$, and so we can apply L'Hôpital's Rule.

$$\lim_{x\to0^+}x^3\ln\left(x\right)=\lim_{x\to0^+}\frac{\dfrac{d}{dx}\left(\ln\left(x\right)\right)}{\dfrac{d}{dx}\left(x^{-3}\right)}=\lim_{x\to0^+}\frac{\dfrac{1}{x}}{-3x^{-4}}=\lim_{x\to0^+}\frac{x^3}{-3}=0$$

Example 1.1.31: Evaluate $\displaystyle\lim_{x\to\infty}\left(1+\frac{1}{x}\right)^x$.

As $x \to \infty$, the base $\left(1+\dfrac{1}{x}\right) \to 1$, and of course the exponent tends to ∞, so this limit produces the indeterminate form 1^∞. To deal with this form, we set the limit equal to a new variable (say y), and then we take the natural logarithm of each side. Then we follow our nose.

$$y = \lim_{x\to\infty}\left(1+\frac{1}{x}\right)^x$$

$$\ln(y) = \ln\left(\lim_{x\to\infty}\left(1+\frac{1}{x}\right)^x\right) = \lim_{x\to\infty}\left(\ln\left(1+\frac{1}{x}\right)^x\right) = \lim_{x\to\infty}\left(x\cdot\ln\left(1+\frac{1}{x}\right)\right)$$

The limit on the right produces the form $\infty \cdot 0$, which is handled using the method of Example 1.1.30.

$$\ln(y) = \lim_{x\to\infty}\frac{\ln\left(1+\dfrac{1}{x}\right)}{\dfrac{1}{x}}$$

Now this limit produces the form $\dfrac{0}{0}$, and so we can apply L'Hôpital's Rule.

$$\ln(y) = \lim_{x\to\infty}\frac{\dfrac{d}{dx}\left(\ln\left(1+\dfrac{1}{x}\right)\right)}{\dfrac{d}{dx}\left(\dfrac{1}{x}\right)} = \lim_{x\to\infty}\frac{\left(1+\dfrac{1}{x}\right)^{-1}\left(-\dfrac{1}{x^2}\right)}{-\dfrac{1}{x^2}} = \lim_{x\to\infty}\left(1+\frac{1}{x}\right)^{-1} = (1+0)^{-1} = 1$$

So we have the equation $\ln(y) = 1$, whose solution is $y = e$. Therefore, our final result is as follows:

$$\lim_{x\to\infty}\left(1+\frac{1}{x}\right)^x = e$$

Give yourself a hand if you recognized at the beginning that this limit is the *definition* of the number e. The indeterminate form 0^0 is handled using the same method as that of the last example.

Example 1.1.32: Evaluate $\displaystyle\lim_{x\to\frac{\pi}{2}^-}\left(\tan(x) - \sec(x)\right)$.

This limit produces the indeterminate form $\infty - \infty$. We'll convert the trig functions to sines and cosines, and then try to manipulate them so that the limit produces some form that we already know how to deal with.

$$\lim_{x \to \frac{\pi}{2}^-} \left(\tan(x) - \sec(x) \right) = \lim_{x \to \frac{\pi}{2}^-} \left(\frac{\sin(x)}{\cos(x)} - \frac{1}{\cos(x)} \right) = \lim_{x \to \frac{\pi}{2}^-} \left(\frac{\sin(x) - 1}{\cos(x)} \right)$$

The limit on the right produces $\frac{0}{0}$, and so we will use L'Hôpital's Rule.

$$\lim_{x \to \frac{\pi}{2}^-} \left(\tan(x) - \sec(x) \right) = \lim_{x \to \frac{\pi}{2}^-} \left(\frac{\frac{d}{dx}\left(\sin(x) - 1 \right)}{\frac{d}{dx}\left(\cos(x) \right)} \right) = \lim_{x \to \frac{\pi}{2}^-} \frac{\cos(x)}{-\sin(x)} = \frac{\cos\left(\frac{\pi}{2}\right)}{-\sin\left(\frac{\pi}{2}\right)} = 0$$

LO 1.1D – Limits and the Behavior of Functions

The full Learning Objective of this section is as follows:

LO 1.1D: Deduce and interpret behavior of functions using limits.

The first item of Essential Knowledge under **LO 1.1D** discusses the application of limits to explain asymptotic and unbounded behavior of functions.

EK 1.1D1: Asymptotic and unbounded behavior of functions can be explained and described using limits.

Bounded and Unbounded Functions

We'll begin by explaining what a bounded function is. A function f is **bounded** on its domain if there exists a real number M such that, for all x in the domain of f, $|f(x)| \leq M$. A classic example is the sine function: for all real x, $|\sin(x)| \leq 1$. On the other hand, a function that is not bounded is called **unbounded**. We've discussed this idea informally in **Prelude: The Functions of Precalculus**, and now we're going to revisit the idea and express it in terms of *limits*.

Example 1.1.33: Discuss the end behavior of $f(x) = -2x^3 + 3x$ in terms of limits.

f is a polynomial function of odd degree with a negative leading coefficient. Its end behavior can be expressed in terms of limits as follows.

$$\lim_{x \to -\infty} \left(-2x^3 + 3x \right) = \infty \quad \text{and} \quad \lim_{x \to \infty} \left(-2x^3 + 3x \right) = -\infty$$

The fact that these limits are *infinite* means that f is *unbounded*. You should graph this function on a calculator to verify that it has the advertised end behavior.

Another example of unboundedness in functions is vertical asymptotes of rational functions. We're going to rephrase the definition of vertical asymptotes in terms of limits.

Vertical Asymptotes

The line $x = c$ is a **vertical asymptote** of a function $f(x)$ if either $\lim_{x \to c^-} f(x) = \pm\infty$ or $\lim_{x \to c^+} f(x) = \pm\infty$.

Example 1.1.34: Discuss vertical asymptotes of $f(x) = \dfrac{1}{x^2 - 3x - 4}$ using limits.

If we factor the denominator, we see that $f(x) = \dfrac{1}{(x+1)(x-4)}$. f is undefined at $x = -1$ and at $x = 4$, and if we attempt to evaluate the limits of f at these two points then we find that we obtain the form $\dfrac{1}{0}$, which signals infinite one-sided limits. These limits are given below.

$$\lim_{x \to -1^-} \frac{1}{(x+1)(x-4)} = \infty \quad \text{and} \quad \lim_{x \to -1^+} \frac{1}{(x+1)(x-4)} = -\infty \Rightarrow x = -1 \text{ is a vertical}$$

asymptote of f.

$$\lim_{x \to 4^-} \frac{1}{(x+1)(x-4)} = -\infty \quad \text{and} \quad \lim_{x \to 4^+} \frac{1}{(x+1)(x-4)} = \infty \Rightarrow x = 4 \text{ is a vertical asymptote}$$

of f.

You should graph this function using a calculator to verify that it not only has the vertical asymptotes that we claim, but also that the one-sided limits are as we claim.

As you can see, vertical asymptotes can be described using *infinite limits*. Next we'll review horizontal asymptotes, which can be described using *limits at infinity*.

Horizontal Asymptotes

The line $y = R$ is a **horizontal asymptote** of a function $f(x)$ if either $\lim\limits_{x \to -\infty} f(x) = R$ or $\lim\limits_{x \to \infty} f(x) = R$.

Example 1.1.35: Find the horizontal asymptote of $f(x) = \dfrac{4 - 3x - 6x^2}{2x^2 + 5}$.

We compute the limits of f at infinity. We'll do this by dividing the numerator and denominator by x^2 inside the limits.

$$\lim_{x \to \infty} \frac{4 - 3x - 6x^2}{2x^2 + 5} = \lim_{x \to \infty} \frac{\dfrac{4}{x^2} - \dfrac{3}{x} - 6}{2 + \dfrac{5}{x^2}} = -3$$

$$\lim_{x \to -\infty} \frac{4 - 3x - 6x^2}{2x^2 + 5} = \lim_{x \to -\infty} \frac{\dfrac{4}{x^2} - \dfrac{3}{x} - 6}{2 + \dfrac{5}{x^2}} = -3$$

Since both infinite limits yield -3, the line $y = -3$ is the only horizontal asymptote of f. Next, we will look at a function that has two horizontal asymptotes.

Example 1.1.36: Find the horizontal asymptotes of $f(x) = \dfrac{3x}{\sqrt{4x^2 + 1}}$.

We'll compute the $\lim\limits_{x \to \infty} \dfrac{3x}{\sqrt{4x^2 + 1}}$ first. We'll do this by dividing the numerator and denominator inside the limit by x. In the denominator, we'll use the fact that $x = \sqrt{x^2}$ whenever $x > 0$.

$$\lim_{x\to\infty}\frac{3x}{\sqrt{4x^2+1}}=\lim_{x\to\infty}\frac{\dfrac{3x}{x}}{\dfrac{\sqrt{4x^2+1}}{\sqrt{x^2}}}=\lim_{x\to\infty}\frac{\dfrac{3x}{x}}{\sqrt{\dfrac{4x^2}{x^2}+\dfrac{1}{x^2}}}=\lim_{x\to\infty}\frac{3}{\sqrt{4+\dfrac{1}{x^2}}}=\frac{3}{\sqrt{4+0}}=\frac{3}{2}$$

Now we'll compute $\displaystyle\lim_{x\to-\infty}\frac{3x}{\sqrt{4x^2+1}}$ using the same method, but this time we need to take care to note that when $x<0$, $x=-\sqrt{x^2}$.

$$\lim_{x\to-\infty}\frac{3x}{\sqrt{4x^2+1}}=\lim_{x\to-\infty}\frac{\dfrac{3x}{x}}{\dfrac{\sqrt{4x^2+1}}{-\sqrt{x^2}}}=-\lim_{x\to-\infty}\frac{\dfrac{3x}{x}}{\sqrt{\dfrac{4x^2}{x^2}+\dfrac{1}{x^2}}}=-\lim_{x\to-\infty}\frac{3}{\sqrt{4+\dfrac{1}{x^2}}}=-\frac{3}{\sqrt{4+0}}=-\frac{3}{2}$$

Based on the limits of f at $\pm\infty$, we conclude that the two horizontal asymptotes of f are $y=-\frac{3}{2}$ and $y=\frac{3}{2}$. Don't believe us? Graph the function on a calculator and verify it for yourself.

Next, we'll review a limit concept that will be particularly useful when we review sequences and series.

EK 1.1D2: Relative magnitudes of functions and their rates of change can be compared using limits.

Relative Magnitudes of Functions

Some functions grow more rapidly than others, but it can sometimes be tricky to compare functions by inspecting their graphs. Consider the following graph:

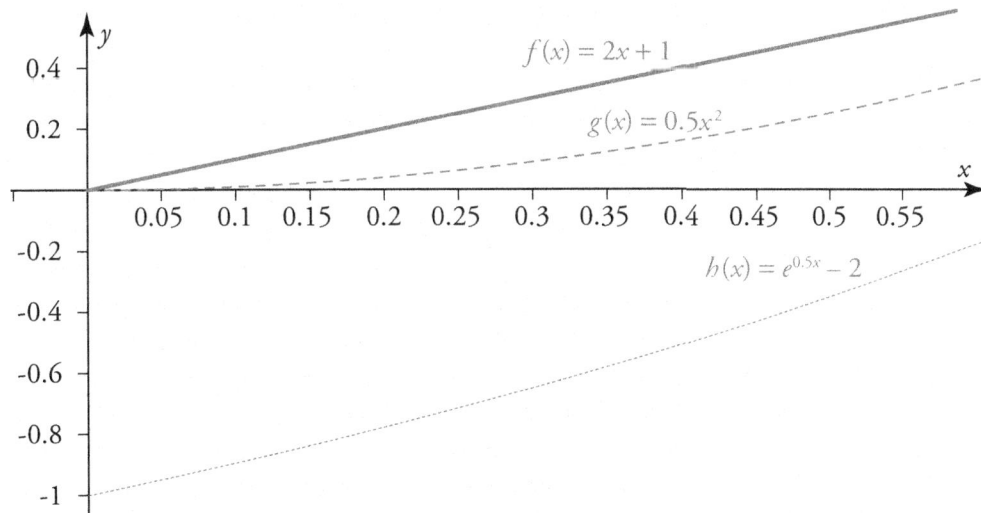

In the short run (that is, on the interval $[0,0.5]$), it *appears* as though the linear function $f(x)=2x+1$ (solid curve) is growing more rapidly than both the quadratic function $g(x)=0.5x^2$ (dashed curve) and the exponential function $h(x)=e^{0.5x}-2$. That is, what we see here is $h(x)<g(x)<f(x)$. However, if we look at the behavior of the same

three functions in the long run (that is, on the interval $[0, 2]$), something very different happens. See the figure below.

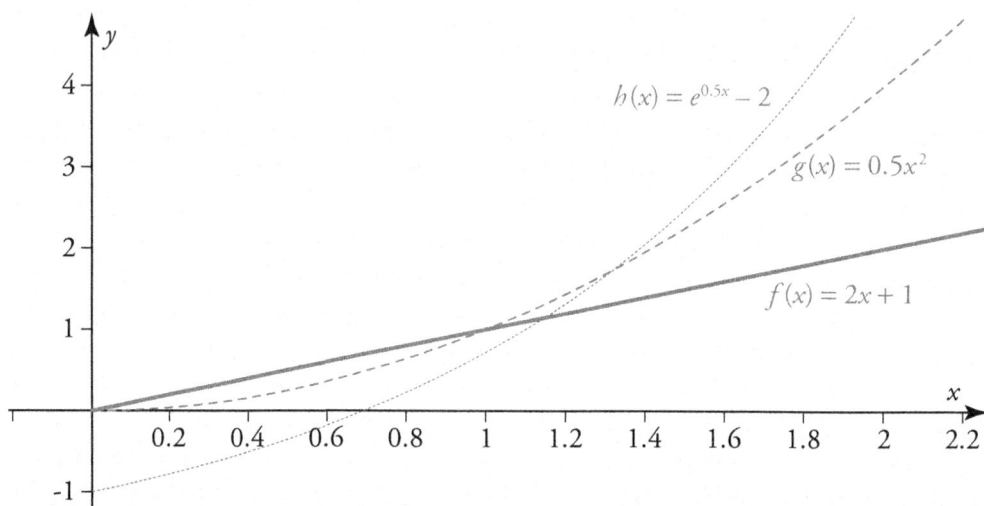

From about $x = 1.5$ onwards, we see a reversal in the previous inequality. That is, we see $f(x) < g(x) < h(x)$. In this example, we only had to make a small adjustment in the viewing window to witness the reversal. But what if we had a different set of functions for which the "long run" occurs over the interval $[0, 10^9]$? We clearly need a more efficient way to determine the relative magnitudes of functions and their rates of change.

This is where limits come in. Suppose f and g are functions, and we wish to see which, if either, function has a larger **relative magnitude**. Relative magnitude refers to the *size of the absolute value* of one function when compared to that of another. Since we have no way of knowing in general what constitutes the "long run" for f and g, we'll consider the longest run imaginable: the interval $[0, \infty)$. Here's the important part: we'll determine which function has the larger relative magnitude by computing $\lim\limits_{x \to \infty} \dfrac{f(x)}{g(x)}$. We make our determination using the following decision rules:

- If $\lim\limits_{x \to \infty} \dfrac{f(x)}{g(x)} = \pm\infty$, then the relative magnitude of f is greater than that of g.

- If $\lim\limits_{x \to \infty} \dfrac{f(x)}{g(x)} = 0$, then the relative magnitude of g is greater than that of f.

Example 1.1.37: Use limits to compare relative magnitudes of $f(x) = 2x + 1$, $g(x) = 0.5x^2$, and $h(x) = e^{0.5x} - 2$.

We'll compare f and g first. We will be using L'Hôpital's Rule.

$$\lim_{x \to \infty} \frac{f(x)}{g(x)} = \lim_{x \to \infty} \frac{2x + 1}{0.5x^2} = \lim_{x \to \infty} \frac{\dfrac{d}{dx}(2x + 1)}{\dfrac{d}{dx}(0.5x^2)} = \lim_{x \to \infty} \frac{2}{x} = 0$$

Based on the above calculation, we see that the relative magnitude of g is greater than that of f. What this means is that in the long run, $g(x)$ grows faster than $f(x)$ does.

Now we'll compare g and h. We'll need to use L'Hôpital's Rule again.

$$\lim_{x\to\infty}\frac{g(x)}{h(x)} = \lim_{x\to\infty}\frac{0.5x^2}{e^{0.5x}-2} = \lim_{x\to\infty}\frac{\frac{d}{dx}\left(0.5x^2\right)}{\frac{d}{dx}\left(e^{0.5x}-2\right)} = \lim_{x\to\infty}\frac{x}{0.5e^{0.5x}}$$

This is still indeterminate. We'll need to use L'Hôpital's Rule one more time.

$$\lim_{x\to\infty}\frac{g(x)}{h(x)} = \lim_{x\to\infty}\frac{\frac{d}{dx}(x)}{\frac{d}{dx}\left(0.5e^{0.5x}\right)}\lim_{x\to\infty}\frac{1}{0.25e^{0.5x}} = 0$$

From this calculation, we see that the relative magnitude of h is greater than that of g.

Putting all of this together, what this means is that the inequality $f(x) < g(x) < h(x)$ will continue to hold in the long run.

Some common types of functions that you'll see in "order of magnitude" questions on the AP Calculus Exam are listed below in *increasing* relative magnitude.

- Logarithmic
- Linear
- Polynomial (order of magnitude *increases* with degree)
- Exponential

What this means is that, in the long run, a linear function will *always* overtake a logarithmic function, a polynomial function of degree 2 will *always* overtake a linear or logarithmic function, and an exponential function will *always* overtake any of the other function types in the list. We can use these facts to compute some limits by inspection.

Example 1.1.38: Determine the value of $\lim_{x\to\infty}\frac{\log_3(x)}{x^3}$ by considering relative magnitudes.

Of course, we *could* use L'Hôpital's Rule to compute this, but we don't have to. This limit is a comparison of the relative magnitudes of $f(x) = \log_3(x)$ and $g(x) = x^3$. Since f is logarithmic, and g is a third degree polynomial, g has the higher relative magnitude. Thus, $\lim_{x\to\infty}\frac{\log_3(x)}{x^3} = 0$.

Relative Magnitudes of Rates of Change of Functions

As you will be reminded when we review **EK 2.1A2** in the next chapter, the rate of change of a function is given by its *derivative*. So if f and g are two differentiable functions, we can compare the relative magnitudes of their rates of change by computing $\lim_{x\to\infty}\frac{f'(x)}{g'(x)}$ and applying the same test. We'll illustrate with an example.

Example 1.1.39: Let $f(x) = 4x^5 + 3$ and $g(x) = 5x^3 + 2$. Use limits to compare the relative magnitudes of the rates of change of f and g.

First we note that $f'(x) = 20x^4$ and $g'(x) = 15x^2$.

$$\lim_{x \to \infty} \frac{f'(x)}{g'(x)} = \lim_{x \to \infty} \frac{20x^4}{15x^2} = \lim_{x \to \infty} \frac{4}{3}x^2 = \infty$$

Based on the above calculation, the relative magnitude of the rate of change of f is greater than the relative magnitude of the rate of change of g.

EU 1.2 – Continuity

LO 1.2A – Understanding continuity of functions

The full Learning Objective of this section is as follows:

LO 1.2A: Analyze functions for intervals of continuity or points of discontinuity.

Loosely speaking, a function is continuous at a point if you can graph the function without picking up your pencil at that point. Here we're going to get more formal with the notion of continuity. In fact, the first item of Essential Knowledge under **LO 1.2A** is the very *definition* of continuity of function at a point.

EK 1.2A1: A function f is continuous at $x = c$ provided that $f(c)$ exists, $\lim_{x \to c} f(x)$ exists, and $\lim_{x \to c} f(x) = f(c)$.

The third condition is the most important one. In order for f to be continuous at $x = c$, it is not enough for $f(c)$ and $\lim_{x \to c} f(x)$ to both exist. We further require that those two numbers be *equal* to each other. We will illustrate the three conditions in **EK 1.2A1** by showing *discontinuous* functions that *fail* to satisfy them.

In the figure below, we see the graph of a function f which is not defined at $x = c$, and as you can see, f is not continuous at $x = c$. This function fails the first condition in the definition of continuity at a point.

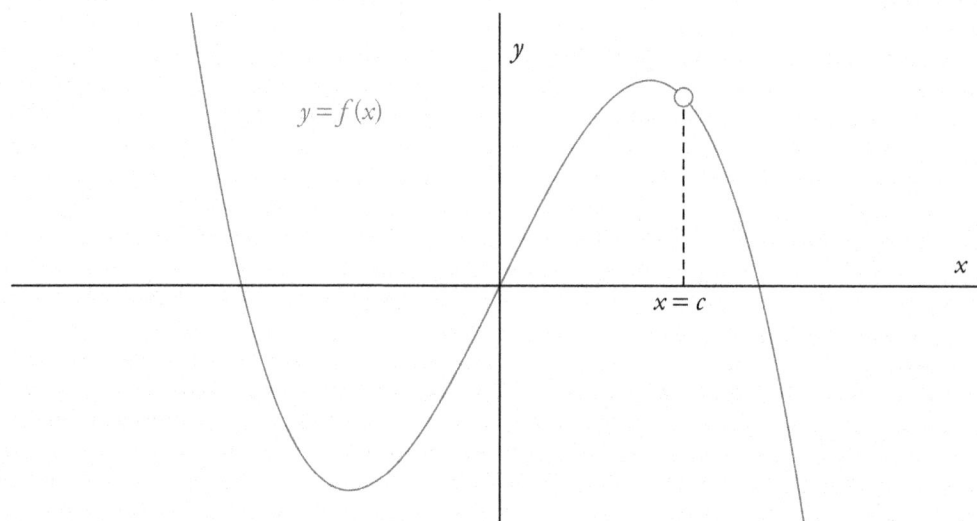

In the next figure, we see the graph of another function that fails to be continuous at $x = c$. This time $f(c)$ exists, but $\lim\limits_{x \to c} f(x)$ does not. This function fails the second condition of the definition.

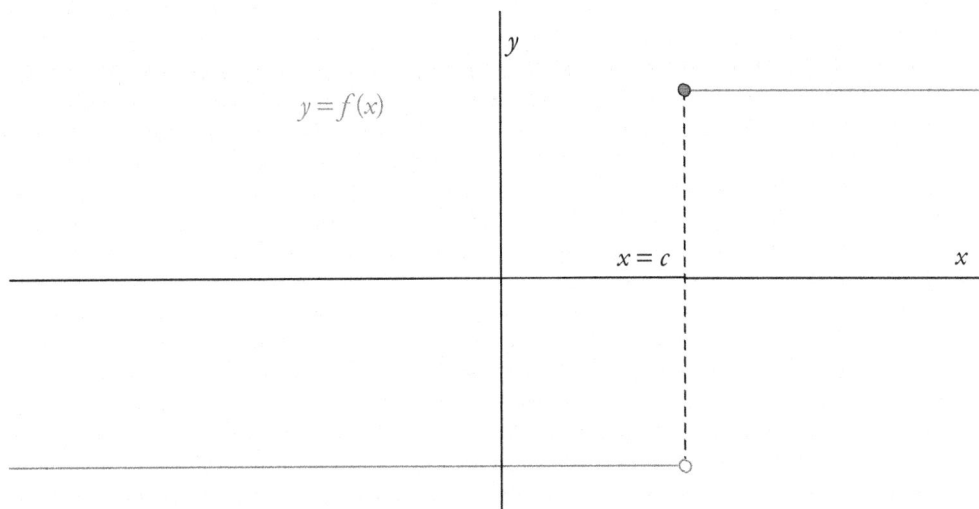

$y = f(x)$

In the next figure, the function f is again discontinuous at $x = c$. But this time, both $f(c)$ and $\lim\limits_{x \to c} f(x)$ exist, so the first two conditions of the definition are satisfied. It's the third condition that is not satisfied, because $\lim\limits_{x \to c} f(x) \neq f(c)$.

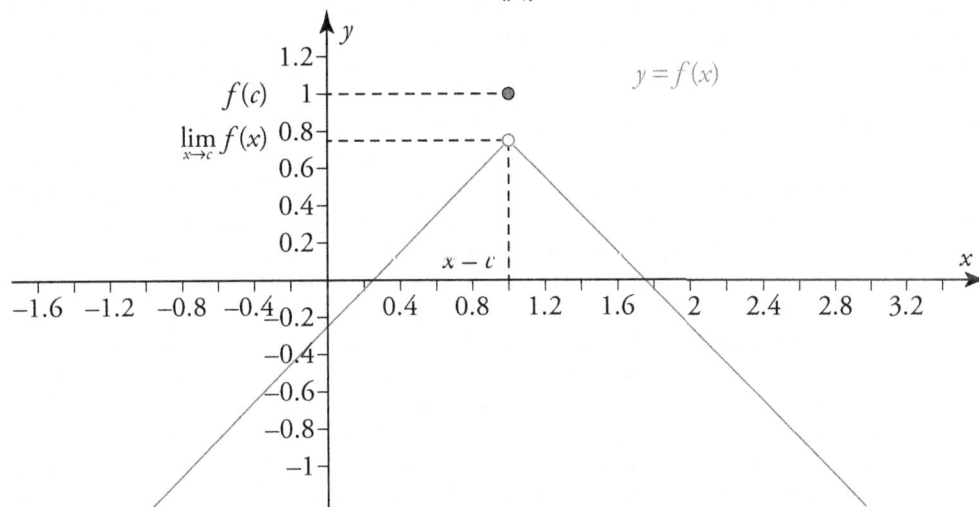

Example 1.2.1: For each function below, determine if it is continuous at $x = 1$. If it is not, state which condition(s) of the definition of continuity at a point are violated by the function at $x = 1$.

$$f(x) = \frac{1}{x + 1}$$

$$f(x) = \frac{1}{x - 1}$$

$$f(x) = (x-1)^2$$

$$f(x) = \begin{cases} x+1 & x \neq 1 \\ 3 & x = 1 \end{cases}$$

$$f(x) = \begin{cases} x+1 & x \leq 1 \\ 2 & x > 1 \end{cases}$$

You may find it helpful to graph these functions.

Since $f(1) = \lim\limits_{x \to 1} f(x) = \dfrac{1}{2}$, f is continuous at $x = 1$.

f is not continuous at $x = 1$. All three conditions are violated: $f(1)$ does not exist, $\lim\limits_{x \to \infty} f(x)$ does not exist, and consequently the two are not equal to each other.

Since $f(1) = \lim\limits_{x \to 1} f(x) = 0$, f is continuous at $x = 1$.

f is not continuous at $x = 1$. The first two conditions *are* satisfied: $f(1) = 3$ and $\lim\limits_{x \to 1} f(x) = 2$. It's the third condition that is violated, because $\lim\limits_{x \to 1} f(x) \neq f(2)$.

Since $f(1) = \lim\limits_{x \to 1} f(x) = 2$, f is continuous at $x = 1$.

Next we'll review specific classes of functions whose continuity can be determined by inspection.

EK 1.2A2: Polynomial, rational, power, exponential, logarithmic, and trigonometric functions are continuous at all points in their domains.

The third condition in the definition of continuity at a point requires that $\lim\limits_{x \to c} f(x) = f(c)$. Another way to say this is that f is continuous at $x = c$ if its limit can be computed by *direct substitution*. In **EK 1.1C1**, we saw that the limits of polynomial, rational, power, exponential, logarithmic, and trigonometric functions can all be computed by direct substitution at any point in their domains.

It therefore follows that all of these functions are continuous at all points in their domains.

Example 1.2.2: For each function below, determine if it is continuous at the given point.

$f(x) = -2x^3 + 6x^2 - x + 3$ at $x = 2$.

$f(x) = \dfrac{x}{2x^2 - 32}$ at $x = 0$

$f(x) = \dfrac{x}{2x^2 - 32}$ at $x = -4$

$f(x) = \tan(2x)$ at $x = \dfrac{\pi}{4}$

$f(x) = \ln(x)$ at $x = 0.5$.

Remember: All we need to do is check whether the given point is in the *domain* of each function. We do not need to compute any limits!

f is continuous at $x = 2$. It is a polynomial function, and so its domain is all real x, and that includes $x = 2$.

f is rational function whose domain is all real x such that $2x^2 - 32 \neq 0$, or $x \neq \pm 4$. Since $x = 0$ is in the domain of f, f is continuous at $x = 0$.

f is not continuous at $x = -4$, because $x = -4$ is not in the domain of f (see part b).

$f\left(\dfrac{\pi}{4}\right) = \tan\left(\dfrac{\pi}{2}\right)$, which is undefined. Since $x = \dfrac{\pi}{4}$ is not in the domain of this trigonometric function, it is not continuous at $x = \dfrac{\pi}{4}$.

f is a logarithmic function whose domain is all positive real numbers, and that includes $x = 0.5$. Thus, f is continuous at $x = 0.5$.

In the next EK, we will *classify* the discontinuities of a function.

EK 1.2A3: Types of discontinuities include removable discontinuities, jump discontinuities, and discontinuities due to vertical asymptote.

Removable Discontinuity

A function f has a **removable discontinuity** at $x = c$ if the discontinuity could be removed by appropriately defining (or perhaps *re*-defining) $f(c)$. We'll illustrate with an example.

Example 1.2.3: Consider the function $f(x) = \dfrac{x^3 - 8}{x - 2}$. This is a rational function, so it is continuous at all points in its domain, which is the set of all real $x \neq 2$. Since $f(2)$ is not defined, f is not continuous at $x = 2$. The graph of f is shown below.

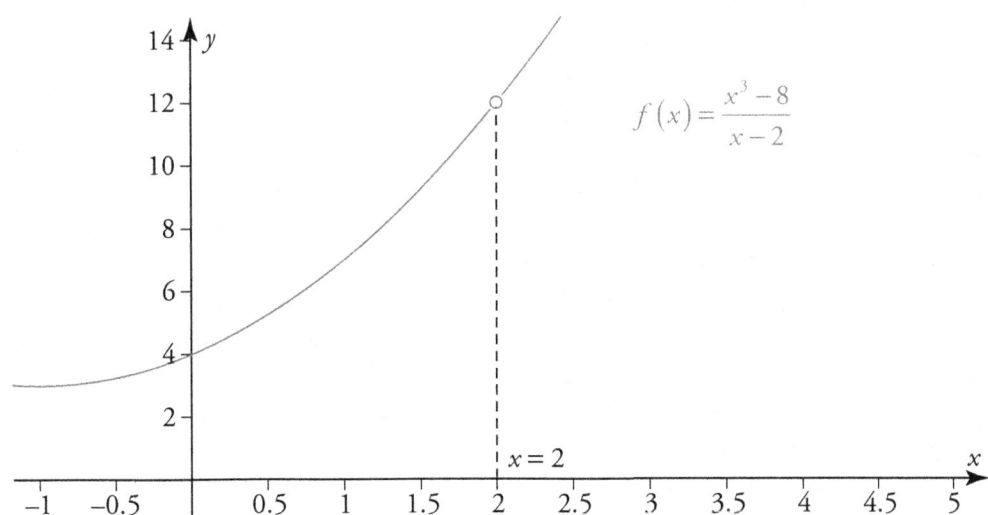

Let's consider the limit of f at $x = 2$.

$$\lim_{x \to 2} f(x) = \lim_{x \to 2} \frac{x^3 - 8}{x - 2} = \lim_{x \to \infty} \frac{(x - 2)(x^2 + 2x + 4)}{x - 2} = \lim_{x \to 2} (x^2 + 2x + 4) = 12$$

$f(2)$ does not exist, but $\lim_{x \to 2} f(x)$ does. Let us define a new function g such that $g(x) = f(x)$ for all $x \neq 2$, and $g(2) = \lim_{x \to 2} f(x) = 12$. We define this function piecewise as follows:

$$g(x) = \begin{cases} \dfrac{x^3 - 8}{x - 2} & x \neq 2 \\ 12 & x = 2 \end{cases}$$

This function *is* continuous at $x = 2$, as you can see from the following graph:

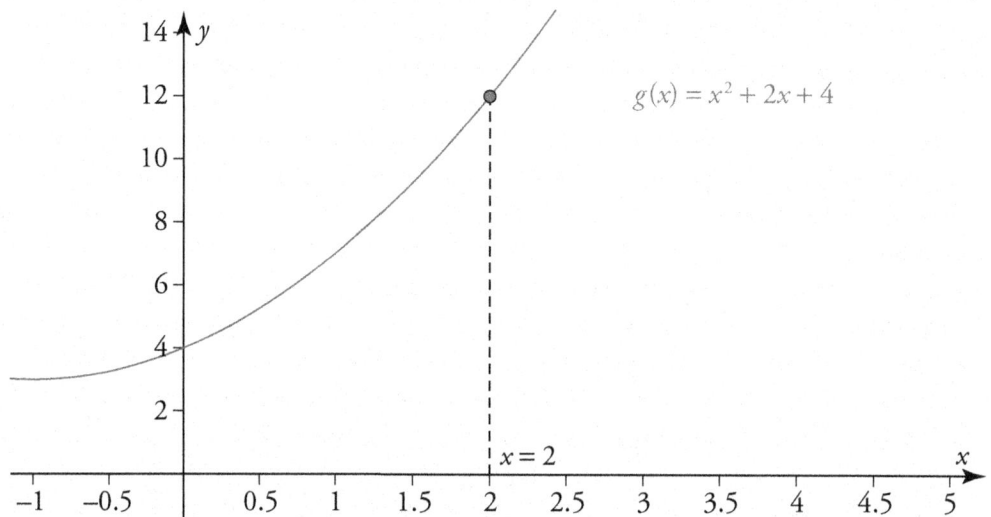

Because we can remove the discontinuity of f by defining $f(2) = 12$, we say that f has a removable discontinuity at $x = 2$.

If a function f has a discontinuity at $x = c$ that *cannot* be removed by defining or re-defining $f(c)$, then f has a **nonremovable discontinuity** at $x = c$. Nonremovable discontinuities come in two varieties: jump discontinuities and vertical asymptotes. We'll review both of them in turn next.

Jump Discontinuity

A jump discontinuity is a nonremovable discontinuity that could only be removed by a vertical line segment of finite length. In terms of limits, f has a jump discontinuity at $x = c$ if $\lim_{x \to c^-} f(x) = R$ and $\lim_{x \to c^+} f(x) = S$, where R and S are distinct real numbers. See the following figure:

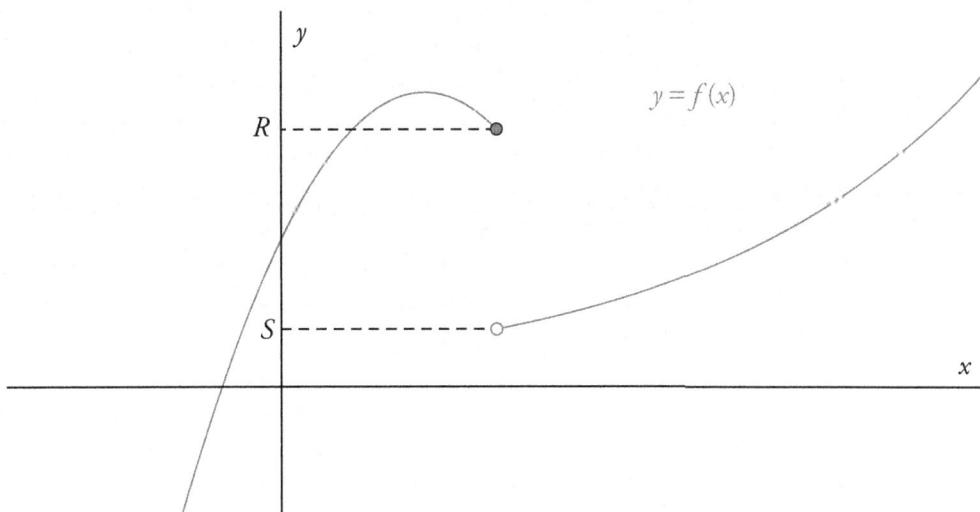

The function in Example 1.1.12 has a jump discontinuity. We recommend that you revisit that example.

Vertical Asymptote

We gave a thorough review of vertical asymptotes in **EK 1.1D1**, so we won't devote more space to it here. We'll simply point out that vertical asymptotes are also nonremovable discontinuities.

Summary

Classification of the discontinuities of a function f can be boiled down to a few simple conditions.

Condition:	Type of Discontinuity:
$f(c)$ is not defined, but $\lim\limits_{x \to c} f(x)$ exists.	Removable Discontinuity
$f(c)$ is not defined and $\lim\limits_{x \to c} f(x)$ does not exist, but both one-sided limits are finite.	Jump Discontinuity
$f(c)$ is not defined, $\lim\limits_{x \to c^-} f(x) = \pm\infty$, and $\lim\limits_{x \to c^+} f(x) = \pm\infty$	Vertical Asymptote

We'll close this section with an example.

Example 1.2.4: Discuss the continuity of the function $f(x) = \dfrac{x^2 + 4x + 3}{x^2 + x - 6}$.

This is a rational function, so its domain is all real x such that $x^2 + x - 6 \neq 0$. Since $x^2 + x - 6 = (x+3)(x-2)$, we see that the domain of f is all real $x \neq -3, 2$. This tells us that f is continuous at all real x except $x = -3$ and $x = 2$. We will examine each discontinuity individually.

The limit of f at $x = 3$ produces the indeterminate form $\dfrac{0}{0}$. We'll use algebraic manipulation (see **EK 1.1C2**) to evaluate this limit.

$$\lim_{x \to -3} \frac{x^2 + 4x + 3}{x^2 + x - 6} = \lim_{x \to -3} \frac{(x+3)(x+1)}{(x+3)(x-2)} = \lim_{x \to -3} \frac{x+1}{x-2} = \frac{2}{5}$$

Since $f(-3)$ does not exist, but $\lim_{x \to -3} f(x)$, f has a removable discontinuity at $x = -3$. The limit of f at $x = 2$ produces the form $\frac{15}{0}$, which is a signal that the one-sided limits are infinite.

$$\lim_{x \to 2^-} \frac{x^2 + 4x + 3}{x^2 + x - 6} = \lim_{x \to 2^-} \frac{(x+3)(x+1)}{(x+3)(x-2)} = \lim_{x \to 2^-} \frac{x+1}{x-2} = -\infty$$

$$\lim_{x \to 2^+} \frac{x^2 + 4x + 3}{x^2 + x - 6} = \lim_{x \to 2^+} \frac{(x+3)(x+1)}{(x+3)(x-2)} = \lim_{x \to 2^+} \frac{x+1}{x-2} = \infty$$

These infinite one-sided limits indicate that f has a vertical asymptote at $x = 2$. A graph of f is displayed below to illustrate these results.

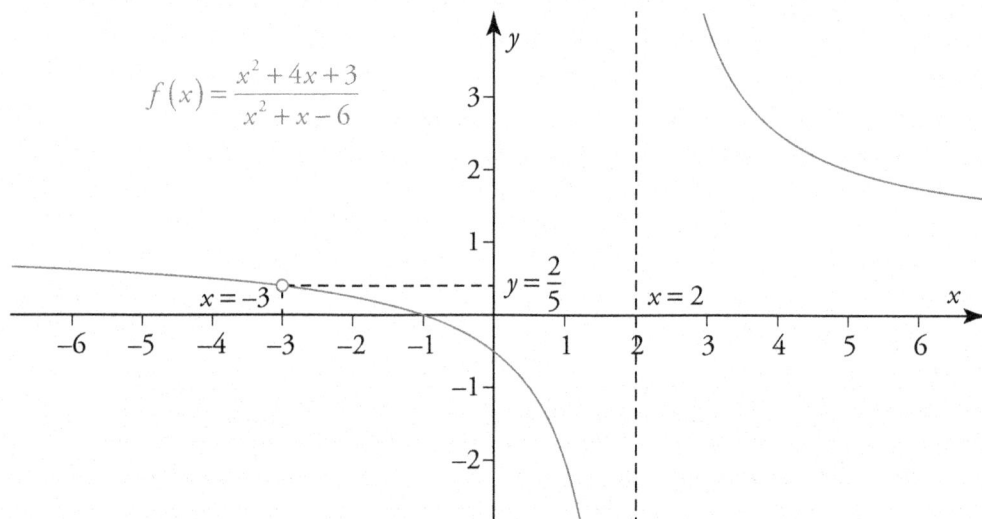

LO 1.2B – Continuity and the Theorems of Calculus

The full Learning Objective of this section is as follows:

LO 1.2B: Determine the applicability of important calculus theorems using continuity.

The first and only item of Essential Knowledge under LO 1.2B shows the importance of continuity property of functions for some important theorems in calculus to hold.

EK 1.2B1: Continuity is an essential condition for theorems such as the Intermediate Value Theorem, the Extreme Value Theorem, and the Mean Value Theorem.

Intermediate Value Theorem

Suppose a function f is continuous at all points in the interval $[a, b]$, and that there is a real number M such that M is between $f(a)$ and $f(b)$. Then there exists at least one c in the interval $[a, b]$ such that $f(c) = M$. In other words, f takes on every value between $f(a)$ and $f(b)$ somewhere in the interval. See the following figure:

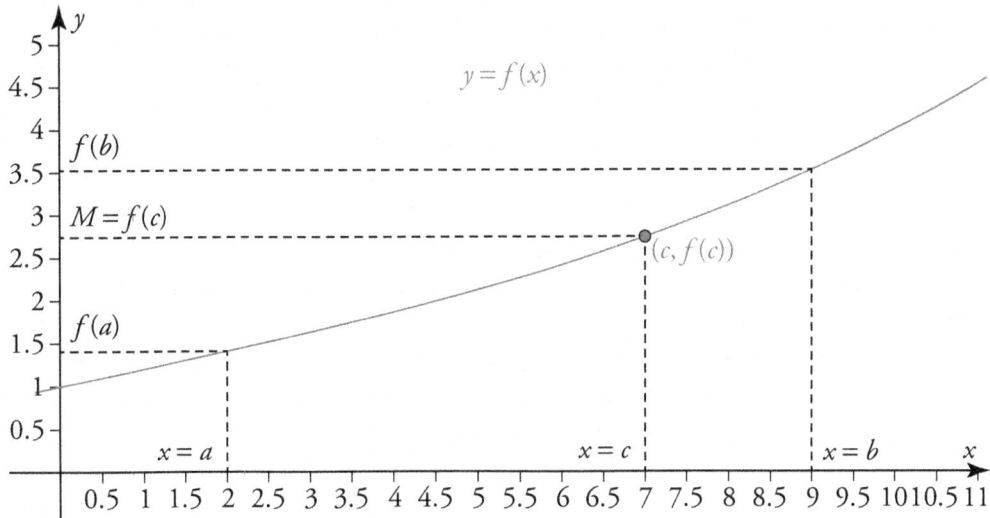

The Intermediate Value Theorem says that if the hypotheses of the theorem are satisfied, then the graph of f *must* cross the horizontal line $y = M$ somewhere in the interval $[a, b]$. Note the importance of *continuity* here. If f were not continuous on $[a, b]$, then there would be no guarantee that f takes on the value M anywhere in $[a, b]$.

We illustrate the importance of the Intermediate Value Theorem with an example.

Example 1.2.5: Show that the equation $x^5 - 2x^3 - 5 = 0$ must have a solution in the interval $[0, 2]$.

This equation is difficult (if not impossible!) to solve using any techniques from precalculus. While we cannot solve the equation, we *can* use the Intermediate Value Theorem to prove the desired result.

Let $f(x) = x^5 - 2x^3 - 5$. This is a polynomial function, so it's continuous everywhere. Note that $f(0) = -5$ and $f(2) = 11$. Since 0 is between $f(0)$ and $f(2)$, there must exist at least one c in $[0, 2]$ such that $f(c) = 0$, which proves that $x^5 - 2x^3 - 5 = 0$ somewhere in $[0, 2]$. See the following figure:

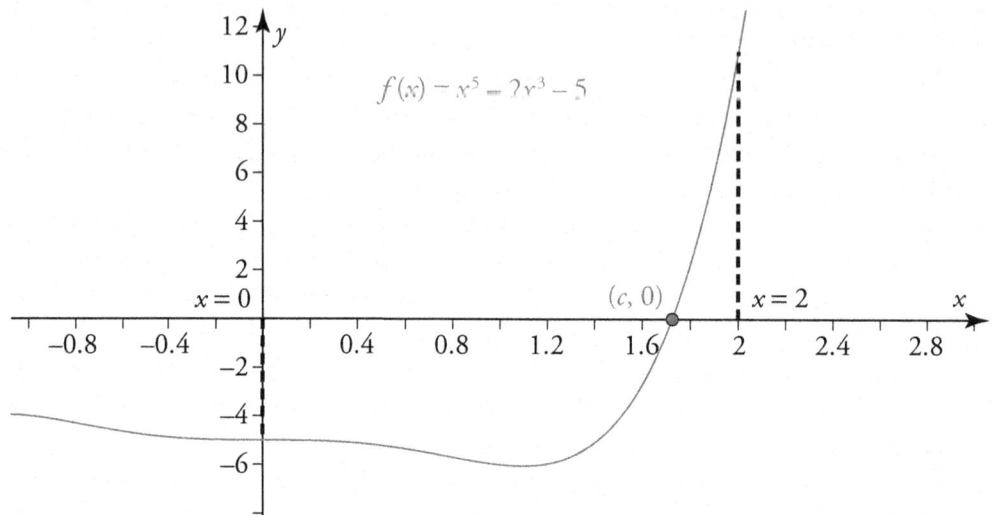

$$f(x) = x^5 - 2x^3 - 5$$

$x = 0$

$(c, 0)$

$x = 2$

Extreme Value Theorem

If a function f is continuous at all points in the interval $[a, b]$, then f must attain a maximum and a minimum in that interval. This means that there exists a c and d in the closed interval $[a, b]$ such that $f(c) \leq f(x)$ and $f(x) \leq f(d)$. These are the minimum and maximum values of f occurring in $[a, b]$ at $x = c$ and $x = d$, respectively. See the following figure:

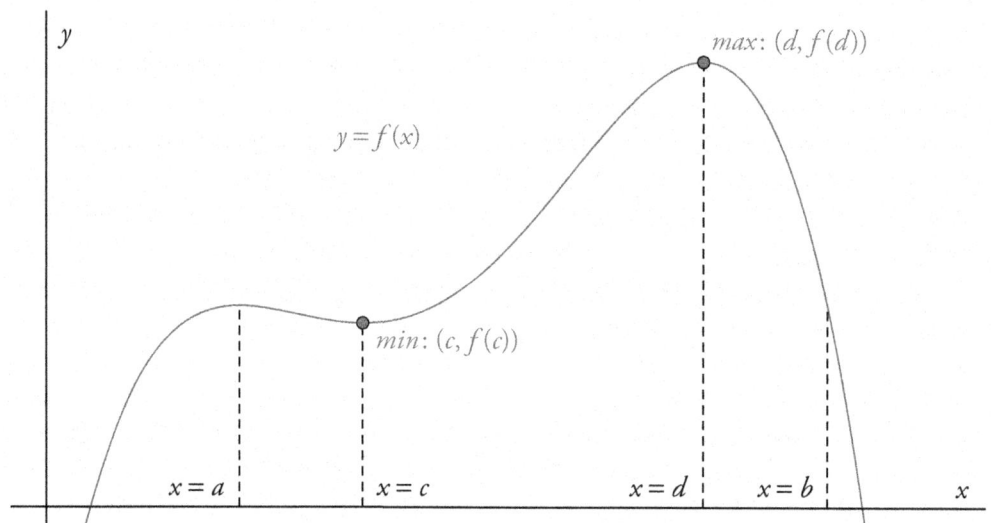

You will learn about absolute and relative maxima/minima and the application of this theorem in detail in section **EK 2.3C3**. For now, just remember that this theorem holds good only for functions that are *continuous* on a closed interval.

Before we get to the next theorem, we note that it refers to differentiability and the derivative, which we will review in the next chapter.

Mean Value Theorem

If a function f is continuous on $[a, b]$ and differentiable on (a, b). Then there exists at least one number c in (a, b) such that $f'(c) = \dfrac{f(b) - f(a)}{b - a}$. See the following figure:

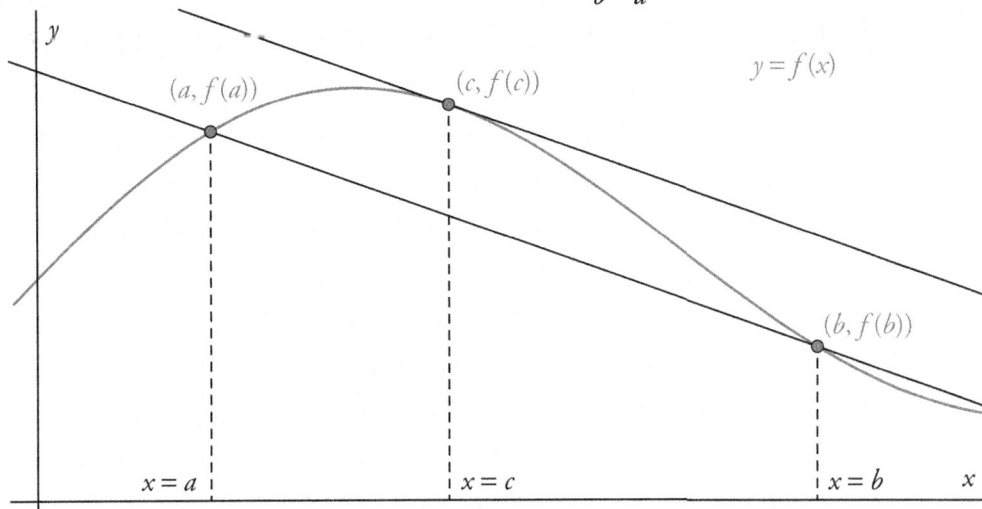

The slope of the line joining the points $(a, f(a))$ and $(b, f(b))$ is $\dfrac{f(b) - f(a)}{b - a}$, and the slope of line tangent to the curve at the point $(c, f(c))$ is $f'(c)$. $f'(c)$ interpreted as the *instantaneous* rate of change of f at $x = c$, and $\dfrac{f(b) - f(a)}{b - a}$ is interpreted as the *mean* rate of change of f over $[a, b]$. What the Mean Value Theorem says is that, if the hypotheses are satisfied, then there must be a point c in (a, b) at which the two slopes are equal. This is equivalent to saying that there must be a point c in (a, b) at which the instantaneous rate of change of f is equal to the mean rate of change of f.

You will learn about this theorem in detail in chapter **EU 2.4**. For now, just remember that this theorem holds good only for functions that are *continuous* on a closed interval.

Chapter 2: Differential Calculus

Chapter 2 Overview _____

Chapter 2 is divided into the following four Enduring Understandings:

- **EU 2.1:** The derivative of a function is defined as the limit of a difference quotient and can be determined using a variety of strategies.
- **EU 2.2:** A function's derivative, which is itself a function, can be used to understand the behavior of the function.
- **EU 2.3:** The derivative has multiple interpretations and applications including those that involve instantaneous rates of change.
- **EU 2.4:** The Mean Value Theorem connects the behavior of a differentiable function over an interval to the behavior of the derivative of that function at a particular point in the interval.

EU 2.1 – The Derivative _____

LO 2.1A – The Derivative as the Limit of a Difference Quotient

Now that you've brushed up on limits, we'll review a special type of limit: the derivative. The full Learning Outcome for this section is as follows:

LO 2.1A: Identify the derivative of a function as the limit of a difference quotient.

The first item of Essential Knowledge pertains to what exactly a *difference quotient* is and what it means.

EK 2.1A1: The difference quotients $\dfrac{f(a+h)-f(a)}{h}$ and $\dfrac{f(x)-f(a)}{x-a}$ express the average rate of change of a function over an interval.

The two expressions given in the above EK are called difference quotients for the simple reason that they are both a quotient of two differences. The denominator of the first expression may not *look* like a difference, but it really is because $h = (a+h) - h$.

The difference quotient $\dfrac{f(a+h)-f(a)}{h}$ expresses the average rate of change of a function f over a *fixed* interval $[a, b]$, where $b = a + h$. We'll illustrate this below.

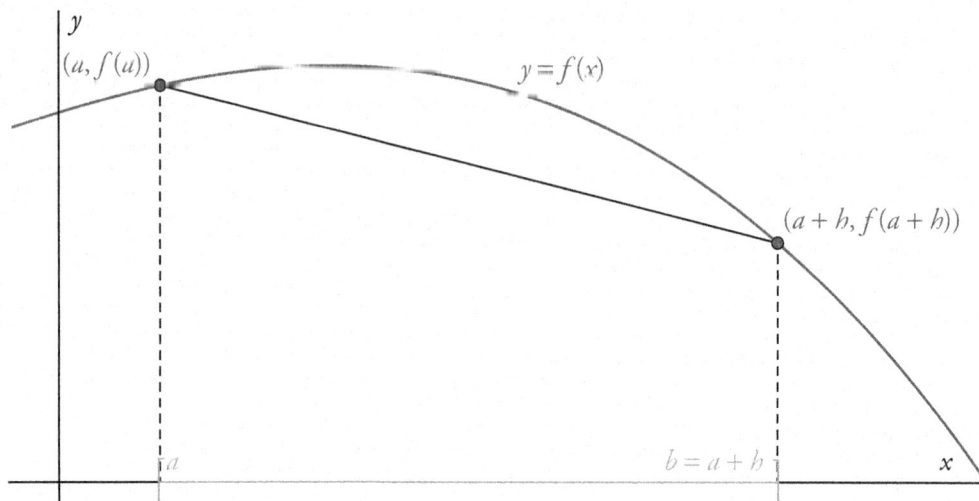

Geometrically, the difference quotient $\dfrac{f(a+h)-f(a)}{h}$ is the slope of the secant line that joins the points $(a, f(a))$ and $(a+h, f(a+h))$.

Example 2.1.1: Find the average rate of change of $f(x) = 2x^2 - 3x$ over the interval $[1, 4]$.

In this case $a = 1$ and $b = a + h = 4$, so $h = b - a = 3$. The average rate of change is then found as follows:

$$\frac{f(4) - f(1)}{3} = \frac{\left(2(4)^2 - 3(4)\right) - \left(2(1)^2 - 3(1)\right)}{3} = 7$$

You're going to get to know this function really well, because we are going to beat it to death in the next several examples.

The difference quotient $\dfrac{f(x) - f(a)}{x - a}$ is different from the previous one in that it expresses the average rate of change of f over the interval $[a, x]$, where x is *variable*. We'll illustrate this below.

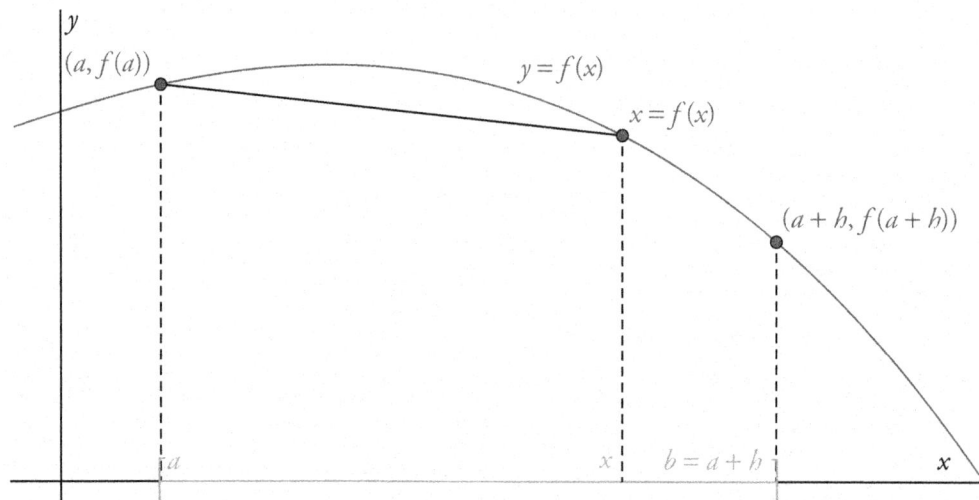

As you may have already guessed, the slope of secant line that joins the points $(a, f(a))$ and $(x, f(x))$ is equal to the difference quotient $\dfrac{f(x) - f(a)}{x - a}$.

Example 2.1.2: Find the average rate of change of $f(x) = 2x^2 - 3x$ over the intervals $[1, 1.5]$, $[1, 2]$, $[1, 2.5]$, $[1, 3]$, and $[1, 3.5]$.

Another way to phrase this problem is this: find the average rate of change of $f(x) = 2x^2 - 3x$ over $[1, x]$ for $x = 1.5, 2, 2.5, 3$, and 3.5.

$$\frac{f(x) - f(1)}{x - 1} = \frac{2x^2 - 3x - \left(2(1)^2 - 3(1)\right)}{x - 1} = \frac{2x^2 - 3x + 1}{x - 1}$$

The numerator is factorable: $2x^2 - 3x + 1 = (2x - 1)(x - 1)$. For $x \neq 1$, we may simplify the difference quotient as follows:

$$\frac{f(x) - f(1)}{x - 1} = \frac{(2x - 1)(x - 1)}{x - 1} = 2x - 1$$

This difference quotient is a *function of* x. We'll finish this off by plugging and chugging.

$$x = 1.5: \qquad \frac{f(1.5) - f(1)}{1.5 - 1} = 2(1.5) - 1 = 2$$

$$x = 2: \qquad \frac{f(2) - f(1)}{2 - 1} = 2(2) - 1 = 3$$

$$x = 2.5: \qquad \frac{f(2.5) - f(1)}{2.5 - 1} = 2(2.5) - 1 = 4$$

$$x = 3: \qquad \frac{f(3) - f(1)}{3 - 1} = 2(3) - 1 = 5$$

$$x = 3.5: \qquad \frac{f(3.5) - f(1)}{3.5 - 1} = 2(3.5) - 1 = 6$$

In the next EK, we will make the transition from *average* rates of change to *instantaneous* rates of change.

EK 2.1A2: The instantaneous rate of change of a function at a point can be expressed by $\lim\limits_{h \to 0} \dfrac{f(a + h) - f(a)}{h}$ or $\lim\limits_{x \to a} \dfrac{f(x) - f(a)}{x - a}$, provided that the limit exists. These are common forms of the definition of the derivative and are denoted $f'(a)$.

Consider again the function in Examples 2.1.1 and 2.1.2. We found the *average* rates of change of f over $[1, x]$ for $x = 1.5, 2, 2.5, 3, 3.5$, and 4. What if we wanted to know the *instantaneous* rate of change at $x = 1$?

In the figure below, we've graphed the function $f(x) = 2x^2 - 3x$. The secants joining the point $(1, 1)$ and the other labeled points are sketched in, as is the tangent line to the graph of f at $(1, 1)$. The slope of this tangent line is the instantaneous rate of change of f at $x = 1$, which we denote $f'(1)$.

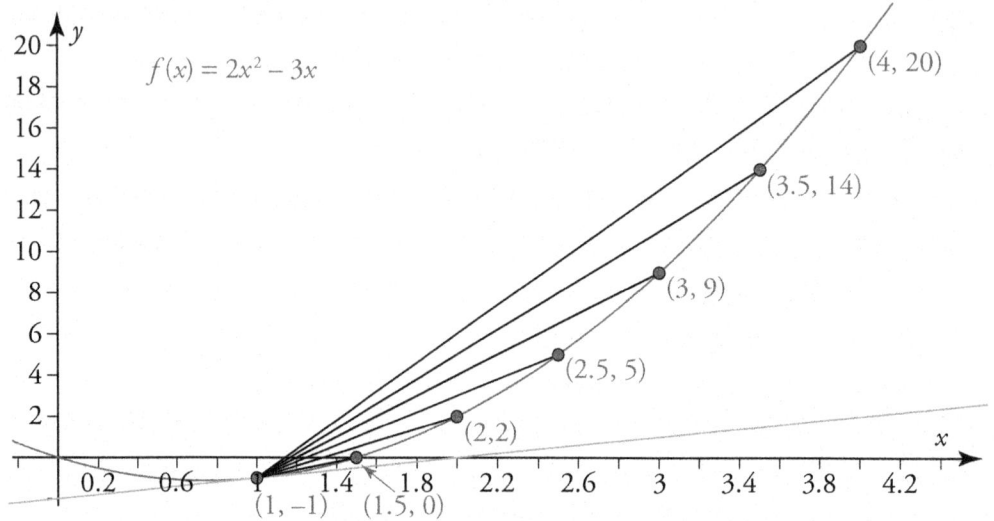

$f(x) = 2x^2 - 3x$

As you can see, the slopes of the secants of f over $[1, x]$ approach the slope of the tangent as x approaches 1 from the right. Mathematically, we express this as

$$\lim_{x \to 1^+} \frac{f(x) - f(1)}{x - 1},$$

or

$$\lim_{h \to 0^+} \frac{f(1 + h) - f(1)}{h}.$$

A similar figure can be drawn to illustrate the limits from the left, as shown below.

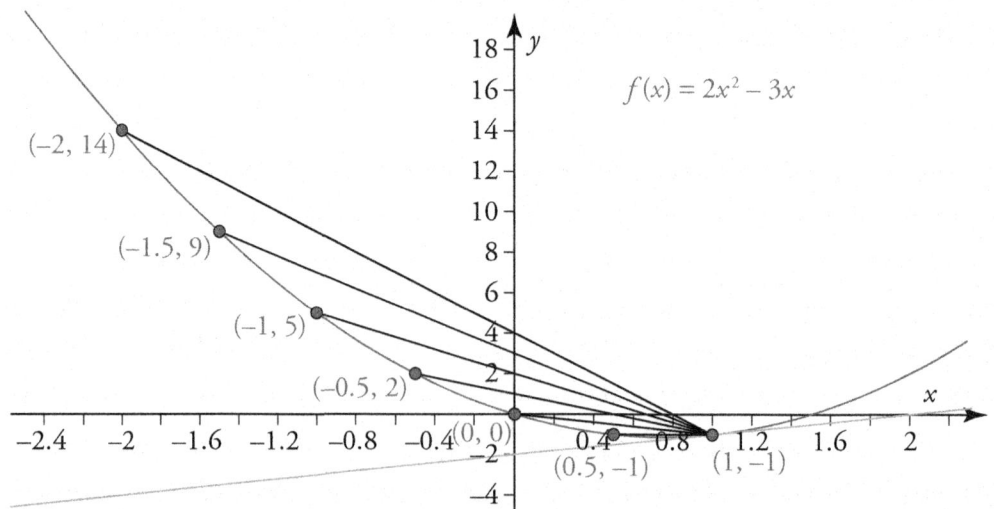

$f(x) = 2x^2 - 3x$

Example 2.1.3: Compute $f'(1)$ using each of the two difference quotients given above.

We'll start with the first difference quotient.

$$f'(1) = \lim_{x \to 1} \frac{f(x) - f(1)}{x - 1}$$

We've already developed and simplified the difference quotient in Example 2.1.2.

$$f'(1) = \lim_{x \to 1} (2x - 1) = 1$$

Now we compute $f'(1)$ with the second difference quotient.

$$f'(1) = \lim_{h \to 0} \frac{f(1+h) - f(1)}{h} = \lim_{h \to 0} \frac{\left(2(1+h)^2 - 3(1+h)\right) - \left(2(1)^2 - 3(1)\right)}{h}$$

$$f'(1) = \lim_{h \to 0} \frac{2h^2 + h}{h} = \lim_{h \to 0} \frac{h(2h+1)}{h} = \lim_{h \to 0} (2h + 1) = 1$$

Notice the phrase "provided the limit exists" in the EK. Sometimes the limit *does not* exist, in which case it is not possible to compute an instantaneous rate of change of a function at a point.

Example 2.1.4: Show that the instantaneous rate of change of $f(x) = |x|$ is not defined at $x = 0$.

We need to show that $\lim_{x \to 0} \frac{|x| - |0|}{x - 0} = \lim_{x \to 0} \frac{|x|}{x}$ does not exist. We do this by considering the left- and right-handed limits separately.

$$\lim_{x \to 0^-} \frac{|x|}{x} = \lim_{x \to 0^-} \frac{-x}{x} = \lim_{x \to 0^-} (-1) = -1$$

$$\lim_{x \to 0^+} \frac{|x|}{x} = \lim_{x \to 0^+} \frac{x}{x} = \lim_{x \to 0^+} 1 = 1$$

Since $\lim_{x \to 0^-} \frac{|x|}{x} \neq \lim_{x \to 0^+} \frac{|x|}{x}$, it follows that $\lim_{x \to 0} \frac{|x|}{x}$ does not exist.

If the limit $\lim_{x \to a} \frac{f(x) - f(a)}{x - a}$ exists, then we say that f is **differentiable** at $x = a$, and we say that f is **not differentiable** otherwise.

In this EK, we reviewed how to find the instantaneous rate of change of a function *at a point*. In the next EK, we will review how to find the instantaneous rate of change of a function *as a function of x*.

EK 2.1A3: The derivative of f is the function whose value at x is $\lim_{h \to 0} \frac{f(x+h) - f(x)}{h}$ provided this limit exists.

Example 2.1.5: Let $f(x) = 2x^2 - 3x$. Find the instantaneous rate of change of f as a function of x.

$$\lim_{h \to 0} \frac{f(x+h) - f(x)}{h} = \lim_{h \to 0} \frac{\left(2(x+h)^2 - 3(x+h)\right) - \left(2x^2 - 3x\right)}{h}$$

$$\lim_{h \to 0} \frac{f(x+h) - f(x)}{h} = \lim_{h \to 0} \frac{\left(2x^2 + 4hx + h^2 - 3x - 3h\right) - \left(2x^2 - 3x\right)}{h}$$

$$\lim_{h \to 0} \frac{f(x+h) - f(x)}{h} = \lim_{h \to 0} \frac{4hx + h^2 - 3h}{h} = \lim_{h \to 0} \frac{h(4x + h - 3)}{h} = \lim_{h \to 0} (4x + h - 3) = 4x - 3$$

The next EK is an easy one. It just introduces some new notation.

EK 2.1A4: For $y = f(x)$, notations for the derivative include $\dfrac{dy}{dx}$, $f'(x)$, and y'.

Example 2.1.6: Let $y = 2x^2 - 3x$. We may denote the result of Example 2.1.5 in each of the following ways:

$$\frac{dy}{dx} = 4x - 3$$

$$f'(x) = 4x - 3$$

$$y' = 4x - 3$$

We may think of $\dfrac{dy}{dx}$ as $\dfrac{d}{dx}(y)$, where $\dfrac{d}{dx}$ is the derivative *operator* that acts on y. In this notation, the result of Example 2.1.5 reads as follows:

$$\frac{d}{dx}\left(2x^2 - 3x\right) = 4x - 3$$

EK 2.1A5: The derivative can be represented graphically, numerically, analytically, and verbally.

As we mentioned while reviewing **E 2.1A2**, the derivative of a function at a point is equal to the slope of the *tangent line* to the graph at that point. This leads us immediately to a graphical representation of the derivative.

Example 2.1.7: In the figure below, both the function $f(x) = 2x^2 - 3x$ and the line tangent to the graph of the function at the point $(1, -1)$ are graphed. Use this information to determine $f'(1)$.

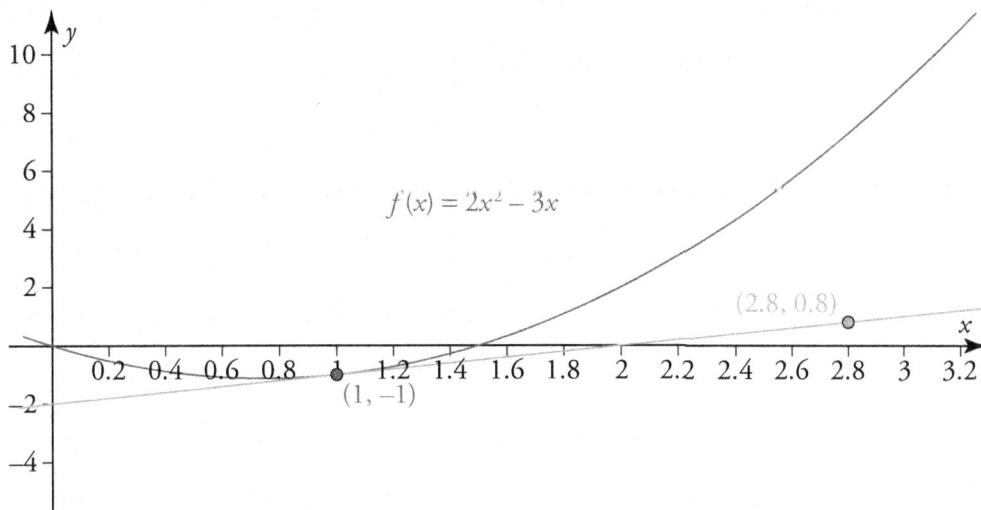

We can determine the slope of the tangent line from the two given points on the line, and this slope is the derivative of f at $x = 1$.

$$f'(1) = \frac{0.8 - (-1)}{2.8 - 1} = 1$$

Note that we did not actually need the picture to answer the question. We could just as well have been given a numerical table of points on the tangent line, like the following:

x	y
1	−1
2.8	0.8

Alternatively, we could have been given a verbal description of the tangent line, which we would then have had to translate into mathematical notation. An example is as follows:

The line tangent to f at the point $(1, -1)$ passes through the point $(2, 8, 0.8)$.

Any of these representations would have yielded the same result: $f'(1) = 1$.

LO 2.1B – Estimation of Derivatives

This Learning Objective is very simply stated.

LO 2.1B: Estimate derivatives.

This LO has but one item of Essential Knowledge.

EK 2.1B1: The derivative at a point can be estimated from information given in tables or graphs.

We will return to our example $f(x) = 2x^2 - 3x$. We've already presented a graphical representation of $f'(1)$ in our discussion of **EK 2.1A2** and **EK 2.1A4**. We'll come in for a closer look this time.

Example 2.1.8: Use the information in the following graph to estimate $f'(1)$.

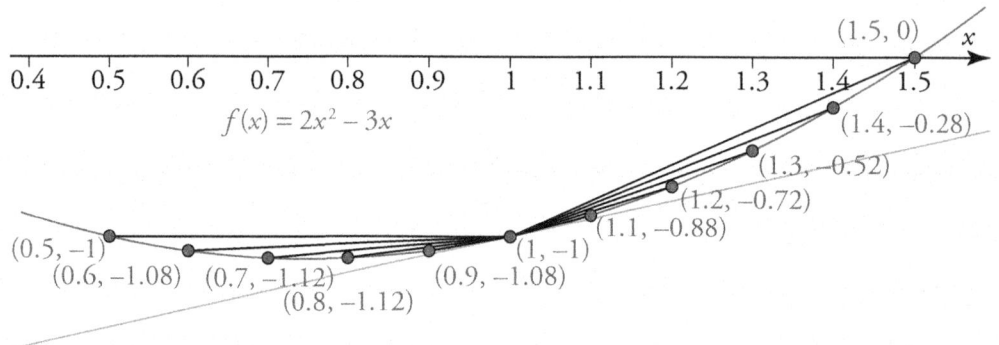
(0.39, 0.81)

From the above graph, we can construct the following table for $\displaystyle\lim_{x\to1^+}\frac{f(x)-f(1)}{x-1}$:

x:	1.5	1.4	1.3	1.2	1.1
$y=f(x)$:	0	−0.28	−0.52	−0.72	−0.88
$\dfrac{f(x)-f(1)}{x-1}$:	2	1.8	1.6	1.4	1.2

We have arranged the values of x so that they approach $x=1$ from above when read from left to right, and we have the estimate $\displaystyle\lim_{x\to1^+}\frac{f(x)-f(1)}{x-1}\approx1.2$. Similarly, we can construct the following table for $\displaystyle\lim_{x\to1^-}\frac{f(x)-f(1)}{x-1}$:

x:	0.5	0.6	0.7	0.8	0.9
$y=f(x)$:	−1	−1.08	−0.42	−1.12	−1.08
$\dfrac{f(x)-f(1)}{x-1}$:	0	0.2	0.4	0.6	0.8

This time we have the estimate $\displaystyle\lim_{x\to1^-}\frac{f(x)-f(1)}{x-1}\approx0.8$. It looks as though it *could* be true that $\displaystyle\lim_{x\to1^-}\frac{f(x)-f(1)}{x-1}=1=\lim_{x\to1^+}\frac{f(x)-f(1)}{x-1}$, and of course we know from previous examples that that *is* true, but we could get better precision on our estimate by moving in closer to $x=1$.

Example 2.1.9: Below is a table of coordinates of points on the graph of $f(x) = 2x^2 - 3x$. Complete the table and use it to estimate $f'(1)$.

x:	0.9	0.99	0.999	1.001	1.01	1.1
$y = f(x)$:	−1.08	−1.0098	−1.000998	−0.998998	−0.9898	−0.88
$\dfrac{f(x) - f(1)}{x - 1}$:						

First, we work out the first three columns to get an estimate of $\displaystyle\lim_{x \to 0^-} \frac{f(x) - f(1)}{x - 1}$.

$$\frac{f(0.9) - f(1)}{0.9 - 1} = \frac{-1.08 - (-1)}{0.9 - 1} = 0.8$$

$$\frac{f(0.99) - f(1)}{0.99 - 1} = \frac{-1.0098 - (-1)}{0.99 - 1} = 0.98$$

$$\frac{f(0.999) - f(1)}{0.999 - 1} = \frac{-1.000998 - (-1)}{0.999 - 1} = 0.998$$

From this we see that $\displaystyle\lim_{x \to 0^-} \frac{f(x) - f(1)}{x - 1} \approx 0.998$.

Next, we'll work out the last three columns to get an estimate of $\displaystyle\lim_{x \to 0^+} \frac{f(x) - f(1)}{x - 1}$.

$$\frac{f(1.1) - f(1)}{1.1 - 1} = \frac{-0.88 - (-1)}{1.1 - 1} = 1.2$$

$$\frac{f(1.01) - f(1)}{1.01 - 1} = \frac{-0.9898 - (-1)}{1.01 - 1} = 1.02$$

$$\frac{f(1.001) - f(1)}{1.001 - 1} = \frac{-0.998998 - (-1)}{1.001 - 1} = 1.002$$

From this, we see that $\displaystyle\lim_{x \to 0^+} \frac{f(x) - f(1)}{x - 1} \approx 1.002$.

Comparing these results with those of the last example, we see that this estimate makes a much more convincing case that $\displaystyle\lim_{x \to 0^-} \frac{f(x) - f(1)}{x - 1} = 1 = \lim_{x \to 0^+} \frac{f(x) - f(1)}{x - 1}$.

LO 2.1C – Calculating Derivatives

In this section, we will review how to get away from using the limit definition of the derivative and, instead, use the differentiation rules that you learned in your calculus course. The LO for this section is short and sweet.

LO 2.1C: Calculate derivatives.

In the first item of Essential Knowledge, we will apply the limit definition of the derivative to a handful of functions.

EK 2.1C1: Direct application of the definition of the derivative can be used to find the derivative of selected functions, including polynomial, power, sine, cosine, exponential, and logarithmic functions.

We will start with a polynomial function.

Example 2.1.10: Let $f(x) = x^3 - 2x + 4$. Compute $f'(x)$.

$$f'(x) = \lim_{h \to 0} \frac{\left((x+h)^3 - 2(x+h) + 4\right) - \left(x^3 - 2x + 4\right)}{h}$$

$$f'(x) = \lim_{h \to 0} \frac{\left(x^3 + 3hx^2 + 3h^2x + h^3 - 2x - 2h + 4\right) - \left(x^3 - 2x + 4\right)}{h}$$

$$f'(x) = \lim_{h \to 0} \frac{3hx^2 + 3h^2x + h^3 - 2h}{h} = \lim_{h \to 0} \frac{h\left(3x^2 + 3hx + h^2 - 2\right)}{h} = \lim_{h \to 0}\left(3x^2 + 3hx + h^2 - 2\right)$$

$$f'(x) = 3x^2 - 2$$

Next, we'll do two examples of power functions.

Example 2.1.11: Let $f(x) = \dfrac{1}{x^2}$. Compute $f'(x)$.

This may not look like a power function, but if we rewrite it as $f(x) = x^{-2}$, we see that it is a power function with a negative exponent.

$$f'(x) = \lim_{h \to 0} \frac{\dfrac{1}{(x+h)^2} - \dfrac{1}{x^2}}{h} = \lim_{h \to 0} \frac{x^2 - (x+h)^2}{hx^2(x+h)^2} = \lim_{h \to 0} \frac{\left(x - (x+h)\right)\left(x + (x+h)\right)}{hx^2(x+h)^2}$$

$$f'(x) = \lim_{h \to 0} \frac{-h(2x+h)}{hx^2(x+h)^2} = \lim_{h \to 0} \frac{-(2x+h)}{x^2(x+h)^2} = -\frac{2x}{x^4}$$

$$f'(x) = -\frac{2}{x^3}$$

Example 2.1.12: Let $f(x) = \sqrt{x}$. Compute $f'(x)$.

If we rewrite this function as $f(x) = x^{1/2}$, we see that it is a power function with a rational exponent.

$$f'(x) = \lim_{h \to 0} \frac{\sqrt{x+h} - \sqrt{x}}{h} = \lim_{h \to 0}\left(\frac{\sqrt{x+h} - \sqrt{x}}{h} \cdot \frac{\sqrt{x+h} + \sqrt{x}}{\sqrt{x+h} + \sqrt{x}}\right)$$

$$f'(x) = \lim_{h \to 0} \frac{x+h-x}{h\left(\sqrt{x+h}+\sqrt{x}\right)} = \lim_{h \to 0} \frac{h}{h\left(\sqrt{x+h}+\sqrt{x}\right)} = \lim_{h \to 0} \frac{1}{\sqrt{x+h}+\sqrt{x}}$$

$$f'(x) = \frac{1}{2\sqrt{x}}$$

In our next example, we will compute the derivative of a trigonometric function.

Example 2.1.13: Let $f(x) = \sin(x)$. Compute $f'(x)$.

$$\lim_{h \to 0} \frac{\sin(x+h) - \sin(x)}{h}$$

Here, we need to use the sum formula $\sin(u+v) = \sin(u)\cos(v) + \cos(u)\sin(v)$.

$$f'(x) = \lim_{h \to 0} \frac{\sin(x)\cos(h) + \cos(x)\sin(h) - \sin(x)}{h}$$

$$f'(x) = \lim_{h \to 0} \left(\frac{\sin(x)\cos(h) - \sin(x)}{h} + \frac{\cos(x)\sin(h)}{h} \right)$$

$$f'(x) = \lim_{h \to 0} \left(\frac{\sin(x)(\cos(h) - 1)}{h} + \frac{\cos(x)\sin(h)}{h} \right)$$

For the purposes of this limit, functions of x are regarded as *constants*, and so we may rewrite the last line as follows:

$$f'(x) = \sin(x) \cdot \lim_{h \to 0} \left(\frac{\cos(h) - 1}{h} \right) + \cos(x) \cdot \lim_{h \to 0} \frac{\sin(h)}{h}$$

Noting that $\lim_{h \to 0} \left(\frac{\cos(h) - 1}{h} \right) = 0$ and $\lim_{h \to 0} \frac{\sin(h)}{h} = 1$, we have the following:
$$f'(x) = \cos(x)$$

As an exercise, you should show that $\frac{d}{dx}\left(\cos(x)\right) = -\sin(x)$. You will need the sum formula $\cos(u+v) = \cos(u)\cos(v) - \sin(u)\sin(v)$.

Next, we'll take the derivative of an exponential function.

Example 2.1.14: Let $f(x) = e^x$. Compute $f'(x)$.

This one is a little tricky to do using the definition of a derivative. You've undoubtedly seen the definition of e that states $e = \lim_{n \to \infty} \left(1 + \frac{1}{n} \right)^n$. Here, we will need a lesser-known alternate definition: e is the unique positive number for which $\lim_{n \to 0} \frac{e^n - 1}{n} = 1$. Now let's get to it.

$$f'(x) = \lim_{h \to 0} \frac{e^{x+h} - e^x}{h} = \lim_{h \to 0} \frac{e^x e^h - e^x}{h} = \lim_{h \to 0} \frac{e^x\left(e^h - 1\right)}{h}$$

As far as the limit is concerned, e^x is a constant, and so we may pull it outside of the limit.

$$f'(x) = e^x \lim_{h \to 0} \frac{e^h - 1}{h}$$

According to the alternate definition of e, $\lim_{h \to 0} \frac{e^h - 1}{h} = 1$. So, we finally obtain our result.

$$f'(x) = e^x$$

We will finish off this EK with an example of differentiation of a logarithmic function.

Example 2.1.15: Let $f(x) = \ln(x)$. Compute $f'(x)$.

This one is also tricky to do with the definition of the derivative, so we'll go through it carefully.

$$f'(x) = \lim_{h \to 0} \frac{\ln(x+h) - \ln(x)}{h}$$

We begin by using the properties of logarithms to simplify the limit.

$$f'(x) = \lim_{h \to 0} \frac{1}{h} \cdot \ln\left(\frac{x+h}{x}\right) = \lim_{h \to 0} \ln\left(1 + \frac{h}{x}\right)^{1/h}$$

Here's where we have to get clever. In the exponent, we will multiply by $\frac{x}{x}$.

$$f'(x) = \lim_{h \to 0} \ln\left(1 + \frac{h}{x}\right)^{x/xh} = \lim_{h \to 0} \ln\left(\left(1 + \frac{h}{x}\right)^{x/h}\right)^{1/x}$$

Notice that in the above limit, $h \to 0 \Rightarrow \frac{h}{x} \to 0$. Let $n = \frac{h}{x}$ and use yet another alternate definition of e: $e = \lim_{n \to 0}(1+n)^{1/n}$. Then we obtain the following:

$$f'(x) = \ln\left(e^{1/x}\right) = \frac{1}{x}$$

Before we move on to the next EK, we will review some basic differentiation rules.

Example 2.1.16: Use the definition of the derivative to show the following:

1. $\dfrac{d}{dx}(c) = 0$, where c is any real number.

2. $\dfrac{d}{dx}(x) = 1$

3. $\dfrac{d}{dx}\left(x^n\right)=nx^{n-1}$, where n is any natural number.

4. $\dfrac{d}{dx}\left(cx^n\right)=cnx^{n-1}$, where n is any natural number and c is any real number.

We'll work these out one at a time.

1. Let $f(x)=c$, where c is a real number. This is a *constant function*. Note that for this function, $f(x+h)=f(x)=c$.

$$f'(x)=\lim_{h\to 0}\frac{f(x+h)-f(x)}{h}=\lim_{h\to 0}\frac{c-c}{h}=\lim_{h\to 0}\frac{0}{h}=0$$

So, the derivative of a constant function is zero.

2. Let $f(x)=x$.

$$f'(x)=\lim_{h\to 0}\frac{x+h-x}{h}=\lim_{h\to 0}\frac{h}{h}=\lim_{h\to 0}1=1$$

3. Let $f(x)=x^n$, where n is a natural number.

$$f'(x)=\lim_{h\to 0}\frac{(x+h)^n-x^n}{h}$$

At this point, we need a result called the *Binomial Theorem*, which says the following:

$$(x+h)^n=x^n+nx^{n-1}h+\frac{n(n-1)}{2}x^{n-2}h^2+\cdots+h^2$$

We'll use this to expand $(x+h)^n$ in the numerator.

$$f'(x)=\lim_{h\to 0}\frac{x^n+nx^{n-1}h+\dfrac{n(n-1)}{2}x^{n-2}h^2+\cdots+h^2-x^n}{h}$$

$$f'(x)=\lim_{h\to 0}\frac{nx^{n-1}h+\dfrac{n(n-1)}{2}x^{n-2}h^2+\cdots+h^2}{h}$$

$$f'(x)=\lim_{h\to 0}\frac{h\left(nx^{n-1}+\dfrac{n(n-1)}{2}x^{n-1}h+\cdots+h^{n-1}\right)}{h}$$

$$f'(x)=\lim_{h\to 0}\left(nx^{n-1}+\frac{n(n-1)}{2}x^{n-1}h+\cdots+h^{n-1}\right)$$

$$f'(x)=nx^{n-1}$$

This result is called the **Power Rule for Derivatives**.

4. Let $f(x) = cx^n$.

$$f'(x) = \lim_{h \to 0} \frac{c(x+h)^n - cx^n}{h} = \lim_{h \to 0} \frac{c\left((x+h)^n - x^n\right)}{h}$$

Recall from **EK 1.1C1** that $\lim_{x \to c} kf(x) = k \lim_{x \to c} f(x)$, provided $\lim_{x \to c} f(x)$ exists. We can use that here.

$$f'(x) = c \lim_{h \to 0} \frac{(x+h)^n - x^n}{h}$$

We'll finish this off by using our result from part (c).

$$f'(x) = cnx^{n-1}$$

In the next EK, we'll break away from the definition of the derivative and, instead, review the familiar differentiation rules.

EK 2.1C2: Specific rules can be used to calculate derivatives for classes of functions, including polynomial, rational, power, exponential, logarithmic, trigonometric, and inverse trigonometric.

We will put the cart before the horse here and simply present the differentiation rules without justifying them. The justification will come in **EK 2.1C3–EK 2.1C5**.

Polynomial Functions

Let $f(x) = a_n x^n + a_{n-1} x^{n-1} + \cdots + a_2 x^2 + a_1 x + a_0$, where n is a natural number, a_0, a_1, a_2, ..., a_{n-1}, a_n are real numbers, and $a_n \neq 0$. Then f is differentiable for all x, and $f'(x)$ is given by the following rule:

$$f'(x) = na_n x^{n-1} + (n-1)a_{n-1} x^{n-1} + \cdots + 2a_2 x + a_1$$

What we have done here is apply the basic differentiation rules from Example 2.1.16 to each term of the polynomial. Term-by-term differentiation will be justified in **EK 2.1C3**.

Example 2.1.17: Let $f(x) = 5x^4 - 3x^3 + 7x^2 + 11x - 8$. Compute $f'(x)$.

Applying the above rule for differentiation of polynomials, we obtain the following:

$$f'(x) = 20x^3 - 9x^2 + 14x + 11$$

Rational Functions

Let $f(x) = \dfrac{p(x)}{q(x)}$, where $p(x)$ and $q(x)$ are polynomial functions. Then f is differentiable for all x for which $q(x) \neq 0$, and $f'(x)$ is given by the following rule:

$$f'(x) = \frac{p'(x)q(x) - p(x)q'(x)}{\left(q(x)\right)^2}$$

This rule will be justified when we review the Quotient Rule in **EK 2.1C3**.

Example 2.1.18: Let $f(x) = \dfrac{x}{x^2 + 1}$. Compute $f'(x)$.

Applying the above rule in conjunction with the previous rule, we obtain the following:

$$f'(x) = \frac{(1)(x^2 + 1) - (x)(2x)}{(x^2 + 1)^2}$$

$$f'(x) = \frac{-x^2 + 1}{(x^2 + 1)^2}$$

Power Functions

When we derived the Power Rule for Derivatives in part (c) of Example 2.1.16, we did so only for powers that are *natural numbers*. We will justify the use of powers that are *real numbers* when we get to **EK 2.1C5**, but for now we simply state the following as a fact.

For any real number n,

$$\frac{d}{dx}\left(x^n\right) = nx^{n-1}.$$

Example 2.1.19: Let $f(x) = x^{-2/3}$. Compute $f'(x)$.

Applying the latest version of the Power Rule for Derivatives, we obtain the following:

$$f'(x) = -\frac{2}{3}x^{-\frac{2}{3}-1} = -\frac{2}{3}x^{-\frac{5}{3}}$$

Exponential Functions

Let $a > 0$ and $a \neq 1$. Then the function $f(x) = a^x$ is differentiated as follows:

$$f'(x) = \frac{d}{dx}\left(a^x\right) = \frac{d}{dx}\left(e^{\ln(a)x}\right) = e^{\ln(a)x} \cdot \frac{d}{dx}\left(\ln(a)x\right)$$

The last step above is justified by the Chain Rule, which we will review in **EK 2.1C4**. Simplifying, we obtain our final result.

$$\frac{d}{dx}\left(a^x\right) = \ln(a)a^x$$

Example 2.1.20: Let $f(x) = 2^x$. Compute $f'(x)$.

Using the above rule, we obtain the following:

$$f'(x) = \ln(2)\,2^x$$

Logarithmic Functions

Let $a > 0$ and $a \neq 1$. Then the function $f(x) = \log_a(x)$ is differentiated as follows:

$$f'(x) = \frac{d}{dx}\left(\log_a(x)\right) = \frac{d}{dx}\left(\frac{\ln(x)}{\ln(a)}\right) = \frac{1}{\ln(a)}\frac{d}{dx}\left(\ln(x)\right)$$

The last step above is justified by a theorem which states that $\frac{d}{dx}\left(kf(x)\right) = kf'(x)$, which we will review when we get to **EK 2.1C3**. Using the result of Example 2.1.15, we obtain the following differentiation rule:

$$\frac{d}{dx}\left(\log_a(x)\right) = \frac{1}{\ln(a)x}$$

Example 2.1.21: Let $f(x) = \log_3(x)$. Compute $f'(x)$.

Using the above rule, we obtain the following:

$$\frac{d}{dx}\left(\log_3(x)\right) = \frac{1}{\ln(3)x}$$

Trigonometric Functions

In Example 2.1.13, we showed that $\frac{d}{dx}\left(\sin(x)\right) = \cos(x)$, and we asked you to show that $\frac{d}{dx}\left(\cos(x)\right) = -\sin(x)$ as an exercise. It is possible to obtain differentiation rules for the other four trigonometric functions using the Quotient Rule, which we will review in **EK 2.1C3**. Here, we'll simply present the rules.

$$\frac{d}{dx}\left(\sec(x)\right) = \sec(x)\tan(x) \qquad\qquad \frac{d}{dx}\left(\csc(x)\right) = -\csc(x)\cot(x)$$

$$\frac{d}{dx}\left(\tan(x)\right) = \sec^2(x) \qquad\qquad \frac{d}{dx}\left(\cot(x)\right) = -\csc^2(x)$$

Inverse Trigonometric Functions

These derivatives can be computed using Implicit Differentiation, which we will review in **EK 2.1C5**. Here, we'll simply present the differentiation rules.

$$\frac{d}{dx}\left(\sin^{-1}(x)\right) = \frac{1}{\sqrt{1-x^2}} \qquad\qquad \frac{d}{dx}\left(\cos^{-1}(x)\right) = -\frac{1}{\sqrt{1-x^2}}$$

$$\frac{d}{dx}\left(\tan^{-1}\right) = \frac{1}{x^2+1} \qquad\qquad \frac{d}{dx}\left(\cot^{-1}(x)\right) = -\frac{1}{x^2+1}$$

$$\frac{d}{dx}\left(\sec^{-1}(x)\right) = \frac{1}{|x|\sqrt{x^2-1}} \qquad\qquad \frac{d}{dx}\left(\csc^{-1}(x)\right) = -\frac{1}{|x|\sqrt{x^2-1}}$$

In the next EK, we will review some differentiation theorems that both justify and extend the differentiation rules that we just reviewed.

EK 2.1C3: Sums, differences, products, and quotients of functions can be differentiated using derivative rules.

The Constant Multiple Rule and the Sum and Difference Rule

Let k be any real number, and let f and g be differentiable functions at x. Then the following two rules hold:

Constant Multiple Rule:

$$\frac{d}{dx}\big(kf(x)\big) = kf'(x)$$

Sum and Difference Rule:

$$\frac{d}{dx}\big(f(x) \pm g(x)\big) = f'(x) \pm g'(x)$$

These two rules are direct consequences of the Constant Multiple and Sum and Difference Rules for limits, which we reviewed in **EK 1.1C1**. Remember: a derivative *is* a limit, so it makes sense that some rules of limits should apply.

These two rules, in conjunction with the basic differentiation rules developed in Example 2.1.16, provide the justification of the differentiation rule for polynomial functions given in **EK 2.1C2**. In particular, the *term-by-term* differentiation in that rule is justified by the Sum and Difference Rule.

We can also use these rules to extend the range of applicability of the other rules developed in the last EK. We'll illustrate with an example.

Example 2.1.22: Let $f(x) = 6x^2 - 3\sin(x) + 4\tan^{-1}(x)$. Compute $f'(x)$.

$$f'(x) = \frac{d}{dx}\Big(6x^2 - 3\sin(x) + 4\tan^{-1}(x)\Big)$$

Apply the Sum and Difference Rule to differentiate term-by-term.

$$f'(x) = \frac{d}{dx}\big(6x^2\big) - \frac{d}{dx}\big(3\sin(x)\big) + \frac{d}{dx}\big(4\tan^{-1}(x)\big)$$

Now apply the Constant Multiple Rule to bring the coefficients outside the derivative operators.

$$f'(x) = 6\frac{d}{dx}\big(x^2\big) - 3\frac{d}{dx}\big(\sin(x)\big) + 4\frac{d}{dx}\big(\tan^{-1}(x)\big)$$

Finally, apply the differentiation rules of the last EK and simplify.

$$f'(x) = 6 \cdot 2x - 3\cos(x) + 4 \cdot \frac{1}{x^2 + 1}$$

$$f'(x) = 12x - 3\cos(x) + \frac{4}{x^2 + 1}$$

For instructional purposes, we have worked this out step by step. However, after you have practiced using these rules you should be able to work out problems like this in a single step.

We'll offer a word of caution before proceeding on to the next differentiation theorem. Take another look at the previous two differentiation rules, side-by-side with their corresponding limit rules from **EK 1.1C1**.

	Limits:	Derivatives:
Constant Multiple Rule:	$\lim_{x \to c} (kf(x)) = kR$	$\frac{d}{dx}(kf(x)) = kf'(x)$
Sum and Difference Rule:	$\lim_{x \to c}(f(x) \pm g(x)) = R + S$	$\frac{d}{dx}(f(x) \pm g(x)) = f'(x) \pm g'(x)$

Looking at these, you might be tempted to think that, since $\lim_{x \to c}(f(x)g(x)) = RS$, it follows that $\frac{d}{dx}(f(x)g(x)) = f'(x)g'(x)$. If so, *you would be wrong*. Not every limit theorem has a direct analog in differential calculus, as we'll demonstrate in the next rule.

The Product Rule

Let f, g be differentiable functions at x. Then, the product fg is also a differentiable function at x, and

$$\frac{d}{dx}(f(x)g(x)) = f'(x)g(x) + f(x)g'(x).$$

If you're happy to accept this at face value, you may want to skip straight to the next example. But if you're just dying to see why the Product Rule looks like it does, then read the following proof:

$$\frac{d}{dx}(f(x)g(x)) = \lim_{h \to 0} \frac{f(x+h)g(x+h) - f(x)g(x)}{h}$$

We're going to be a little sneaky here and both subtract and add $f(x)g(x+h)$ in the numerator.

$$\frac{d}{dx}(f(x)g(x)) = \lim_{h \to 0} \frac{f(x+h)g(x+h) - f(x)g(x+h) + f(x)g(x+h) - f(x)g(x)}{h}$$

Now we're going to split this into two limits as follows:

$$\frac{d}{dx}(f(x)g(x)) =$$
$$\lim_{h \to 0} \frac{f(x+h)g(x+h) - f(x)g(x+h)}{h} + \lim_{h \to 0} \frac{f(x)g(x+h) - f(x)g(x)}{h}$$

We'll evaluate the first limit on the right side by factoring $g(x+h)$ out of the numerator, and then applying the Product Rule for limits.

$$\lim_{h \to 0} \left(\frac{f(x+h) - f(x)}{h} \cdot g(x+h) \right) = f'(x) g(x+0) = f'(x) g(x)$$

We'll evaluate the second limit on the right side by factoring $f(x)$ out of the numerator, and then applying the Constant Rule for limits (remember: $f(x)$ is a *constant* as far as this limit is concerned).

$$\lim_{h \to 0} \left(f(x) \cdot \frac{g(x+h) - g(x)}{h} \right) = f(x) g'(x)$$

Putting these results together, we have the Product Rule.

$$\frac{d}{dx} \left(f(x) g(x) \right) = f'(x) g(x) + f(x) g'(x)$$

We'll illustrate the Product Rule with an example.

Example 2.1.23: Compute $\frac{d}{dx} \left(x^2 \tan(x) \right)$.

$x^2 \tan(x)$ is a product of two functions that we know how to differentiate: x^2 and $\tan(x)$.

$$\frac{d}{dx} \left(x^2 \tan(x) \right) = \frac{d}{dx} \left(x^2 \right) \cdot \tan(x) + x^2 \cdot \frac{d}{dx} \left(\tan(x) \right)$$

$$\frac{d}{dx} \left(x^2 \tan(x) \right) = 2x \tan(x) + x^2 \sec^2(x)$$

Just as with the Product Rule, the Quotient Rule for derivatives also does not look like its analog for limits.

The Quotient Rule

Let f, g be differentiable functions at x, and let $g(x) \neq 0$. Then the quotient $\frac{f}{g}$ is also differentiable at x, and

$$\frac{d}{dx} \left(\frac{f(x)}{g(x)} \right) = \frac{f'(x) g(x) - f(x) g'(x)}{\left(g(x) \right)^2}$$

You should recognize this from our differentiation rule for rational functions in the last EK. It turns out that the Quotient Rule applies to *all* quotients of differentiable functions, not just quotients of polynomials. To see why the Quotient Rule looks the way it does, see the following proof. If you're not interested in the proof, you may skip straight to the next example.

$$\frac{d}{dx}\left(\frac{f(x)}{g(x)}\right) = \lim_{h \to 0} \frac{\dfrac{f(x+h)}{g(x+h)} - \dfrac{f(x)}{g(x)}}{h} = \lim_{h \to 0} \frac{f(x+h)g(x) - f(x)g(x+h)}{hg(x)g(x+h)}$$

Next we'll use a clever trick. We'll subtract and add $f(x)g(x)$ in the numerator.

$$\frac{d}{dx}\left(\frac{f(x)}{g(x)}\right) = \lim_{h \to 0} \frac{f(x+h)g(x) - f(x)g(x) + f(x)g(x) - f(x)g(x+h)}{hg(x)g(x+h)}$$

$$\frac{d}{dx}\left(\frac{f(x)}{g(x)}\right) = \lim_{h \to 0} \frac{(f(x+h) - f(x))g(x) - f(x)(g(x+h) - g(x))}{hg(x)g(x+h)}$$

$$\frac{d}{dx}\left(\frac{f(x)}{g(x)}\right) = \lim_{h \to 0} \frac{\left(\dfrac{f(x+h) - f(x)}{h}\right)g(x) - f(x)\left(\dfrac{g(x+h) - g(x)}{h}\right)}{g(x)g(x+h)}$$

Now apply the Quotient Rule for limits, and simplify.

$$\frac{d}{dx}\left(\frac{f(x)}{g(x)}\right) = \frac{f'(x)g(x) - f(x)g'(x)}{g(x)g(x+0)} = \frac{f'(x)g(x) - f(x)g'(x)}{(g(x))^2}$$

We will now justify the differentiation rules for the secant and tangent functions that were given in the last EK.

Example 2.1.24: Prove the following:

1. $\dfrac{d}{dx}(\sec(x)) = \sec(x)\tan(x)$

2. $\dfrac{d}{dx}(\tan(x)) = \sec^2(x)$

We will make use of the identities $\sec(x) = \dfrac{1}{\cos(x)}$ and $\tan(x) = \dfrac{\sin(x)}{\cos(x)}$.

$$\frac{d}{dx}(\sec(x)) = \frac{d}{dx}\left(\frac{1}{\cos(x)}\right) = \frac{\dfrac{d}{dx}(1) \cdot \cos(x) - 1 \cdot \dfrac{d}{dx}(\cos(x))}{\cos^2(x)}$$

$$\frac{d}{dx}(\sec(x)) = \frac{0 \cdot \cos(x) - 1 \cdot (-\sin(x))}{\cos^2(x)} = \frac{\sin(x)}{\cos^2(x)} = \frac{1}{\cos(x)} \cdot \frac{\sin(x)}{\cos(x)}$$

$$\frac{d}{dx}(\sec(x)) = \sec(x)\tan(x)$$

$$\frac{d}{dx}\big(\tan(x)\big) = \frac{d}{dx}\left(\frac{\sin(x)}{\cos(x)}\right) = \frac{\frac{d}{dx}\big(\sin(x)\big)\cdot\cos(x) - \sin(x)\cdot\frac{d}{dx}\big(\cos(x)\big)}{\cos^2(x)}$$

$$\frac{d}{dx}\big(\tan(x)\big) = \frac{\cos(x)\cdot\cos(x) - \sin(x)\cdot\big(-\sin(x)\big)}{\cos^2(x)} = \frac{\cos^2(x) + \sin^2(x)}{\cos^2(x)} = \frac{1}{\cos^2(x)}$$

$$\frac{d}{dx}\big(\tan(x)\big) = \sec^2(x)$$

The Quotient Rule can be used to differentiate non-trigonometric functions as well.

Example 2.1.25: Compute $\dfrac{d}{dx}\left(\dfrac{\ln(x)}{x}\right)$.

$$\frac{d}{dx}\left(\frac{\ln(x)}{x}\right) = \frac{\frac{d}{dx}\big(\ln(x)\big)\cdot x - \ln(x)\cdot\frac{d}{dx}(x)}{x^2} = \frac{\frac{1}{x}\cdot x - \ln(x)\cdot 1}{x^2}$$

$$\frac{d}{dx}\left(\frac{\ln(x)}{x}\right) = \frac{1 - \ln(x)}{x^2}$$

In the next EK, we will learn how to differentiate *composite functions*.

EK 2.1C4: The Chain Rule provides a way to differentiate composite functions.

The Chain Rule

If g is differentiable at x, and if f is differentiable at $g(x)$, then the composite function $f\circ g$ is differentiable at x, and

$$\frac{d}{dx}\big((f\circ g)(x)\big) = f'\big(g(x)\big)g'(x).$$

The derivative of a composite function at x is the derivative of the *outer* function evaluated at the *inner* function of x times the derivative of the *inner* function evaluated at x.

If we let $u = g(x)$ and $y = f(u)$, then we get the following alternate formulation of the Chain Rule:

$$\frac{dy}{dx} = \frac{dy}{du}\cdot\frac{du}{dx}$$

We will illustrate this with several examples.

Example 2.1.26: Compute $\dfrac{d}{dx}\big(\sin^2(x)\big)$.

If we rewrite $\sin^2(x)$ as $(\sin(x))^2$, then we can see that it is the composition $f(g(x))$, where $u = g(x) = x$ and $y = f(u) = u^2$.

We'll begin by applying the Power Rule for Derivatives to the outer function.

$$\frac{d}{dx}\left(\sin^2(x)\right) = \frac{d}{dx}\left(\left(\sin(x)\right)^2\right) = 2\left(\sin(x)\right) \cdot \frac{d}{dx}\left(\sin(x)\right)$$

Note the $\frac{d}{dx}\left(\sin(x)\right)$ on the right side. That is required by the Chain Rule. Next, we'll evaluate that derivative and simplify.

$$\frac{d}{dx}\left(\sin^2(x)\right) = 2\left(\sin(x)\right) \cdot \cos(x) = 2\sin(x)\cos(x)$$

Example 2.1.27: Compute $\frac{d}{dx}\left(\sin\left(x^2\right)\right)$.

This time we have $\sin(x^2)$, which looks like the composition $f(g(x))$, where $u = g(x) = x^2$, and $y = f(u) = \sin(u)$.

$$\frac{d}{dx}\left(\sin\left(x^2\right)\right) = \cos\left(x^2\right) \cdot \frac{d}{dx}\left(x^2\right) = \cos\left(x^2\right) \cdot 2x = 2x\cos\left(x^2\right)$$

Example 2.1.28: Compute $\frac{d}{dx}\left(e^{\tan(x)}\right)$.

$e^{\tan(x)}$ looks like the composition $f(g(x))$, where $u = g(x) = \tan(x)$, and $y = f(u) = e^u$.

$$\frac{d}{dx}\left(e^{\tan(x)}\right) = e^{\tan(x)} \cdot \frac{d}{dx}\left(\tan(x)\right) = e^{\tan(x)}\sec^2(x)$$

The Chain Rule can also be extended to *multiply composite* functions. We do it by repeated application of the Chain Rule. For instance, if $y = f(u)$, $u = g(v)$, and $v = h(x)$ are all differentiable, then the doubly composite function $f(g(h(x)))$ is also differentiable, and we'll apply the Chain Rule twice to obtain

$$\frac{d}{dx}\left(f\left(g\left(h(x)\right)\right)\right) = f'\left(g\left(h(x)\right)\right)g'\left(h(x)\right)h'(x).$$

Alternatively, we may express this as

$$\frac{dy}{dx} = \frac{dy}{du} \cdot \frac{du}{dv} \cdot \frac{dv}{dx}.$$

Example 2.1.29: Compute $\frac{d}{dx}\left(\cos^3(5x)\right)$.

$\cos^3(5x)$ looks like the composition $f(g(h(v)))$, where $v = h(x) = 5x$, $u = g(v) = \cos(v)$, and $y = f(u) = u^3$.

$$\frac{d}{dx}\left(\cos^3(5x)\right) = 3\cos^2(5x) \cdot \frac{d}{dx}\left(\cos(5x)\right)$$

$$\frac{d}{dx}\left(\cos^3(5x)\right) = 3\cos^2(5x) \cdot \left(-\sin(5x)\right) \cdot \frac{d}{dx}(5x)$$

$$\frac{d}{dx}\left(\cos^3(5x)\right) = 3\cos^2(5x) \cdot \left(-\sin(5x)\right) \cdot 5$$

$$\frac{d}{dx}\left(\cos^3(5x)\right) = -15\cos^2(5x)\sin(5x)$$

At this point, it would be good to gather all of our differentiation rules in one place. We will do this in a table on the next page. In each entry in the table, we'll assume that u is a differentiable function of x and that c, k, and n are real numbers.

Elementary Differentiation Rules

1. $\dfrac{d}{dx}(kf(x)) = kf'(x)$

2. $\dfrac{d}{dx}(f(x) \pm g(x)) = f'(x) \pm g'(x)$

3. $\dfrac{d}{dx}(f(x)g(x)) = f'(x)g(x) + f(x)g'(x)$

4. $\dfrac{d}{dx}\left(\dfrac{f(x)}{g(x)}\right) = \dfrac{f'(x)g(x) - f(x)g'(x)}{\left(g(x)\right)^2}$

5. $\dfrac{d}{dx}(c) = 0$

6. $\dfrac{d}{dx}\left(u^n\right) = nu^{n-1}u'$

7. $\dfrac{d}{dx}\left(e^u\right) = e^u u'$

8. $\dfrac{d}{dx}\left(a^u\right) = \ln(a)a^u u', \ (a > 0, a \neq 1)$

9. $\dfrac{d}{dx}(\ln(u)) = \dfrac{u'}{u}$

10. $\dfrac{d}{dx}\left(\log_a(u)\right) = \dfrac{u'}{\ln(a)u}, \ (a > 0, a \neq 1)$

11. $\dfrac{d}{dx}(\sin(u)) = \cos(u)u'$

12. $\dfrac{d}{dx}(\cos(u)) = -\sin(u)u'$

13. $\dfrac{d}{dx}(\tan(u)) = \sec^2(u)u'$

14. $\dfrac{d}{dx}(\cot(u)) = -\csc^2(u)u'$

15. $\dfrac{d}{dx}(\sec(u)) = \sec(u)\tan(u)u'$

16. $\dfrac{d}{dx}\big(\csc(u)\big) = -\csc(u)\cot(u)u'$

17. $\dfrac{d}{dx}\big(\sin^{-1}(u)\big) = \dfrac{u'}{\sqrt{1-u^2}}$

18. $\dfrac{d}{dx}\big(\cos^{-1}(u)\big) = -\dfrac{u'}{\sqrt{1-u^2}}$

19. $\dfrac{d}{dx}\big(\tan^{-1}(u)\big) = \dfrac{u'}{u^2+1}$

20. $\dfrac{d}{dx}\big(\cot^{-1}(u)\big) = -\dfrac{u'}{u^2+1}$

21. $\dfrac{d}{dx}(\sec^{-1}(u)) = \dfrac{u'}{|u|\sqrt{u^2-1}}$

22. $\dfrac{d}{dx}\big(\csc^{-1}(u)\big) = -\dfrac{u'}{|u|\sqrt{u^2-1}}$

Now, we will move on to a direct application of the Chain Rule: Implicit Differentiation.

EK 2.1C5: The Chain Rule is the basis for implicit differentiation.

Suppose y depends on x in an unknown way. Suppose further that y is differentiable at x, and that $f(y)$ is also a differentiable function. To compute $\dfrac{d}{dx}\big(f(y)\big)$, we must apply the Chain Rule because y is actually a function of x, and so $f(y)$ is actually a composite function.

$$\frac{d}{dx}\big(f(y)\big) = f'(y)y'$$

Example 2.1.30: Suppose that y depends on x, and that y is differentiable at x. Compute $\dfrac{d}{dx}\big(3y^5\big)$.

Applying the Chain Rule as above, we get the following:

$$\frac{d}{dx}\big(3y^5\big) = 15y^4 y'$$

Implicit Differentiation

Suppose that x and y are related by an equation that cannot be explicitly solved for y. We then say that x and y are *implicitly* related. We compute $\dfrac{dy}{dx}$ using Implicit Differentiation, as follows:

1. Differentiate both sides of the equation with respect to x, applying the Chain Rule as appropriate.
2. Algebraically solve for y' in terms of x and y.

Example 2.1.31: Find y' if $3x^2 + xy + 5y = 4$.

We'll differentiate both sides with respect to x. Note that we will have to use the Product Rule on the second term on the left side.

$$\frac{d}{dx}\left(3x^2 + xy + 5y\right) = \frac{d}{dx}(4)$$

$$6x + \frac{d}{dx}(x)\cdot y + x\cdot\frac{d}{dx}(y) + 5y' = 0$$

$$6x + y + xy' + 5y' = 0$$

Now, we'll solve for y'.

$$xy' + 5y' = -6x - y$$

$$(x+5)y' = -6x - y$$

$$y' = \frac{-6x - y}{x+5}$$

Implicit Differentiation can also be used to justify the differentiation rules for the inverse trigonometric functions that were presented in **EK 2.1C2**.

Example 2.1.32: Prove the following:

1. $\dfrac{d}{dx}\left(\sin^{-1}(x)\right) = \dfrac{1}{\sqrt{1-x^2}}$

2. $\dfrac{d}{dx}\left(\csc^{-1}(x)\right) = -\dfrac{1}{|x|\sqrt{x^2-1}}$

1. Let $y = \sin^{-1}(x)$, then take the sine of both sides.

$$\sin(y) = \sin\left(\sin^{-1}(x)\right)$$

$$\sin(y) = x$$

Note that if we think of y as an angle in a right triangle, then $\sin(y) = \dfrac{x}{1} = \dfrac{opp}{hyp}$. We'll illustrate this as follows:

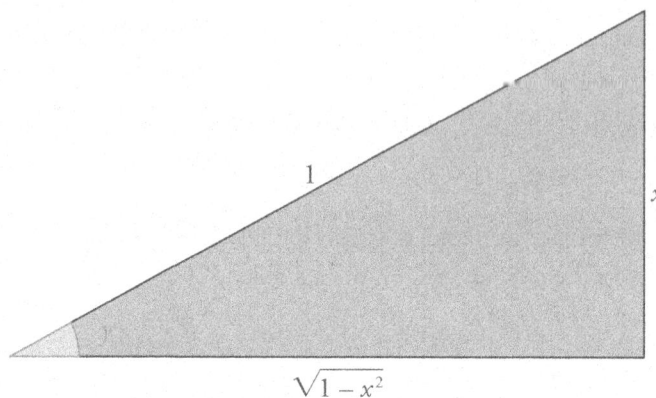

Differentiating implicitly, we obtain the following:

$$\frac{d}{dx}\left(\sin\left(y\right)\right) = \frac{d}{dx}\left(x\right)$$

$$\cos\left(y\right)y' = 1$$

$$y' = \frac{1}{\cos\left(y\right)}$$

Referring back to the triangle, we see that $\cos\left(y\right) = \dfrac{adj}{hyp} = \dfrac{\sqrt{1-x^2}}{1} = \sqrt{1-x^2}$, so we have the following:

$$y' = \frac{d}{dx}\left(\sin^{-1}\left(x\right)\right) = \frac{1}{\sqrt{1-x^2}}$$

2. Let $y = \csc^{-1}\left(x\right)$, then take the cosecant of both sides.

$$\csc\left(y\right) = \csc\left(\csc^{-1}\left(x\right)\right)$$

$$\frac{1}{\sin\left(y\right)} = x$$

$$\sin\left(y\right) = \frac{1}{x}$$

Now take the inverse sine of both sides.

$$\sin^{-1}\left(\sin\left(y\right)\right) = \sin^{-1}\left(\frac{1}{x}\right)$$

$$y = \sin^{-1}\left(\frac{1}{x}\right)$$

Differentiate using the Chain Rule.

$$y' = \frac{1}{\sqrt{1-\left(\frac{1}{x}\right)^2}}\frac{d}{dx}\left(\frac{1}{x}\right) = \frac{1}{\sqrt{1-\left(\frac{1}{x}\right)^2}}\cdot\left(-\frac{1}{x^2}\right) = -\frac{1}{x^2\sqrt{1-\frac{1}{x^2}}}$$

Noting that $|x| = \sqrt{x^2}$ and $x^2 = |x|^2$, we simplify as follows:

$$y' = -\frac{1}{|x|\sqrt{x^2}\sqrt{1-\frac{1}{x^2}}} = -\frac{1}{|x|\sqrt{x^2\left(1-\frac{1}{x^2}\right)}}$$

$$y' = \frac{d}{dx}\left(\csc^{-1}(x)\right) = -\frac{1}{|x|\sqrt{x^2-1}}$$

Suppose now that we wanted to differentiate $f(x) = \frac{x^2\sin(x)}{\sqrt[3]{x^2+1}}$. We *could* use the Quotient Rule, Product Rule, and Chain Rule to get it done, but that would be both messy and tedious. We could make this a lot easier by using properties of logarithms. This is exactly what we do when we use Logarithmic Differentiation, which is a variant of Implicit Differentiation.

Logarithmic Differentiation

We will list out the steps before turning to an example. Let $y = f(x)$.
1. Take the natural logarithm of both sides, and use the properties of logarithms to simplify as necessary.
2. Implicitly differentiate both sides, using the rule $\frac{d}{dx}\left(\ln(u)\right) = \frac{u'}{u}$.
3. Algebraically solve for y'.

Example 2.1.33: Let $f(x) = \frac{x^2\sin(x)}{\sqrt[3]{x^2+1}}$. Compute $f'(x)$.

We'll use Logarithmic Differentiation. First, we take the natural logarithm of both sides and apply the properties of logarithms to the right side.

$$\ln\left(f(x)\right) = \ln\left(\frac{x^2\sin(x)}{\sqrt[3]{x^2+1}}\right)$$

$$\ln\left(f(x)\right) = \ln\left(x^2\right) + \ln\left(\sin(x)\right) - \ln\left(\left(x^2+1\right)^{1/3}\right)$$

$$\ln\left(f(x)\right) = 2\ln(x) + \ln\left(\sin(x)\right) - \frac{1}{3}\ln\left(x^2+1\right)$$

Next, we'll differentiate implicitly.

$$\frac{d}{dx}\left(\ln\left(f\left(x\right)\right)\right)=\frac{d}{dx}\left(2\ln\left(x\right)+\ln\left(\sin\left(x\right)\right)-\frac{1}{3}\ln\left(x^2+1\right)\right)$$

$$\frac{f'\left(x\right)}{f\left(x\right)}=\frac{2}{x}+\frac{\cos\left(x\right)}{\sin\left(x\right)}-\frac{2x}{3\left(x^2+1\right)}$$

$$f'\left(x\right)=\left(\frac{2}{x}+\frac{\cos\left(x\right)}{\sin\left(x\right)}-\frac{2x}{3\left(x^2+1\right)}\right)f\left(x\right)$$

$$f'\left(x\right)=\left(\frac{2}{x}+\cot\left(x\right)-\frac{2x}{3\left(x^2+1\right)}\right)\frac{x^2\sin\left(x\right)}{\sqrt[3]{x^2+1}}$$

Now, as we promised in **EK 2.1C2**, we will show that the Power Rule for Derivatives works when the power n is *any* real number.

Example 2.1.34: Prove that $\frac{d}{dx}\left(x^n\right)=nx^{n-1}$, where n is any real number.

We'll use Logarithmic Differentiation. Let $y=x^n$.

$$\ln\left(y\right)=\ln\left(x^n\right)=n\ln\left(x\right)$$

$$\frac{d}{dx}\left(\ln\left(y\right)\right)=\frac{d}{dx}\left(n\ln\left(x\right)\right)$$

$$\frac{y'}{y}=\frac{n}{x}$$

$$y'=n\frac{y}{x}=n\frac{x^n}{x}$$

$$y'=\frac{d}{dx}\left(x^n\right)=nx^{n-1}$$

Next, we will turn our attention to invertible functions whose inverse functions cannot be explicitly written down. For instance, consider the function $f(x)=x^3+4x-4$, which is graphed below.

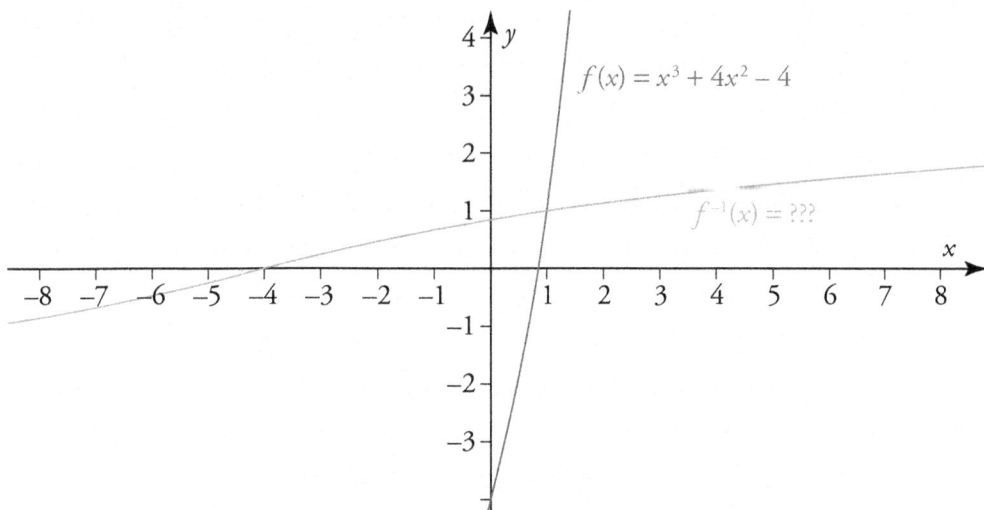

The graph of f passes the horizontal line test, so it is one-to-one, and therefore has an inverse function f^{-1}. As you can see, f^{-1} has been graphed as well. But what *is* this inverse function? To find it, you would have to do the following:

1. Let $y = f(x)$: $y = x^3 + 4x - 4$
2. Swap x and y: $x = y^3 + 4y - 4$
3. Solve for $y = f^{-1}(x)$:   ??????????

The equation in step 2 is too difficult to solve for y, so we cannot get an explicit formula for $f^{-1}(x)$. However, we *can* say something about the derivative of f^{-1} (if it exists).

EK 2.1C6: The Chain Rule can be used to find the derivative of an inverse function, provided the derivative of that inverse function exists.

Let $y = f^{-1}(x)$. Compose both sides of this equation with f.

$$f(y) = f\left(f^{-1}(x)\right)$$

$$f(y) = x$$

Now implicitly differentiate both sides. Note that we are assuming that both f and f^{-1} are differentiable.

$$\frac{d}{dx}\left(f(y)\right) = \frac{d}{dx}(x)$$

$$f'(y)\,y' = 1$$

$$y' = \frac{1}{f'(y)}$$

Recalling that $y = f^{-1}(x)$ and $y' = (f^{-1})'(x)$, we have our result.

$$\left(f^{-1}\right)'(x) = \frac{1}{f'\left(f^{-1}(x)\right)}$$

Example 2.1.35: Let $f(x) = x^3 + 4x - 4$.

1. Evaluate $f(2)$.
2. Evaluate $(f^{-1})'(12)$.

The first part is easy, but essential.

1. $f(2) = 2^3 + 4(2) - 4 = 8 + 8 - 4 = 12$ So the point $(2, 12)$ is on the graph of $y = f(x)$. That means that the point $(12, 2)$ is on the graph of $y = f^{-1}(x)$, which means that $f^{-1}(12) = 2$. We will use this in part b.

2. Use the result that we just derived for the derivative of an inverse function.

$$\left(f^{-1}\right)'(12) = \frac{1}{f'\left(f^{-1}(12)\right)} = \frac{1}{f'(2)}$$

Now $f'(x) = 3x^2 + 4$, so $f'(2) = 16$. We'll use this information to finish off the problem.

$$\left(f^{-1}\right)'(12) = \frac{1}{16}$$

We will close our review of **LO 2.1C** with a BC topic.

EK 2.1C7: (BC) Methods for calculating derivatives of real-valued functions can be extended to vector-valued functions, parametric functions, and functions in polar coordinates.

We will take these one at a time.

Vector-Valued Functions

Let $\vec{r}(t) = (x(t), (y(t)))$ be a vector-valued function. If x and y are both differentiable at t, then $\vec{r}$ is also differentiable at t, and

$$\vec{r}'(t) = \left(x'(t), y'(t)\right).$$

Example 2.1.36: Let $\vec{r}(t) = (t^3 e^t, \cos(6t))$. Compute $\vec{r}'(t)$.

Here $x(t) = t^3 e^t$, and $y(t) = \cos(6t)$. We'll use the Product Rule to differentiate x.

$$x'(t) = \frac{d}{dt}\left(t^3 e^t\right) = \frac{d}{dt}\left(t^3\right) \cdot e^t + t^3 \cdot \frac{d}{dt}\left(e^t\right) = 3t^2 e^t + t^3 e^t$$

Now, we'll use the Chain Rule to differentiate y.

$$y'(t) = \frac{d}{dt}\left(\cos(6t)\right) = -\sin(6t) \cdot \frac{d}{dt}(6t) = -6\sin(6t)$$

Putting these results together, we obtain the following:

$$\vec{r}'(t) = \left(3t^2 e^t + t^3 e^t, -6\sin(6t)\right)$$

Parametric Functions

Suppose a smooth curve is given by the parametric equations $x = f(t)$ and $y = g(t)$ for $a \leq t \leq b$. If $f'(t) \neq 0$, then the derivative at (x, y) is

$$\frac{dy}{dx} = \frac{dy/dt}{dx/dt} = \frac{g'(t)}{f'(t)}.$$

This is called the **parametric form of the derivative.**

Example 2.1.37: Let $x = t^2 + 3t$ and $y = -2t + 6$ for $1 \leq t \leq 5$.

1. Find (x, y) when $t = 2$.
2. Find $\dfrac{dy}{dx}$ when $t = 2$.

1. $x(2) = 2^2 + 3(2) = 10$ and $y(2) = -2(2) + 6 = 2$. So when $t = 2$, $(x, y) = (10, 2)$.
2. Use the above result for the parametric form of the derivative.

$$\frac{dy}{dx} = \frac{\frac{d}{dt}(-2t + 6)}{\frac{d}{dt}(t^2 + 3t)} = \frac{-2}{2t + 3}$$

$$\left.\frac{dy}{dx}\right|_{t=2} = \frac{-2}{2(2) + 3} = -\frac{2}{7}$$

Polar Coordinates

First recall that polar coordinates (r, θ) are related to Cartesian coordinates (x, y) by the equations $x = r\cos(\theta)$ and $y = r\sin(\theta)$. If $r = f(\theta)$ is a polar function, then the rectangular coordinates (x, y) of the points on the graph of the function are given by $x = f(\theta)\cos(\theta)$ and $y = f(\theta)\sin(\theta)$. We can compute $\dfrac{dx}{d\theta}$ and $\dfrac{dy}{d\theta}$ using the Product Rule.

$$\frac{dx}{d\theta} = \frac{d}{d\theta}\left(f(\theta)\cos(\theta)\right) = -f(\theta)\sin(\theta) + f'(\theta)\cos(\theta)$$

$$\frac{dy}{d\theta} = \frac{d}{d\theta}\left(f(\theta)\sin(\theta)\right) = f(\theta)\cos(\theta) + f'(\theta)\sin(\theta)$$

Now we're ready to talk about derivatives of polar graphs.

Let $r = f'(\theta)$ be a differentiable polar function. Then if $\dfrac{dx}{d\theta} \neq 0$, $\dfrac{dy}{dx}$ at the point with polar coordinates $(r, \theta) = (f(\theta), \theta)$ is computed as follows:

$$\frac{dy}{dx} = \frac{\dfrac{dy}{d\theta}}{\dfrac{dx}{d\theta}} = \frac{f(\theta)\cos(\theta) + f'(\theta)\sin(\theta)}{-f(\theta)\sin(\theta) + f'(\theta)\cos(\theta)}.$$

This is called the **polar form of the derivative**.

Example 2.1.37: Let $r = 2 - 2\cos(\theta)$. Find $\dfrac{dy}{dx}$ when $\theta = \dfrac{\pi}{6}$.

Use the above result for the polar form of the derivative.

$$\frac{dy}{dx} = \frac{\left(2 - 2\cos(\theta)\right)\cos(\theta) + \dfrac{d}{d\theta}\left(2 - 2\cos(\theta)\right)\sin(\theta)}{-\left(2 - 2\cos(\theta)\right)\sin(\theta) + \dfrac{d}{d\theta}\left(2 - 2\cos(\theta)\right)\cos(\theta)}$$

$$\frac{dy}{dx} = \frac{\left(2 - 2\cos(\theta)\right)\cos(\theta) + 2\sin(\theta)\sin(\theta)}{-\left(2 - 2\cos(\theta)\right)\sin(\theta) + 2\sin(\theta)\cos(\theta)}$$

$$\frac{dy}{dx} = \frac{2\cos(\theta) - 2\cos^2(\theta) + 2\sin^2(\theta)}{-2\sin(\theta) + 4\sin(\theta)\cos(\theta)} = \frac{2\cos(\theta) - 2\cos(2\theta)}{-2\sin(\theta) + 2\sin(2\theta)}$$

$$\frac{dy}{dx} = \frac{\cos(\theta) - \cos(2\theta)}{-\sin(\theta) + \sin(2\theta)}$$

$$\left.\frac{dy}{dx}\right|_{\theta = \frac{\pi}{6}} = \frac{\cos\left(\dfrac{\pi}{6}\right) - \cos\left(\dfrac{\pi}{3}\right)}{-\sin\left(\dfrac{\pi}{6}\right) + \sin\left(\dfrac{\pi}{3}\right)} = \frac{\dfrac{\sqrt{3}}{2} - \dfrac{1}{2}}{-\dfrac{1}{2} + \dfrac{\sqrt{3}}{2}}$$

$$\left.\frac{dy}{dx}\right|_{\theta = \frac{\pi}{6}} = 1$$

LO 2.1D – Higher Order Derivatives

In **EK 2.1A3** we reviewed the fact that when you differentiate a function $f(x)$ to obtain its derivative $f'(x)$, you get *another function*. So you may ask, "Can't we take the derivative of f' as well?" The answer, of course, is "YES, provided f' is differentiable," and this is what gives rise to higher order derivatives. That leads us to the LO for this section.

LO 2.1D: Determine higher order derivatives.

The **order** of a derivative refers to the number of times a function is differentiated. If a function f is differentiated once to obtain the f', then f' is called the **first derivative**, or **derivative of order 1**, of f. In the first item of Essential Knowledge of this section, we will look at second derivatives.

EK 2.1D1: Differentiating f' produces the second derivative f'', provided the derivative of f' exists; repeating this process produces higher order derivatives of f.

f'' is a **derivative of order 2**. There are no new differentiation rules to be learned here. In order to obtain f'', all you do is apply the usual differentiation rules to f'. So we'll get right to the examples.

Example 2.1.38: Let $f(x) = e^{3x}$. Compute $f''(x)$.

We'll need to compute $f'(x)$ first. Note that because f is a composite function, we'll need the Chain Rule.

$$f'(x) = \frac{d}{dx}\big(f(x)\big) = \frac{d}{dx}\big(e^{3x}\big) = e^{3x} \cdot \frac{d}{dx}(3x) = e^{3x} \cdot 3 = 3e^{3x}$$

Now compute $f''(x)$, again using the Chain Rule.

$$f''(x) = \frac{d}{dx}\big(f'(x)\big) = \frac{d}{dx}\big(3e^{3x}\big) = 3e^{3x} \cdot \frac{d}{dx}(3x) = 3e^{3x} \cdot 3 = 9e^{3x}$$

Example 2.1.39: Let $f(x) = \sec(x)$. Compute $f''(x)$.

We'll get $f'(x)$ first. This is an elementary derivative.

$$f'(x) = \frac{d}{dx}\big(f(x)\big) = \frac{d}{dx}\big(\sec(x)\big) = \sec(x)\tan(x)$$

Now we'll compute $f''(x)$. We'll need the Product Rule.

$$f''(x) = \frac{d}{dx}\big(f'(x)\big) = \frac{d}{dx}\big(\sec(x)\tan(x)\big) = \frac{d}{dx}\big(\sec(x)\big) \cdot \tan(x) + \sec(x) \cdot \frac{d}{dx}\big(\tan(x)\big)$$

$$f''(x) = \sec(x)\tan(x) \cdot \tan(x) + \sec(x) \cdot \sec^2(x)$$

$$f''(x) = \sec(x)\tan(x) + \sec^3(x)$$

Example 2.1.40: Let $f(x) = \dfrac{1}{x^2 + 1}$. Compute $f''(x)$.

We'll write the function as $f(x) = (x^2 + 1)^{-1}$ and use the Chain Rule to compute $f'(x)$.

$$f'(x) = \frac{d}{dx}\big(f(x)\big) = \frac{d}{dx}\Big(\big(x^2 + 1\big)^{-1}\Big) = -\big(x^2 + 1\big)^{-2} \cdot \frac{d}{dx}\big(x^2 + 1\big) = -\frac{1}{\big(x^2 + 1\big)^2} \cdot 2x$$

$$f'(x) = -\frac{2x}{\left(x^2+1\right)^2}$$

Now we'll use the Quotient Rule and the Chain Rule to compute $f''(x)$.

$$f''(x) = \frac{d}{dx}\left(f'(x)\right) = \frac{d}{dx}\left(-\frac{2x}{\left(x^2+1\right)^2}\right) = -\frac{\frac{d}{dx}(2x)\cdot\left(x^2+1\right)^2 - 2x\cdot\frac{d}{dx}\left(\left(x^2+1\right)^2\right)}{\left(\left(x^2+1\right)^2\right)^2}$$

$$f''(x) = -\frac{2\cdot\left(x^2+1\right)^2 - 2x\cdot 2\left(x^2+1\right)\cdot\frac{d}{dx}\left(x^2+1\right)}{\left(x^2+1\right)^4}$$

$$f''(x) = -\frac{2\cdot\left(x^2+1\right)^2 - 2x\cdot 2\left(x^2+1\right)\cdot(2x)}{\left(x^2+1\right)^4} = -\frac{2\left(x^2+1\right)^2 - 8x^2\left(x^2+1\right)}{\left(x^2+1\right)^4}$$

We'll clean this up by factoring and reducing.

$$f''(x) = -\frac{2\left(x^2+1\right)\left(x^2+1-4x^2\right)}{\left(x^2+1\right)^4} = -\frac{2\left(1-3x^2\right)}{\left(x^2+1\right)^3}$$

$$f''(x) = \frac{2\left(3x^2-1\right)}{\left(x^2+1\right)^3}$$

There's no reason that we have to stop differentiating with the second derivative. We can go to higher orders, and that's where the next EK takes us.

EK 2.1D2: Higher order derivatives are represented with a variety of notations. For $y = f(x)$, notations for the second derivative include $\dfrac{d^2 y}{dx^2}$, $f''(x)$, and y''. Higher order derivatives can be denoted $\dfrac{d^n y}{dx^n}$ or $f^{(n)}(x)$.

Note that if $y = f(x)$, then as the EK says a natural alternate notation for $f''(x)$ is y''. We can also use the operator notation. Recall that $y' = \dfrac{d}{dx}(y) = \dfrac{dy}{dx}$. Continuing with this line of thought, we get the following:

$$y'' = \frac{d}{dx}(y') = \frac{d}{dx}\left(\frac{d}{dx}(y)\right)$$

We'll combine the two derivative operators into one by borrowing exponent notation as follows:

$$y'' = \frac{d^2}{dx^2}(y) = \frac{d^2 y}{dx^2}$$

Note that these superscripts aren't *really* exponents. d isn't some quantity that can be multiplied or squared. However, since the combination of the two $\frac{d}{dx}$ operators formally resembles multiplication, we use exponent notation to simplify them.

Derivatives of higher orders use similar notation. The third derivative of $y = f(x)$ can be written as $f'''(x)$ or y''' or $\frac{d^3 y}{dx^3}$. As you can imagine, the string of primes in the superscript can get awkwardly long after a while. The standard convention is to ditch the prime notation after the third derivative, and instead indicate the order of the derivative in a superscript in parentheses. For instance, for the seventh derivative, instead of $f'''''''(x)$, we would write $f^{(7)}(x)$ or $y^{(7)}$ or $\frac{d^7 y}{dx^7}$.

Example 2.1.41: Compute $\frac{d^2}{dx^2}\left(x^2 \cos(x)\right)$.

We'll use the Product Rule to compute $f'(x)$.

$$\frac{d}{dx}\left(x^2 \cos(x)\right) = \frac{d}{dx}\left(x^2\right) \cdot \cos(x) + x^2 \cdot \frac{d}{dx}\left(\cos(x)\right) = 2x \cdot \cos(x) + x^2 \cdot \left(-\sin(x)\right)$$

$$\frac{d}{dx}\left(x^2 \cos(x)\right) = 2x \cos(x) - x^2 \sin(x)$$

Now we'll compute the second derivative, using the Product Rule on each term above. We'll do it one term at a time.

$$\frac{d}{dx}\left(2x \cos(x)\right) = \frac{d}{dx}(2x) \cdot \cos(x) + 2x \cdot \frac{d}{dx}\left(\cos(x)\right) = 2 \cos(x) - 2x \sin(x)$$

$$\frac{d}{dx}\left(x^2 \sin(x)\right) = \frac{d}{dx}\left(x^2\right) \cdot \sin(x) + x^2 \cdot \frac{d}{dx}\left(\sin(x)\right) = 2x \sin(x) + x^2 \cos(x)$$

Now, we combine these to get $\frac{d^2}{dx^2}\left(x^2 \cos(x)\right)$.

$$\frac{d^2}{dx^2}\left(x^2 \cos(x)\right) = \frac{d}{dx}\left(2x \cos(x) - x^2 \sin(x)\right)$$

$$\frac{d^2}{dx^2}\left(x^2 \cos(x)\right) = 2 \cos(x) - 2x \sin(x) - \left(2x \sin(x) + x^2 \cos(x)\right)$$

$$\frac{d^2}{dx^2}\left(x^2 \cos(x)\right) = \left(2 - x^2\right)\cos(x) - 4x \sin(x)$$

We'll wrap up our examples with one on Implicit Differentiation.

Example 2.1.42: Let $x^2 + y^2 = 1$. Compute y'' in terms of x and y only.

First, we find y'.

$$\frac{d}{dx}\left(x^2 + y^2\right) = \frac{d}{dx}(1)$$

$$2x + 2\,yy' = 0$$

$$y' = -\frac{x}{y}$$

Now, we compute y''. We'll need to use the Quotient Rule.

$$y'' = \frac{d}{dx}\left(y'\right) = \frac{d}{dx}\left(-\frac{x}{y}\right) = -\frac{\frac{d}{dx}(x)\cdot y - x\cdot\frac{d}{dx}(y)}{y^2} = -\frac{y - xy'}{y^2}$$

Since the problem requires us to express y'' in terms of x and y only, we can't leave our answer in terms of x, y, and y'. We'll substitute $-\frac{x}{y}$ in for y' and simplify.

$$y'' = -\frac{y - x\left(-\dfrac{x}{y}\right)}{y^2} = -\frac{y + \dfrac{x^2}{y}}{y^2}\cdot\frac{y}{y} = -\frac{y^2 + x^2}{y^3}$$

From the equation relating x and y, we know that $x^2 + y^2 = 1$, so we finish this off as follows:

$$y'' = -\frac{1}{y^3}$$

Example 2.1.43: Let $f(x) = \sin(2x)$. Compute $f^{(5)}(x)$.

We'll work our way up to $f^{(5)}(x)$, one derivative at a time. We'll need to use the Chain Rule every step of the way.

$$f'(x) = \frac{d}{dx}\left(f(x)\right) = \frac{d}{dx}\left(\sin(2x)\right) = \cos(2x)\cdot\frac{d}{dx}(2x) = 2\cos(2x)$$

$$f''(x) = \frac{d}{dx}\left(f'(x)\right) = \frac{d}{dx}\left(2\cos(2x)\right) = 2\left(-\sin(2x)\right)\cdot\frac{d}{dx}(2x) = -4\sin(2x)$$

$$f'''(x) = \frac{d}{dx}\left(f''(x)\right) = \frac{d}{dx}\left(-4\sin(2x)\right) = -4\cos(2x)\cdot\frac{d}{dx}(2x) = -8\cos(2x)$$

$$f^{(4)}(x) = \frac{d}{dx}\left(f'''(x)\right) = \frac{d}{dx}\left(-8\cos(2x)\right) = -8\left(-\sin(2x)\right) \cdot \frac{d}{dx}(2x) = 16\sin(2x)$$

$$f^{(5)}(x) = \frac{d}{dx}\left(f^{(4)}(x)\right) = \frac{d}{dx}\left(16\sin(2x)\right) = 16\cos(2x) \cdot \frac{d}{dx}(2x) = 32\cos(2x)$$

EU 2.2 – Derivatives and the Behavior of Functions _____

LO 2.2A –Derivatives and Properties of a Function

The full Learning Objective of this section is as follows:

LO 2.2A: Use derivatives to analyze properties of a function.

The first item of Essential Knowledge under LO 2.2A covers the information that can be extracted about a function f by looking at f' and f''.

EK 2.2A1: First and second derivatives of a function can provide information about the function and its graph including intervals of increase or decrease, local (relative) and global (absolute) extrema, intervals of upward or downward concavity, and points of inflection.

If the graph of a function was always available, you could tell at a glance where the function increases or decreases, where there are "peaks" or "valleys" or whether the concavity is upward or downward. More often than not, though, you are looking at an algebraic representation of a function such as $f(x) = x^3 - 6x + 2$. How will you know where this function goes up or down, or whether it peaks in a given interval such as $[-2, 2]$? This is where derivatives spring into action.

This EK covers a lot of ground, so we'll focus on one issue at a time.

Intervals of Increase or Decrease

Let f be a differentiable function defined on an open interval (a, b).

1. f is **increasing on (a, b)** if $f'(x) > 0$ for all x in (a, b).
2. f is **decreasing on (a, b)** if $f'(x) < 0$ for all x in (a, b).

If f is either decreasing or increasing for all x in an open interval (a, b), then we say that f is **strictly monotonic on (a, b)**.

Example 2.2.1: Determine the open intervals on which $f(x) = x^3 - 6x + 2$ is increasing and on which it is decreasing.

We'll find the zeros of $f'(x)$ and check its sign in between each zero.

$$f'(x) = 3x^2 - 6 = 0 \Rightarrow x = \pm\sqrt{2}$$

Our next move is to test $f'(x)$ on the intervals $(-\infty, -\sqrt{2})$, $(-\sqrt{2}, \sqrt{2})$, and $(\sqrt{2}, \infty)$. We'll do this in a table.

Interval/Point	Test Value x	$f'(x)$	Behavior of f
$(-\infty, -\sqrt{2})$	$x = -2$	$f'(-2) = 6 > 0$	Increasing
$(-\sqrt{2}, \sqrt{2})$	$x = 0$	$f'(0) = -6 < 0$	Decreasing
$(\sqrt{2}, \infty)$	$x = 2$	$f'(-2) = 6 > 0$	Increasing

So, $f(x)$ is increasing on $(-\infty, -\sqrt{2})$ and $(\sqrt{2}, \infty)$, and decreasing on $(-\sqrt{2}, \sqrt{2})$.

In Example 2.2.1, we relied on finding the zeros of f' to identify the open intervals to be tested. This works when f is a polynomial function, which is differentiable everywhere. But what if f is *not* differentiable everywhere? What if f isn't even *defined* everywhere? We must take these things into account, and we do in the following two examples.

Example 2.2.2: Determine the open intervals on which $f(x) = |x|$ is increasing and on which it is decreasing.

Recall from Example 2.1.4 that $f(x) = |x|$ is not differentiable at $x = 0$. For $x \neq 0$, we may differentiate f piecewise as follows.

$$f'(x) = \frac{d}{dx}(|x|) = \frac{d}{dx}\left(\begin{cases} x & x \geq 0 \\ -x & x < 0 \end{cases}\right) = \begin{cases} 1 & x > 0 \\ -1 & x < 0 \end{cases}$$

As you can see, $f'(x)$ *never* equals zero. However, it is clear that $f'(x) > 0$ for all x in $(0, \infty)$, and $f'(x) < 0$ for all x in $(-\infty, 0)$. So f is increasing on $(0, \infty)$, and decreasing on $(-\infty, 0)$.

Example 2.2.3: Determine the open intervals on which $f(x) = \frac{1}{x}$ is increasing and on which it is decreasing.

$f'(x) = -\frac{1}{x^2}$, and this function has no zeros. However, neither f nor f' is defined at $x = 0$, so we should test f' on the intervals $(-\infty, 0)$ and $(0, \infty)$. We can tell by inspection that $f'(x) < 0$ for all x in each of these intervals, so f is decreasing on both $(-\infty, 0)$ and $(0, \infty)$.

The points that we used to identify the open intervals on which to test f' in these examples are important enough to give a definition for, which we do below.

Let f be defined at $x = c$. c is a **critical number** of f if either $f'(c) = 0$ or $f'(c)$ does not exist.

In Example 2.2.1, we found that the critical numbers of $f(x) = x^3 - 6x + 2$ are $x = \pm\sqrt{2}$, and in Example 2.2.2, we found that the critical number of $f(x) = |x|$ is $x = 0$. Looking back at Example 2.2.3, we see that $x = 0$ is *not* a critical number of $f(x) = \frac{1}{x}$ because f is not defined there.

Local (Relative) Extrema

We will start with a definition. Let a function f be defined on an open interval (a, b) that contains c.

1. If $f(c) \geq f(x)$ for all x in (a, b), f has a **relative maximum** at the point $(c, f(c))$.
2. If $f(c) \leq f(x)$ for all x in (a, b), f has a **relative minimum** at the point $(c, f(c))$.

In the figure below, f has a relative maximum at $(c_1, f(c_1))$ and a relative minimum at $(c_2, f(c_2))$. Note that a relative maximum is *not* necessarily the highest point on the graph, nor is a relative minimum necessarily the lowest. A relative maximum is the highest point in a local neighborhood, and likewise a relative minimum is the lowest point in a local neighborhood.

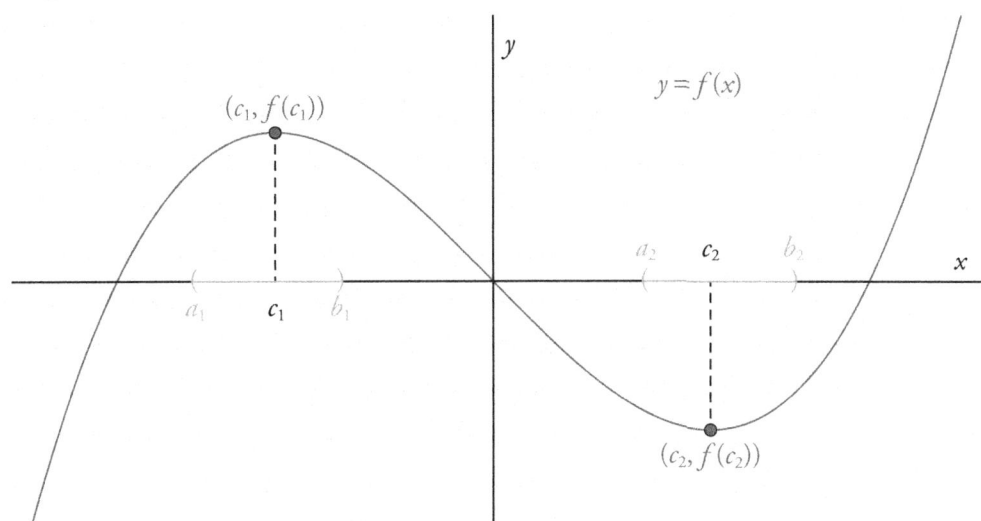

Relative maxima and relative minima are referred to collectively as **relative extrema**, and they are the peaks and valleys of the graph of a function. It's easy to spot the relative extrema of a function by looking at its graph, but what if we don't have the graph? What if all we have is a formula for $f(x)$? Not to worry. There's a theorem that states that *relative extrema only occur at critical numbers*. This doesn't guarantee that *every* critical number corresponds to a relative extremum, but it does guarantee that we needn't bother looking anywhere else.

Suppose we've found all of the critical numbers of a function f. Now what? How do we test them? That's where the First Derivative Test comes in.

The First Derivative Test

Let c be a critical number of a function f, let f be continuous on an open interval I containing c, and let f be differentiable everywhere in I, except possibly at $x = c$. Then we can find the relative extrema of f as follows:

1. If $f'(x)$ changes sign from negative to positive at $x = c$, then f has a relative minimum at $(c, f(c))$.
2. If $f'(x)$ changes sign from positive to negative at $x = c$, then f has a relative maximum at $(c, f(c))$.

3. If $f'(x)$ does not change sign at $x = c$, then f has no relative extremum at $(c, f(c))$. We'll illustrate this with two examples.

Example 2.2.4: Find and classify all relative extrema of $f(x) = x^3 - 6x + 2$.

In Example 2.2.1, we found that the critical numbers of this function are $x = \pm\sqrt{2}$. We'll expand the table that we used in that example to include the critical numbers.

Interval/Point	Test Value x	$f'(x)$	Behavior of f
$(-\infty, -\sqrt{2})$	$x = -2$	$f'(-2) = 6 > 0$	Increasing
$x = -\sqrt{2}$	N/A	$f'(-\sqrt{2}) = 0$	Relative Maximum
$(-\sqrt{2}, \sqrt{2})$	$x = 0$	$f'(0) = -6 < 0$	Decreasing
$x = \sqrt{2}$	N/A	$f'(-\sqrt{2}) = 0$	Relative Minimum
$(\sqrt{2}, \infty)$	$x = 2$	$f'(-2) = 6 > 0$	Increasing

The nice thing about this format is that the First Derivative Test is built right into the table.

- $f'(x) > 0$ on $(-\infty, -\sqrt{2})$, and $f'(x) < 0$ on $(-\sqrt{2}, \sqrt{2})$, so f has a relative maximum at $(-\sqrt{2}, f(-\sqrt{2})) = (-\sqrt{2}, 4\sqrt{2} + 2)$.
- $f'(x) < 0$ on $(-\sqrt{2}, \sqrt{2})$, and $f'(x) > 0$ on $(\sqrt{2}, \infty)$, so f has a relative minimum at $(\sqrt{2}, f(\sqrt{2})) = (\sqrt{2}, -4\sqrt{2} + 2)$.

We'll illustrate this in the following figure:

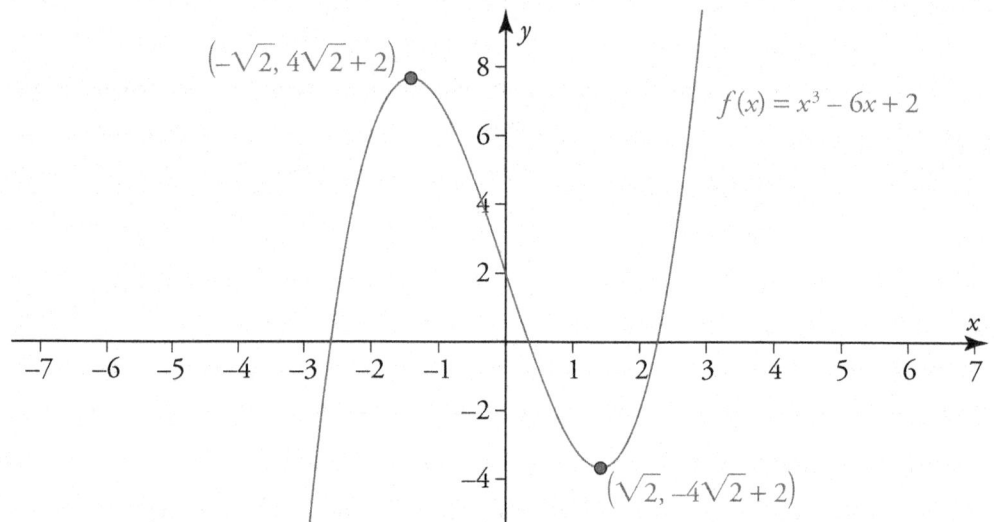

Example 2.2.5: Find and classify all relative extrema of $f(x) = |x|$.

In Example 2.2.2, we found that the one and only critical number of f is $x = 0$. We'll make a table similar to that of the last example.

Interval/Point	Test Value x	$f'(x)$	Behavior of f
$(-\infty, 0)$	$x = -1$	$f'(-1) = -1 < 0$	Decreasing
$x = 0$	N / A	$f'(0)$ is undefined	Relative Minimum
$(0, \infty)$	$x = 1$	$f'(1) = 1 > 0$	Increasing

So f has a relative minimum at $(0, f(0)) = (0, 0)$. See the following figure:

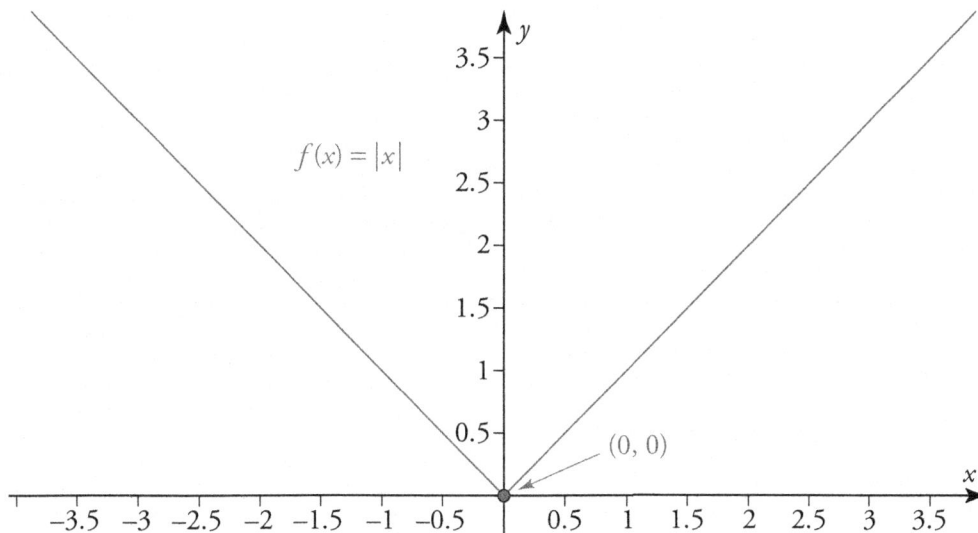

Global (Absolute) Extrema

Let f be defined on an interval I containing c.

1. $f(c)$ is an **absolute maximum** of f on I if $f(c) \geq f(x)$ for all x in I.
2. $f(c)$ is an **absolute minimum** of f on I if $f(c) \leq f(x)$ for all x in I.

These definitions are more general than the definitions of relative extrema in two important ways.

- The interval I here need not be open. It could be open, closed, or neither.
- The number c need not be a critical number of f.

Not every function has an absolute maximum or an absolute minimum on every interval. For instance, $f(x) = x^2$ has an absolute minimum but no absolute maximum on $[0, \infty)$, and it has neither on $(0, \infty)$. See the following figures:

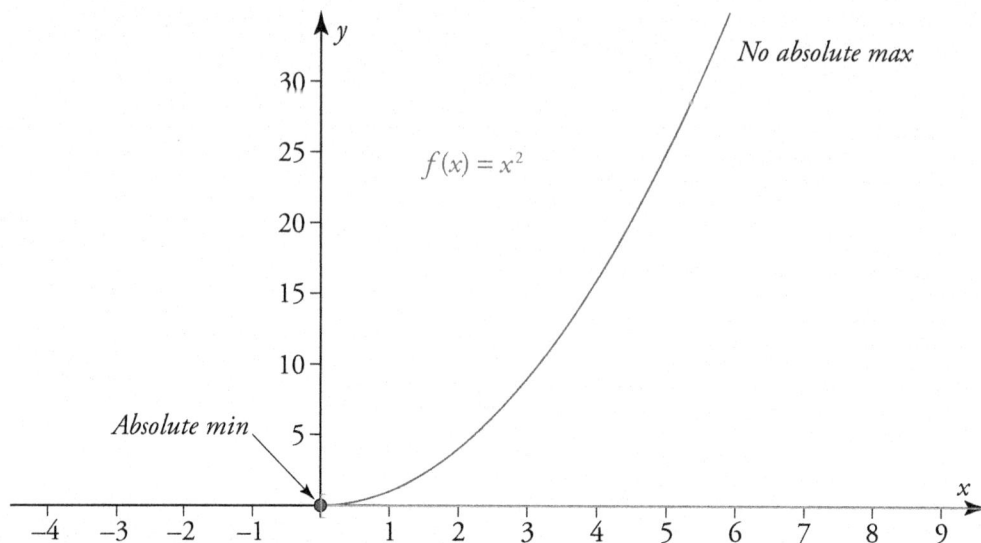

f(x) = x²

Absolute min

No absolute max

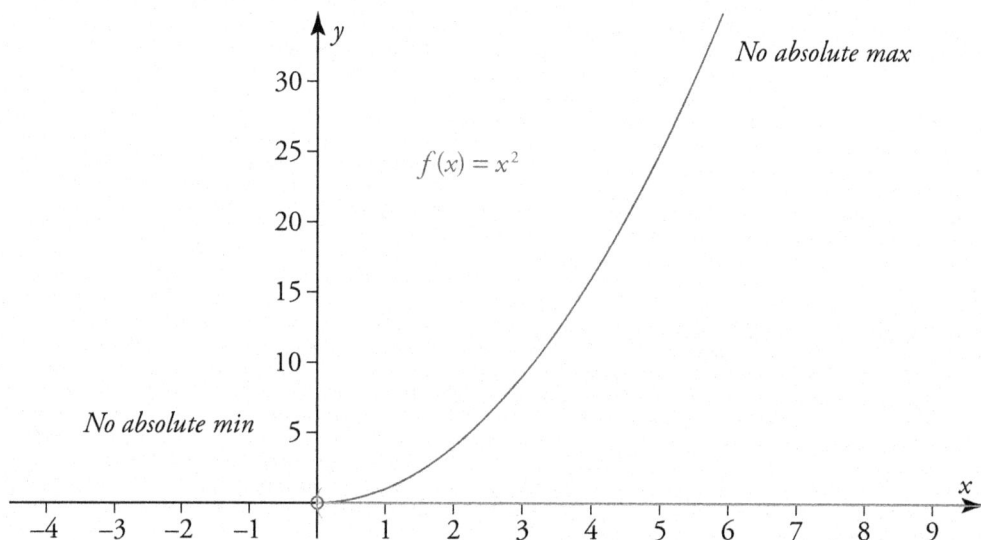

f(x) = x²

No absolute min

No absolute max

There is an existence theorem that tells us when a function absolutely, positively *must* have absolute extrema. We review this next.

The Extreme Value Theorem

If f is continuous on a closed interval $[a, b]$, then f has both an absolute minimum and an absolute maximum on $[a, b]$.

This tells us *when* we can find absolute extrema, but it does not tell us *how to find them*. We'll turn to this next.

If f is continuous on $[a, b]$, then its absolute extrema will occur

1. at a critical number c inside (a, b), or
2. at one of the endpoints of $[a, b]$.

Let's look at a couple of examples.

Example 2.2.6: Find and classify the absolute extrema of $f(x) = x^3 - 6x + 2$ on $[-2, 2]$.

Since f is continuous and the interval is closed, there is guaranteed to be an absolute maximum and an absolute minimum. We already know from Example 2.2.1 that the critical numbers of f are $x = \pm\sqrt{2}$. We need to evaluate f at each critical point and at each endpoint. The largest and smallest of these function values are, respectively, the absolute maximum and absolute minimum of f on $[-2, 2]$.

- $f(-2) = (-2)^3 - 6(-2) + 2 = 6$
- $f(-\sqrt{2}) = (-\sqrt{2})^3 - 6(-\sqrt{2}) + 2 = 4\sqrt{2} + 2 \approx 7.66$
- $f(\sqrt{2}) = (\sqrt{2})^3 - 6(\sqrt{2}) + 2 = -4\sqrt{2} + 2 \approx -3.66$
- $f(2) = 2^3 - 6(2) + 2 = -2$

So on $[-2, 2]$, f has an absolute maximum of $4\sqrt{2} + 2$ and an absolute minimum of $-4\sqrt{2} + 2$. We'll illustrate this below.

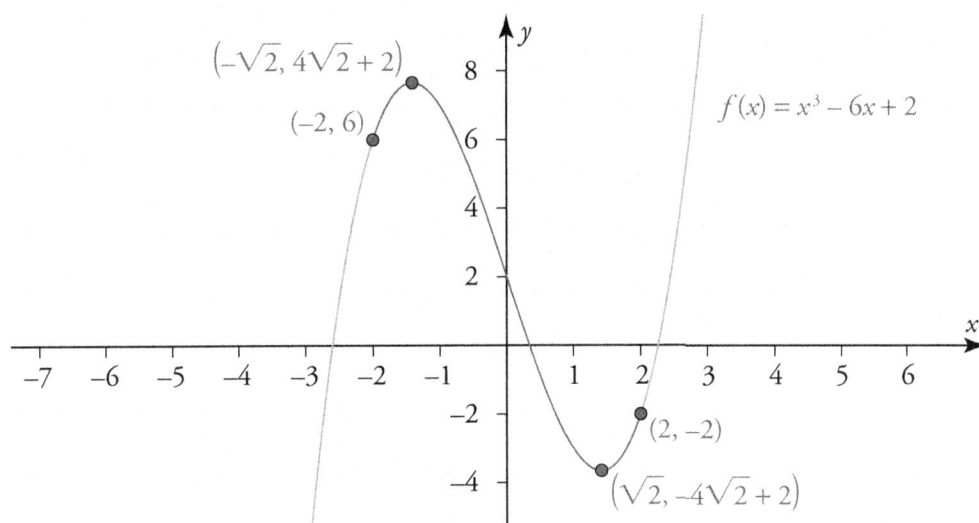

If the interval I changes, then the absolute extrema on I may change. We demonstrate this by tweaking our previous example.

Example 2.2.7: Find and classify the absolute extrema of $f(x) = x^3 - 6x + 2$ on $[-2, 3]$.

Again, the critical numbers of f are $x = \pm\sqrt{2}$. We will evaluate f at each critical number and at each endpoint of the interval.

- $f(-2) = (-2)^3 - 6(-2) + 2 = 6$
- $f(-\sqrt{2}) = (-\sqrt{2})^3 - 6(-\sqrt{2}) + 2 = 4\sqrt{2} + 2 \approx 7.66$
- $f(\sqrt{2}) = (\sqrt{2})^3 - 6(\sqrt{2}) + 2 = -4\sqrt{2} + 2 \approx -3.66$
- $f(3) = 3^3 - 6(3) + 2 = 11$

As you can see, by changing the interval from $[-2, 2]$ to $[-2, 3]$, the absolute minimum of f remained the same at $-4\sqrt{2} + 2$, but the absolute maximum changed to 11. We'll illustrate this below.

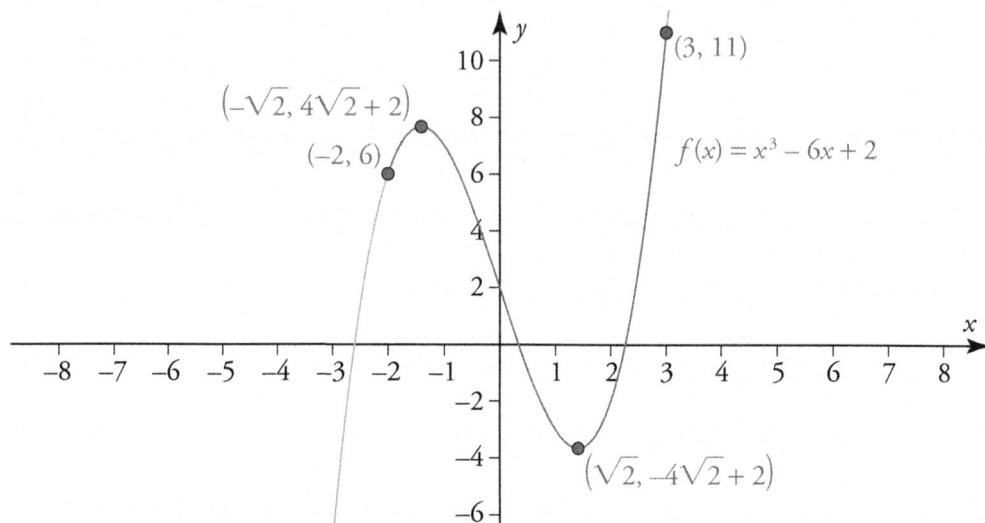

The graph shows $f(x) = x^3 - 6x + 2$ with labeled points $(-\sqrt{2}, 4\sqrt{2} + 2)$, $(-2, 6)$, $(3, 11)$, and $(\sqrt{2}, -4\sqrt{2} + 2)$.

Note that although there is no longer an *absolute* maximum at the point $(-\sqrt{2}, 4\sqrt{2} + 2)$, there is still a *relative* maximum there.

Intervals of Upward and Downward Concavity

Let f be a *twice* differentiable function defined on an open interval (a, b).

1. f is **concave upward on (a, b)** if $f''(x) > 0$ for all x in (a, b).
2. f is **concave downward on (a, b)** if $f''(x) < 0$ for all x in (a, b).

Note that $f''(x) > 0$ implies that f' is *increasing*, and $f''(x) < 0$ implies that f' is *decreasing*.

Example 2.2.8: Find the open intervals on which $f(x) = x^3 - 6x + 2$ is concave upward and on which it is concave downward.

$$f'(x) = 3x^2 - 6 = 0$$

$$f''(x) = 6x = 0 \Rightarrow x = 0$$

Now we'll test $f''(x)$ on the intervals $(-\infty, 0)$ and $(0, \infty)$.

Interval	Test Value x	$f''(x)$	Behavior of f
$(-\infty, 0)$	$x = -1$	$f''(-1) = -6 < 0$	Concave Downward
$(0, \infty)$	$x = 1$	$f''(1) = 6 > 0$	Concave Upward

Points of Inflection

Let f be a continuous function on an open interval containing c, and let f be differentiable at c. The point $(c, f(c))$ is a **point of inflection** of the graph of f if the concavity of f changes at this point. We'll illustrate this in the following figure:

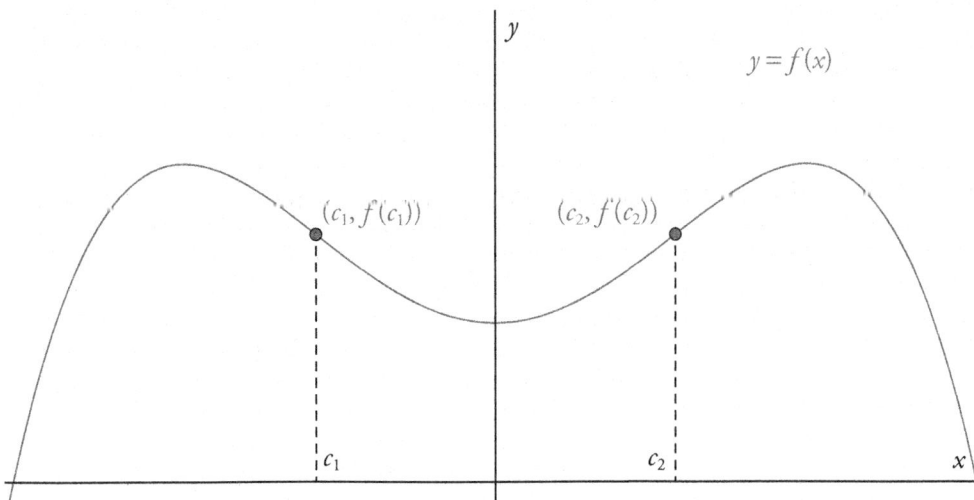

In the figure, the function f is concave downward on $(-\infty, c_1)$, concave upward on (c_1, c_2), and concave downward again on (c_2, ∞). Consequently, the points $(c_1, f(c_1))$ and $(c_2, f(c_2))$ are points of inflection of the graph of f.

Points of inflection occur only at points for which either $f''(x) = 0$ or $f''(x)$ does not exist. In other words, *points of inflection occur only at critical numbers of f'*.

Example 2.2.9: Refer back to Example 2.2.8. Since $f(x) = x^3 - 6x + 2$ is continuous and differentiable at $x = 0$, and since f changes concavity from downward to upward at $x = 0$, the point $(0, 2)$ is a point of inflection of the graph of f. We'll illustrate this in the following figure:

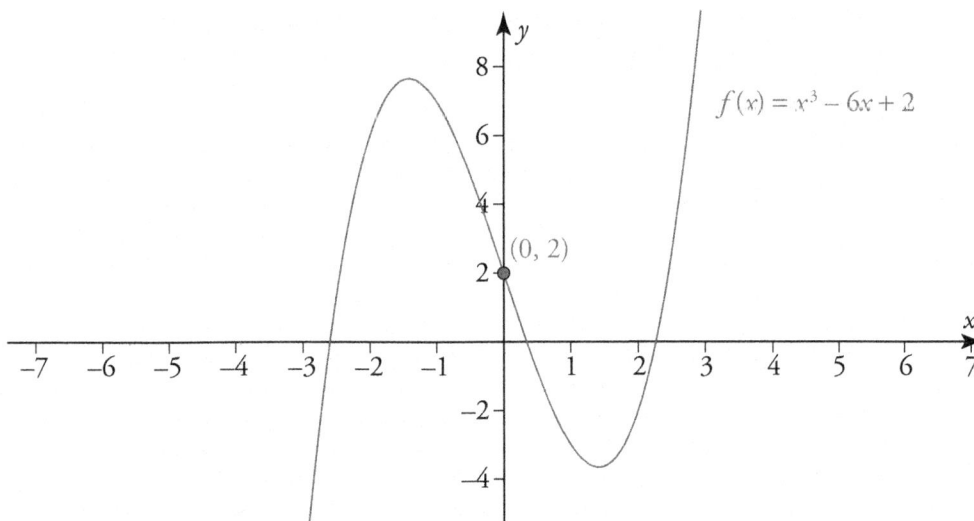

Next, we'll review an alternative to the First Derivative Test for classifying the critical numbers of a function f. It turns out that the second derivative can tell us something about relative extrema. Take another look at the following figure from Example 2.2.4:

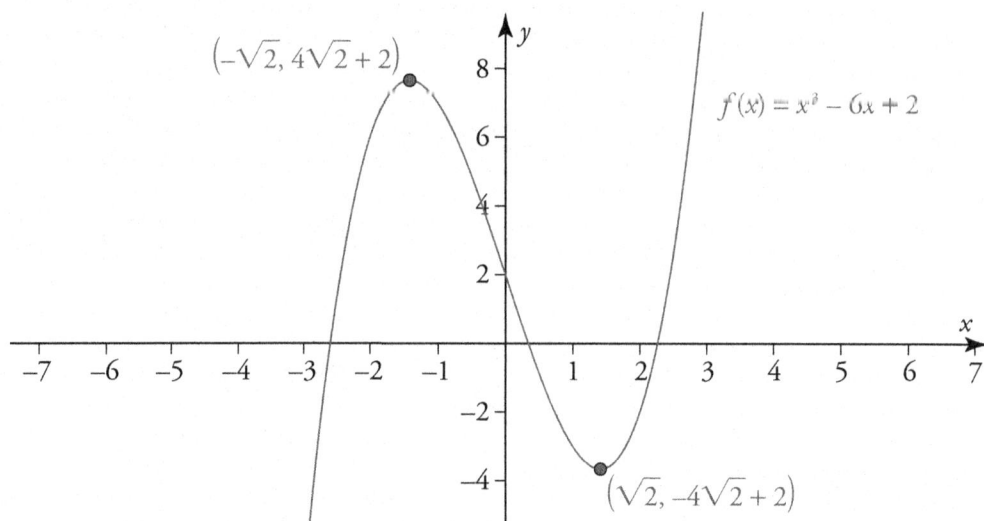

$(-\sqrt{2}, 4\sqrt{2} + 2)$

$f(x) = x^3 - 6x + 2$

$(\sqrt{2}, -4\sqrt{2} + 2)$

Observe that the point $(-\sqrt{2}, 4\sqrt{2} + 2)$ is a relative maximum of f, and f is concave downward at that point, which means that $f''(-\sqrt{2})$ is *negative*. Also, the point $(\sqrt{2}, -4\sqrt{2} + 2)$ is a relative minimum of f, and f is concave upward at that point, which means that $f''(\sqrt{2})$ is *positive*. We generalize these observations in the Second Derivative Test.

The Second Derivative Test

Let c be a critical number of a function f, and let f be twice differentiable on an open interval containing c. Then the following statements hold:

1. If $f''(c) > 0$, then f has a relative minimum at the point $(c, f(c))$.
2. If $f''(c) < 0$, then f has a relative maximum at the point $(c, f(c))$.
3. If $f''(c) = 0$, then the second derivative yields no information about the point $(c, f(c))$. Use the First Derivative Test.

The advantage that the Second Derivative Test has over the First Derivative Test is that fewer computations are involved. When applying the First Derivative Test, you must test *two* points, one on *each side* of the critical number c, while for the Second Derivative Test, you need only to test the critical number c itself. The drawback of the Second Derivative Test is that it fails if $f''(c) = 0$, whereas the First Derivative Test *never* fails.

Example 2.2.10: Refer back to Examples 2.2.4 and 2.2.8. Show by explicit calculation that $f''(-\sqrt{2}) < 0$ and that $f''(\sqrt{2}) > 0$, as required by the Second Derivative Test.

In Example 2.2.8, we computed $f''(x) = 6x$. Evaluating this at each of the critical numbers, we obtain the following:

$$f''\left(-\sqrt{2}\right) = 6\left(-\sqrt{2}\right) = -6\sqrt{2} < 0$$

$$f''\left(\sqrt{2}\right) = 6\sqrt{2} > 0$$

Our next EK is closely associated with this one.

EK 2.2A2: Key features of functions and their derivatives can be identified and related to their graphical, numerical, and analytical representations.

The "key features" mentioned here are precisely those that were mentioned in **EK 2.2A1**: intervals of increase or decrease, relative and absolute extrema, intervals of upward or downward concavity, and points of inflection. We just reviewed at length how to identify these key features and relate them to the *analytical* representation of a function and its first and second derivatives, so we will consider that covered. Here we will focus on how to relate these features to a function's graphical and numerical representations.

Graphical Representation of a Function

The "key features" listed above are much easier to spot when a function is represented graphically than they are when the same function is represented analytically. Since there are no new concepts to present here, we'll get right to an example.

Example 2.2.11: Consider the following graph of a function $y = f(x)$.

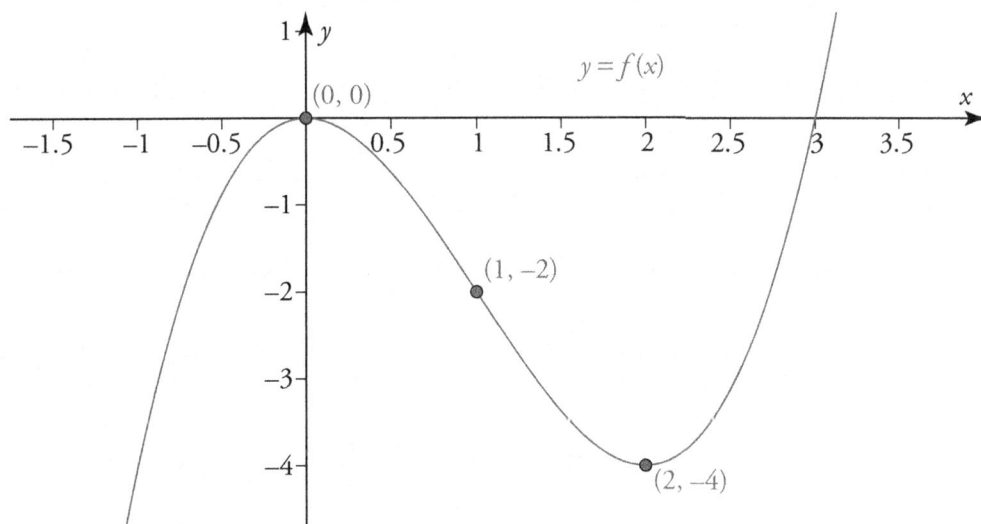

Use the graph to answer the following questions:
1. On what open interval(s) is f increasing? Decreasing?
2. At what point(s) does f have a relative maximum? Relative minimum?
3. On what open interval(s) is f concave upward? Downward?
4. At what point(s) does the graph of f have a point of inflection?

We'll answer these questions by inspecting the graph.
1. f is increasing on $(-\infty, 0)$ and $(2, \infty)$ and decreasing on $(0, 2)$.
2. f has a relative maximum at the point $(0, 0)$ and a relative minimum at the point $(2, -4)$.
3. f is concave downward on $(-\infty, 1)$ and concave upward on $(1, \infty)$.
4. The graph of f has a point of inflection at the point $(1, -2)$.

Numerical Representation of a Function

Again, there are no new concepts here, so we'll go straight to an example

Example 2.2.12: Consider the function f that is twice differentiable for all real x and that satisfies the criteria given in the following table. Neither f' nor f'' has any zeros that are not shown in the table.

x:	–6	–4	–2	0	2	4	6
$f(x)$:	8.2	13.8	9.8	1	–7.8	–11.8	–6.2
$f'(x)$:	6	0	–3.6	–4.8	–3.6	0	6
$f''(x)$:	–3.6	–2.4	–1.2	0	1.2	2.4	3.6

1. On what open interval(s) is f increasing? Decreasing?
2. At what point(s) does f have a relative maximum? Relative minimum?
3. On what open interval(s) is f concave upward? Downward?
4. At what point(s) does the graph of f have a point of inflection?

We'll answer these questions by inspecting the table.

1. Since $f'(-6) > 0$ and $f'(6) > 0$, and since the only zeros of f are at $x = \pm 4$, f is increasing on the intervals $(-\infty, -4)$ and $(4, \infty)$. By similar reasoning, since $f'(x) < 0$ at $x = -2, 0$, and 2, f is decreasing on the interval $(-4, 4)$.

2. Since f changes from increasing to decreasing at $x = -4$, the point $(-4, 13.8)$ is a relative maximum of f. Since f changes from decreasing to increasing at $x = 4$, the point $(-4, -11.8)$ is a relative minimum of f.

3. Since $f''(x) < 0$ for $x = -6, -4$, and -2, and since the only zero of f'' is at $x = 0$, f is concave downward on the interval $(-\infty, 0)$. By similar reasoning, since $f''(x) > 0$ for $x = 2, 4$, and 6, f is concave upward on the interval $(0, \infty)$.

4. Since f changes from concave downward to concave upward at $x = 0$, the point $(0, 1)$ is a point of inflection of the graph of f.

EK 2.2A3: Key features of the graphs of f, f', and f'' are related to one another.

Here, we summarize the connections between f, f', and f''.

If:	**Then:**
f is increasing . . .	. . . $f'(x) > 0$
f is decreasing . . .	. . . $f'(x) < 0$
f has a relative max or min . . .	. . . $f'(x) = 0$ or $f'(x)$ does not exist.
f is concave upward . . .	. . . $f''(x) > 0$ and f' is increasing.
f is concave downward . . .	. . . $f''(x) < 0$ and f' is decreasing.
The graph of f has an inflection point . . .	. . . $f''(x) = 0$ or $f''(x)$ does not exist.

We will use these connections in the following example:

Example 2.2.13: In the following figure, three functions are graphed. One of the functions is $f(x)$, one is $f'(x)$, and one is $f''(x)$. Determine which function is which. The grid lines on the x-axis are spaced 0.5 units apart.

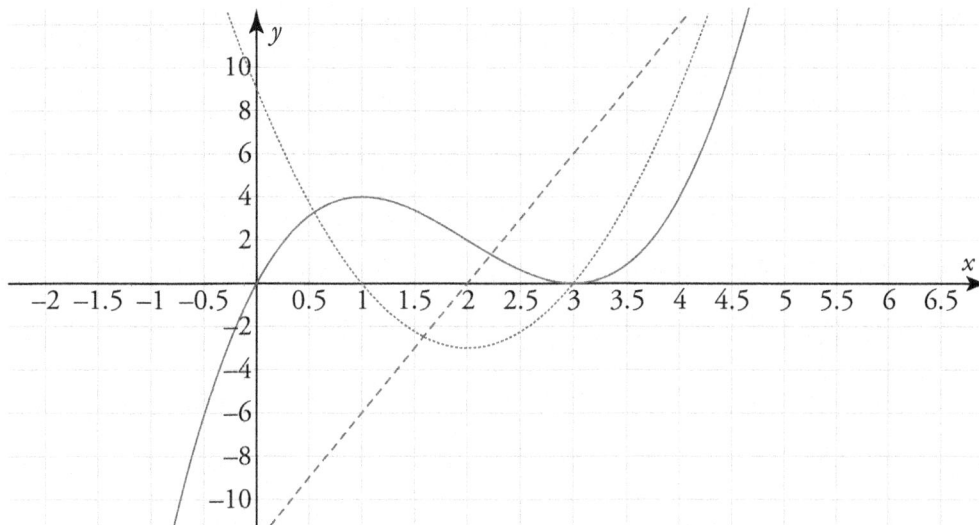

The function whose graph is given by the solid curve has relative extrema at $x = 1$ and at $x = 3$, so the derivative of this function should either be zero or undefined at these points. The function whose graph is given by the dashed curve is defined but not zero at each of these points, but the function whose graph is given by the dotted curve is zero at both points. What this means is that the function given by the dotted curve is the derivative of the function given by the solid curve.

The function whose graph is given by the dotted curve has a relative minimum at $x = 2$. The function whose graph is given by the dashed curve is zero at this point, so this function is the derivative of the function given by the dotted curve.

Considering all of this together, $f(x)$ is given by the solid curve, $f'(x)$ is given by the dotted curve, and $f''(x)$ is given by the dashed curve.

We'll close this section by reviewing what first and second derivatives can tell us about *polar* graphs.

EK 2.2A4: (BC) For a curve given by a polar equation $r = f(\theta)$, derivatives of r, x, and y with respect to θ and first and second derivatives of y with respect to x can provide information about the curve.

We'll organize our review according to the various interesting properties of polar graphs: slope, concavity, horizontal tangents, vertical tangents, and tangents at the pole.

Slope

Just as with functions given in Cartesian coordinates, the slope of a function given in polar coordinates is given by $\dfrac{dy}{dx}$, where $y = r\sin(\theta) = f(\theta)\sin(\theta)$ and $x = r\cos(\theta) = f(\theta)\cos(\theta)$. As we reviewed in **EK 2.1C7**, the polar form of the derivative is given as follows:

$$\frac{dy}{dx} = \frac{\dfrac{dy}{d\theta}}{\dfrac{dx}{d\theta}} = \frac{f(\theta)\cos(\theta) + f'(\theta)\sin(\theta)}{-f(\theta)\sin(\theta) + f'(\theta)\cos(\theta)}$$

Example 2.2.14: Refer back to Example 2.1.37. We showed that for $r = 2 - 2\cos(\theta)$, the value of $\dfrac{dy}{dx}$ when $\theta = \dfrac{\pi}{6}$ is 1. This means that the slope of $r = 2 - 2\cos(\theta)$ when $\theta = \dfrac{\pi}{6}$ is 1. We'll illustrate this in the following graph:

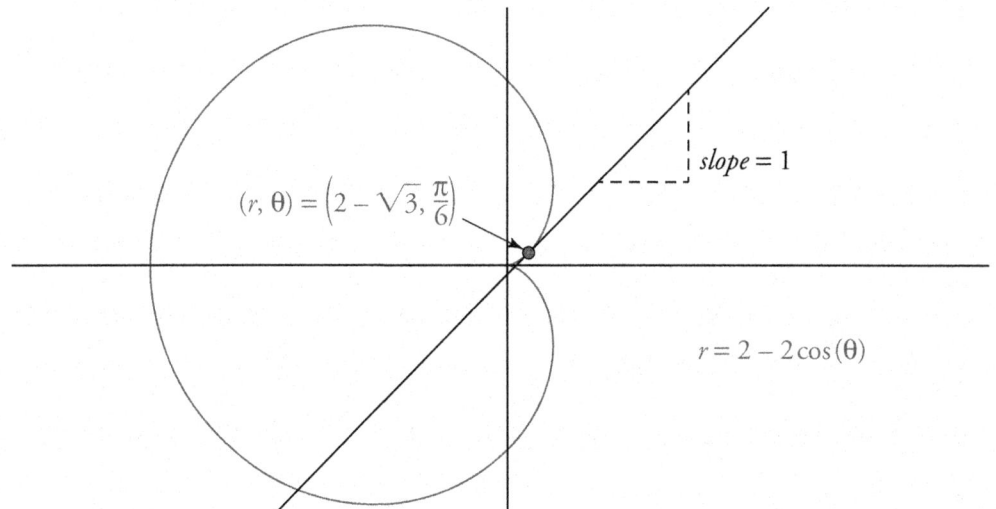

In the graph, the point $(r, \theta) = \left(2 - \sqrt{3}, \dfrac{\pi}{6}\right)$ has been plotted, as has the line tangent to the graph of the polar function at that point. The slope of the tangent line is 1.

Concavity

As you might expect, the concavity of a polar graph is determined by the second derivative $\dfrac{d^2 y}{dx^2}$. We need a formula for the second derivative in polar coordinates.

$$\frac{d^2 y}{dx^2} = \frac{d\left(\dfrac{dy}{dx}\right)}{dx} = \frac{\dfrac{d}{d\theta}\left(\dfrac{dy}{dx}\right)}{\dfrac{dx}{d\theta}}$$

Recall that $\dfrac{dy}{dx} = \dfrac{dy/d\theta}{dx/d\theta}$, so we need to apply the Quotient Rule to the numerator in the above expression. If we do that, we obtain the following result:

$$\frac{d^2 y}{dx^2} = \frac{\dfrac{d^2 y}{d\theta^2} \cdot \dfrac{dx}{d\theta} - \dfrac{dy}{d\theta} \cdot \dfrac{d^2 x}{d\theta^2}}{\left(\dfrac{dx}{d\theta}\right)^3}$$

It is not worth the effort to express this in terms of θ and $f(\theta)$, so we'll leave it in this form.

Example 2.2.15: Show analytically that the polar function $r = 2 - 2\cos(\theta)$ is concave upward at $\theta = \dfrac{\pi}{6}$.

What we need to show is that $\left.\dfrac{d^2 y}{dx^2}\right|_{\theta = \frac{\pi}{6}} > 0$. We show the necessary derivatives below, and we evaluate them at $\theta = \dfrac{\pi}{6}$. Note that not every step in the derivation is shown. You should try to fill in the missing steps yourself.

- $\dfrac{dx}{d\theta} = \dfrac{d}{d\theta}\big((2 - 2\cos(\theta))\cos(\theta)\big) = 2\sin(2\theta) - 2\sin(\theta)$

- $\left.\dfrac{dx}{d\theta}\right|_{\theta = \frac{\pi}{6}} = 2\sin\left(\dfrac{\pi}{3}\right) - 2\sin\left(\dfrac{\pi}{6}\right) = \sqrt{3} - 1$

- $\dfrac{d^2 x}{d\theta^2} = \dfrac{d}{d\theta}\big(2\sin(2\theta) - 2\sin(\theta)\big) = 4\cos(2\theta) - 2\cos(\theta)$

- $\left.\dfrac{d^2 x}{d\theta^2}\right|_{\theta = \frac{\pi}{6}} = 4\cos\left(\dfrac{\pi}{3}\right) - 2\cos\left(\dfrac{\pi}{6}\right) = 2 - \sqrt{3}$

- $\dfrac{dy}{d\theta} = \dfrac{d}{d\theta}\big((2 - 2\cos(\theta))\sin(\theta)\big) = 2\cos(\theta) - 2\cos(2\theta)$

- $\left.\dfrac{dy}{d\theta}\right|_{\theta = \frac{\pi}{6}} = 2\cos\left(\dfrac{\pi}{6}\right) - 2\cos\left(\dfrac{\pi}{3}\right) = \sqrt{3} - 1$

- $\dfrac{d^2 y}{d\theta^2} = \dfrac{d}{d\theta}\big(2\cos(\theta) - 2\cos(2\theta)\big) = -2\sin(\theta) + 4\sin(2\theta)$

- $\left.\dfrac{d^2 y}{d\theta^2}\right|_{\theta = \frac{\pi}{6}} = -2\sin\left(\dfrac{\pi}{6}\right) + 4\sin\left(\dfrac{\pi}{3}\right) = -1 + 2\sqrt{3}$

Now we have all of the pieces that we need to compute $\dfrac{d^2 y}{dx^2}$ at $\theta = \dfrac{\pi}{6}$.

$$\left.\frac{d^2 y}{dx^2}\right|_{\theta = \frac{\pi}{6}} = \frac{\left(2\sqrt{3} - 1\right)\left(\sqrt{3} - 1\right) - \left(\sqrt{3} - 1\right)\left(2 - \sqrt{3}\right)}{\left(\sqrt{3} - 1\right)^3}$$

$$\frac{d^2y}{dx^2}\bigg|_{\theta=\frac{\pi}{6}} = \frac{3}{\sqrt{3}} \cdot 1 > 0$$

So the polar function $r = 2 - 2\cos(\theta)$ is indeed concave upward at $\theta = \dfrac{\pi}{6}$. You can ascertain this visually by looking at the graph in Example 2.2.14.

Horizontal Tangents

The graph of a polar function $r = f(\theta)$ has a horizontal tangent wherever the slope of the graph is zero. This occurs wherever the conditions $\dfrac{dy}{d\theta} = 0$ and $\dfrac{dx}{d\theta} \neq 0$ are both satisfied.

Example 2.2.16: Find all points on the graph of $r = 2 - 2\cos(\theta)$ at which the line tangent to the graph is horizontal.

We've already presented $\dfrac{dy}{d\theta}$ and $\dfrac{dx}{d\theta}$ in Example 2.2.15. We need to find all solutions of the equation $\dfrac{dy}{d\theta} = 0$ in the interval $[0, 2\pi)$. We will need the trig identity $\cos(2\theta) = 2\cos^2(\theta) - 1$.

$$\frac{dy}{d\theta} = 2\cos(\theta) - 2\cos(2\theta) = 0$$

$$2\cos(\theta) - 2\left(2\cos^2(\theta) - 1\right) = 0$$

$$2\cos^2(\theta) - \cos(\theta) - 1 = 0$$

$$\left(2\cos(\theta) + 1\right)\left(\cos(\theta) - 1\right) = 0$$

Setting each factor equal to zero and solving, we obtain the following:

$$2\cos(\theta) + 1 = 0 \Rightarrow \cos(\theta) = -\frac{1}{2} \Rightarrow \theta = \frac{2\pi}{3}, \frac{4\pi}{3}$$

$$\cos(\theta) - 1 = 0 \Rightarrow \cos(\theta) = 1 \Rightarrow \theta = 0$$

Of these three solutions, we reject $\theta = 0$, because $\dfrac{dx}{d\theta}\bigg|_{\theta=0} = 0$. The polar coordinates at which the function has horizontal tangents are thus $\left(3, \dfrac{2\pi}{3}\right)$ and $\left(3, \dfrac{4\pi}{3}\right)$.

Vertical Tangents

The graph of a polar function $r = f(\theta)$ has a vertical tangent wherever the slope of the graph is infinite. This occurs wherever the conditions $\dfrac{dx}{d\theta} = 0$ and $\dfrac{dy}{d\theta} \neq 0$ are both satisfied.

Example 2.2.16: Find all points on the graph of $r = 2 - 2\cos(\theta)$ at which the line tangent to the graph is vertical.

We need to find all solutions of the equation $\dfrac{dx}{d\theta} = 0$ in the interval $[0, 2\pi)$. We will need the trig identity $\sin(2\theta) = 2\sin(\theta)\cos(\theta)$.

$$\frac{dx}{d\theta} = 2\sin(2\theta) - 2\sin(\theta)$$

$$2\sin(\theta)\cos(\theta) - \sin(\theta) = 0$$

$$\sin(\theta)\big(2\cos(\theta) - 1\big) = 0$$

Setting each factor equal to zero and solving, we obtain the following:

$$\sin(\theta) = 0 \Rightarrow \theta = 0, \pi$$

$$2\cos(\theta) - 1 = 0 \Rightarrow \cos(\theta) = \frac{1}{2} \Rightarrow \theta = \frac{\pi}{3}, \frac{5\pi}{3}$$

Of these four solutions, we reject $\theta = 0$, because $\left.\dfrac{dy}{d\theta}\right|_{\theta=0} = 0$. The polar coordinates at which the function has vertical tangents are thus $\left(1, \dfrac{\pi}{3}\right)$, $(4, \pi)$, and $\left(1, \dfrac{5\pi}{3}\right)$.

Tangents at the Pole

Suppose the graph of a polar function $r = f(\theta)$ passes through the pole when $\theta = \alpha$. That means that $f(\alpha) = 0$. Plugging this condition into the polar form of the derivative, we obtain the following:

$$\left.\frac{dy}{dx}\right|_{\theta=\alpha} = \frac{f(\alpha)\cos(\alpha) + f'(\alpha)\sin(\alpha)}{-f(\alpha)\sin(\alpha) + f'(\alpha)\cos(\alpha)} = \frac{f'(\alpha)\sin(\alpha)}{f'(\alpha)\cos(\alpha)}$$

If $f'(\alpha) \neq 0$, then we obtain the following result:

$$\left.\frac{dy}{dx}\right|_{\theta=\alpha} = \tan(\alpha)$$

This means that the line $\theta = \alpha$ is tangent to the graph of $r = f(\theta)$, provided that the conditions $f(\alpha) = r\big|_{\theta=\alpha} = 0$ and $f'(\alpha) = \left.\dfrac{dr}{d\theta}\right|_{\theta=\alpha} \neq 0$ are both satisfied.

Example 2.2.17: Show that the function $r = 2 - 2\cos(\theta)$ has no tangents at the pole.

To find any potential tangents at the pole, we set $r = 0$ and solve for θ.

$$r = 2 - 2\cos(\theta) = 0 \Rightarrow \cos(\theta) = 1 \Rightarrow \theta = 0$$

Next we need to check the condition $\dfrac{dr}{d\theta} \neq 0$.

$$\frac{dr}{d\theta} = 2\sin(\theta) \Rightarrow \frac{dr}{d\theta}\bigg|_{\theta=0} \sin(0) = 0$$

Since the only candidate for a tangent at the pole does not satisfy the last condition, this function has no tangent at the pole.

The following figure shows the graph of $r = 2 - 2\cos(\theta)$ with all of its horizontal and vertical tangents sketched in.

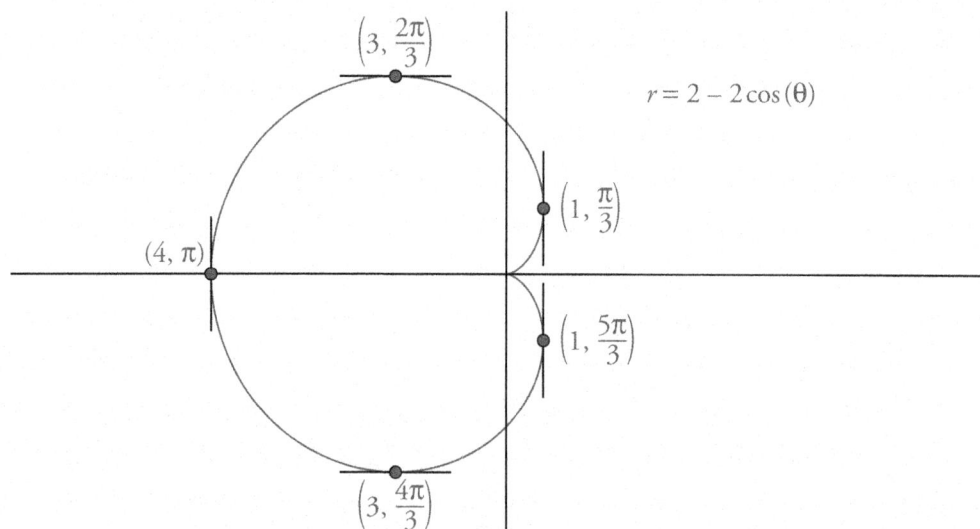

LO 2.2B –Differentiability and Continuity

The full statement of the LO for this section is given below.

LO 2.2B: Recognize the connection between differentiability and continuity.

In the first EK of this section, we reviewed the fact that continuity does not imply differentiability.

EK 2.2B1: A continuous function may fail to be differentiable at a point in its domain.

We've already seen an example of this in Examples 2.2.2 and 2.2.5, in which we looked at the function $f(x) = |x|$. This function is continuous at $x = 0$, because $\lim\limits_{x \to 0} |x| = |0| = 0$. However, as we showed in Example 2.2.2, $f'(0)$ does not exist for this function, so f is not differentiable at $x = 0$. We'll show one more example.

Example 2.2.18: Consider the function f defined below.

$$f(x) = \begin{cases} -x+2 & x < 1 \\ x^2 & x \geq 1 \end{cases}$$

Show that f is both continuous and not differentiable at $x = 1$.

First, we show continuity. We consider $\lim\limits_{x \to 1} f(x)$ from each side.

$$\lim_{x \to 1^-} f(x) = \lim_{x \to 1^-} (-x+2) = 1$$

$$\lim_{x \to 1^+} f(x) = \lim_{x \to 1^+} (x^2) = 1$$

Since the one-sided limits both exist and are equal to 1, we have $\lim\limits_{x \to 1} f(x) = 1$. Also, since $f(1) = 1^2 = 1$, we have $\lim\limits_{x \to 1} f(x) = f(1)$, and, thus, f is continuous at $x = 1$.

Now we show non-differentiability. We consider $\lim\limits_{x \to 1} \dfrac{f(x) - f(1)}{x - 1}$ from each side.

$$\lim_{x \to 1^-} \frac{f(x) - f(1)}{x - 1} = \lim_{x \to 1^-} \frac{-x+2-1}{x-1} = \lim_{x \to 1^-} \frac{-(x-1)}{x-1} = \lim_{x \to 1^-} (-1) = -1$$

$$\lim_{x \to 1^+} \frac{f(x) - f(1)}{x - 1} = \lim_{x \to 1^+} \frac{x^2-1}{x-1} = \lim_{x \to 1^+} \frac{(x-1)(x+1)}{x-1} = \lim_{x \to 1^+} (x+1) = 2$$

Since the one-sided limits do not agree, $\lim\limits_{x \to 1} \dfrac{f(x) - f(1)}{x - 1}$ does not exist, and f is thus not differentiable at $x = 1$. We show the graph of f in the following figure:

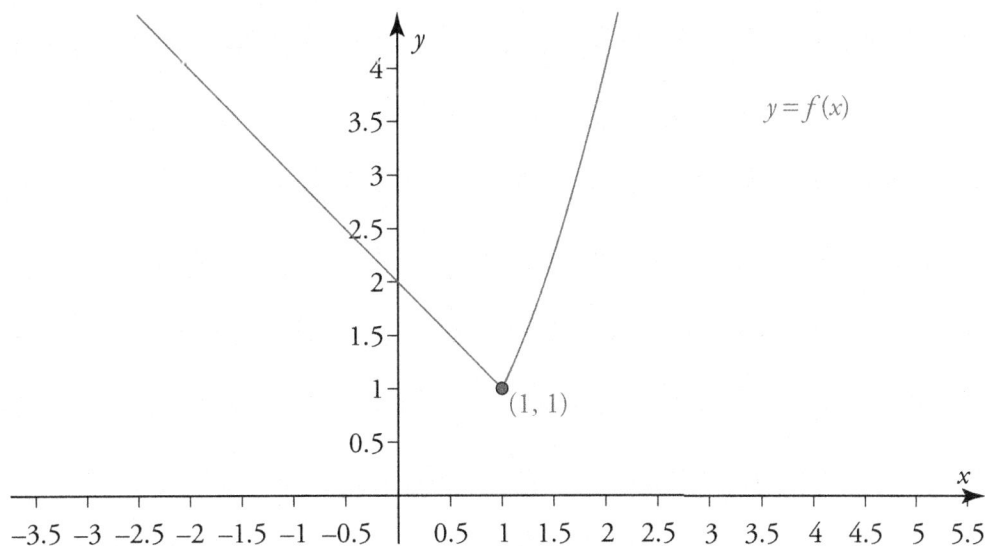

Notice that the graph comes to a sharp point at $x = 1$, just as the graph of $f(x) = |x|$ does at $x = 0$. This sharp point is a sure sign that f is not differentiable at that point, because the slope of the graph changes abruptly there.

We just established by means of a counterexample that shows that the statement "continuity implies differentiability" is *false*. It turns out that the converse of this statement, namely "differentiability implies continuity", is *true*, as we will review in the next EK.

EK 2.2B2: If a function is differentiable at a point, then it is continuous at that point.

Suppose a function f is differentiable at $x = c$. Then $\lim\limits_{x \to c} \dfrac{f(x) - f(c)}{x - c}$ exists, and we refer to its value as $f'(c)$. The limit $\lim\limits_{x \to c}(x - c) = 0$ obviously exists, and so using the Product Rule for Limits, we obtain the desired result.

$$\lim\limits_{x \to c}\left(\frac{f(x) - f(c)}{x - c} \cdot (x - c) \right) = f'(c) \cdot 0$$

$$\lim\limits_{x \to c}\left(f(x) - f(c) \right) = 0$$

$$\lim\limits_{x \to c} f(x) = f(c)$$

This last line means that f is continuous at $x = c$ *by definition*. Thus, differentiability implies continuity.

EU 2.3 – Interpretations and Applications of the Derivative _____

LO 2.3A – Interpretations of the Derivative

The full statement of the LO for this chapter is as follows:

LO 2.3A: Interpret the meaning of a derivative within a problem.

The items of Essential Knowledge for this Learning Objective focus on interpreting the derivative in real world situations. Quantities in the real world have *units of measurement*, and our first EK focuses on that.

EK 2.3A1: The unit for $f'(x)$ is the unit for f divided by the unit for x.

It is easy to see why this is if we consider that $y = f(x)$, and that we can write $f'(x) = \dfrac{dy}{dx}$.

$$\frac{dy}{dx} \begin{matrix} \leftarrow \textit{Units of } y \\ \leftarrow \textit{Units of } x \end{matrix}$$

Example 2.3.1: If the position of a particle (in meters, *m*) is given as a function of time (in seconds, *s*) by $x(t)$, then the velocity of the particle, which is given by $v(t) = x'(t)$, has units of $\dfrac{m}{s}$.

Example 2.3.2: If the cost C (in dollars, \$) of producing x units of a product is given by the function $C(x)$, then the marginal cost of production, which is given by $M(x) = C'(x)$, has units of $\dfrac{\$}{unit}$.

EK 2.3A2: The derivative of a function can be interpreted as the instantaneous rate of change with respect to its independent variable.

Simply stated, if $y = f(x)$, then $\dfrac{dy}{dx} = f'(x)$ can be interpreted as the rate of change of y (a.k.a. f) with respect to x.

Example 2.3.3: If the position of a particle is given as a function of time by $x(t)$, then the velocity $v(t) = x'(t)$ of the particle is the rate of change of particle's position with respect to time.

Example 2.3.4: If the cost C of producing x units of a product is given by the function $C(x)$, then the marginal cost of production $M(x) = C'(x)$ is the rate of change of the production cost with respect to the number of units produced.

LO 2.3B – The Tangent Line Problem

The tangent line problem was one of the original motivations for the development of differential calculus, and so it's no surprise that it is one of the Learning Objectives for the exam. The full statement of the LO is as follows:

LO 2.3B: Solve problems involving the slope of a tangent line.

The first item of Essential Knowledge is one that we already touched on in **EK 2.1A2**.

EK 2.3B1: The derivative at a point is the slope of the tangent to a graph at that point on the graph.

As they say, a picture is worth a thousand words, so here's a picture.

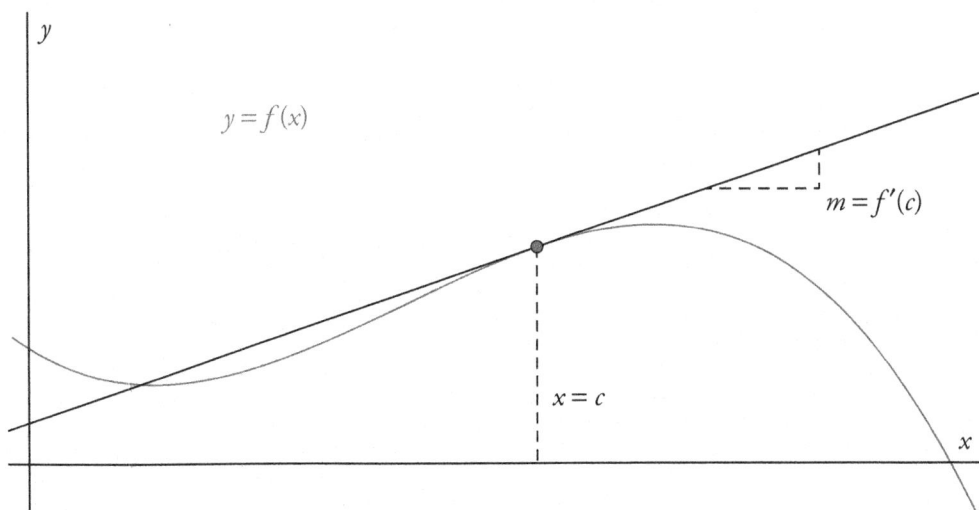

In the figure, m is the slope of the tangent line to the graph of $y = f(x)$, and it is equal to $f'(c)$.

Example 2.3.5: Find the equation of the tangent line to the graph of $f(x) = x^3 - 2x - 1$ at $x = 2$.

First note that the y-coordinate of the point of tangency is $f(2) = 3$. Now use the derivative to find the slope m of the tangent line. $f'(x) = 3x^2 - 2$, and so $m = f'(2) = 10$. The equation of the line in point-slope form is $y = 10(x - 2) + 3$, and in slope-intercept form it is $y = 10x - 17$.

In the next EK, we review an important application of tangent lines: the tangent line approximation.

EK 2.3B2: The tangent line is the graph of a locally linear approximation of the function near the point of tangency.

As you saw in Example 2.3.5, the point-slope form of the equation of the tangent line to the graph of a differentiable function $y = f(x)$ at $x = c$ is $y = f'(c)(x - c) + f(c)$. The equation of this line defines a function that is *exactly* equal to $f(x)$ *at* $x = c$, and that is *approximately* equal to $f(x)$ *near* $x = c$. We will refer to this function as $T(x)$, where the T stands for "tangent". We will illustrate this in the following figure:

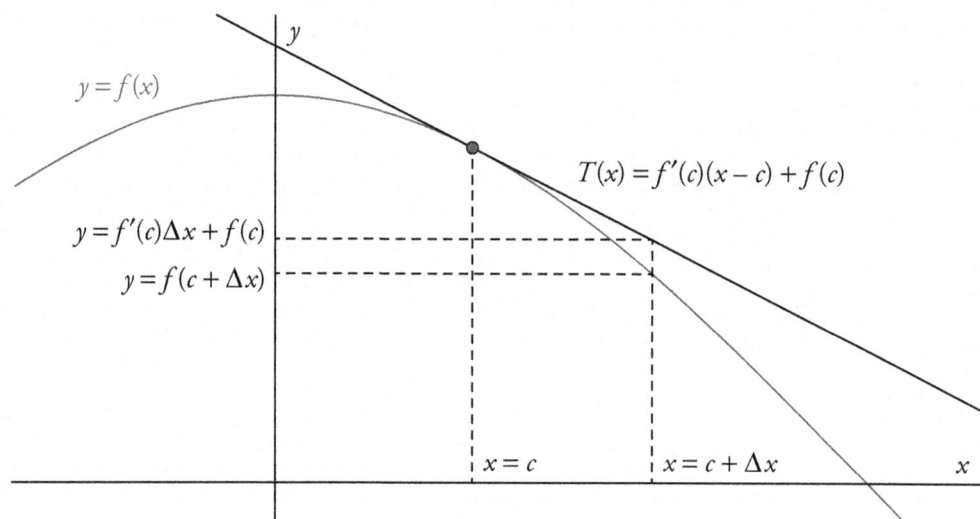

At a point $x = c + \Delta x$, the absolute value of the difference between $f(x)$ and $T(x)$ is $|T(c + \Delta x) - f(c + \Delta x)| = |f'(c)\Delta x + f(c) - f(c + \Delta x)|$. If we regard $T(x)$ as an approximation to $f(x)$, then this difference is the error in that approximation. The error goes to zero as $\Delta x \to 0$, and it is small for small values of Δx.

Example 2.3.6: Find the tangent line approximation $T(x)$ for $f(x) = \ln(x)$ at $x = 1$, and use the approximation to complete the following table:

x:	1	1.001	1.01	1.1
$f(x)$:	0	0.00099950	0.00995033	0.09531018
$T(x)$:				
$\|T(x) - f(x)\|$:				

First note that $f(1) = 0$, and $f'(x) = \dfrac{1}{x}$, so $f'(1) = 1$. We need these to find the tangent line approximation $T(x)$.

$$T(x) = f'(1)(x-1) + f(1)$$

$$T(x) = x - 1$$

Now we have what we need to fill in the table.

x:	1	1.001	1.01	1.1
$f(x)$:	0	0.00099950	0.00995033	0.09531018
$T(x)$:	0	0.001	0.01	0.1
$\|T(x) - f(x)\|$:	0	0.00000050	0.00004967	0.00468982

Notice how the error $\left| T(x) - f(x) \right|$ gets larger as x gets larger.

LO 2.3C – Motion, Related Rates, and Optimization

Here, we will continue with real world applications of derivatives, as the LO states.

LO 2.3C: Solve problems involving related rates, optimization, rectilinear motion, (BC) and planar motion.

We will begin by reviewing motion along a straight line.

EK 2.3C1: The derivative can be used to solve rectilinear motion problems involving position, speed, velocity, and acceleration.

If the position x of a particle (as measured relative to the origin of some coordinate system) is given as a function of time by $x(t)$, then the particle's speed, velocity, and acceleration are given by the following functions of time:

Velocity:	$v(t) = x'(t)$
Speed:	$\|v(t)\| = \|x'(t)\|$
Acceleration:	$a(t) = v'(t) = x''(t)$

Example 2.3.7: The position (in m) of a particle at time $t \geq 0\,s$ is given by $x(t) = t^2 e^{-t}$. Find the position, velocity, speed, and acceleration of the particle at $t = 3$. Round your answers to the nearest thousandth.

Position:

$$x(2) = 3^2 \cdot e^{-3} \approx 0.448\ m$$

Velocity:

$$v(t) = x'(t) = \frac{d}{dt}\left(t^2 e^{-t}\right) = 2te^{-t} - t^2 e^{-t} = \left(2t - t^2\right)e^{-t}$$

$$v(3) = \left(2(3) - 3^2\right)e^{-3} = -3e^{-3} \approx -0.149\ \frac{m}{s}$$

Speed:

$$\left|v(3)\right| \approx \left|-0.149\ \frac{m}{s}\right| = 0.149\ \frac{m}{s}$$

Acceleration:

$$a(t) = v'(t) = \frac{d}{dt}\left(\left(2t - t^2\right)e^{-t}\right) = \left(2 - 2t\right)e^{-t} - \left(2t - t^2\right)e^{-t} = \left(2 - 4t + t^2\right)e^{-t}$$

$$a(3) = \left(2 - 4(3) + 3^2\right)e^{-3} = -e^{-3} \approx -0.050\ \frac{m}{s^2}$$

Next, we will turn our attention to problems involving related rates.

EK 2.3C2: The derivative can be used to solve related rates problems, that is, finding a rate at which one quantity is changing by relating it to other quantities whose rates of change are known.

Related rates problems are practical applications of Implicit Differentiation. The independent variable is almost always time t, and the quantities involved in the problem are implicit functions of t. To solve a related rates problem, you need to know the value of each quantity in the problem at a particular instant, the rates of change of all but one of the quantities with respect to t, and an equation that relates all of the quantities together. We'll illustrate with an example.

Example 2.3.8: A conical tank of radius $R = 5\ ft$ and height $H = 12\ ft$ is being filled with water at the rate of $10\ \frac{ft^3}{min}$. At what rate is the surface of the water rising when the water is $8\ ft$ high, as in the following figure? Round your answer to the nearest thousandth.

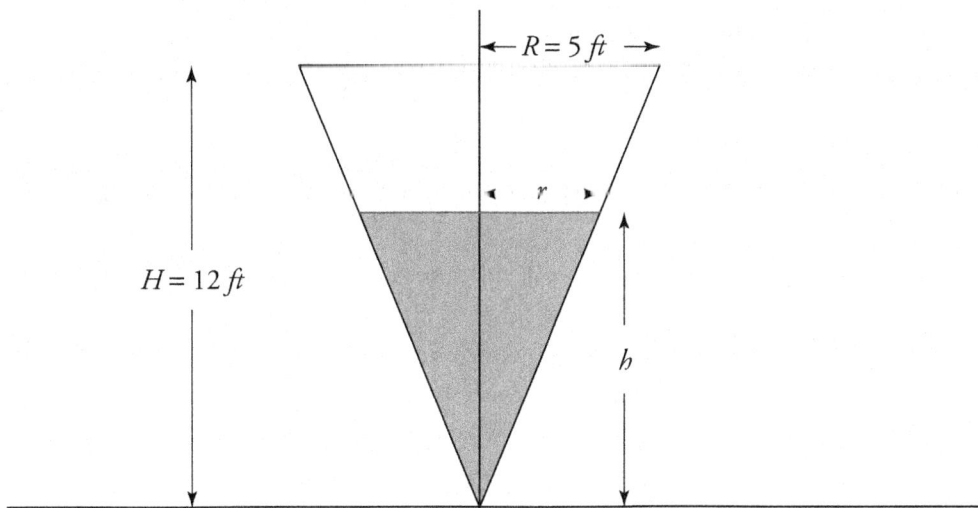

Taking our cue from the figure, we'll let the height and radius of the water be h and r, respectively. These quantities are related to the volume V of the tank by the formula for the volume of a cone, which is $V = \frac{1}{3}\pi r^2 h$. We were given that $\frac{dV}{dt} = 10 \frac{ft^3}{min}$, and we were asked to find $\frac{dh}{dt}$ when $h = 8\,ft$. We were not given any information about r, but it turns out that we don't need it. We can use similar triangles to find r in terms of h.

$$\frac{12}{5} = \frac{h}{r} \Rightarrow r = \frac{5}{12}h$$

Plugging this into the volume formula, we obtain the following:

$$V = \frac{1}{3}\pi\left(\frac{5}{12}h\right)^2 h$$

$$V = \frac{25}{432}\pi h^3$$

Now differentiate both sides with respect to t, plug in the given data, and then solve for $\frac{dh}{dt}$.

$$\frac{dV}{dt} = \frac{25}{144}\pi h^2 \frac{dh}{dt}$$

$$10 = \frac{25}{144}\pi (8)^2 \frac{dh}{dt}$$

$$\frac{dh}{dt} = \frac{1440}{1600\pi} \approx 0.286 \frac{ft}{min}$$

As you saw throughout **EU 2.2**, the derivative is particularly useful for finding maxima and minima of functions. This can be exploited to solve problems of interest in the real world, as we review in the next EK.

EK 2.3C3: The derivative can be used to solve optimization problems, that is, finding a maximum or minimum value of a function over a given interval.

In any optimization problem, you will either be given or required to derive a **primary equation**, which relates the quantity to be maximized or minimized to other variables. Since you do not yet know how to deal with functions of several variables, you also be given one or more **constraint equations**, with which you can reduce the quantity to be optimized to a function of a *single* variable. We will illustrate with an example.

Example 2.3.9: A poster is to be printed on a rectangular sheet that has an area of $18 \, ft^2$. The upper and lower margins are to have a width of $9 \, in$, and the right and left margins are to have a width of $6 \, in$. The rectangular printed area is to be centered on the sheet. Find the dimensions of the sheet that will maximize the printed area.

Let the dimensions of the sheet be x and y. Clearly, we must have $x, y > 0$. Since the area of the sheet was given in ft^2, we should convert the margin widths to feet: $9 \, in = \dfrac{3}{4} \, ft$ and $6 \, in = \dfrac{1}{2} \, ft$. The following figure is a sketch of the poster with the dimensions and margins labeled, and the printed area shaded.

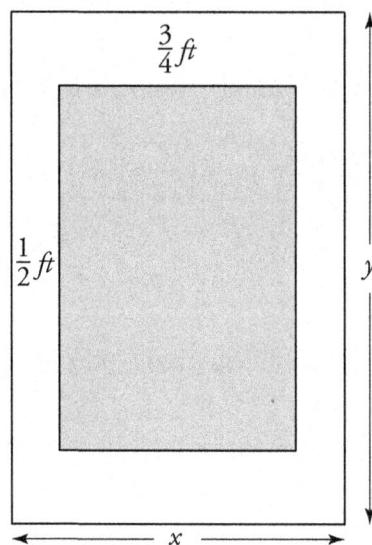

Let the printed area be A. This is the quantity to be maximized. The base of the printed area has a length of $x - 2 \cdot \dfrac{1}{2} = x - 1$, and the height of the printed area has a length of $y - 2 \cdot \dfrac{3}{4} = y - \dfrac{3}{2}$. Since the printed area is a rectangle, we'll use the formula for the area of a rectangle to get our primary equation.

$$A = (x-1)\left(y - \frac{3}{2}\right) \quad \textbf{primary equation}$$

This gives A as a function of two variables, so we need a constraint to reduce this to a function of a single variable. The constraint will come from the fact that the total area of the sheet is 18 ft^3.

$$18 = xy \quad \textbf{constraint equation}$$

Solving the constraint equation for y yields $y = \frac{18}{x}$. We'll plug this into the primary equation to get A as a function of x alone.

$$A = (x-1)\left(\frac{18}{x} - \frac{3}{2}\right)$$

$$A = \frac{39}{2} - \frac{3}{2}x - \frac{18}{x}$$

Now we find the critical numbers of A by setting $\frac{dA}{dx} = 0$.

$$\frac{dA}{dx} = -\frac{3}{2} + \frac{18}{x^2} = 0 \Rightarrow x = \pm 2\sqrt{3}$$

We will reject the negative critical number, since $x > 0$. So the only critical number of A is $x = 2\sqrt{3}$. We will verify that this corresponds to a maximum (as opposed to a minimum) by applying the Second Derivative Test.

$$\frac{d^2A}{dx^2} = -\frac{36}{x^3}$$

$$\left.\frac{d^2A}{dx^2}\right|_{x=2\sqrt{3}} = -\frac{36}{\left(2\sqrt{3}\right)^3} = -\frac{\sqrt{3}}{2} < 0$$

According to the Second Derivative Test, A has a relative maximum at $x = 2\sqrt{3}$. Since this is the only critical number, the graph of A vs x has no other turning points, and so A has an *absolute* maximum at $x = 2\sqrt{3}$.

So, the dimensions of the poster that maximize the printed area are $x = 2\sqrt{3}$ *ft* and $y = \frac{18}{2\sqrt{3}} = 3\sqrt{3}$ *ft*.

In the last EK of this LO, we will return to motion. Only this time, we will not restrict ourselves to motion along a straight line.

EK 2.3C4: (BC) Derivatives can be used to determine velocity, speed, and acceleration for a particle moving along curves given by parametric or vector-valued functions.

Here, we'll consider the motion of a particle in the plane. The x- and y-coordinates of the particle are given as functions of the parameter t (time): $x(t), y(t)$. The position of the particle is given as a vector-valued function of which these parametric functions are the components: $\vec{r}(t) = (x(t), y(t))$. The velocity, speed, and acceleration of the particle are given as follows:

Velocity:	$\vec{v}(t) = \vec{r}'(t) = \left(x'(t), y'(t)\right)$
Speed:	$\left|\vec{v}(t)\right| = \left|\vec{r}'(t)\right| = \sqrt{\left(x'(t)\right)^2 + \left(y'(t)\right)^2}$
Acceleration:	$\vec{a}(t) = \vec{v}'(t) = \vec{r}''(t) = \left(x''(t), y''(t)\right)$

Example 2.3.10: The position (in m) of a particle in the plane for $t \geq 0\,s$ is given by the vector-valued function $\vec{r}(t) = (-\sin(2t), \cos(2t))$. Find the position, velocity, speed, and acceleration of the particle at $t = \dfrac{\pi}{12}$. Round your answers to the nearest thousandth.

Position:

$$\vec{r}\left(\frac{\pi}{12}\right) = \left(-\sin\left(\frac{\pi}{6}\right), \cos\left(\frac{\pi}{6}\right)\right) = \left(-\frac{1}{2}, \frac{\sqrt{3}}{2}\right) \approx (-0.5, 0.866)\,m$$

Velocity:

$$\vec{v}(t) = \vec{r}'(t) = \left(-2\cos(2t), -2\sin(2t)\right)$$

$$\vec{v}\left(\frac{\pi}{12}\right) = \vec{r}'\left(\frac{\pi}{12}\right) = \left(-2\cos\left(\frac{\pi}{6}\right), -2\sin\left(\frac{\pi}{6}\right)\right) = \left(-\sqrt{3}, -1\right) \approx (-1.732, -1)\frac{m}{s}$$

Speed:

$$\left|\vec{v}(t)\right| = \sqrt{\left(-\sqrt{3}\right)^2 + (-1)^2} = 2\frac{m}{s}$$

Acceleration:

$$\vec{a}(t) = \vec{v}'(t) = \left(4\sin(2t), -4\cos(2t)\right)$$

$$\vec{a}\left(\frac{\pi}{12}\right) = \vec{v}'\left(\frac{\pi}{12}\right) = \left(4\sin\left(\frac{\pi}{6}\right), -4\cos\left(\frac{\pi}{6}\right)\right) = \left(2, -2\sqrt{3}\right) \approx (2, -3.464)\frac{m}{s^2}$$

LO 2.3D – Rates of Change in Applied Contexts

This LO focuses on applications of the derivative as a *rate of change*.

LO 2.3D: Solve problems involving rates of change in applied contexts.

This LO has just one item of Essential Knowledge.

EK 2.3D1: The derivative can be used to express information about rates of change in applied contexts.

You've already seen this in the context of motion. Velocity is the rate of change of position with respect to time, and acceleration is the rate of change of velocity with respect to time. Here, we'll do an example from a different context altogether.

Example 2.3.11: A manufacturer has an annual inventory cost of C (in dollars), which depends on q, the order size (in units) when the inventory is replenished. The inventory cost is given by the function $C(q) = 1.030000q^{-1} + 7.1q$. Find the rate of change of inventory cost with respect to order size when the order size is 410. Round your answer to the nearest cent per unit.

First, compute the derivative of C with respect to q.

$$C'(q) = -1{,}030{,}000q^{-2} + 7.1$$

Now compute $C'(410)$.

$$C'(410) = -1{,}030{,}000(410)^{-2} + 7.1 = \$0.97 \, / \, unit$$

LO 2.3E – Verifying Solutions to Differential Equations

As the section title states, we will be reviewing how to verify solutions to differential equations here. Note that we will not have to *solve* any differential equations in this chapter. That is reserved for Chapter 3 on Integration. The LO for this section is stated below.

LO 2.3E: Verify solutions to differential equations.

We'll start off by reviewing the most obvious difference between algebraic equations and differential equations. Solutions to algebraic equations are *numbers*, whereas solutions to differential equations are *functions*, or families thereof.

EK 2.3E1: Solutions to differential equations are functions or families of functions.

Consider the differential equation $\dfrac{dy}{dx} = 2x$. You can easily verify that the function $y = x^2$ is a solution to this equation. However, this isn't the only solution. $y = x^2 + 1$ also works. So do $y = x^2 + 2$, $y = x^2 - 8$, and $y = x^2 + \dfrac{\pi^2}{3e^5}$. In fact, for *any* real number C, the function $y = x^2 + C$ is a solution to the differential equation. The solution set is, thus, a *family of functions*. We will sketch a few representatives of this family in the following figure:

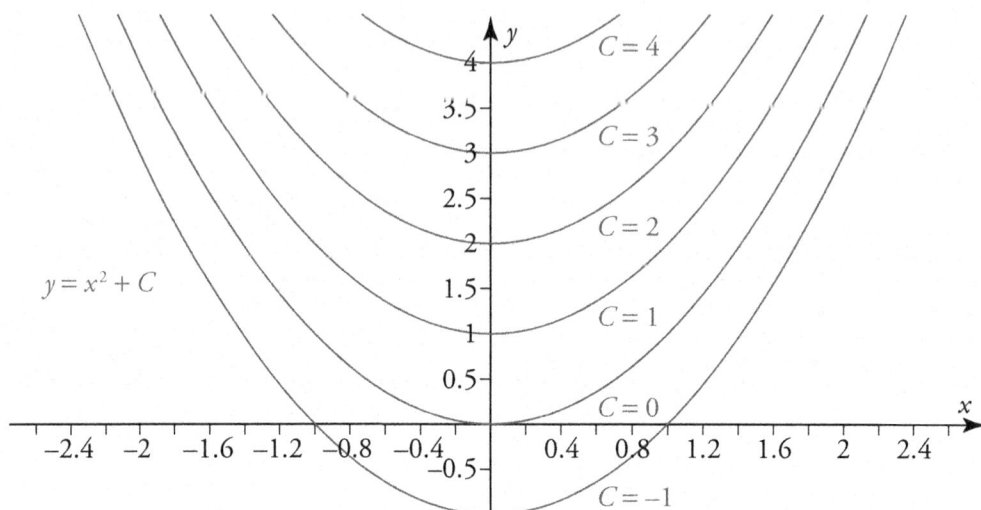

A family of solutions such as this is referred to as a *general solution* to a differential equation. Sometimes a differential equation comes with an *initial condition*, in which case the solution is just one function, which we will call the *particular solution* to the differential equation.

Continuing with our example, suppose now that we wish to know the solution of the differential equation $\frac{dy}{dx} = 2x$, subject to the initial condition $y = 3$ when $x = 1$. We know that the *family* of functions $y = x^2 + C$ satisfies the *equation*, but only *one* member of this family satisfies *both* the equation *and* the initial condition. To find out which member, we'll plug the initial condition into the general solution. That is, we'll plug $y = 3$ and $x = 1$ into $y = x^2 + C$, and solve for C. When we do this, we get: $3 = 1^2 + C \Rightarrow C = 2$. Thus, the particular solution is $y = x^2 + 2$. We'll illustrate this in the following figure:

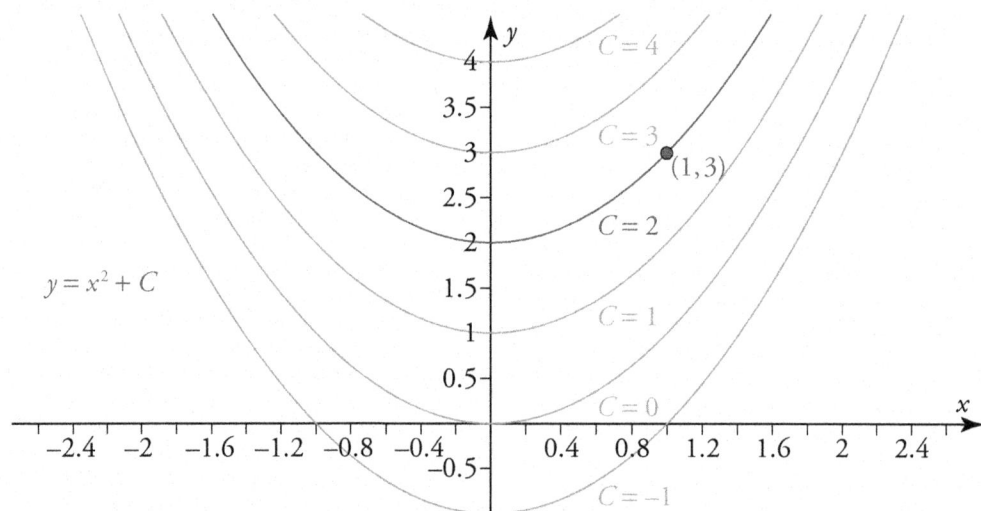

In the next item of Essential Knowledge, we will get to what this Learning Objective is really all about.

EK 2.3E2: Derivatives can be used to verify that a function is a solution to a given differential equation.

Even though we have not yet reviewed how to *solve* differential equations, you can at least *verify* that a given function is a solution of a given differential equation by plugging the function into the equation. This verification will necessarily involve taking derivatives, as we will demonstrate in the next example.

Example 2.3.11: Verify that the family of functions $y = C_1 e^{-2x} + C_2 e^{-3x}$ satisfies the differential equation $\dfrac{d^2 y}{dx^2} + 5\dfrac{dy}{dx} + 6y = 0$.

We'll begin by taking the first and second derivatives of y.

$$\frac{dy}{dx} = -2C_1 e^{-2x} - 3C_2 e^{-3x}$$

$$\frac{d^2 y}{dx^2} = 4C_1 e^{-2x} + 9C_2 e^{-3x}$$

Plugging these into the differential equation yields the following:

$$4C_1 e^{-2x} + 9C_2 e^{-3x} + 5\left(-2C_1 e^{-2x} - 3C_2 e^{-3x}\right) + 6\left(C_1 e^{-2x} + C_2 e^{-3x}\right) = 0$$

$$\left(4 - 10 + 6\right)C_1 e^{-2x} + \left(9 - 15 + 6\right)C_2 e^{-3x} = 0$$

$$0 = 0$$

Thus, $y = C_1 e^{-2x} + C_2 e^{-3x}$ satisfies the equation.

LO 2.3F – Estimating Solutions to Differential Equations

We're still not ready to review how to solve differential equations, but we're about to take one step closer to it with this Learning Objective.

LO 2.3F: Estimate solutions to differential equations.

In this LO, we'll review two approaches to estimating differential equations: one qualitative, and one quantitative. We will begin with the qualitative approach.

EK 2.3F1: Slope fields provide visual clues to the behavior of solutions to first order differential equations.

During the exam, you may be asked to deal with slope fields for differential equations of the form $\dfrac{dy}{dx} = F\left(x, y\right)$. A slope field is a field of line segments in the xy-plane such that at each point (x, y), the slope of the line segment at that point is equal to the slope of the solution curve that passes through that point. These slopes can be calculated from the differential equation when the function $F(x,y)$ is known, because the slope of the solution curve is given by $\dfrac{dy}{dx}$. We'll turn now to an example.

Example 2.3.12: The slope field for the differential equation $\dfrac{dy}{dx} = xy$ is shown below.

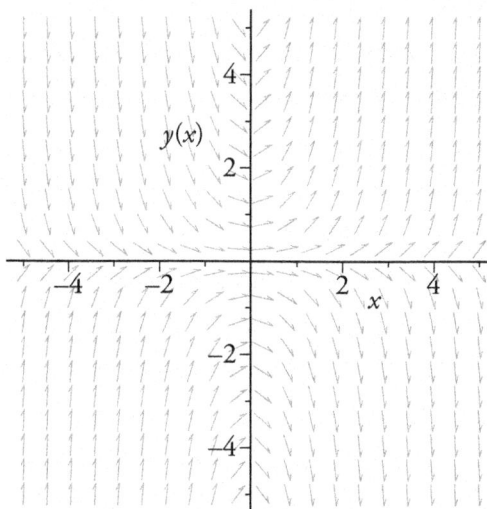

Even though you don't know the general solution to the equation, you can infer from this plot that the solutions are concave upward when $y > 0$, and concave downward when $y < 0$. If we plot the solution curve that passes through the point $(0, 2)$, you can see this more clearly.

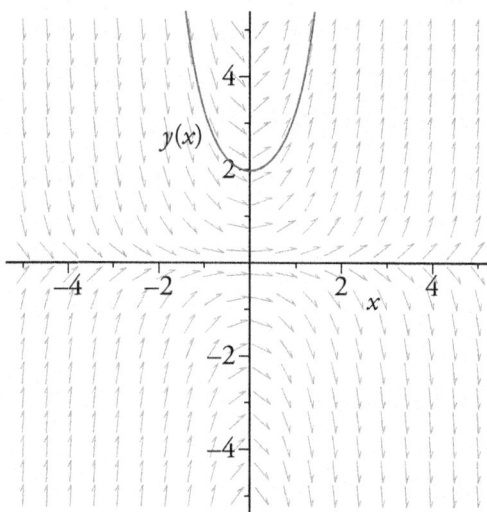

And now, we'll turn our attention to estimating solutions to differential equations quantitatively.

EK 2.3F2: (BC) For differential equations, Euler's method provides a procedure for approximating a solution or a point on a solution curve.

Just as in **EK 2.3F1**, we restrict our attention to differential equations of the form $\frac{dy}{dx} = F(x, y)$. Euler's method approximates a particular solution by the following procedure:

Given a *step size h* and a point (x_0, y_0) through which the particular solution passes, generate a sequence of points (x_1, y_1), (x_2, y_2), ..., (x_n, y_n), ... as follows:

$$x_n = x_{n-1} + h \qquad\qquad y_n = y_{n-1} + hF\left(x_{n-1}, y_{n-1}\right)$$

This method approximates a particular solution curve to the differential equation by a *sequence of line segments*, as illustrated below.

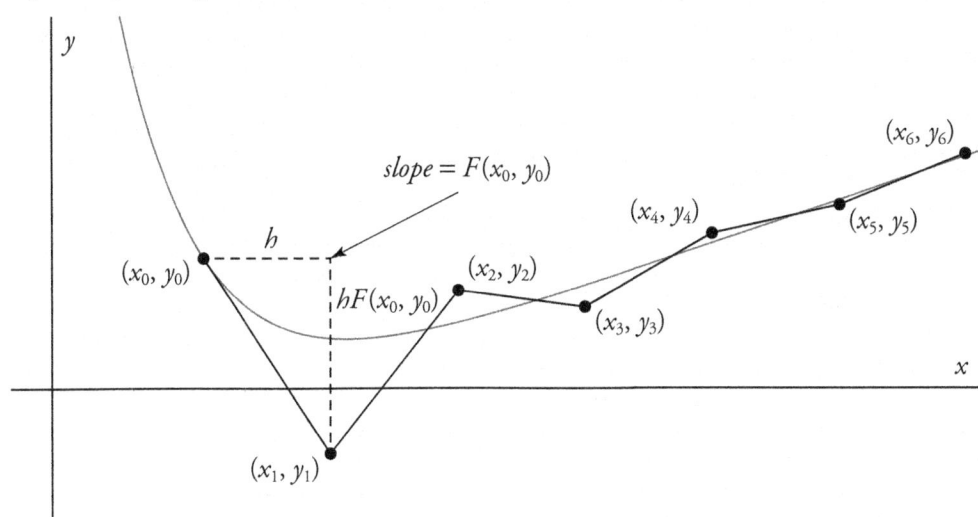

In the figure, the smooth curve is the exact solution curve, and the sequence of line segments is the Euler approximation. The slope of the nth line segment is $F(x_{n-1}, y_{n-1})$. The slope of the first line segment has been labeled. The smaller the step size h, the better the approximation. We will now illustrate with an example.

Example 2.3.13: Consider the particular solution to the differential equation $\frac{dy}{dx} = 4x - 5y$ that passes through the point $(2, 1)$. Using a step size of 0.1, find the Euler approximation of this solution at $x = 3$. Round your answer to the nearest thousandth.

We were given $(x_0, y_0) = (2, 1)$ and $h = 0.1$. It will take 10 steps of size 0.1 to go from $x = 2$ to $x = 3$. We present our results in the following table:

n	x_n	y_n
0	2	1
1	2.1	1.3
2	2.2	1.49
3	2.3	1.625
4	2.4	1.7325
5	2.5	1.82625
6	2.6	1.913125
7	2.7	1.9965625
8	2.8	2.07828125
9	2.9	2.159140625
10	3	2.239570313

Rounded to the nearest thousandth, the solution at $x_{10} = 3$ is approximately $y_{10} \approx 2.240$.

EU 2.4 – The Mean Value Theorem

LO 2.4A – Applying the Mean Value Theorem

The final Learning Objective pertains to a Theorem that we mentioned briefly in **EK 1.2B1**: The Mean Value Theorem.

LO 2.4A: Apply the Mean Value Theorem to describe the behavior of a function over an interval.

There is only one item of Essential Knowledge for this LO.

EK 2.4A1: If a function f is continuous over the interval $[a, b]$ and differentiable over the interval (a, b), the Mean Value Theorem guarantees a point within that open interval where the instantaneous rate of change equals the average rate of change over the interval.

In the following figure, the slope of the secant line that passes through the points $(a, f(a))$ and $(b, f(b))$ is equal to the average (a.k.a. *mean*) rate of change of f over $[a, b]$, which is given by $\dfrac{f(b) - f(a)}{b - a}$. The slope of the tangent line to the graph of f at the point $(c, f(c))$ is equal to the instantaneous rate of change of f at $x = c$, which is given by $f'(c)$.

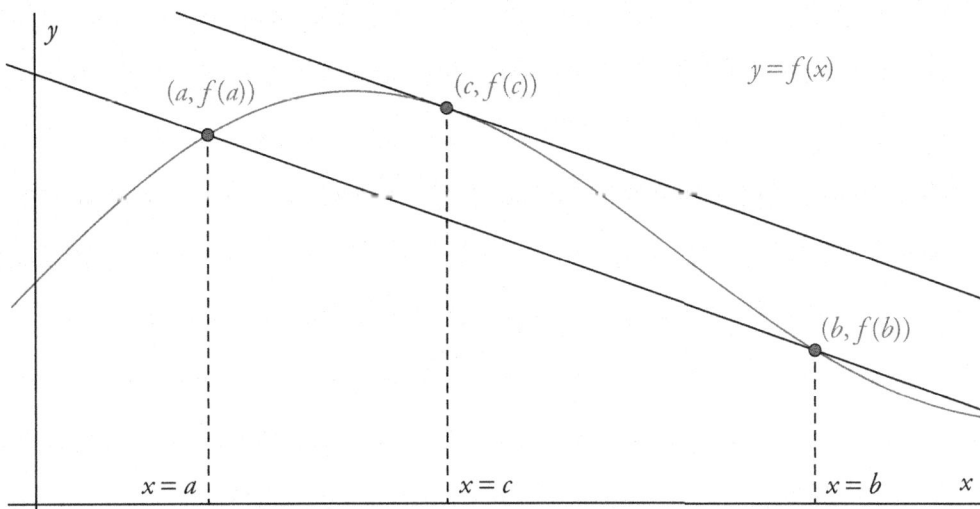

The Mean Value Theorem guarantees that if its hypotheses are satisfied, then there must exist a point c in (a, b) such that the secant line is parallel to the tangent line at $x = c$. Since parallel lines have equal slopes, we can express the conclusion of the Mean Value Theorem as follows:

$$f'(c) = \frac{f(b) - f(a)}{b - a}$$

Example 2.4.1: Let $f(x) = x^3 - 2x$ on the interval $[-1, 1]$. Find all values of c in the interval $(-1, 1)$ such that the instantaneous rate of change of f at $x = c$ equals the average rate of change of f over $[-1, 1]$.

Since $f'(x) = 3x^2 - 2$, we equate $f'(c)$ to the average rate of change of f, and then solve for c.

$$f'(c) = \frac{f(1) - f(-1)}{1 - (-1)}$$

$$3c^2 - 2 = -1$$

$$c = \pm \frac{1}{\sqrt{3}}$$

Both of these values of c are in the interval $(-1, 1)$.

Chapter 3: Integral Calculus

Chapter 3 Overview

Chapter 3 is divided into the following five Enduring Understandings:

- **EU 3.1:** Antidifferentiation is the inverse process of differentiation.
- **EU 3.2:** The definite integral of a function over an interval is the limit of a Riemann sum over that interval and can be calculated using a variety of strategies.
- **EU 3.3:** The Fundamental Theorem of Calculus, which has two distinct formulations, connects differentiation and integration.
- **EU 3.4:** The definite integral of a function over an interval is a mathematical tool with many interpretations and applications involving accumulation.
- **EU 3.5:** Antidifferentiation is an underlying concept involved in solving separable differential equations. Solving separable differential equations involves determining a function or relation given its rate of change.

EU 3.1 – Antidifferentiation

LO 3.1A – Antiderivatives

In Chapter 2, you had a thorough review of differentiation and its applications. In this section, we will review the *reverse* operation: antidifferentiation. The full Learning Objective of this section is as follows:

LO 3.1A: Recognize antiderivatives of basic functions.

The first item of Essential Knowledge under LO 3.1A is the very definition of an antiderivative.

EK 3.1A1: An antiderivative of a function *f* is a function *g* whose derivative is *f*.

What this means is that antidifferentiation and differentiation are inverses of each other. *Almost*. To see why the two operations are almost inverses and not exactly inverses, consider the following:

Example 3.1.1: x^4 is an antiderivative of $4x^3$, because $\frac{d}{dx}\left(x^4\right) = 4x^3$.

Notice that we say "*an* antiderivative" and not "*the* antiderivative". That is because antiderivatives are *not unique*. Consider the following:

Example 3.1.2: $x^4 + 1$ is an antiderivative of $4x^3$, because $\dfrac{d}{dx}(x^4+1)=4x^3$.

See? $4x^3$ has more than one antiderivative. Here's another one.

Example 3.1.3: $x^4 - 7$ is an antiderivative of $4x^3$, because $\dfrac{d}{dx}\left(x^4-7\right)=4x^3$.

We could go on forever generating examples like this, but rather than do that we will simply point out that, for any real number C, *anything* of the form $x^4 + C$ is an antiderivative of $4x^3$, because $\dfrac{d}{dx}\left(x^4+C\right)=4x^3$.

Next, we will review the standard notation for antiderivatives. Suppose that a function g is an antiderivative of a function f, i.e.: $g'(x)=f(x)$. We can write this as follows:

$$\frac{dg(x)}{dx}=f(x)$$

Alternatively, this can be written in differential form as follows:

$$dg(x)=f(x)dx$$

Next, we will antidifferentiate both sides. The symbol for this operation is the antidifferentiation sign: $\int$.

$$\int dg(x)=\int f(x)dx$$

Since antidifferentiation and differentiation are (almost!) inverses of each other, the antiderivative of the differential of $g(x)$ yields $g(x)$ itself.

$$g(x)=\int f(x)dx$$

You may be wondering, "Where does the $+ C$ come in?" The answer is that it would come in after computing the right-hand side of the equation, which we would only be able to do if we knew the function $f(x)$.

To see this notation in action, we will now return to our previous examples: antiderivatives of $4x^3$.

Example 3.1.4: $\int 4x^3 dx=x^4+C$

The second item of Essential Knowledge under **LO 3.1A** pertains to the operational rules by which antiderivatives are computed.

At the beginning of this section, we said that antidifferentiation is the reverse of differentiation. What this means is that every differentiation rule can be "turned around" to yield an antidifferentiation rule. This idea is expressed as an item of Essential Knowledge as follows:

EK 3.1A2: Differentiation rules provide the foundation for finding antiderivatives.

This means that, with every derivative, you get a free antiderivative. We will explore this by revisiting the rules for differentiation.

Rules for Scalar Multiples and Sums and Differences of Functions

First, let us recall the Scalar Multiple Rule and the Sum and Difference Rule for Derivatives. If k is a real number, and f and g are differentiable functions, then we have the following:

$$\frac{d}{dx}\big(kf(x)\big) = k\frac{d}{dx}\big(f(x)\big) = kf'(x)$$

$$\frac{d}{dx}\big(f(x) \pm g(x)\big) = \frac{d}{dx}\big(f(x)\big) \pm \frac{d}{dx}\big(g(x)\big) = f'(x) \pm g'(x)$$

We can obtain antiderivative rules from these. If F and G are antiderivatives for f and g, respectively, then we can write the following:

$$\frac{d}{dx}\big(kF(x)\big) = k\frac{d}{dx}\big(F(x)\big) = kF'(x) = kf(x)$$

$$\frac{d}{dx}\big(F(x) \pm G(x)\big) = \frac{d}{dx}\big(F(x)\big) \pm \frac{d}{dx}\big(G(x)\big) = F'(x) \pm G'(x) = f(x) \pm g(x)$$

We can now use the definition of antiderivatives to turn these rules around as follows:

$$\frac{d}{dx}\big(kF(x)\big) = kf(x) \Rightarrow \int kf(x)\,dx = k\int f(x)\,dx$$

$$\frac{d}{dx}\big(F(x) \pm G(x)\big) = f(x) \pm g(x) \Rightarrow \int \big(f(x) \pm g(x)\big)\,dx = \int f(x)\,dx \pm \int g(x)\,dx$$

Thus, we have obtained the Scalar Multiple Rule and the Sum and Difference Rule for Antiderivatives from their corresponding rules for antiderivatives.

Rules for Constants

Next, recall for any real number C, the derivative of $f(x) = C$ is zero. This makes sense, because the derivative is interpreted as a *rate of change*, and since a constant function does not change at all, its rate of change is zero. We can obtain an antidifferentiation rule from this as follows:

$$\frac{d}{dx}(C) = 0 \Rightarrow \int 0\,dx = C$$

Next, recall that for any nonzero real number k, the derivative of $f(x) = kx$ is equal to k. This makes sense, because the derivative is interpreted as a *slope*, and in this case f is a linear function with slope k. We can obtain an antidifferentiation rule from this as follows:

$$\frac{d}{dx}(kx) = k \Rightarrow \int k\,dx = kx + C$$

Combining the last two rules, we can find the antiderivative of any constant function.

Power Rule

Next, recall the Power Rule for Derivatives. Let n be a real number. Then the Power Rule is as given below.

$$\frac{d}{dx}\left(x^n\right) = nx^{n-1}$$

Let us use our knowledge of derivatives to compute $\frac{1}{n+1}x^{n+1}$, $n \neq 1$.

First, we use the Scalar Multiple Rule to bring the constant factor $\frac{1}{n+1}$ outside of the derivative operator.

$$\frac{d}{dx}\left(\frac{1}{n+1}x^{n+1}\right) = \frac{1}{n+1}\left(\frac{d}{dx}\left(x^{n+1}\right)\right)$$

Second, we apply the Power Rule for Derivatives and simplify.

$$\frac{d}{dx}\left(\frac{1}{n+1}x^{n+1}\right) = \frac{1}{n+1} \cdot (n+1)x^{n+1-1} = x^n$$

Using the definition of the antiderivative, we can turn this around to obtain the Power Rule for Antiderivatives as follows:

$$\frac{d}{dx}\left(\frac{1}{n+1}x^{n+1}\right) = x^n \Rightarrow \int x^n\,dx = \frac{1}{n+1}x^{n+1} + C, \ (n \neq -1)$$

Let us look at some examples of computing antiderivatives using some or all of the above three rules. First, we will look at a straightforward example using only the Power Rule for Antiderivatives.

Example 3.1 5: $\int x^8\,dx = \frac{1}{9}x^9 + C$

In the next example, we will use the Scalar Multiple Rule for Antiderivatives to bring it outside of the antidifferentiation sign, and then apply the Power Rule.

Example 3.1.6: $\int 8x^2\,dx = 8\int x^2\,dx = 8 \cdot \frac{1}{3}x^3 + C = \frac{8}{3}x^3 + C$

In the next example, we antidifferentiate a function with four terms. We will use the Sum and Difference Rules, and then we will use the Scalar Multiple Rule and the Power Rule to finish it off.

Example 3.1.7: $\int \left(3 - 8x^3 + 5x^4 - x^5\right)dx = \int 3\,dx - \int 8x^3\,dx + \int 5x^4\,dx - \int x^5\,dx$

$$\int\left(3-8x^{3}+5x^{4}-x^{5}\right)dx = \int 3\,dx - 8\int x^{3}\,dx + 5\int x^{4}\,dx - \int x^{5}\,dx$$

$$\int\left(3-8x^{3}+5x^{4}-x^{5}\right)dx = 3x - 8\cdot\frac{1}{4}x^{4} + 5\cdot\frac{1}{5}x^{5} - \frac{1}{6}x^{6} + C$$

$$\int\left(3-8x^{3}+5x^{4}-x^{5}\right)dx = 3x - 2x^{4} + x^{5} - \frac{1}{6}x^{6} + C$$

With practice, you will be able to quickly compute these antiderivatives without devoting a separate step to applying each rule.

Sometimes a function must be rewritten using the rules of exponents before it can be antidifferentiated. Consider the following examples:

Example 3.1.8: $\int\dfrac{1}{t^{3}}\,dt = \int t^{-3}\,dt = \dfrac{1}{-3+1}t^{-3+1} + C = -\dfrac{1}{2}t^{-2} + C$

We started by rewriting $\frac{1}{t^{3}}$ using a negative exponent, and then we applied the Power Rule. Typically, if a function is written without negative exponents, then its antiderivative will also be written without negative exponents. So we finish this off as follows:

$$\int\frac{1}{t^{3}}\,dt = -\frac{1}{2t^{2}} + C$$

Example 3.1.9: $\int 14\sqrt[4]{y^{3}}\,dy = 14\int y^{\frac{3}{4}}\,dy = 14\cdot\dfrac{1}{\frac{3}{4}+1}y^{\frac{3}{4}+1} + C = 14\cdot\dfrac{4}{7}y^{\frac{7}{4}} + C = 8y^{\frac{7}{4}} + C$

We started by bringing the scalar factor outside the antidifferentiation sign and rewriting the radical using a rational exponent, and then we applied the Power Rule and simplified. Typically, if a function is written without rational exponents, then the final form of the antiderivative will also be written without rational exponents. So we conclude as follows:

$$\int 14\sqrt[4]{y^{3}}\,dy = 8\sqrt[4]{y^{7}} + C \text{ or } 8y\sqrt[4]{y^{3}} + C$$

Next, we will review the antidifferentiation rules for the elementary functions of precalculus, namely the trigonometric, inverse trigonometric, logarithmic, and exponential functions.

Rules for Trigonometric Functions

We can obtain antidifferentiation rules from the differentiation rules of trigonometric functions as follows:

$$\frac{d}{dx}\left(\sin(x)\right) = \cos(x) \Rightarrow \int\cos(x)\,dx = \sin(x) + C$$

$$\frac{d}{dx}\left(-\cos(x)\right) = \sin(x) \Rightarrow \int\sin(x)\,dx = -\cos(x) + C$$

$$\frac{d}{dx}\big(\sec(x)\big) = \sec(x)\tan(x) \Rightarrow \int \sec(x)\tan(x)\,dx = \sec(x) + C$$

$$\frac{d}{dx}\big(-\csc(x)\big) = \csc(x)\cot(x) \Rightarrow \int \csc(x)\cot(x)\,dx = -\csc(x) + C$$

$$\frac{d}{dx}\big(\tan(x)\big) = \sec^2(x) \Rightarrow \int \sec^2(x)\,dx = \tan(x) + C$$

$$\frac{d}{dx}\big(-\cot(x)\big) = \csc^2(x) \Rightarrow \int \csc^2(x)\,dx = -\cot(x) + C$$

Example 3.1.10: $\int \big(3\cos(x) + 2\csc(x)\cot(x)\big)dx = 3\int \cos(x)\,dx + 2\int \csc(x)\cot(x)\,dx$

$$\int \big(3\cos(x) + 2\csc(x)\cot(x)\big)dx = 3\sin(x) - 2\cot(x) + C$$

Rules for Inverse Trigonometric Functions

We can also obtain antidifferentiation rules from the differentiation rules for inverse trigonometric functions (for $a > 0$) as follows:

$$\frac{d}{dx}\left(\arcsin\left(\frac{x}{a}\right)\right) = \frac{1}{\sqrt{a^2 - x^2}} \Rightarrow \int \frac{1}{\sqrt{a^2 - x^2}}\,dx = \arcsin\left(\frac{x}{a}\right) + C$$

$$\frac{d}{dx}\left(\frac{1}{a}\arctan\left(\frac{x}{a}\right)\right) = \frac{1}{x^2 + a^2} \Rightarrow \int \frac{1}{x^2 + a^2}\,dx = \frac{1}{a}\arctan\left(\frac{x}{a}\right) + C$$

$$\frac{d}{dx}\left(\frac{1}{a}\operatorname{arcsec}\left(\frac{|x|}{a}\right)\right) = \frac{1}{x\sqrt{x^2 - a^2}} \Rightarrow \int \frac{1}{x\sqrt{x^2 - a^2}}\,dx = \frac{1}{a}\operatorname{arcsec}\left(\frac{|x|}{a}\right) + C$$

Example 3.1.11: $\int \frac{1}{\sqrt{9 - x^2}}\,dx = \int \frac{1}{\sqrt{3^2 - x^2}}\,dx = \arcsin\left(\frac{x}{3}\right) + C$

Example 3.1.12: $\int \frac{2}{x^2 + 5}\,dx = 2\int \frac{1}{x^2 + \sqrt{5}^2}\,dx = \frac{2}{\sqrt{5}}\arctan\left(\frac{x}{\sqrt{5}}\right) + C$

Rule for Logarithmic Functions

Earlier we reviewed the Power Rule for Antiderivatives, which is stated as follows:

$$\int x^n\,dx = \frac{1}{n+1}x^{n+1} + C \ \ (n \neq -1)$$

Notice that this rule is valid for any n *except* -1. So what happens when $n = -1$? Can we just not antidifferentiate x^{-1}? It turns out that we can, thanks to the differentiation rule for the natural logarithm.

$$\frac{d}{dx}\big(\ln|x|\big) = \frac{1}{x} \Rightarrow \int \frac{1}{x}\,dx = \ln|x| + C$$

Recall that $x^{-1} = \dfrac{1}{x}$, so this antidifferentiation rule really does answer the question that we posed above.

Example 3.1.13: $\displaystyle\int 4x^{-1}dx = 4\int \frac{1}{x}dx = 4\ln|x| + C$

Rules for Exponential Functions

We will conclude our review of antidifferentiation rules involving elementary functions by discussion exponential functions. Let $a > 0$, $a \neq 1$.

$$\frac{d}{dx}\left(e^x\right) = e^x \Rightarrow \int e^x dx = e^x + C$$

$$\frac{d}{dx}\left(\frac{1}{\ln(a)}a^x\right) = a^x \Rightarrow \int a^x dx = \frac{1}{\ln(a)}a^x + C$$

Example 3.1.14: $\displaystyle\int\left(5e^x - 3\cdot 2^x\right)dx = 5\int e^x dx - 3\int 2^x dx$

$$\int\left(5e^x - 3\cdot 2^x\right)dx = 5e^x - \frac{3}{\ln(2)}2^x + C$$

Example 3.1.15: $\displaystyle\int e^{x-3}dx = \int e^x e^{-3}dx = e^{-3}\int e^x dx = e^{-3}e^x + C = e^{x-3} + C$

EU 3.2 – Definite Integration

LO 3.2A(a) – Riemann Sums

The subject of this EU is definite integration, but before we get to that we need to review some prerequisites, namely partitions and Riemann sums. The full Learning Objective for this section is stated below.

LO 3.2A(a): Interpret the definite integral as the limit of a Riemann sum.

There is only one item of Essential Knowledge for this section, and it is as follows:

EK 3.2A1: A Riemann sum, which requires a partition of an interval I, is the sum of products, each of which is the value of the function at a point in a subinterval multiplied by the length of that subinterval of the partition.

We will start with partitions. A partition is just a prescription for dividing a closed interval $I = [a, b]$ into subintervals. In more formal terms, a partition Δ of a closed interval I into n subintervals is of the form

$$a = x_0 < x_1 < \cdots < x_i < \cdots < x_{n-1} < x_n = b.$$

In this partition, the ith subinterval is $[x_{i-1}, x_i]$, and its length is $\Delta x_i = x_i - x_{i-1}$. For a uniform partition, each subinterval is of equal length: $\Delta x_i = \dfrac{b-a}{n}$. If the lengths of any two subintervals are different, then the partition is nonuniform.

Uniform Partition of [a, b]

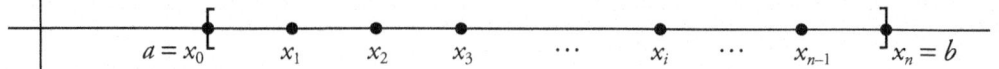

Nonuniform Partition of [a, b]

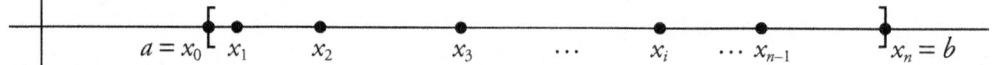

Example 3.2.1: The interval $[0, 2]$ is partitioned into 4 subintervals such that $x_i = \dfrac{i}{2}$. Is this partition uniform or nonuniform?

The partition is uniform if each subinterval $[x_{i-1}, x_i]$ has the same length $\Delta x_i = \dfrac{2-0}{4} = \dfrac{1}{2}$. We will check this.

$$\Delta x_i = x_i - x_{i-1} = \frac{i}{2} - \frac{i-1}{2} = \frac{1}{2}$$

Thus, the partition is uniform. You can see this in the figure below.

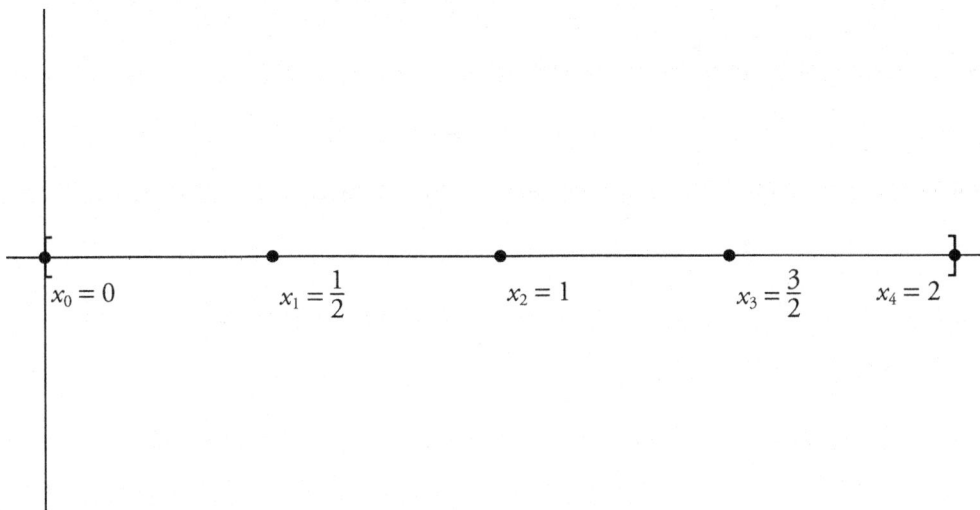

$x_0 = 0$ $x_1 = \dfrac{1}{2}$ $x_2 = 1$ $x_3 = \dfrac{3}{2}$ $x_4 = 2$

Example 3.2.2: The interval $[0, 2]$ is partitioned into 4 subintervals such that $x_i = \dfrac{i^2}{8}$. Is this partition uniform or nonuniform?

The partition is uniform if each subinterval $[x_{i-1}, x_i]$ has the same length $\Delta x_i = \dfrac{2-0}{4} = \dfrac{1}{2}$. We will check this.

$$\Delta x_i = x_i - x_{i-1} = \frac{i^2}{8} - \frac{(i-1)^2}{8} = \frac{i^2 - (i^2 - 2i + 1)}{8}$$

$$\Delta x_i = \frac{2i-1}{8}$$

In this example, Δx_i is not constant. It depends on i. Thus, the partition is nonuniform. You can see this in the figure below.

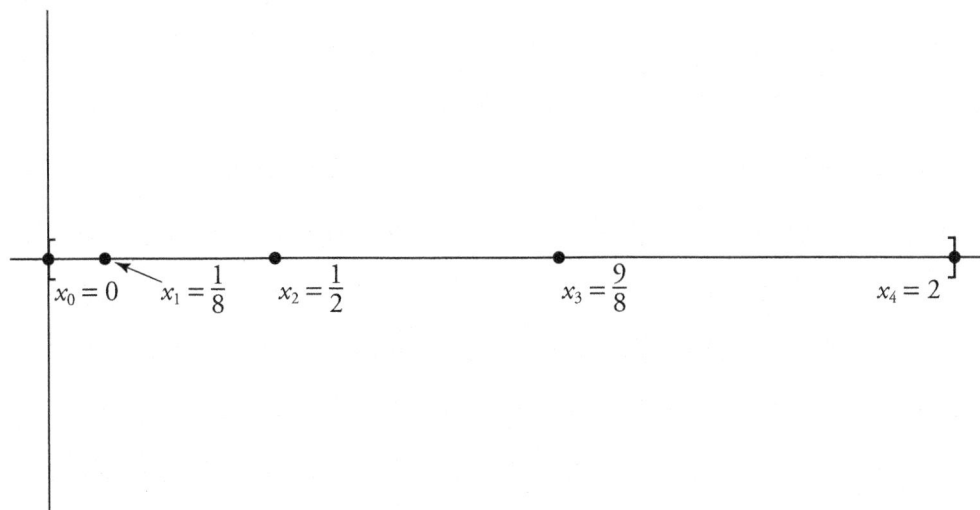

$x_0 = 0$ $x_1 = \dfrac{1}{8}$ $x_2 = \dfrac{1}{2}$ $x_3 = \dfrac{9}{8}$ $x_4 = 2$

Now we are ready to define Riemann sums. If f is defined on a closed interval $[a, b]$, and Λ is a partition of $[a, b]$ of the form given above, and if c_i is any point in the ith subinterval, then the sum

$$\sum_{i=1}^{n} f(c_i) \Delta x_i$$

is called a **Riemann sum of f for the partition Δ**. As you can see, it is a sum of products. Each product is $f(c_i)$, the value of f at c_i in the ith subinterval, multiplied by Δx_i, the length of the ith subinterval.

Two special Riemann sums are the left Riemann sum and the right Riemann sum. The left Riemann sum is obtained when c_i is the left endpoint of the ith subinterval, that is, when $c_i = x_{i-1}$. The right Riemann sum is obtained when c_i is the right endpoint of the ith subinterval, that is, when $c_i = x_i$.

Left Riemann Sum:

$$\sum_{i=1}^{n} f(x_{i-1}) \Delta x_i$$

Right Riemann Sum:

$$\sum_{i=1}^{n} f(x_i) \Delta x_i$$

Example 3.2.3: Compute the right and left Riemann sums of $f(x) = x^2 - 3x$ for the partition of $[0, 2]$ into 4 subintervals defined by $x_i = \dfrac{i}{2}$.

As noted in Example 3.2.1, $\Delta x_i = \dfrac{1}{2}$, and the points of the partition are $x_0 = 0$, $x_1 = \dfrac{1}{2}$, $x_2 = 1$, $x_3 = \dfrac{3}{2}$, and $x_4 = 2$.

First, we will compute the left Riemann sum.

$$\sum_{i=1}^{4} f(x_{i-1}) \Delta x_i = f(0) \cdot \frac{1}{2} + f\left(\frac{1}{2}\right) \cdot \frac{1}{2} + f(1) \cdot \frac{1}{2} + f\left(\frac{3}{2}\right) \cdot \frac{1}{2}$$

$$\sum_{i=1}^{4} f(x_{i-1}) \Delta x_i = 0 \cdot \frac{1}{2} + \left(-\frac{5}{4}\right) \cdot \frac{1}{2} + (-2) \cdot \frac{1}{2} + \left(-\frac{9}{4}\right) \cdot \frac{1}{2} = -\frac{11}{4}$$

And now we will compute the right Riemann sum.

$$\sum_{i=1}^{4} f(x_i) \Delta x_i = f\left(\frac{1}{2}\right) \cdot \frac{1}{2} + f(1) \cdot \frac{1}{2} + f\left(\frac{3}{2}\right) \cdot \frac{1}{2} + f(2) \cdot \frac{1}{2}$$

$$\sum_{i=1}^{4} f(x_i) \Delta x_i = \left(-\frac{5}{4}\right) \cdot \frac{1}{2} + (-2) \cdot \frac{1}{2} + \left(-\frac{9}{4}\right) \cdot \frac{1}{2} + (-2) \cdot \frac{1}{2} = -\frac{15}{4}$$

Example 3.2.4: Compute the left and right Riemann sums of $f(x) = \sqrt{x}$ for the partition $[0, 2]$ into 4 subintervals defined by $x_i = \dfrac{i^2}{8}$.

As noted in Example 3.2.2, $\Delta x_i = \dfrac{2i - 1}{8}$, and the points of the partition are $x_0 = 0$, $x_1 = \dfrac{1}{8}$, $x_2 = \dfrac{1}{2}$, $x_3 = \dfrac{9}{8}$, and $x_4 = 2$.

The left Riemann sum is as follows:

$$\sum_{i=1}^{4} f(x_{i-1})\Delta x_i = f(0) \cdot \frac{1}{8} + f\left(\frac{1}{8}\right) \cdot \frac{3}{8} + f\left(\frac{1}{2}\right) \cdot \frac{5}{8} + f\left(\frac{9}{8}\right) \cdot \frac{7}{8}$$

$$\sum_{i=1}^{4} f(x_{i-1})\Delta x_i = 0 \cdot \frac{1}{8} + \sqrt{\frac{1}{8}} \cdot \frac{3}{8} + \sqrt{\frac{1}{2}} \cdot \frac{5}{8} + \sqrt{\frac{9}{8}} \cdot \frac{7}{8} = \frac{17\sqrt{2}}{16}$$

The right Riemann sum is as follows:

$$\sum_{i=1}^{4} f(x_i)\Delta x_i = f\left(\frac{1}{8}\right) \cdot \frac{1}{8} + f\left(\frac{1}{2}\right) \cdot \frac{3}{8} + f\left(\frac{9}{8}\right) \cdot \frac{5}{8} + f(2) \cdot \frac{7}{8}$$

$$\sum_{i=1}^{4} f(x_i)\Delta x_i = \sqrt{\frac{1}{8}} \cdot \frac{1}{8} + \sqrt{\frac{1}{2}} \cdot \frac{3}{8} + \sqrt{\frac{9}{8}} \cdot \frac{5}{8} + \sqrt{2} \cdot \frac{7}{8} = \frac{25\sqrt{2}}{16}$$

LO 3.2A(b) – Riemann Integration

We will now turn our attention to definite integration. In **EU 3.1**, we reviewed antidifferentiation. When the *antiderivative* of $f(x)$ is computed, the result is a family of functions, called antiderivatives of $f(x)$. When the *definite integral* of $f(x)$ is computed, the result (if it exists) is a *number*. That number is the limit of a Riemann sum. The full Learning Objective for this section is as follows:

LO 3.2A(b): Express the limit of a Riemann sum in integral notation.

The first item of Essential Knowledge for this section is given below.

EK 3.2A2: The definite integral of a continuous function f over the interval $[a, b]$, denoted by $\int_a^b f(x)\,dx$, is the limit of Riemann sums as the width of the subintervals approach 0. That is, $\int_a^b f(x)\,dx = \lim_{\max \Delta x_i \to 0} \sum_{i=1}^{n} f(x_i^*)\Delta x_i$, where x_i^* is a value in the ith subinterval, n is the number of subintervals, and $\max \Delta x_i$ is the width of the largest subinterval. Another form of the definition is $\int_a^b f(x)\,dx = \lim_{n \to \infty} \sum_{i=1}^{n} f(x_i^*)\Delta x_i$, where $\Delta x_i = \frac{b-a}{n}$ and x_i^* is a value in the ith subinterval.

That is an awful lot of information packed into one EK, so we will carefully review each part.

First, if $[a, b]$ is partitioned into n subintervals, then whether the partition is uniform (as in Example 3.2.1) or nonuniform (as in Example 3.2.2), then (at least one) of the subintervals is the longest one. The length of this subinterval is denoted as $\max \Delta x_i$.

Example 3.2.5: Find $\max \Delta x_i$ for the partition of $[0, 2]$ into 4 subintervals defined by $x_i = \frac{i}{2}$.

As we saw in Example 3.2.1, each subinterval has the same length: $\Delta x_i = \frac{1}{2}$. Thus, $\max \Delta x_i = \frac{1}{2}$.

Example 3.2.6: Find $\max \Delta x_i$ for the partition of $[0, 2]$ into 4 subintervals defined by $x_i = \frac{i^2}{8}$.

As we saw in Example 3.2.2, the length of the ith subinterval is $\Delta x_i = \frac{2i-1}{8}$. Δx_i is an increasing function of i, so the longest subinterval is the fourth one. Thus, $\max \Delta x_i = x_4 - x_3 = \frac{4^2}{8} - \frac{3^2}{8} = \frac{7}{8}$.

Now, we are ready to move on to definite integrals. Let Δ be a partition of $[a, b]$ into n subintervals, and let x_i^* be *any* point in $[x_{i-1}, x_i]$, and let f be defined on $[a, b]$. If the limit

$$\lim_{\max \Delta x_i \to 0} \sum_{i=1}^{n} f\left(x_i^*\right) \Delta x_i$$

exists for *all* partitions Δ, then f is said to be **integrable** on $[a, b]$. If that happens, then the above limit is called the definite integral of f from a to b, and is denoted as follows:

$$\int_a^b f(x)\, dx = \lim_{\max \Delta x_i \to 0} \sum_{i=1}^{n} f\left(x_i^*\right) \Delta x_i$$

So the definite integral is a *limit of a Riemann sum*. Consequently, it is sometimes referred to as the Riemann integral. There is an interesting theorem that relates integrability and continuity. Simply stated: *continuity implies integrability*. So any function f that is continuous on $[a, b]$ is also integrable on $[a, b]$, and so $\int_a^b f(x)\, dx$ exists.

Taking the limit as $\max \Delta x_i \to 0$ makes intuitive sense. If the length of the *longest* subinterval approaches zero, then the length of *every* subinterval approaches zero. As intuitive as this may be however, it is not convenient to compute with. The quantity whose limit we are computing depends on the variable n, so it would be nice if the limit referred to n as well. We can make that happen by noting that, as the length of each subinterval of the partition approaches zero, then the number n of subintervals must *increase* if the subintervals are to fill up all of $[a, b]$. More precisely,

$$\max \Delta x_i \to 0 \Rightarrow n \to \infty.$$

So we will re-formulate our definition of the definite integral as follows:

$$\int_a^b f(x)\, dx = \lim_{n \to \infty} \sum_{i=1}^{n} f\left(x_i^*\right) \Delta x_i$$

The statement of the EK says only that we should know how to use this definition for *uniform* partitions of $[a, b]$ into n subintervals, for which $\Delta x_i = \dfrac{b-a}{n}$. It should be noted that a quick, easy rule of thumb for calculating definite integrals in this way is to always let x_i^* be the *right endpoint* of the ith subinterval. For uniform partitions, we have the following:

i	$x_i^* =$ **Right endpoint of** $[x_{i-1}, x_i]$
1	$x_1^* = a + \Delta x_i = a + \dfrac{b-a}{n}$
2	$x_2^* = a + 2\Delta x_i = a + \dfrac{2(b-a)}{n}$
3	$x_3^* = a + 3\Delta x_i = a + \dfrac{3(b-a)}{n}$
$\vdots$	$\vdots$
i	$x_i^* = a + i\Delta x_i = a + \dfrac{i(b-a)}{n}$

So for a uniform partition of $[a, b]$ into n subintervals, the right endpoint of the ith subinterval is located at $x_i^* = a + \dfrac{i(b-a)}{n}$.

EK 3.2A3: The information in a definite integral can be translated into the limit of a related Riemann sum, and the limit of a Riemann sum can be written as a definite integral.

This is a continuation of **EK3.2A2**. We will start by reviewing the translation of a definite integral into the limit of a related Riemann sum via two examples.

Example 3.2.7: Translate the definite integral $\int_0^4 (x^2 + 2x)\, dx$ into the limit of Riemann sum.

We will use a uniform partition, so that $\Delta x_i = \dfrac{4-0}{n} = \dfrac{4}{n}$, and we will let x_i^* be the right endpoint of the ith subinterval, so $x_i^* = 0 + i\dfrac{4-0}{n} = \dfrac{4i}{n}$. We apply the definition of the definite integral as follows:

$$\int_0^4 (x^2 + 2x)\, dx = \lim_{n \to \infty} \sum_{i=1}^n \left(\left(\frac{4i}{n}\right)^2 + 2\left(\frac{4i}{n}\right) \right)\frac{4}{n}$$

Example 3.2.8: Translate the definite integral $\int_{-1}^2 3x^2\, dx$.

We will use a uniform partition, so that $\Delta x_i = \dfrac{2-(-1)}{n} = \dfrac{3}{n}$, and we will let x_i^* be the right endpoint of the ith subinterval, so $x_i^* = -1 + i\dfrac{2-(-1)}{n} = -1 + \dfrac{3i}{n}$. We apply the definition of the definite integral as follows:

$$\int_{-1}^{2} 3x^2\,dx = \lim_{n\to\infty} \sum_{i=1}^{n} 3\left(-1+\frac{3i}{n}\right)^2 \frac{3}{n}$$

It should be noted that the EK *does not* require that you *calculate* definite integrals as limits of Riemann sums.

Now we will review the other half of **EK 3.2A3**, which is writing the limit of a Riemann sum as a definite integral. The limit of a Riemann sum will be presented in the following form:

$$\lim_{n\to\infty} \sum_{i=1}^{n} f\left(x_i^*\right)\Delta x_i$$

We will need to be able to extract from this expression $f(x)$, a, and b so that we can write down an expression of the following form:

$$\int_{a}^{b} f(x)\,dx$$

We will illustrate how to do this with an example.

Example 3.2.9: Write

$$\lim_{n\to\infty} \sum_{i=1}^{n} \left(-5\left(2+\frac{3i}{n}\right)^2 + 6\left(2+\frac{3i}{n}\right)\right)\frac{3}{n}$$

as a definite integral of the form $\int_{a}^{b} f(x)\,dx$.

The factor of $\dfrac{3}{n}$ is Δx_i, which is equal to $\dfrac{b-a}{n}$. So upon comparison of these two expressions, we can already tell that the interval of integration has a length of $b - a = 3$. In parentheses, we see a function whose argument is $x_i^* = 2 + \dfrac{3i}{n}$. The function, which we'll call f, is given by $f(x) = -5x^2 + 6x$. This is the integrand of the definite integral. As we reviewed earlier, $x_i^* = a + i\Delta x_i$. Comparing this to $x_i^* = 2 + \dfrac{3i}{n}$, we see that $a = 2$. Since we already determined that $b - a = 3$, this implies that $b = 5$. We now know everything we need to know in order to write the limit of this Riemann sum as a definite integral.

$$\int_a^b f(x)\,dx = \int_2^5 \left(-5x^2 + 6x\right)dx$$

LO 3.2B – Approximate Definite Integration

The full Learning Objective for this section is stated below.

LO 3.2B: Approximate a definite integral.

The first item of Essential Knowledge for this section is given below.

EK 3.2B1: Definite integrals can be approximated for functions that are represented graphically, numerically, algebraically, and verbally.

This actually contains a lot of stuff. We will review each of the four items in the list one at a time. We will begin with a function that is represented graphically. Consider the following example:

Example 3.2.10: Consider the function $f(x)$ given by the following graph:

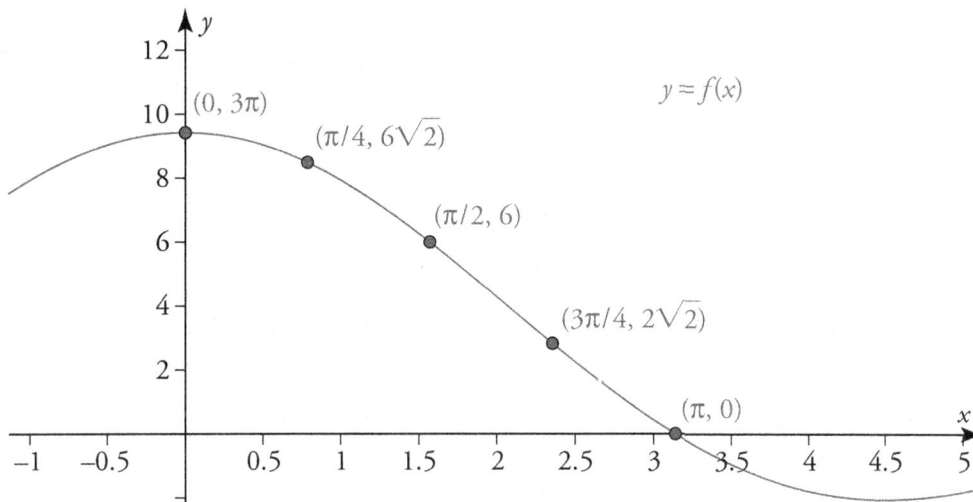

Approximate $\displaystyle\int_0^\pi f(x)\,dx$ using a left Riemann sum with a uniform partition of $[0, \pi]$ into four subintervals. Round your answer to the nearest tenth.

The length of each subinterval is $\Delta x_i = \dfrac{\pi - 0}{4} = \dfrac{\pi}{4}$, and the points in the partition are $x_0^* = 0,\ x_1^* = \dfrac{\pi}{4},\ x_2^* = \dfrac{\pi}{2},\ x_3^* \dfrac{3\pi}{4}$, and $x_4^* = \pi$. Recall that the left Riemann sum uses the left endpoint of each subinterval. So, the integral is approximated as follows:

$$\int_0^\pi f(x)\,dx \approx \sum_{i=1}^4 f\left(x_{i-1}^*\right)\Delta x_i$$

$$\int_0^\pi f(x)\,dx \approx f\left(x_0^*\right)\Delta x_1 + f\left(x_1^*\right)\Delta x_2 + f\left(x_2^*\right)\Delta x_3 + f\left(x_3^*\right)\Delta x_4$$

$$\int_0^\pi f(x)\,dx \approx f(0)\cdot\frac{\pi}{4} + f\left(\frac{\pi}{4}\right)\cdot\frac{\pi}{4} + f\left(\frac{\pi}{2}\right)\cdot\frac{\pi}{4} + f\left(\frac{3\pi}{4}\right)\cdot\frac{\pi}{4}$$

$$\int_0^\pi f(x)\,dx \approx 3\pi\cdot\frac{\pi}{4} + 6\sqrt{2}\cdot\frac{\pi}{4} + 6\cdot\frac{\pi}{4} + 2\sqrt{2}\cdot\frac{\pi}{4}$$

$$\int_0^\pi f(x)\,dx \approx 21.0$$

Next, we will look at a function that is represented numerically.

Example 3.2.11: Consider a continuous function $y = f(x)$ which has solution points (x, y) given by the following table:

x	$y = f(x)$
1	2
2	5
3	8
4	6
5	3
6	5

Approximate $\displaystyle\int_0^6 f(x)\,dx$ using a right Riemann sum with a uniform partition of $[0, 6]$ into six subintervals.

The length of each subinterval is $\Delta x_i = \dfrac{6-0}{6} = 1$, and the points in the partition are $x_0^* = 0$, $x_1^* = 1$, $x_2^* = 2$, $x_3^* = 3$, $x_4^* = 4$, $x_5^* = 5$, and $x_6^* = 6$. Recall that the right Riemann sum uses the right endpoint of each subinterval. So, the integral is approximated as follows:

$$\int_0^6 f(x)\,dx \approx \sum_{i=1}^6 f\left(x_i^*\right)\Delta x_i$$

$$\int_0^6 f(x)\,dx \approx f\left(x_1^*\right)\Delta x_1 + f\left(x_2^*\right)\Delta x_2 + f\left(x_3^*\right)\Delta x_3 + f\left(x_4^*\right)\Delta x_4 + f\left(x_5^*\right)\Delta x_5 + f\left(x_6^*\right)\Delta x_6$$

$$\int_0^6 f(x)\,dx \approx f(1)\cdot 1 + f(2)\cdot 1 + f(3)\cdot 1 + f(4)\cdot 1 + f(5)\cdot 1 + f(6)\cdot 1$$

$$\int_{0}^{6} f(x)\,dx \approx 2\cdot 1 + 5\cdot 1 + 8\cdot 1 + 6\cdot 1 + 3\cdot 1 + 5\cdot 1$$

$$\int_{0}^{6} f(x)\,dx \approx 29$$

The next item listed in **EK 3.2B1** is functions that are represented algebraically. However, we already reviewed how to approximate definite integrals of such functions in Examples 3.2.3 and 3.2.4, so we will not repeat that discussion here. The final item in the list is functions that are represented verbally. We illustrate this with an example.

Example 3.2.12: Consider the function f, where $f(x)$ is equal to the natural base e raised to the power of the negative square of x. Approximate $\int_{0}^{1} f(x)\,dx$ using a left Riemann sum with a uniform partition of $[0, 1]$ into four subintervals. Round your answer to the nearest thousandth.

The function described in the problem statement is $f(x) = e^{-x^2}$, and no elementary antiderivative exists for it. The length of each subinterval is $\Delta x_i = \dfrac{1-0}{4} = \dfrac{1}{4}$, and the points in the partition are $x_0^* = 0, x_1^* = \dfrac{1}{4}, x_2^* = \dfrac{1}{2}, x_3^* = \dfrac{3}{4}$, and $x_4^* = 1$. Recall that the left Riemann sum uses the left endpoint of each subinterval. So the integral is approximated as follows:

$$\int_{0}^{1} f(x)\,dx \approx \sum_{i=1}^{4} f\left(x_i^*\right)\Delta x_i$$

$$\int_{0}^{1} f(x)\,dx \approx f\left(x_1^*\right)\Delta x_1 + f\left(x_2^*\right)\Delta x_2 + f\left(x_3^*\right)\Delta x_3 + f\left(x_4^*\right)\Delta x_4$$

$$\int_{0}^{1} f(x)\,dx \approx f(0)\cdot\frac{1}{4} + f\left(\frac{1}{4}\right)\cdot\frac{1}{4} + f\left(\frac{1}{2}\right)\cdot\frac{1}{4} + f\left(\frac{3}{4}\right)\cdot\frac{1}{4}$$

$$\int_{0}^{1} f(x)\,dx \approx e^{-0^2}\cdot\frac{1}{4} + e^{-\left(\frac{1}{4}\right)^2}\cdot\frac{1}{4} + e^{-\left(\frac{1}{2}\right)^2}\cdot\frac{1}{4} + e^{-\left(\frac{3}{4}\right)^2}\cdot\frac{1}{4}$$

$$\int_{0}^{1} f(x)\,dx \approx 0.822$$

While the first item of Essential Knowledge pertains to different *representations of functions*, the second pertains to different *methods of approximation*.

EK 3.2B2: Definite integrals can be approximated using a left Riemann sum, a right Riemann sum, a midpoint Riemann sum, or a trapezoidal sum; approximations can be computed using either uniform or nonuniform partitions.

We already reviewed left and right Riemann sums in Examples 3.2.3 and 3.2.4. In Example 3.2.3, we computed right and left Riemann sums of $f(x) = x^2 - 3x$ for a *uniform* partition of $[0, 2]$ into 4 subintervals. What we were really doing there was approximating $\int_0^2 (x^2 - 3x)dx$. In Example 3.2.4, we computed right and left Riemann sums of $f(x) = \sqrt{x}$ for a *nonuniform* partition of $[0, 2]$ into 4 subintervals. In that example, we were approximating $\int_0^2 \sqrt{x}dx$. So we will consider those items covered. We will move on to the midpoint Riemann sum. Let f be integrable on the interval $[a, b]$, and let $[a, b]$ be partitioned by $a = x_0 < x_1 < x_2 < \cdots < x_n = b$. The midpoint Riemann sum is given as follows:

$$\sum_{i=1}^n f\left(\frac{x_{i-1} + x_i}{2}\right)\Delta x_i$$

Note that $\frac{x_{i-1} + x_i}{2}$ is the *midpoint* of the ith subinterval $[x_{i-1}, x_i]$, which is why this sum is named the midpoint Riemann sum. Just as with the left and right Riemann sums, Δx_i is the length of the ith subinterval of the partition. We will illustrate this with an example.

Example 3.2.13: Approximate $\int_0^2 (x^2 - 3x)dx$ using a midpoint Riemann sum with a uniform partition of $[0, 2]$ into four subintervals.

For this partition, $\Delta x_i = \frac{2-0}{4} = \frac{1}{2}$ and $x_0 = 0$, $x_1 = \frac{1}{2}$, $x_2 = 1$, $x_3 = \frac{3}{2}$, and $x_4 = 2$. So the integral is approximated as follows:

$$\int_0^2 \left(x^2 - 3x\right)dx \approx \sum_{i=1}^4 f\left(\frac{x_{i-1} + x_i}{2}\right)\Delta x_i$$

$$\int_0^2 \left(x^2 - 3x\right)dx \approx f\left(\frac{x_0 + x_1}{2}\right)\Delta x_1 + f\left(\frac{x_1 + x_2}{2}\right)\Delta x_2 + f\left(\frac{x_2 + x_3}{2}\right)\Delta x_3 + f\left(\frac{x_3 + x_4}{2}\right)\Delta x_4$$

$$\int_0^2 \left(x^2 - 3x\right)dx \approx f\left(\frac{1}{4}\right)\cdot\frac{1}{2} + f\left(\frac{3}{4}\right)\cdot\frac{1}{2} + f\left(\frac{5}{4}\right)\cdot\frac{1}{2} + f\left(\frac{7}{4}\right)\cdot\frac{1}{2}$$

$$\int_0^2 \left(x^2 - 3x\right)dx \approx -\frac{11}{16}\cdot\frac{1}{2} - \frac{27}{16}\cdot\frac{1}{2} - \frac{35}{16}\cdot\frac{1}{2} - \frac{35}{16}\cdot\frac{1}{2} = -\frac{27}{8}$$

We will close this section by reviewing trapezoidal sums. Let f be integrable on $[a, b]$. Using the same partition as before, the trapezoidal sum is given as follows:

$$\frac{b-a}{2n}\left[f\left(x_0\right) + 2f\left(x_1\right) + 2f\left(x_2\right) + \cdots + 2f\left(x_{n-1}\right) + 2f\left(x_n\right)\right]$$

We will see why this is called the *trapezoidal* sum when we connect definite integrals to areas in the next section. For now, we will take this formula as a given. Let us look at an example.

Example 3.2.14: Approximate $\int_0^2 (x^2 - 3x)dx$ using a trapezoidal sum with a uniform partition of $[0, 2]$ into four subintervals.

For this partition, $x_0 = 0$, $x_1 = \frac{1}{2}$, $x_2 = 1$, $x_3 = \frac{3}{2}$, and $x_4 = 2$. So the integral is approximated as follows:

$$\int_0^2 \left(x^2 - 3x\right)dx \approx \frac{2-0}{2(4)}\left[f\left(x_0\right) + 2f\left(x_1\right) + 2f\left(x_2\right) + \cdots + 2f\left(x_{n-1}\right) + 2f\left(x_n\right)\right]$$

$$\int_0^2 \left(x^2 - 3x\right)dx \approx \frac{1}{4}\left[f(0) + 2f\left(\frac{1}{2}\right) + 2f(1) + 2f\left(\frac{3}{2}\right) + f(2)\right]$$

$$\int_0^2 \left(x^2 - 3x\right)dx \approx \frac{1}{4}\left[0 + 2\left(-\frac{5}{4}\right) + 2(-2) + 2\left(-\frac{9}{4}\right) - 2\right] = -\frac{13}{4}$$

LO 3.2C: Areas and Properties of Definite Integrals

The full Learning Objective for this section is as follows:

LO 3.2C: Calculate a definite integral using areas and properties of definite integrals.

This short sentence actually says a lot, as we will see as we review the items of Essential Knowledge for this section.

EK 3.2C1: In some cases, a definite integral can be evaluated by using geometry and the connection between the definite integral and area.

This is a big one. Here, we will review the connection between a definite integral and an area. Let f be integrable on the interval $[a, b]$, *and* let $f(x) \geq 0$ on $[a, b]$. Consider the function f shown in the figure below.

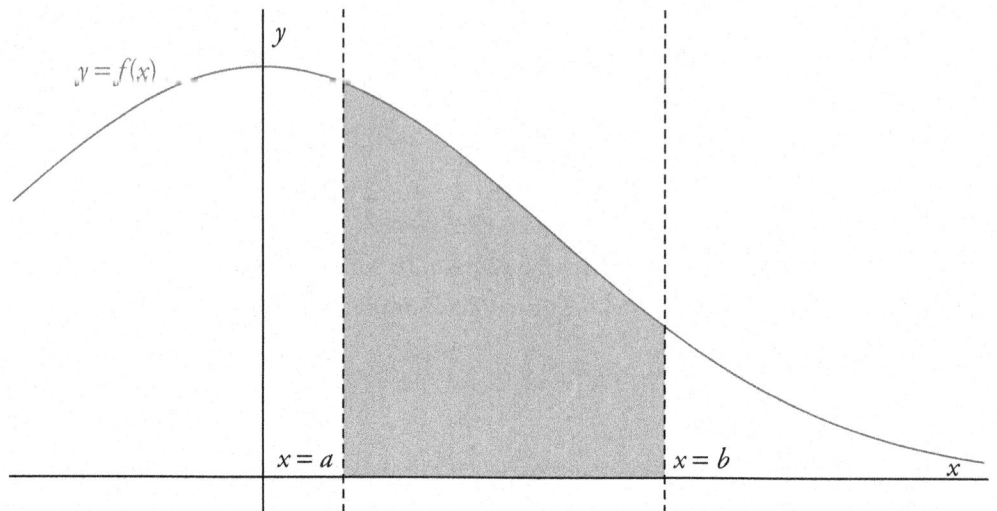

As you can see, f is continuous (and therefore integrable) on the interval $[a, b]$. Furthermore, $f(x) \geq 0$ on $[a, b]$. The area A of the shaded region bounded by $x = a$, $x = b$, the y-axis, and the graph of f can be approximated by the areas of four *rectangles*, as shown below.

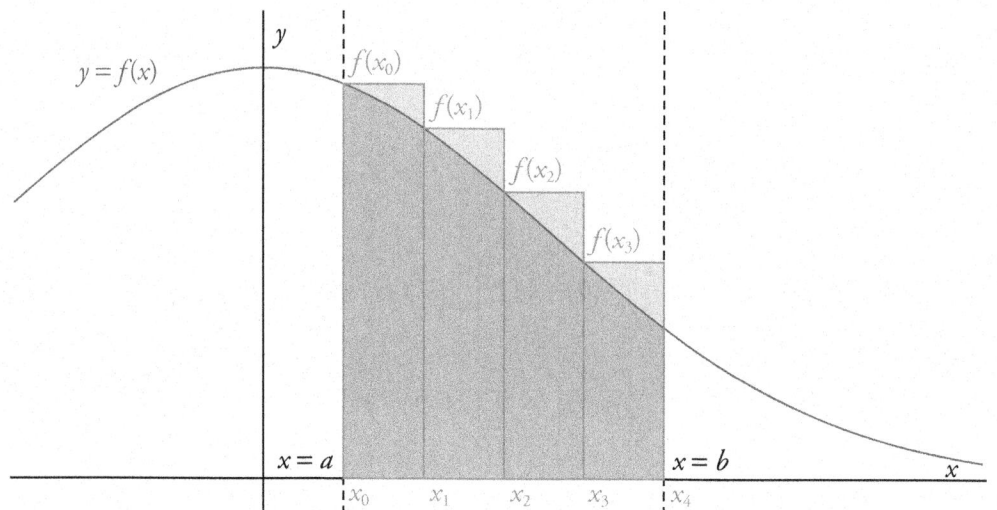

These rectangles were drawn such that the lengths of their bases are equal. In other words, the partition $a = x_0 < x_1 < x_2 < x_3 < x_4 = b$ is *uniform*, so that for each subinterval, $\Delta x_i = \dfrac{b-a}{4}$. From left to right the height of each rectangle is, respectively, $f(x_0)$, $f(x_1)$, $f(x_2)$, and $f(x_3)$. Let us write out an explicit expression for our approximation of the area A. It is the sum of the areas the four rectangles. Recalling that the area of a rectangle is equal to the product of the length of its base and the length of its height, we obtain the following:

$$A \approx f(x_0)\Delta x_1 + f(x_1)\Delta x_2 + f(x_2)\Delta x_3 + f(x_3)\Delta x_4$$

$$A \approx \sum_{i=1}^{4} f\left(x_{i-1}\right) \Delta x_i$$

Now we notice something significant. The right side of the last line is none other than the left Riemann sum of f with a uniform partition of $[a, b]$ into four subintervals. Recall that this is sum *also* approximates $\int_a^b f(x)\,dx$. Next, we approximate A with eight rectangles.

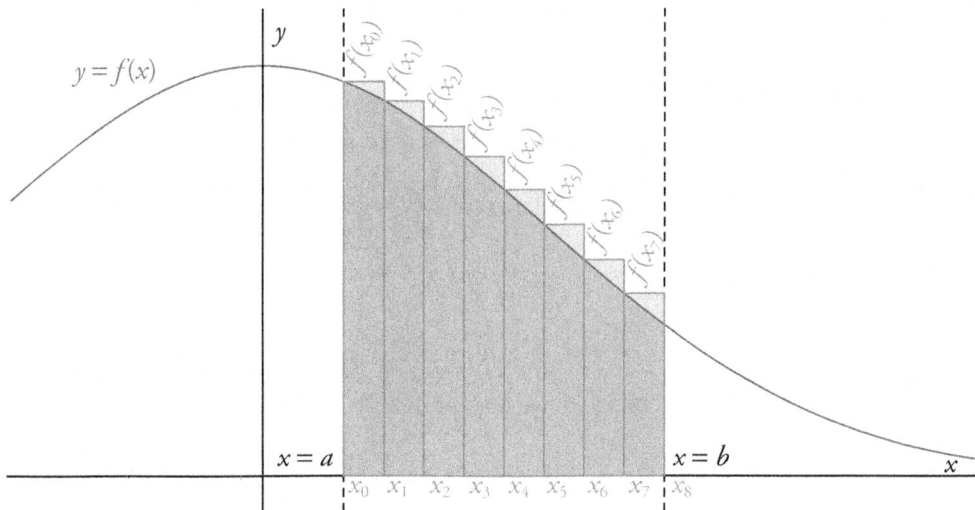

Again, the rectangles were drawn such that the lengths of their bases are equal, but this time we have the uniform partition $a = x_0 < x_1 < x_2 < x_3 < x_4 < x_5 < x_6 < x_7 < x_8 = b$ with $\Delta x_i = \dfrac{b-a}{8}$. From left to right, the height of each rectangle is $f(x_0)$, $f(x_1)$, $f(x_2)$, $f(x_3)$, $f(x_4)$, $f(x_5)$, $f(x_6)$, and $f(x_7)$. Explicitly writing out our approximation, we obtain the following:

$$A \approx f\left(x_0\right)\Delta x_1 + f\left(x_1\right)\Delta x_2 + f\left(x_2\right)\Delta x_3 + f\left(x_3\right)\Delta x_4 +$$
$$f\left(x_4\right)\Delta x_5 + f\left(x_5\right)\Delta x_6 + f\left(x_6\right)\Delta x_7 + f\left(x_7\right)\Delta x_8$$

$$A \approx \sum_{i=1}^{8} f\left(x_{i-1}\right) \Delta x_i$$

Not only is A approximated by this left Riemann sum, but also the approximation has *improved*. You can see this visually by comparing the last two figures: the excess area of the rectangles has decreased as number of rectangles has increased. In fact, as the number n of rectangles approaches infinity, the excess area approaches *zero*. In other words, A is *exactly* equal to $\lim\limits_{n\to\infty}\sum\limits_{i=1}^{n}f\left(x_{i-1}\right)\Delta x_i$. In fact, even for nonuniform partitions of $[a, b]$, and using *any* point x_i^* in the ith subinterval of the partition, we have the following result:

$$A = \lim\limits_{n\to\infty}\sum\limits_{i=1}^{n}f\left(x_i^*\right)\Delta x_i$$

The quantity on the right side of that equation should look familiar. It is the alternative definition of $\int_{a}^{b}f\left(x\right)dx$ from **EK 3.2A2**.

So, we are finally able to make the connection between areas and definite integrals. If f is integrable on the interval $[a, b]$, and if $f(x) \geq 0$ on $[a, b]$, then the area A of the region bounded by $x = a$, $x = b$ the y-axis, and the graph of f is given as follows:

$$A = \int_{a}^{b}f\left(x\right)dx$$

You might be wondering, "Why do we require that $f(x) \geq 0$ on $[a, b]$?" Good question! We have a good answer: let $f(x) = x^3$ and let $[a, b] = [-1, 1]$. As you can (and should) easily verify, $\int_{-1}^{1}x^3 dx = 0$. However, as you can see in the following figure, the area of the region bounded by $x = -1$, $x = 1$ the y-axis, and the graph of f is clearly not zero.

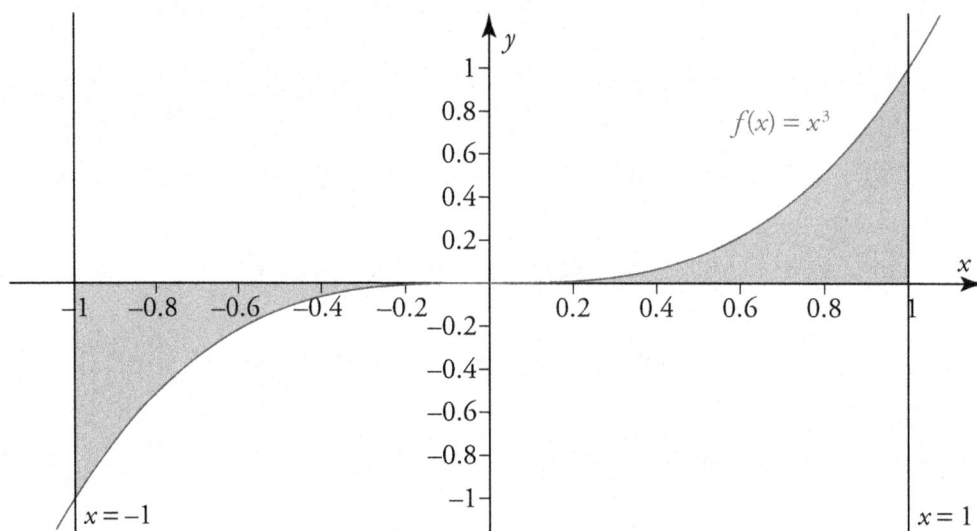

We have said a lot about the connection between areas and definite integrals, but we have yet to address the actual statement of this EK. We will remedy that now with some examples.

Example 3.2.15: Consider the function $f(x) = \sqrt{9 - x^2}$, which is graphed below.

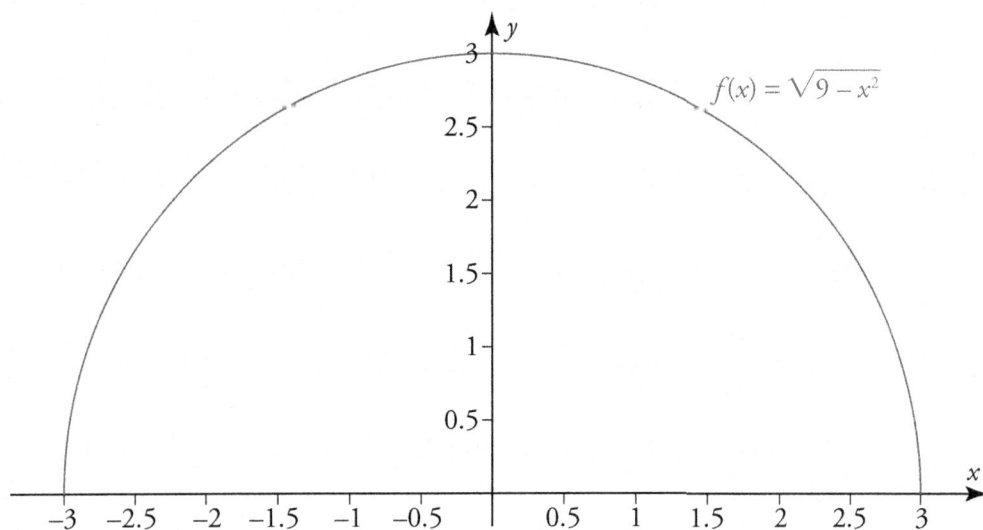

$$f(x) = \sqrt{9 - x^2}$$

Evaluate $\displaystyle\int_{-3}^{3} f(x)\,dx$.

Now we *could* integrate this using trigonometric substitution. *But we do not have to.*

Notice that on $[-3, 3]$, f is continuous *and* $f(x) \geq 0$. Based on the discussion preceding this example, we see then that $\displaystyle\int_{-3}^{3} f(x)\,dx$ is equal to the area of the region bounded by $x = -3$, $x = 3$ the y-axis, and the graph of f. But that region is a *semicircle of radius 3*, and we know that the area of a semicircle of radius r is $\frac{1}{2}\pi r^2$. Therefore, we obtain the following:

$$\int_{-3}^{3} f(x)\,dx = \frac{1}{2}\pi \cdot 3^2 = \frac{9\pi}{2}$$

Example 3.2.16: Consider the function f, which is defined piecewise as follows:

$$f = \begin{cases} 1 & 0 \leq x < 3 \\ x - 2 & 3 \leq x \leq 5 \end{cases}$$

The graph of this function is shown below.

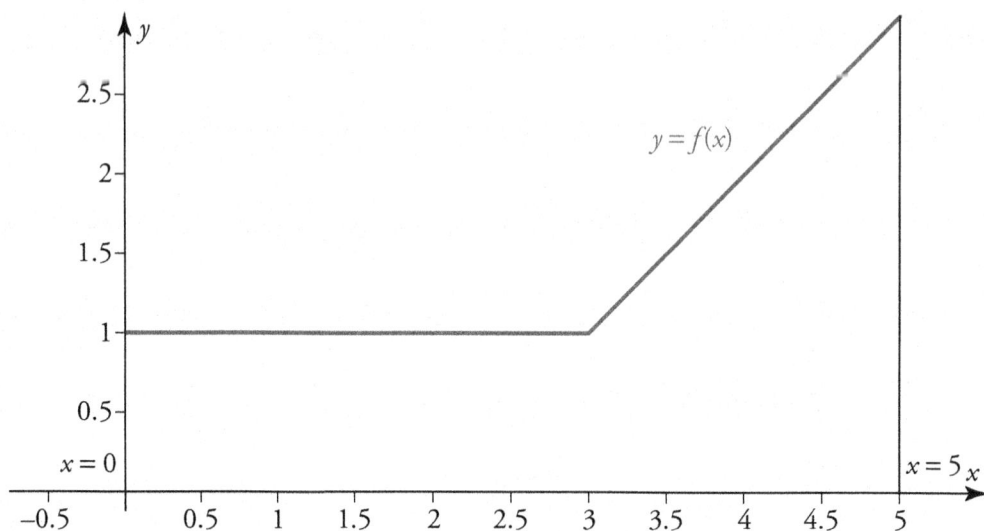

Evaluate $\int_0^5 f(x)\,dx$.

Since f is continuous (and therefore integrable) on $[0,5]$, and $f(x) \geq 0$ on $[0,5]$, it follows that $\int_0^5 f(x)\,dx$ is equal to the area A of the region bounded by $x = 0$, $x = 5$, the y-axis, and the graph of f. We will split this region into two subregions, as shown below.

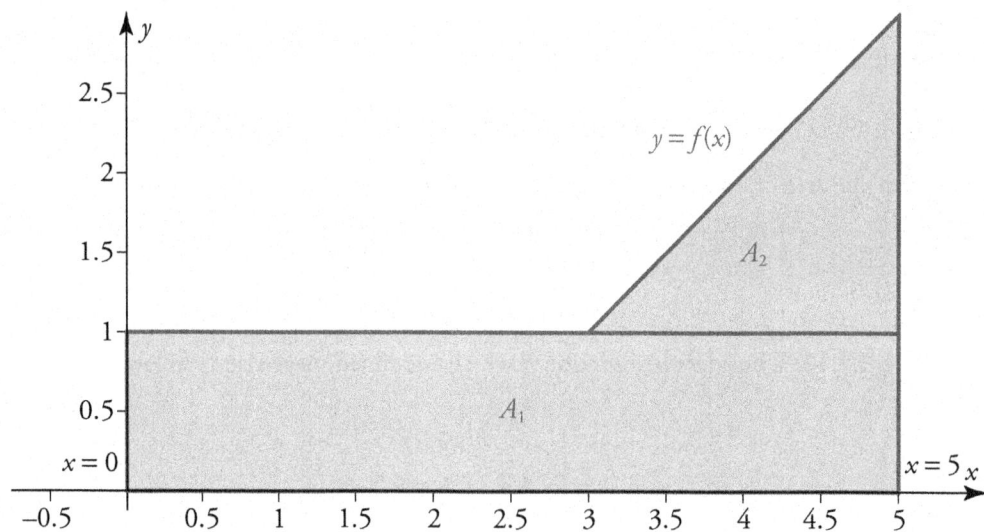

The first subregion, whose area is labeled A_1, is a rectangle with base $b_1 = 5$ and height $h_1 = 1$, so $A_1 = b_1 h_1 = (5)(1) = 5$. The second subregion, whose area is labeled A_2, is a triangle with base $b_2 = 2$ and height $h_2 = 2$, so $A_2 = \frac{1}{2}b_2 h_2 = \frac{1}{2}(2)(2) = 2$. Combining these results, we obtain the following:

$$\int\limits_0^5 f(x)\,dx = A = A_1 + A_2 = 5 + 2 = 7$$

We close our discussion of **EK 3.2C1** by fulfilling our earlier promise to explain what's so *trapezoidal* about the trapezoidal sum. Recall that a trapezoid is a quadrilateral in which *precisely* one pair of sides is parallel.

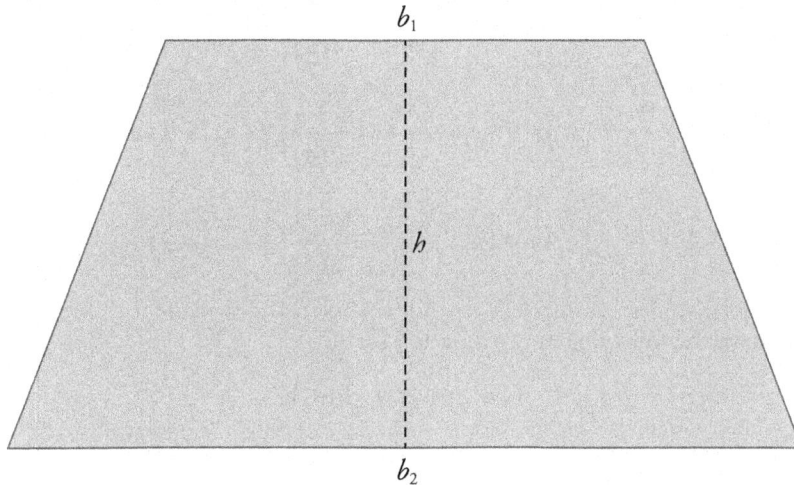

The trapezoid has bases b_1 and b_2, height h, and area $A = \dfrac{1}{2}(b_1 + b_2)h$.

Consider a function f that is continuous on $[a,b]$ such that $f(x) \geq 0$ on $[a,b]$, and consider the region bounded by $x = a$, $x = b$, the y-axis, and the graph of f. We will approximate the area A of region, only this time with *trapezoids*.

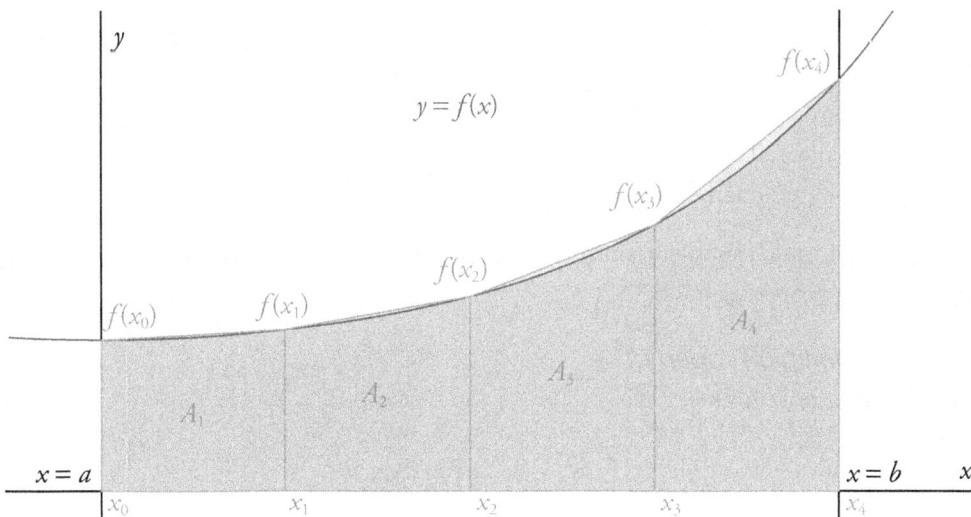

In the above figure we have used four trapezoids. The partition $a = x_0 < x_1 < x_2 < x_3 < x_4 = b$ is uniform, and so the height of each trapezoid is $\Delta x_i = \dfrac{b-a}{4}$. The bases of the ith

trapezoid are $f(x_{i-1})$ and $f(x_i)$. So the expressions for the areas of the four trapezoids are as follows:

$$A_1 = \frac{1}{2}\big[f(x_0)+f(x_1)\big]\frac{b-a}{4} = \frac{b-a}{2(4)}\big[f(x_0)+f(x_1)\big]$$

$$A_2 = \frac{1}{2}\big[f(x_1)+f(x_2)\big]\frac{b-a}{4} = \frac{b-a}{2(4)}\big[f(x_1)+f(x_2)\big]$$

$$A_3 = \frac{1}{2}\big[f(x_2)+f(x_3)\big]\frac{b-a}{4} = \frac{b-a}{2(4)}\big[f(x_2)+f(x_3)\big]$$

$$A_4 = \frac{1}{2}\big[f(x_3)+f(x_4)\big]\frac{b-a}{4} = \frac{b-a}{2(4)}\big[f(x_3)+f(x_4)\big]$$

So, we have the following approximation for A:

$$A \approx A_1 + A_2 + A_3 + A_4$$

$$A \approx \frac{b-a}{2(4)}\big[f(x_0)+f(x_1)+f(x_1)+f(x_2)+f(x_2)+f(x_3)+f(x_3)+f(x_4)\big]$$

$$A \approx \frac{b-a}{2(4)}\big[f(x_0)+2f(x_1)+2f(x_2)+2f(x_3)+f(x_4)\big]$$

Generalizing this to n trapezoids, we obtain the trapezoidal sum:

$$A \approx \frac{b-a}{2n}\big[f(x_0)+2f(x_1)+2f(x_2)+\cdots+2f(x_{n-1})+2f(x_n)\big].$$

EK 3.2C2: Properties of definite integrals include the integral of a constant times a function, the integral of the sum of two functions, reversal of the limits of integration, and the integral of a function over adjacent intervals.

This EK is much simpler than the previous one. Let f, g be integrable on $[a, b]$. Then the following properties hold:

Constant Multiple Property

Let k be a real number.

$$\int_a^b kf(x)\,dx = k\int_a^b f(x)\,dx$$

Sum Property

$$\int_a^b \left[f(x) + g(x) \right] dx = \int_a^b f(x)\, dx + \int_a^b g(x)\, dx$$

Reversal of Limits Property

$$\int_a^b f(x)\, dx = -\int_b^a f(x)\, dx$$

Adjacent Interval Property

Let c be in the interval (a, b).

$$\int_a^b f(x)\, dx = \int_a^c f(x)\, dx + \int_c^b f(x)\, dx$$

We will illustrate these properties with examples.

Example 3.2.17: Let $\int_{-1}^4 f(x)dx = 6,\ \int_4^7 f(x)dx = -3,$ and $\int_{-1}^4 g(x)dx = -7$. Compute the following:

1. $\int_{-1}^4 5f(x)\, dx$

2. $\int_4^{-1} f(x)\, dx$

3. $\int_{-1}^4 \left[f(x) + g(x) \right] dx$

4. $\int_{-1}^7 f(x)\, dx$

We will use the four properties given above to evaluate these.

1. Use the Constant Multiple Property.

$$\int_{-1}^4 5f(x)\, dx = 5\int_{-1}^4 f(x)\, dx = 5(6) = 30$$

2. Use the Sum Property.

$$\int_{-1}^4 \left[f(x) + g(x) \right] dx = \int_{-1}^4 f(x)\, dx + \int_{-1}^4 g(x)\, dx = 6 + (-7) = -1$$

3. Use the Reversal of Limits Property.

$$\int_4^{-1} f(x)\, dx = -\int_{-1}^4 f(x)\, dx = -6$$

4. Use the Adjacent Interval Property.

$$\int_{-1}^{7} f(x)\,dx = \int_{-1}^{4} f(x)\,dx + \int_{4}^{7} f(x)\,dx = 6 + (-3) = 3$$

Difference Rule

We can obtain a useful corollary of the Constant Multiple and Sum Properties as follows:

$$\int_{a}^{b} \left[f(x) - g(x) \right] dx = \int_{a}^{b} \left[f(x) + \left(-1g(x) \right) \right] dx$$

$$\int_{a}^{b} \left[f(x) - g(x) \right] dx = \int_{a}^{b} f(x)\,dx + \int_{a}^{b} -1g(x)\,dx$$

$$\int_{a}^{b} \left[f(x) - g(x) \right] dx = \int_{a}^{b} f(x)\,dx - \int_{a}^{b} g(x)\,dx$$

Zero Interval Rule

Suppose f is continuous at $x = c$. Then using the Reversal of Limits Property with $a = b = c$, we obtain the following corollary:

$$\int_{c}^{c} f(x)\,dx = -\int_{c}^{c} f(x)\,dx$$

Adding $\int_{c}^{c} f(x)\,dx$ to both sides, we obtain the following:

$$2\int_{c}^{c} f(x)\,dx = 0$$

$$\int_{c}^{c} f(x)\,dx = 0$$

Example 3.2.18: Let f be continuous at $x = 3$, and let $\int_{-1}^{4} f(x)\,dx = 6$ and $\int_{-1}^{4} g(x)\,dx = -7$. Compute the following:

1. $\int_{-1}^{4} \left[f(x) - g(x) \right] dx$

2. $\int_{3}^{3} f(x)\,dx$

We will use the two new rules to compute these integrals.

1. Use the Difference Rule.

$$\int_{-1}^{4}\left[f(x)-g(x)\right]dx = \int_{-1}^{4}f(x)\,dx - \int_{-1}^{4}g(x)\,dx = 6-(-7)=13$$

2. Use the Zero Interval Rule.

$$\int_{3}^{3}f(x)\,dx = 0$$

EK 3.2C3: The definition of the definite integral may be extended to functions with removable or jump discontinuities.

As we reviewed in **EK 3.2A2**, continuity implies integrability. The converse is not true. That is, if a function f is integrable on $[a,b]$, it does *not* imply that f is continuous on $[a,b]$. There are two cases covered on the exam: removable discontinuities and jump discontinuities.

Removable Discontinuities

If f is continuous on $[a,b]$ except for a removable discontinuity at a point $x=c$ in $[a,b]$, then f is integrable on $[a,b]$. Furthermore,

$$\int_{a}^{b}f(x)\,dx = \int_{a}^{b}g(x)\,dx,$$

where

$$g(x)-\begin{cases} f(x) & x \neq c \\ \lim_{x\to c}f(x) & x=c \end{cases}.$$

That is, if g is a continuous function on *all* of $[a,b]$ that is obtained by removing the removable discontinuity of f at c, then the two functions have the same integral on $[a,b]$. We will illustrate with an example.

Example 3.2.19: Compute $\displaystyle\int_{0}^{2}\frac{x^2-1}{x-1}\,dx$.

The function $f(x)=\dfrac{x^2-1}{x-1}$ is discontinuous at $x=1$, which is in $[0,2]$. However, this discontinuity is removable, and $\displaystyle\lim_{x\to 1}\frac{x^2-1}{x-1}=\lim_{x\to 1}\frac{(x-1)(x+1)}{x-1}=\lim_{x\to 1}(x+1)=2$. Define the function g as follows:

$$g(x)=\begin{cases} f(x) & x \neq 1 \\ 2 & x=1 \end{cases}$$

The graph of this function is shown below.

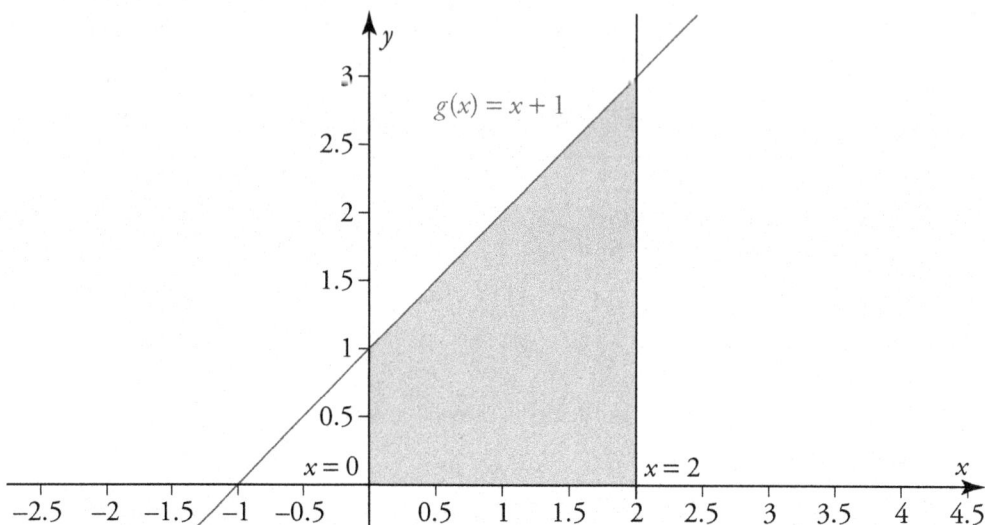

So $\int_0^2 \dfrac{x^2-1}{x-1}\,dx = \int_0^2 g(x)\,dx$, and the integral on the right side can be evaluated by using geometric formulas, as we reviewed in **EK 3.2C1**. If we cut the shaded region into two subregions at $y = 1$, then we obtain a rectangular region of area 2 and a triangular region of area 2. So we obtain the following as our final result:

$$\int_0^2 \frac{x^2-1}{x-1}\,dx = \int_0^2 g(x)\,dx = 2+2 = 4$$

This case can also be extended to functions that have *finitely many* removable discontinuities in $[a,b]$.

Example 3.2.20: Compute $\displaystyle\int_0^4 \frac{x^3+x^2-6x}{x^2-2x}\,dx$.

The function $f(x) = \dfrac{x^3+x^2-6x}{x^2-2x}$ has removable discontinuities at $x=0$ and $x=2$, both of which are in $[0,4]$. Furthermore,

$$\lim_{x\to 0}\frac{x^3+x^2-6x}{x^2-2x} = \lim_{x\to 0}\frac{x(x+3)(x-2)}{x(x-2)} = \lim_{x\to 0}(x+3) = 3,$$

and

$$\lim_{x\to 2}\frac{x^3+x^2-6x}{x^2-2x} = \lim_{x\to 2}\frac{x(x+3)(x-2)}{x(x-2)} = \lim_{x\to 2}(x+3) = 5.$$

Define the function g as follows:

$$g(x) = \begin{cases} f(x) & x \neq 0, 2 \\ 3 & x = 0 \\ 5 & x = 2 \end{cases}$$

The graph of this function is shown below.

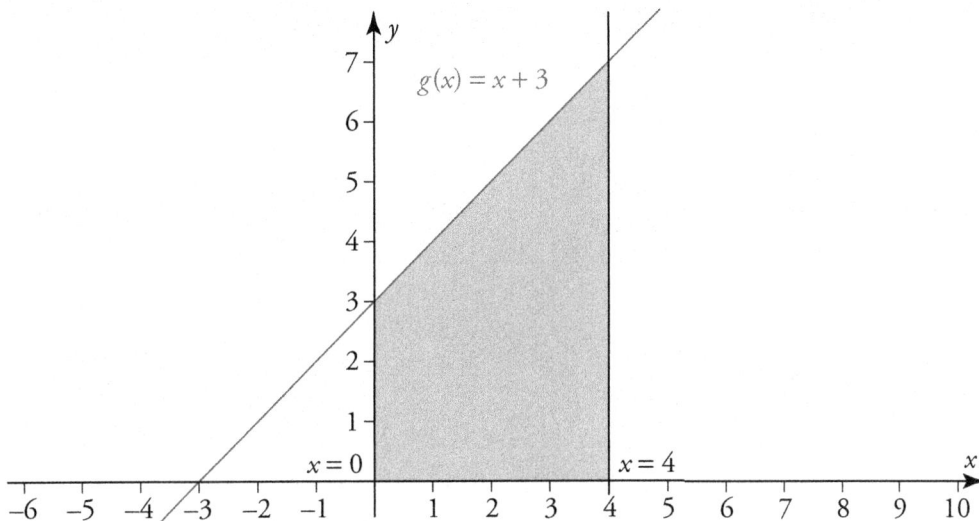

The shaded region can be cut at $y = 3$ into a rectangular region of area 12 and a triangular region of area 8. So we have the following:

$$\int_0^4 \frac{x^3 + x^2 - 6x}{x^2 - 2x}\,dx = \int_0^4 g(x)\,dx = 12 + 8 = 20$$

Jump Discontinuities

If f is continuous on $[a, b]$, except for a *finite* jump discontinuity at $x = c$ in (a, b), then f is integrable on $[a, b]$. Furthermore,

$$\int_a^b f(x)\,dx = \int_a^c f(x)\,dx + \int_c^b f(x)\,dx$$

Example 3.2.21: Compute $\displaystyle\int_{-2}^2 \left(\frac{|x|}{x} + 3 \right) dx$.

The graph of the function $f(x) = \dfrac{|x|}{x} + 3$ is shown below.

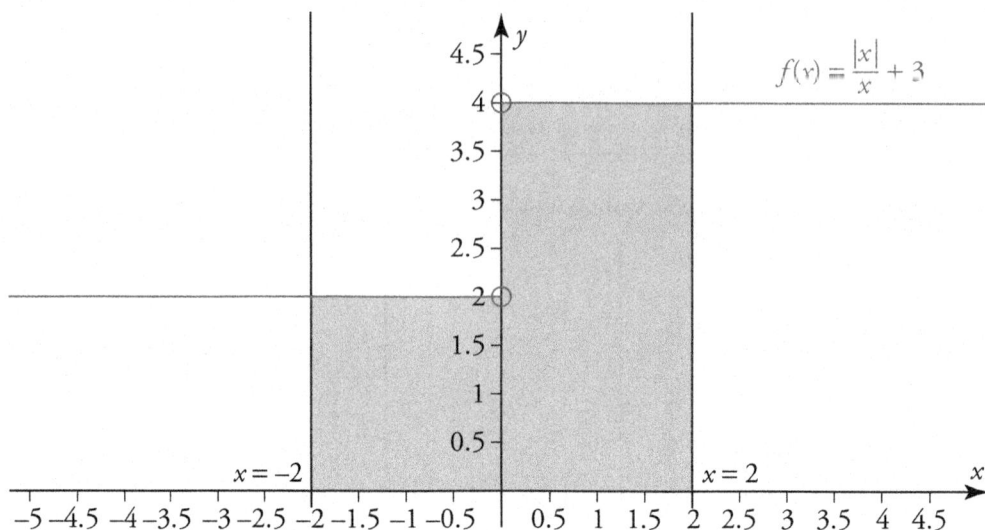

f has a finite jump discontinuity at $x = 0$, so we compute the integral as follows:

$$\int_{-2}^{2} \left(\frac{|x|}{x} + 3 \right) dx = \int_{-2}^{0} \left(\frac{|x|}{x} + 3 \right) dx + \int_{0}^{2} \left(\frac{|x|}{x} + 3 \right) dx$$

For the first integral on the right side, the integrand has a removable discontinuity at $x = 0$, and likewise for the second integral on the right side. Using the formula for the area of a rectangle, we see that $\int_{-2}^{0} \left(\frac{|x|}{x} + 3 \right) dx = 4$ and $\int_{0}^{2} \left(\frac{|x|}{x} + 3 \right) dx = 8$. So we obtain the following for the total integral:

$$\int_{-2}^{2} \left(\frac{|x|}{x} + 3 \right) dx = 4 + 8 = 12$$

This case can also be extended to functions that have *finitely many* jump discontinuities in (a, b).

Example 3.2.22: Compute $\int_{1}^{4} [\![x]\!]\, dx$, where $f(x) = [\![x]\!]$ is the greatest integer function.

The graph of $f(x) = [\![x]\!]$ is shown below.

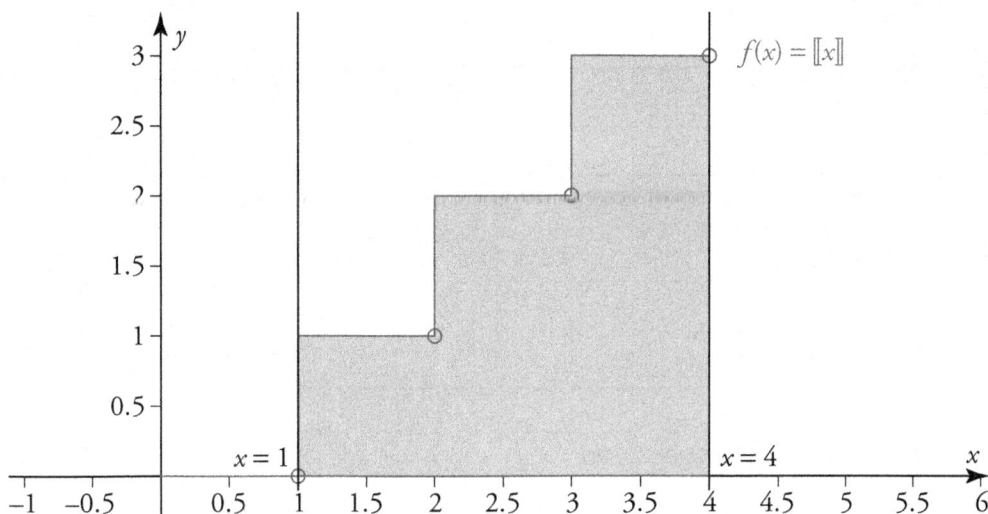

f has two finite jump discontinuities in $(1, 4)$, so we compute the integral as follows:

$$\int\limits_1^4 [\![x]\!]\, dx = \int\limits_1^2 [\![x]\!]\, dx + \int\limits_2^3 [\![x]\!]\, dx + \int\limits_3^4 [\![x]\!]\, dx$$

Each of the three integrals on the right side has a removable discontinuity at its left endpoint. Using the formula for the area of a rectangle, we see that $\int\limits_1^2 [\![x]\!]\, dx = 1$, $\int\limits_2^3 [\![x]\!]\, dx = 2$, and $\int\limits_3^4 [\![x]\!]\, dx = 3$. So we obtain the following for the total integral:

$$\int\limits_1^4 [\![x]\!]\, dx = 1 + 2 + 3 = 6$$

LO 3.2D: (BC) Improper Integrals

The full Learning Objective is given below.

LO 3.2D: (BC) Evaluate an improper integral or show that an improper integral diverges.

First, we will classify improper integrals, and then we will learn how to evaluate them, if they converge.

EK 3.2D1: (BC) An improper integral is an integral that has one or both limits infinite or has an integrand that is unbounded in the interval of integration.

As stated in the EK, there are two broad classes of improper integrals. We will review them one at a time.

Integrals with One or More Infinite Limits of Integration

In each of the three forms that follow, we assume that f is continuous on its interval of integration.

Form	Example
$\displaystyle\int_a^\infty f(x)\,dx$	$\displaystyle\int_1^\infty \frac{1}{x}\,dx$
$\displaystyle\int_{-\infty}^b f(x)\,dx$	$\displaystyle\int_{-\infty}^1 e^x\,dx$
$\displaystyle\int_{-\infty}^\infty f(x)\,dx$	$\displaystyle\int_{-\infty}^\infty \frac{1}{x^2+1}\,dx$

Integrands That Are Unbounded

Each of these forms have two *finite* limits of integration, so the integral looks like $\displaystyle\int_a^b f(x)\,dx$.

Form	Example
f is continuous on $[a,b)$ and unbounded at $x=b$.	$\displaystyle\int_{-1}^0 \frac{1}{x^2}\,dx$
f is continuous on $(a,b]$ and unbounded at $x=a$.	$\displaystyle\int_0^1 \frac{1}{\sqrt{x}}\,dx$
f is unbounded at a point c in (a,b) and continuous on the rest of $[a,b]$.	$\displaystyle\int_0^\pi \sec(x)\tan(x)\,dx$

EK 3.2D2: (BC) Improper integrals can be determined using limits of definite integrals.

None of the functions listed in **EK 3.2D1** are Riemann integrable. That's why the integrals are called *improper*. However, there *is* a method by which some of them (not all!) can be evaluated. The method involves using a dummy variable and taking a limit. We will now review how to use this method for each type of improper integral.

Integrals with One or More Infinite Limits of Integration

Once again, for each of these forms we assume that f is continuous on its entire interval of integration. The prescriptions for dealing with each of the forms are reviewed below. We start with the first two forms.

$$\int_a^\infty f(x)\,dx = \lim_{b\to\infty}\int_a^b f(x)\,dx$$

$$\int_{-\infty}^b f(x)\,dx = \lim_{a\to-\infty}\int_a^b f(x)\,dx$$

For the first two forms, we substitute a dummy variable for the infinite limit of integration, then we integrate using the Fundamental Theorem of Calculus (to be reviewed in **EU 3.3**), and then we take the limit of the resulting expression. If the limit exists and equals a real number L, then we say that the improper integral *converges* to L. If the limit does not exist, then we say that the improper integral *diverges*. We conclude with the third form.

$$\int_{-\infty}^{\infty} f(x)\,dx = \int_{-\infty}^{c} f(x)\,dx + \int_{c}^{\infty} f(x)\,dx$$

For the third form, we choose a real number c at which to break up the interval of integration (which is the entire real number line), then we apply the Adjacent Interval Property to split the improper integral into two improper integrals. On the right side, the first improper integral is of the second form, and the second improper integral is of the first form. They are evaluated according to the prescriptions given above. If *either* of the improper integrals on the right side diverges, then the entire improper integral diverges. If both improper integrals on the right side converge, then the entire improper integral converges to the sum of their values.

We will now go over some examples. Since this is a BC topic, we will assume that you are familiar with the AP topics of the Fundamental Theorem of Calculus. If you need to review this topic, it is discussed in **EK 3.3B2**.

Example 3.2.23: Determine whether $\displaystyle\int_{1}^{\infty} \frac{1}{x}\,dx$ converges or diverges. If it converges, state its value.

This improper integral is of the first form reviewed above.

$$\int_{1}^{\infty} \frac{1}{x}\,dx = \lim_{b \to \infty} \int_{1}^{b} \frac{1}{x}\,dx = \lim_{b \to \infty} \Big[\ln(x)\Big]_{1}^{b} = \lim_{b \to \infty} \Big[\ln(b) - \ln(1)\Big] = \infty$$

This improper integral diverges.

Example 3.2.24: Determine whether $\displaystyle\int_{-\infty}^{1} e^{x}\,dx$ converges or diverges. If it converges, state its value.

This improper integral is of the second form reviewed above.

$$\int_{-\infty}^{1} e^{x}\,dx = \lim_{a \to -\infty} \Big[e^{x}\Big]_{a}^{1} = \lim_{a \to -\infty} \Big[e^{1} - e^{a}\Big] = e - 0 = e$$

This improper integral converges to e.

Example 3.2.25: Determine whether $\displaystyle\int_{-\infty}^{\infty} \frac{1}{x^2 + 1}\,dx$ converges or diverges. If it converges, state its value.

This improper integral is of the third form reviewed above. We choose to split the interval of integration at $c = 0$.

$$\int_{-\infty}^{\infty} \frac{1}{x^2+1}\, dx = \int_{-\infty}^{0} \frac{1}{x^2+1}\, dx + \int_{0}^{\infty} \frac{1}{x^2+1}\, dx$$

The two improper integrals on the right side are of the second and first forms, respectively.

We evaluate the first improper integral on the right side.

$$\int_{-\infty}^{0} \frac{1}{x^2+1}\, dx = \lim_{a\to-\infty} \int_{a}^{0} \frac{1}{x^2+1}\, dx = \lim_{a\to-\infty} \left[\tan^{-1}(x) \right]_{a}^{0} = \lim_{a\to-\infty} \left[\tan^{-1}(0) - \tan^{-1}(a) \right]$$

$$\int_{-\infty}^{0} \frac{1}{x^2+1}\, dx = 0 - \left(-\frac{\pi}{2}\right) = \frac{\pi}{2}$$

And now we evaluate the improper second integral on the right side.

$$\int_{0}^{\infty} \frac{1}{x^2+1}\, dx = \lim_{b\to\infty} \int_{0}^{b} \frac{1}{x^2+1}\, dx = \lim_{b\to\infty} \left[\tan^{-1}(x) \right]_{0}^{b} = \lim_{b\to\infty} \left[\tan^{-1}(b) - \tan^{-1}(0) \right]$$

$$\int_{0}^{\infty} \frac{1}{x^2+1}\, dx = \frac{\pi}{2} - 0 = \frac{\pi}{2}$$

Since both improper integrals on the right side converge, the entire improper integral converges. We combine the above two results to obtain its value.

$$\int_{-\infty}^{\infty} \frac{1}{x^2+1}\, dx = \frac{\pi}{2} + \frac{\pi}{2} = \pi$$

Integrands That Are Unbounded

Once again, these integrals all look like $\int_{a}^{b} f(x)\, dx$, where a and b are real numbers. The prescriptions for dealing with each of the three forms are reviewed below, starting with the first two forms.

If f is continuous on $[a, b)$ and unbounded at $x = b$, then

$$\int_{a}^{b} f(x)\, dx = \lim_{t\to b^-} \int_{a}^{c} f(x)\, dx.$$

If f is continuous on $(a, b]$ and unbounded at $x = a$, then

$$\int_{a}^{b} f(x)\, dx = \lim_{t\to a^+} \int_{c}^{b} f(x)\, dx.$$

For the first two forms, we substitute a dummy variable for the limit of integration at which the integrand is unbounded, then we integrate using the Fundamental Theorem of Calculus, and then we take the limit of the resulting expression. We are only interested in one-sided limits because the only way to approach b from *within* $[a, b]$ is from the left, and the only way to approach a from *within* $[a, b]$ is from the right. If the limit exists and equals a real number L, then we say that the improper integral *converges* to L. If the limit does not exist, then we say that the improper integral *diverges*. We will wrap this up with the third form.

If f is unbounded at a point c in (a, b) and continuous on the rest of $[a, b]$, then

$$\int_a^b f(x)\,dx = \int_a^c f(x)\,dx + \int_c^b f(x)\,dx.$$

For the third form we break up the interval of integration at $x = c$, then we apply the Adjacent Interval Property to split the improper integral into two improper integrals. On the right side, the two integrals are of the first and second forms, respectively. They are evaluated according to the prescriptions given above. If *either* of the improper integrals on the right side diverges, then the entire improper integral diverges. If both improper integrals on the right side converge, then the entire improper integral converges to the sum of their values.

We will illustrate these prescriptions with some examples.

Example 3.2.26: Determine whether $\displaystyle\int_{-1}^0 \frac{1}{x^2}\,dx$ converges or diverges. If it converges, state its value.

The integrand is unbounded at the right endpoint $x = 0$, so this is of the first form.

$$\int_{-1}^0 \frac{1}{x^2}\,dx = \lim_{t \to 0^-} \int_{-1}^t x^{-2}\,dx = -\lim_{t \to 0^-}\left[\frac{1}{x}\right]_{-1}^t = -\lim_{t \to 0^-}\left[\frac{1}{t} - \frac{1}{-1}\right] = \infty$$

This improper integral diverges.

Example 3.2.27: Determine whether $\displaystyle\int_0^1 \frac{1}{\sqrt{x}}\,dx$ converges or diverges. If it converges, state its value.

The integrand is unbounded at the left endpoint $x = 0$, so this is of the second form.

$$\int_0^1 \frac{1}{\sqrt{x}}\,dx = \lim_{t \to 0^+} \int_t^1 x^{-1/2}\,dx = 2\lim_{t \to 0^+}\left[\sqrt{x}\right]_t^1 = 2\lim_{t \to 0^+}\left[\sqrt{1} - \sqrt{t}\right] = 2$$

This improper integral converges to 2.

Example 3.2.28: Determine whether $\displaystyle\int_0^\pi \sec(x)\tan(x)\,dx$ converges or diverges. If it converges, state its value.

The integrand is unbounded at $x = \dfrac{\pi}{?}$, which is in the interval $(0, \pi)$, so this is of the third form. We will split the interval of integration at $\dfrac{\pi}{2}$.

$$\int_0^{\pi} \sec(x)\tan(x)\,dx = \int_0^{\pi/2} \sec(x)\tan(x)\,dx + \int_{\pi/2}^{\pi} \sec(x)\tan(x)\,dx$$

The two integrals on the right side are of the first form and second form, respectively. We will work with the first of these next.

$$\int_0^{\pi/2} \sec(x)\tan(x)\,dx = \lim_{t \to \frac{\pi}{2}^-} \int_0^{t} \sec(x)\tan(x)\,dx = \lim_{t \to \frac{\pi}{2}^-} \Big[\sec(x)\Big]_0^{t}$$

$$\int_0^{\pi/2} \sec(x)\tan(x)\,dx = \lim_{t \to \frac{\pi}{2}^-} \Big[\sec(t) - \sec(0)\Big] = \infty$$

Since the first improper integral on the right side diverges, there is no need to evaluate the second. The entire improper integral diverges.

As one final example, we will show you how to tackle an integral that is *doubly improper*. This type could show up on the exam.

Example 3.2.29: Determine whether $\displaystyle\int_0^{\infty} \dfrac{1}{\sqrt[3]{x}}\,dx$ converges or diverges. If it converges, state its value.

This integral is doubly improper because the integrand is unbounded at the lower limit of integration, *and* the upper limit of integration is infinite. We isolate these two issues by splitting the doubly improper integral up into *two* improper integrals, each with only one issue. We'll split up the interval of integration at $x = 1$.

$$\int_0^{\infty} \dfrac{1}{\sqrt[3]{x}}\,dx = \int_0^{1} \dfrac{1}{\sqrt[3]{x}}\,dx + \int_1^{\infty} \dfrac{1}{\sqrt[3]{x}}\,dx$$

If either of the two integrals on the right side diverges, then the entire integral diverges. We'll begin by evaluating the first integral on the right side.

$$\int_0^{1} \dfrac{1}{\sqrt[3]{x}}\,dx = \lim_{t \to 0^+} \int_t^{1} x^{-1/3}\,dx = \dfrac{3}{2}\lim_{t \to 0^+} x^{2/3}\Big|_t^{1} = \dfrac{3}{2}\lim_{t \to 0^+}\left(1^{2/3} - t^{2/3}\right) = \dfrac{3}{2}$$

Since this integral converges, we move on to the second integral on the right side.

$$\int_1^{\infty} \dfrac{1}{\sqrt[3]{x}}\,dx = \lim_{b \to \infty} \int_1^{b} x^{-1/3}\,dx = \dfrac{3}{2}\lim_{b \to \infty} x^{2/3}\Big|_1^{b} = \dfrac{3}{2}\lim_{b \to \infty}\left(b^{2/3} - 1^{2/3}\right) = \infty$$

Since the second improper integral on the right side diverges, the entire improper integral diverges.

EU 3.3 – The Fundamental Theorem of Calculus _____

LO 3.3A – Functions Defined by Integrals

The full Learning Objective is as follows:

LO 3.3A: Analyze functions defined by an integral.

This doesn't tell us much, but the three items of Essential Knowledge will expand on it for us.

EK 3.3A1: The definite integral can be used to define new functions; for example,

$$f(x) = \int_0^x e^{-t^2} dt.$$

Integrals such as the example given in the statement of the EK depend on the value of x in the upper limit of integration. Furthermore, since a definite integral is a limit of a Riemann sum, and since limits are *unique* (if they exist), it follows that for any given real number x, the integral expression returns *at most one* real number. That is precisely what a *function* does, and so the integral expression really does define a function of x.

EK 3.3A2: If f is a continuous function on the interval $[a, b]$, then $\dfrac{d}{dx}\left(\int_a^x f(t)\,dt\right) = f(x)$, where x is between a and b.

We will illustrate this simple result with an example.

Example 3.3.1: Compute $\dfrac{d}{dx}\left(\int_1^x \sin(t)\,dt\right)$.

The function $\sin(x)$ is continuous for *all* real numbers, so it is continuous on all closed intervals $[1, b]$, where x is between 1 and b. So, we'll apply the result.

$$\frac{d}{dx}\left(\int_1^x \sin(t)\,dt\right) = \sin(x)$$

The next EK is a bit more involved.

EK 3.3A3: Graphical, numerical, analytical, and verbal representations of a function f provide information about the function $g(x) = \int_a^x f(t)\,dt$.

We will review this EK through examples.

Example 3.3.2: f is a piecewise linear function defined on $[0, 5]$. It consists of line segments that join the following list of points:

x:	0	1	2	3	4	5
$f(x)$:	0	2	0	0	3	0

A graph of f is shown below. The grid lines are 0.5 units apart.

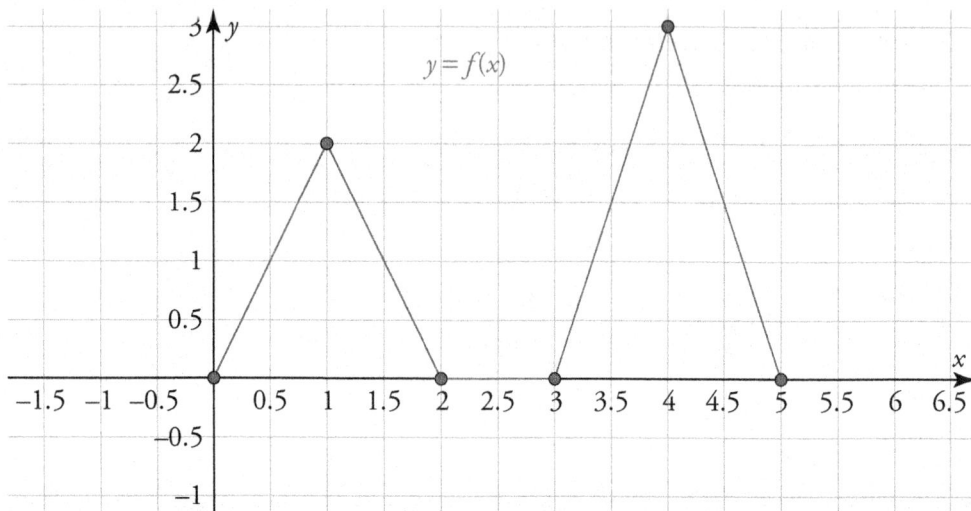

Let $g(x) = \int_0^x f(t)\,dt$. Use the given information about f to fill in the following table:

x:	0	1	2	3	4	5
g(x):						

In this example, the representation of f is a combination of graphical, numerical, and verbal. We evaluate g as follows:

$$g(0) = \int_0^0 f(t)\,dt = 0, \text{ by the Zero Interval Rule.}$$

To evaluate g at the remaining points, we'll use what we learned in **EK 3.2C1** about the connection between definite integrals and geometry.

On $[0, 1]$, the graph of f and the x-axis form a triangle of base 1 and height 2.

$$g(1) = \int_0^1 f(t)\,dt = \frac{1}{2}(1)(2) = 1$$

On $[1, 2]$, the graph of f and the x-axis again form a triangle of base 1 and height 2.

$$g(2) = \int_0^2 f(t)\,dt = \int_0^1 f(t)\,dt + \int_1^2 f(t)\,dt = 1 + \frac{1}{2}(1)(2) = 2$$

On $[2, 3]$, the graph of f is on the x-axis. You can think of the figure as a rectangle of zero area.

$$g(3) = \int_0^3 f(t)\,dt = \int_0^2 f(t)\,dt + \int_2^3 f(t)\,dt = 2 + 0 = 2$$

On [3,4], the graph of f and the x-axis form a triangle of base 1 and height 3.

$$g(4) = \int_0^4 f(t)\,dt = \int_0^3 f(t)\,dt + \int_3^4 f(t)\,dt = 2 + \frac{1}{2}(1)(3) = \frac{7}{2}$$

Finally, on $[4,5]$, the graph of f and the x-axis again form a triangle of base 1 and height 3.

$$g(5) = \int_0^5 f(t)\,dt = \int_0^4 f(t)\,dt + \int_4^5 f(t)\,dt = \frac{7}{2} + \frac{1}{2}(1)(3) = 5$$

We now have everything we need to fill in the table.

x:	0	1	2	3	4	5
$g(x)$:	0	1	2	2	$\frac{7}{2}$	5

Example 3.3.3: Let $f(x) = \cos(x^2)$ be defined on the interval $\left(-\sqrt{2\pi}, \sqrt{2\pi}\right)$. Find the open intervals on which the function $g(x) = \displaystyle\int_{-\sqrt{2\pi}}^{x} f(t)\,dt$ is concave upward and on which it is concave downward.

This time, f is represented analytically. There's no easy way to integrate f, but fortunately we don't have to do that. That's because we don't need the function g itself, but rather, its second derivative.

$$g'(x) = \frac{d}{dx}\left(\int_{-\sqrt{2\pi}}^{x} \cos(t^2)\,dt \right) = \cos(x^2)$$

$$g''(x) = \frac{d}{dx}\left(\cos(x^2) \right) = -2x\sin(x^2)$$

Setting $g''(x) = 0$, we get the following:

$$-2x = 0 \Rightarrow x = 0$$

$$\sin(x^2) = 0 \Rightarrow x^2 = 0, \pi \Rightarrow x = 0, \pm\sqrt{\pi}$$

We will test $g''(x)$ on the intervals $\left(-\sqrt{2\pi}, -\sqrt{\pi}\right)$, $\left(-\sqrt{\pi}, 0\right)$, $\left(0, \sqrt{\pi}\right)$, and $\left(\sqrt{\pi}, \sqrt{2\pi}\right)$.

Interval	Test Value x	$g''(x)$	Behavior of g
$\left(-\sqrt{2\pi},-\sqrt{\pi}\right)$	$x = -\sqrt{\dfrac{3\pi}{2}}$	$g''\left(-\sqrt{\dfrac{3\pi}{2}}\right) = 2\sqrt{\dfrac{3\pi}{2}} > 0$	Concave Upward
$\left(-\sqrt{\pi},0\right)$	$x = -\sqrt{\dfrac{\pi}{2}}$	$g''\left(-\sqrt{\dfrac{\pi}{2}}\right) = -2\sqrt{\dfrac{\pi}{2}} < 0$	Concave Downward
$\left(0,\sqrt{\pi}\right)$	$x = \sqrt{\dfrac{\pi}{2}}$	$g''\left(\sqrt{\dfrac{\pi}{2}}\right) = 2\sqrt{\dfrac{\pi}{2}} < 0$	Concave Upward
$\left(\sqrt{\pi},\sqrt{2\pi}\right)$	$x = \sqrt{\dfrac{3\pi}{2}}$	$g''\left(\sqrt{\dfrac{3\pi}{2}}\right) = -2\sqrt{\dfrac{3\pi}{2}} < 0$	Concave Downward

LO 3.3B(a,b): Antiderivatives and Definite Integrals

The Learning Objectives for this section are short and sweet.

LO 3.3B(a): Calculate antiderivatives.

LO 3.3B(b): Evaluate definite integrals.

We will review these two Learning Objectives together, because the items of Essential Knowledge under them are intertwined.

EK 3.3B1: The function defined by $F(x) = \int_{a}^{x} f(t)\,dt$ is an antiderivative of f.

To demonstrate this, we simply take the derivative of F, using the result from **EK 3.3A2**.

$$F'(x) = \frac{d}{dx}\left(\int_{a}^{x} f(t)\,dt\right) = f(x)$$

Since $F' = f$, this means that F is an antiderivative of f *by definition*.

The next EK is the most important result in a course in single variable calculus. It is the *Fundamental Theorem of Calculus*. This theorem connects definite integration with antidifferentiation.

EK 3.3B2: If f is continuous on the interval $[a,b]$ and F is an antiderivative of f, then $\int_{a}^{b} f(x)\,dx = F(b) - F(a)$.

The Fundamental Theorem of Calculus states that a definite integral can be evaluated by evaluating an antiderivative F of the integrand f at the right and left endpoints of the interval of integration and subtracting the results. Any antiderivative of f will do. Recall that if $F(x)$ is an antiderivative of $f(x)$, then so is $F(x) + C$. But when this alternate form of the antiderivative is used in the right side of the Fundamental Theorem, the two arbitrary constants cancel under subtraction. The upshot of this is that we need not concern ourselves with these constants when finding antiderivatives for use in the Fundamental Theorem.

Let's look at some examples, starting with one that you have seen several times already. Consider the integral $\int_0^2 (x^2 - 3x)\,dx$. We have approximated this integral four times in previous examples. The results are summarized below.

Example	Approximation Used	Result
3.2.3	Left Riemann Sum	$\int_0^2 (x^2 - 3x)\,dx \approx -\dfrac{11}{4} = -2.75$
	Right Riemann Sum	$\int_0^2 (x^2 - 3x)\,dx \approx -\dfrac{15}{4} = -3.75$
3.2.13	Midpoint Riemann Sum	$\int_0^2 (x^2 - 3x)\,dx \approx -\dfrac{27}{8} = -3.375$
3.2.14	Trapezoidal Sum	$\int_0^2 (x^2 - 3x)\,dx \approx -\dfrac{13}{4} = -3.25$

These approximations seem to be fairly close together. They were obtained using a partition of $[0, 2]$ into four subintervals. The agreement between these approximations would improve as the number n of subintervals increases, and as $n \to \infty$, all four approximations would converge to the *same* value, which is the *exact* value of the integral. But what exact value would that be? The Fundamental Theorem of Calculus will answer that for us.

Example 3.3.6: Evaluate $\int_0^2 (x^2 - 3x)\,dx$.

An antiderivative of $x^2 - 3x$ is $\dfrac{1}{3}x^3 - \dfrac{3}{2}x^2$. We are to evaluate this at 2 and at 0, and then subtract the results. We write this as follows:

$$\int_0^2 (x^2 - 3x)\,dx = \left[\frac{1}{3}x^3 - \frac{3}{2}x^2 \right]\Bigg|_0^2$$

The vertical bar on the right side is known as an *evaluation bar*. It tells us to evaluate the quantity in brackets at the upper limit, and at the lower limit, and then subtract.

$$\int_0^2 (x^2 - 3x)\,dx = \left[\frac{1}{3}2^3 - \frac{3}{2}2^2 \right] - \left[\frac{1}{3}0^3 - \frac{3}{2}0^2 \right] = -\frac{10}{3} - 0 = -\frac{10}{3} = -3.\overline{3}$$

So at last, we have the exact value of $\int_0^2 (x^2 - 3x)\,dx$. Let's work out a couple more.

Example 3.3.7: Evaluate $\displaystyle\int_0^{\pi/4} \cos(x)\,dx$.

$$\int_0^{\pi/4} \cos(x)\,dx = \sin(x)\Big|_0^{\pi/4} = \sin\left(\frac{\pi}{4}\right) - \sin(0) = \frac{\sqrt{2}}{2} - 0 = \frac{\sqrt{2}}{2}$$

Example 3.3.8: Evaluate $\displaystyle\int_{\ln(2)}^{\ln(5)} e^x\,dx$.

$$\int_{\ln(2)}^{\ln(5)} e^x\,dx = e^x\Big|_{\ln(2)}^{\ln(5)} = e^{\ln(5)} - e^{\ln(2)} = 5 - 2 = 3$$

EK 3.3B3: The notation $\int f(x)\,dx = F(x) + C$ means that $F'(x) = f(x)$, and $\int f(x)\,dx$ is called an *indefinite integral* of the function f.

The statement of this EK introduces a new term: *indefinite integral*. This is a synonym for antiderivative, and indefinite integration is synonymous with antidifferentiation. Henceforth, we will refer to the function $f(x)$ in an indefinite integral as an *integrand*, and to the arbitrary constant C as a *constant of integration*.

EK 3.3B4: Many functions do not have closed form antiderivatives.

As the EK states, there exist functions $f(x)$ whose antiderivatives cannot be expressed in terms of the elementary functions of precalculus. One such example is $f(x) = e^{-x^2}$. No combination of techniques of antidifferentiation will allow you to express $\int e^{-x^2}\,dx$ in terms of elementary functions.

We note however that the existence of an indefinite integral for a function f in terms of elementary functions on an interval $[a, b]$ has *nothing to do* with whether or not f is integrable on $[a, b]$.

EK 3.3B5: Techniques for finding antiderivatives include algebraic manipulation such as long division and completing the square, substitution of variables, **(BC)** integration by parts, and nonrepeating linear partial fractions.

We'll review these out of order by starting with substitution of variables.

Substitution of Variables

You learned to differentiate composite functions using the Chain Rule, which is stated as follows for differentiable functions f and g.

$$\frac{d}{dx}\big(f(g(x))\big) = f'(g(x))g'(x)$$

If we let $u = g(x)$ and $y = f(u)$, then the Chain Rule takes on a particularly simple form.

$$\frac{dy}{dx} = \frac{dy}{du} \cdot \frac{du}{dx}$$

It turns out that we can get an antidifferentiation rule from this. Suppose we have an indefinite integral of the following form:

$$\int f\big(g(x)\big)g'(x)\,dx$$

We tackle this by making a substitution of variables. Explicitly, we let $u = g(x)$. The differential of this is $du = g'(x)\,dx$. This transforms the integral as follows:

$$\int f\big(g(x)\big)g'(x)\,dx = \int f(u)\,du$$

The above change of variable is normally called a *u*-substitution. The trick is to choose *u* wisely so that the resulting integral can be computed using one or more of the rules that we reviewed above. As a rule of thumb, look for a composite function in the integrand, and then set *u* equal to the inner part of the composite.

Example 3.3.9: Evaluate $\int 2x\sin\big(x^2\big)dx$.

Let $u = x^2$. Then $du = 2x\,dx$.

$$\int 2x\sin\big(x^2\big)dx = \int \sin\big(x^2\big)\cdot 2x\ dx = \int \sin(u)\,du = -\cos(u)+C$$

After the integration is complete, we undo the substitution and switch back to the original variable.

$$\int 2x\sin\big(x^2\big)dx = -\cos\big(x^2\big)+C$$

Example 3.3.10: Evaluate $\int \dfrac{e^{\sqrt{x}}}{\sqrt{x}}dx$.

Let $u = \sqrt{x} = x^{1/2}$. Then $du = \dfrac{1}{2}x^{-1/2}dx = \dfrac{1}{2\sqrt{x}}dx$. Notice that this time, we do not exactly have the differential du in the integrand, as the factor of 2 in the denominator is missing. However, if we multiply the differential by 2, then we obtain $\dfrac{1}{\sqrt{x}}dx = 2\ du$.

$$\int \frac{e^{\sqrt{x}}}{\sqrt{x}}\,dx = \int e^{\sqrt{x}}\cdot\frac{1}{\sqrt{x}}\,dx = \int e^u\,du = e^u +C = e^{\sqrt{x}}+C$$

Here's a tricky one:

Example 3.3.11: Evaluate $\int \dfrac{x}{\big(1+x\big)^2}dx$.

According to our rule of thumb, we should let $u = 1 + x$, since that is what's inside of the composite function in the integrand. Then we have $du = dx$, but we still have to do something with the x in the numerator. Taking a closer look at the substitution, we see that if $u = 1 + x$, then $x = u - 1$. We are now ready to do our substitution of variables.

$$\int \frac{x}{(1+x)^2} \, dx = \int \frac{u-1}{u^2} \, du = \int \left(u^{-1} - u^{-2} \right) du = \ln|u| + \frac{1}{u} + C$$

Now, undo the substitution.

$$\int \frac{x}{(1+x)^2} \, dx = \ln|1+x| + \frac{1}{1+x} + C$$

Our rule of thumb for choosing u is just that: a rule of thumb. It is not always applicable, and we must sometimes be creative when choosing u, as in the following example:

Example 3.3.12: Evaluate $\int \sin(x)\cos(x)\,dx$.

There is no composite function in this integrand. However, the change of variable $u = \sin(x)$ is still a useful choice, because $du = \cos(x)\,dx$, and as you can see, that is also in the integrand. We proceed as follows:

$$\int \sin(x)\cos(x)\,dx = \int u\,du = \frac{1}{2}u^2 + C = \frac{1}{2}\sin^2(x) + C$$

Example 3.3.13: Evaluate $\int \tan^2(x)\sec^2(x)\,dx$.

This time there are *two* composite functions in the integrand: $\tan^2(x)$ and $\sec^2(x)$. We could try to let u equal the inner part of one of them, but which one? Based on the previous examples, we know that whatever we choose for u, we need for the *derivative* of u (or something close to it) to also be in the integrand. If we let $u = \sec(x)$, then $du = \sec(x)\tan(x)\,dx$, which we do not have in the integrand. However, if we let $u = \tan(x)$, then $du = \sec^2(x)\,dx$, which *is* in the integrand. The second choice is the clear favorite.

$$\int \tan^2(x)\sec^2(x)\,dx = \int u^2\,du = \frac{1}{3}u^3 + C = \frac{1}{3}\tan^3(x) + C$$

u-substitution allows us to evaluate integrals of $\tan(x)$, $\cot(x)$, $\sec(x)$, and $\csc(x)$, as we will show below.

More Rules for Trigonometric Functions

Recall the identities $\tan(x) = \dfrac{\sin(x)}{\cos(x)}$ and $\cot(x) = \dfrac{\cos(x)}{\sin(x)}$. We can use these identities to integrate $\tan(x)$ and $\cot(x)$ as follows:

Tangent:

$$\int \tan(x)\,dx = \int \frac{\sin(x)}{\cos(x)}\,dx$$

Let $u = \cos(x)$. Then $du = -\sin(x)\,dx$, or $\sin(x)\,dx = -du$, and we obtain the following:

$$\int \tan(x)\,dx = \int \frac{1}{\cos(x)} \sin(x)\,dx = -\int \frac{1}{u}\,du = -\ln|u| + C = -\ln|\cos(x)| + C$$

Cotangent:

$$\int \cot(x)\,dx = \int \frac{\cos(x)}{\sin(x)}\,dx$$

Let $u = \sin(x)$. Then $du = \cos(x)\,dx$, and we obtain the following:

$$\int \cot(x)\,dx = \int \frac{1}{\sin(x)} \cos(x)\,dx = \int \frac{1}{u}\,du = \ln|u| + C = \ln|\sin(x)| + C$$

To obtain the integration rules for $\sec(x)$ and $\csc(x)$, we need to be a little craftier. We will start with $\int \sec(x)\,dx$. Multiply the integrand by $\dfrac{\sec(x) + \tan(x)}{\sec(x) + \tan(x)}$. Trust me. It is going to work.

$$\int \sec(x)\,dx = \int \sec(x) \cdot \frac{\sec(x) + \tan(x)}{\sec(x) + \tan(x)}\,dx = \int \frac{\sec^2(x) + \sec(x)\tan(x)}{\sec(x) + \tan(x)}\,dx$$

Now, let $u = \sec(x) + \tan(x)$. Then $du = (\sec(x)\tan(x) + \sec^2(x))\,dx$, which is precisely what we have in the numerator of the integrand.

$$\int \sec(x)\,dx = \int \frac{du}{u} = \ln|u| + C = \ln|\sec(x) + \tan(x)| + C$$

Next, we will do a similar trick to evaluate $\int \csc(x)\,dx$. We will multiply the integrand by $\dfrac{\csc(x) + \cot(x)}{\csc(x) + \cot(x)}$.

$$\int \csc(x)\,dx = \int \csc(x) \cdot \frac{\csc(x) + \cot(x)}{\csc(x) + \cot(x)}\,dx = \int \frac{\csc^2(x) + \csc(x)\cot(x)}{\csc(x) + \cot(x)}\,dx$$

Let $u = \csc(x) + \cot(x)$. Then $du = (-\csc(x)\cot(x) - \csc^2(x))\,dx$, or $(\csc(x)\cot(x) + \csc^2(x))\,dx = -du$.

$$\int \csc(x)\,dx = -\int \frac{du}{u} = -\ln|\csc(x) + \cot(x)| + C$$

Here is a tricky one:

Example 3.3.14: $\displaystyle\int \frac{1}{1 + \sin(x)}\,dx$

This integral is not elementary, and there is no obvious u-substitution that will help. Instead, we will multiply the numerator and denominator the integrand by the *Pythagorean conjugate* of the denominator.

$$\int \frac{1}{1+\sin(x)}dx = \int \frac{1}{1+\sin(x)} \cdot \frac{1-\sin(x)}{1-\sin(x)}dx = \int \frac{1-\sin(x)}{1-\sin^2(x)}dx = \int \frac{1-\sin(x)}{\cos^2(x)}dx$$

$$\int \frac{1}{1+\sin(x)}dx = \int \left(\frac{1}{\cos^2(x)} - \frac{\sin(x)}{\cos^2(x)} \right)dx = \int \left(\sec^2(x) - \sec(x)\tan(x) \right)dx$$

$$\int \frac{1}{1+\sin(x)}dx = \tan(x) - \sec(x) + C$$

Long Division

Consider the following indefinite integral:

$$\int \frac{x^3 + x^2 - 6x + 5}{x+3}dx$$

This does not appear to match any of the antidifferentiation rules that we reviewed in **EU 3.1**. However, notice that the integrand is an improper rational expression. If your precalculus instincts are sharp, then you will have a strong urge to use long division when you see this, and that is precisely what we will do.

$$
\begin{array}{r}
x^2 - 2x \\
x+3\overline{)x^3 + x^2 - 6x + 5} \\
-\left(x^3 + 3x^2\right) \\
\overline{-2x^2 - 6x + 5} \\
-\left(2x^2 - 6x\right) \\
\overline{5}
\end{array}
$$

This means that we can rewrite the integrand as the sum of a polynomial and a proper rational expression: $x^2 - 2x + \dfrac{5}{x+3}$. We will split up the integral as follows:

$$\int \frac{x^3 + x^2 - 6x + 5}{x+3}dx = \int \left(x^2 - 2x\right)dx + \int \frac{5}{x+3}dx$$

The first integral on the right side is elementary.

$$\int \left(x^2 - 2x\right)dx = \frac{1}{3}x^3 - x^2$$

We won't worry about the $+C$ until the end. The second integral is not quite elementary, but it is close. It almost looks like the elementary integral $\int \dfrac{dx}{x}$, and it will take on that form after a u-substitution. Let $u = x + 3$. Then $du = dx$, and we obtain the following:

$$\int \frac{5}{x+3}\,dx = 5\int \frac{du}{u} = 5\ln|u| = 5\ln|x+3|$$

We then combine these results to get our final answer.

$$\int \frac{x^3 + x^2 - 6x + 5}{x+3}\,dx = \frac{1}{3}x^3 - x^2 + 5\ln|x+3| + C$$

Completing the Square

Consider the following indefinite integral:

$$\int \frac{1}{x^2 + 2x + 5}\,dx$$

This is not an elementary integral, but with an appropriate substitution we may be able to transform it into one. Looking through the antidifferentiation rules that we reviewed in **EU 3.1**, we see a possible lookalike: $\int \frac{1}{x^2 + a^2}\,dx$. Why this one? Because the integral we are considering has a constant in the numerator and a second degree polynomial in the denominator, and this is the only form that we reviewed that has that.

The big difference between the two denominators is that $x^2 + 2x + 5$ and $x^2 + a^2$ is that the former has a first degree term, while the latter does not. To fix this, we will complete the square. What number must we add to $x^2 + 2x$ to obtain a perfect square trinomial? To answer that, take one-half of the first degree coefficient, and square the result: $\left(\frac{2}{2}\right)^2 = 1$. So, we will add and subtract 1, and then factor and simplify.

$$x^2 + 2x + 5 = x^2 + 2x + 1 + 5 - 1 = (x+1)^2 + 4$$

Now, we rewrite the integral as follows:

$$\int \frac{1}{x^2 + 2x + 5}\,dx = \int \frac{1}{(x+1)^2 + 4}\,dx$$

We're almost there. To make this look like one of our antidifferentiation rules, let $u = x + 1$. Then $du = dx$.

$$\int \frac{1}{x^2 + 2x + 5}\,dx = \int \frac{1}{u^2 + 4}\,du = \frac{1}{2}\tan^{-1}\left(\frac{u}{2}\right) + C = \frac{1}{2}\tan^{-1}\left(\frac{x+1}{2}\right) + C$$

Integration by Parts (BC)

Recall the Product Rule for differentiable functions u and v:

$$\frac{d(uv)}{dx} = \frac{du}{dx}v + u\frac{dv}{dx}$$

Writing this in differential form, we have the following:

$$d(uv) = v\,du + u\,dv$$

Now, integrate both sides.

$$\int d(uv) = \int v\,du + \int u\,dv$$

$$uv = \int v\,du + \int u\,dv$$

Solving this for $\int u\,dv$, we obtain the rule for integration by parts as it is commonly stated.

$$\int u\,dv = uv - \int v\,du$$

When an integral of the form $\int u\,dv$ is too formidable to evaluate directly, one applies this rule in the hopes that $\int v\,du$ is not as difficult. We will illustrate this with several examples.

Example 3.3.15: Evaluate $\int xe^x\,dx$.

This is not an elementary form, and a *u*-substitution won't remedy that (try it). Instead, we will integrate by parts. What we do is we *choose* part of the integrand to be *u*, and then the rest of the integrand (including the differential) is *dv*. We will let $u = x$, and so $dv = e^x dx$. Note that once we choose *u*, we do *not* get to choose *dv*. Now, from these parts, we calculate the remaining parts. We find *du* by taking the differential of *u*, and we find *v* by integrating *dv*.

$$u = x \Rightarrow du = dx$$

$$dv = e^x dx \Rightarrow v = \int e^x dx = e^x$$

We need not worry about the constant of integration until we are finished. We now apply the rule for integration by parts.

$$\int xe^x dx = xe^x - \int e^x dx$$

The integral on the right side is elementary. We will evaluate it, add the constant of integration, and then we're done.

$$\int xe^x dx = xe^x - e^x + C$$

In this example, we chose to let $u = x$. Why that choice? Why not let $u = e^x$? Give it a try. You will see that if you use that choice, you get $\frac{1}{2}x^2 e^x - \frac{1}{2}\int x^2 e^x dx$. The integral gets *more* complicated, not less so. At this point you may be wondering, "Is there a way to anticipate this?" It turns out that there is. The mnemonic device LIATE is helpful for remembering which type of factor to choose for u in integration by parts. It goes like this:

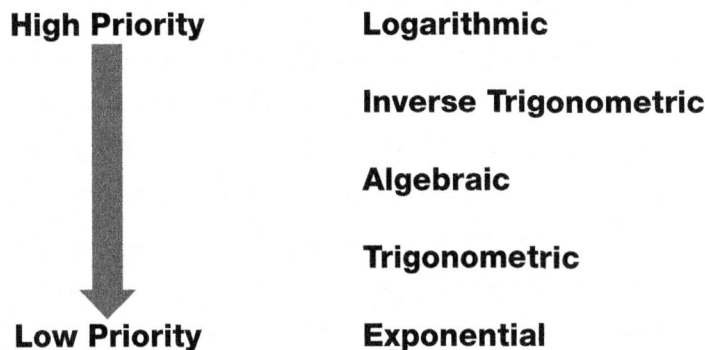

High Priority	**Logarithmic**
	Inverse Trigonometric
	Algebraic
	Trigonometric
Low Priority	**Exponential**

In Example 3.3.15, there were two factors: x and e^x. x is an *algebraic* factor, and e^x is an *exponential* factor. Since algebraic factors are of higher priority than exponential factors, we let $u = x$.

Sometimes, integration by parts must be applied *more than once*. We illustrate this in our next example.

Example 3.3.16: Evaluate $\int x^2 \sin(x)\, dx$.

This integrand has two factors: x^2, which is algebraic, and $\sin(x)$, which is trigonometric. Since algebraic factors are higher on the pecking order than trigonometric factors, let $u = x^2$, and so $dv = \sin(x)\, dx$. Calculate the other parts.

$$u = x^2 \Rightarrow du = 2x\ dx$$

$$dv = \sin(x)\, dx \Rightarrow v = \int \sin(x)\, dx = -\cos(x)$$

Now, apply the rule for integration by parts.

$$\int x^2 \sin(x)\, dx = -x^2 \cos(x) - \int -2x \cos(x)\, dx$$

$$\int x^2 \sin(x)\, dx = -x^2 \cos(x) + 2\int x \cos(x)\, dx$$

The integral on the right side is still not elementary. In fact, we must integrate by parts *again*. However, we are making progress. In the integral on the left the algebraic factor is of degree 2, whereas in the integral on the right it is of degree 1. Let $u = x$, and so $dv = \cos(x)\, dx$. We now calculate the other parts.

$$u = x \Rightarrow du = dx$$

$$dv = \cos(x)\,dx \Rightarrow v = \int \cos(x)\,dx = \sin(x)$$

Now, we apply the rule for integration by parts a second time.

$$\int x^2 \sin(x)\,dx = -x^2 \cos(x) + 2\left(x\sin(x) - \int \sin(x)\,dx\right)$$
$$= -x^2 \cos(x) + 2x\sin(x) - 2\int \sin(x)\,dx$$

At last, the integral in the last term on the right side is elementary. We finish this off by evaluating it.

$$\int x^2 \sin(x)\,dx = -x^2 \cos(x) + 2x\sin(x) + 2\cos(x) + C$$

It should be noted that LIATE is just a rule of thumb, not a mathematical theorem. It is a good jump-off point for simple integrations by parts, but for more complicated ones you may have to be more creative. To see this, let's look at another example.

Example 3.3.16: Evaluate $\int \dfrac{xe^x}{(1+x)^2}\,dx$.

Here, the algebraic factor in the integrand is $\dfrac{x}{(1+x)^2}$, and according to LIATE, that should be our choice for u. That choice *will not work* here (try it). Instead, let $u = xe^x$ and let $dv = \dfrac{1}{(1+x)^2}\,dx$. Then we calculate the other parts as follows:

$$u = xe^x \Rightarrow du = (1+x)e^x\,dx$$

$$dv = \frac{1}{(1+x)^2}\,dx \Rightarrow v = -\frac{1}{1+x}$$

Applying the rule for integration by parts, we obtain the following:

$$\int \frac{xe^x}{(1+x)^2}\,dx = -\frac{xe^x}{1+x} - \int \left(-\frac{1}{1+x}\right)(1+x)e^x\,dx = -\frac{xe^x}{1+x} + \int e^x\,dx = -\frac{xe^x}{1+x} + e^x + C$$

We can simplify this a little.

$$\int \frac{xe^x}{(1+x)^2}\,dx = \left(-\frac{x}{1+x} + 1\right)e^x + C = \frac{e^x}{1+x} + C$$

Nonrepeating Linear Partial Fractions (BC)

This is a method for integrating rational functions. The steps of the method are as follows:

1. If the integrand is an improper rational function, use long division to transform it into the sum of a polynomial and a proper rational function.

2. For each factor of the form $(ax + b)$ in the denominator of the integrand, include in the partial fraction decomposition a term of the form $\dfrac{A}{ax+b}$, where A is a constant to be determined. That is, for an integrand of the form $\dfrac{p(x)}{\left(a_1 x + b_1\right)\left(a_2 x + b_2\right)\cdots\left(a_n x + b_n\right)}$, where $p(x)$ is a polynomial whose degree is less than n, the partial fraction decomposition looks like this:

$$\frac{p(x)}{\left(a_1 x + b_1\right)\left(a_2 x + b_2\right)\cdots\left(a_n x + b_n\right)} = \frac{A_1}{a_1 x + b_1} + \frac{A_2}{a_2 x + b_2} + \cdots + \frac{A_n}{a_n x + b_n}$$

3. Use the above equation to determine the constants A_1, A_2, ..., A_n. Since the EK does not require us to deal with either repeated linear factors or factors of higher degree in the denominator of the integrand, we will not consider them here. We illustrate the above method with two examples.

Example 3.3.17: Evaluate $\displaystyle\int \frac{x}{x^2 - 5x + 6}\, dx$.

Since the integrand is a proper rational function, we can skip the first step in the method. The denominator can be written as a product of linear factors: $(x - 2)(x - 3)$. We set up the partial fraction decomposition of the integrand as follows:

$$\frac{x}{(x-2)(x-3)} = \frac{A_1}{x-2} + \frac{A_2}{x-3}$$

Multiplying both sides by $(x - 2)(x - 3)$, we obtain the following:

$$x = A_1\left(x - 3\right) + A_2\left(x - 2\right)$$

The easiest way to find A_1 and A_2 is to plug in $x = 2$ and $x = 3$, respectively. We find that $x = 2 \Rightarrow A_1 = -2$ and $x = 3 \Rightarrow A_2 = 3$. So we can rewrite the integral as follows:

$$\int \frac{x}{x^2 - 5x + 6}\, dx = \int\left(\frac{-2}{x-2} + \frac{3}{x-3}\right) dx = -2\int \frac{1}{x-2}\, dx + 3\int \frac{1}{x-3}\, dx$$

The two integrals on the right side can be easily evaluated with a substitution of variables by letting $u = x - 2$ and $u = x - 3$, respectively. The final result is given below.

$$\int \frac{x}{x^2 - 5x + 6}\, dx = -2\ln\left|x - 2\right| + 3\ln\left|x - 3\right| + C$$

Next, we'll look at an example of integrating a rational function of *sines and cosines*.

Example 3.3.18: Evaluate $\displaystyle\int \frac{\sin(x)}{\cos^2(x) + \cos(x)}\, dx$.

This integral requires a substitution of variables. Using our rule of thumb for choosing u, we let $u = \cos(x)$, so that $du = -\sin(x)\, dx$, or $\sin(x)\, dx = -du$.

$$\int \frac{\sin(x)}{\cos^2(x)+\cos(x)}dx = -\int \frac{1}{u^2+u}du$$

Now our integrand is a rational function of u, and its denominator can be written as a product of two linear factors: $u(u+1)$. We find the partial fraction decomposition as follows:

$$\frac{1}{u(u+1)} = \frac{A_1}{u} + \frac{A_2}{u+1}\cdots$$

$$1 = A_1(u+1) + A_2 u$$

We can easily find A_1 and A_2 by plugging in $x = 0$ and $x = -1$, respectively. Doing so, we find that $x = 0 \Rightarrow A_1 = 1$ and $x = -1 \Rightarrow A_2 = -1$. This gives us the following:

$$\int \frac{\sin(x)}{\cos^2(x)+\cos(x)}dx = -\int\left(\frac{1}{u} - \frac{1}{u+1}\right)du = -\ln|u| + \ln|u+1| + C$$

Finally, we undo the substitution and use the rules of logs to simplify the result.

$$\int \frac{\sin(x)}{\cos^2(x)+\cos(x)}dx = \ln\left|\frac{\cos(x)+1}{\cos(x)}\right| + C$$

We wrap up this section by reviewing how our techniques can be used with *definite* integration. We start with an example of *u*-substitution.

Example 3.3.19: Evaluate $\displaystyle\int_0^{\pi/4} \sin^3(x)\cos(x)dx$.

This can be integrated with the substitution $u = \sin(x)$, so that $du = \cos(x)\,dx$. The limits of integration are on x, not u, so it would be wrong to rewrite the integral as $\displaystyle\int_0^{\pi/4} u^3\,du$. We will use the substitution to transform the limits of integration into limits on u, as follows:

$$x = 0 \Rightarrow u = \sin(0) = 0$$

$$x = \frac{\pi}{4} \Rightarrow u = \sin\left(\frac{\pi}{4}\right) = \frac{\sqrt{2}}{2}$$

The correct way to rewrite the integral is as follows:

$$\int_0^{\pi/4} \sin^3(x)\cos(x)dx = \int_0^{\sqrt{2}/2} u^3\,du = \frac{1}{4}\left(u^4\right)\Big|_0^{\sqrt{2}/2} = \frac{1}{4}\left(\left(\frac{\sqrt{2}}{2}\right)^4 - 0^4\right) = \frac{1}{16}$$

And now we will do an example of definite integration with integration by parts.

Example 3.3.20: Evaluate $\int_0^{1/2} \sin^{-1}(x)\,dx$.

Following our mnemonic device LIATE, we let $u = \sin^{-1}(x)$, and so $dv = dx$. We calculate the other two parts as follows:

$$u = \sin^{-1}(x) \Rightarrow du = \frac{1}{\sqrt{1-x^2}}\,dx$$

$$dv = dx \Rightarrow v = x$$

Applying the rule for integration by parts, we obtain the following:

$$\int_0^{1/2} \sin^{-1}(x)\,dx = x\sin^{-1}(x)\Big|_0^{1/2} - \int_0^{1/2} \frac{x}{\sqrt{1-x^2}}\,dx$$

Notice that, since we are doing a definite integral, both terms on the right side must be evaluated from $x = 0$ to $x = \frac{1}{2}$. The integral on the right side requires a u-substitution. Let $u = 1 - x^2$. Then we have $du = -2x\,dx \Rightarrow x\,dx = -\frac{1}{2}\,du$, and for the limits of integration we have $x = 0 \Rightarrow u = 1$ and $x = \frac{1}{2} \Rightarrow u = \frac{3}{4}$.

$$\int_0^{1/2} \sin^{-1}(x)\,dx = x\sin^{-1}(x)\Big|_0^{1/2} + \frac{1}{2}\int_1^{3/4} \frac{1}{\sqrt{u}}\,du = x\sin^{-1}(x)\Big|_0^{1/2} - \frac{1}{2}\int_{3/4}^1 u^{-1/2}\,du$$

$$\int_0^{1/2} \sin^{-1}(x)\,dx = x\sin^{-1}(x)\Big|_0^{1/2} - \frac{1}{2}\cdot 2\left(u^{1/2}\right)\Big|_{3/4}^1 = \left(\frac{1}{2}\sin^{-1}\left(\frac{1}{2}\right) - 0\sin^{-1}(0)\right) - \left(1^{1/2} - \left(\frac{3}{4}\right)^{1/2}\right)$$

$$\int_0^{1/2} \sin^{-1}(x)\,dx = \frac{\pi}{12} - 1 + \frac{\sqrt{3}}{2}$$

This concludes our review of integration techniques. A summary of all of the elementary integration rules is given on the next page for your reference.

Elementary Antidifferentiation Rules _____

1. $\int kf(x)\,dx = k\int f(x)\,dx$

2. $\int (f(x) \pm g(x))\,dx = \int f(x)\,dx \pm \int g(x)\,dx$

3. $\int 0\,dx = C$

4. $\int k\,dx = kx + C$

5. $\int x^n\,dx = \frac{1}{n+1}x^n + C \;\; (n \neq -1)$

6. $\int \dfrac{1}{x}\,dx = \ln|x| + C$

7. $\int e^x\,dx = e^x + C$

8. $\int a^x\,dx = \dfrac{1}{\ln(a)}\,a^x + C \ \ (a > 0, a \neq 1)$

9. $\int \sin(x)\,dx = -\cos(x) + C$

10. $\int \cos(x)\,dx = \sin(x) + C$

11. $\int \sec^2(x)\,dx = \tan(x) + C$

12. $\int \csc^2(x)\,dx = -\cot(x) + C$

13. $\int \sec(x)\tan(x)\,dx = \sec(x) + C$

14. $\int \csc(x)\cot(x)\,dx = -\csc(x) + C$

15. $\int \dfrac{1}{\sqrt{a^2 - x^2}}\,dx = \arcsin\left(\dfrac{x}{a}\right) + C \ \ (a > 0)$

16. $\int \dfrac{1}{x^2 + a^2}\,dx = \dfrac{1}{a}\arctan\left(\dfrac{x}{a}\right) + C \ \ (a > 0)$

17. $\int \dfrac{1}{x\sqrt{x^2 - a^2}}\,dx = \dfrac{1}{a}\operatorname{arcsec}\left(\dfrac{|x|}{a}\right) + C$

18. $\int \tan(x)\,dx = -\ln|\cos(x)| + C$

19. $\int \cot(x)\,dx = \ln|\sin(x)| + C$

20. $\int \sec(x)\,dx = \ln|\sec(x) + \tan(x)| + C$

21. $\int \csc(x)\,dx = -\ln|\csc(x) + \cot(x)| + C$

22. $\int f(g(x))g'(x)\,dx = \int f(u)\,du$

23. $\int u\,dv = uv - \int v\,du$

EU 3.4 – Applications of Integration _____

LO 3.4A – Interpreting Definite Integrals

The complete Learning Objective is stated below.

LO 3.4A: Interpret the meaning of a definite integral within a problem.

That statement is quite open to interpretation, so we'll need to go to the EKs to bring it into sharper focus.

EK 3.4A1: A function defined as an integral represents an accumulation of a rate of change.

An easy way to understand this is through an application, so we'll look at this in the context of motion along the x-axis. If the position (a.k.a. **displacement**) of the particle at time $t \geq 0$ is given by $x(t)$, then the velocity of the particle is given by $v(t) = x'(t)$. We can turn this last statement around by expressing displacement as the integral of velocity, which is the *rate of change* of displacement with respect to time.

$$x(t) = \int_0^t v(\lambda)\, d\lambda + x(0)$$

Here, λ is a dummy variable, and $x(0)$ is the position of the particle at $t = 0$. The interpretation here is that the displacement at time t is found by summing via integration the accumulated rate of change (in this case, *velocity*) of the particle over the interval $[0, t]$.

The next EK is closely related to this one.

EK 3.4A2: The definite integral of the rate of change of a quantity over an interval gives the net change of that quantity over that integral.

We'll continue with our motion example. The function $x(t) = \int_0^t v(\lambda)\, d\lambda + x(0)$ is not exactly a definite integral, because the upper limit of integration is a variable. However, if we choose to look at this function at $t = T$, where T is constant, then we obtain a definite integral.

$$x(T) = \int_0^T v(\lambda)\, d\lambda + x(0) \Rightarrow \int_0^T v(\lambda)\, d\lambda = x(T) - x(0)$$

As you can see from the right side, the definite integral of the velocity over the fixed interval $[0, T]$ is equal to the net change in displacement over that same interval.

The final EK for this LO covers some familiar ground.

EK 3.4A3: The limit of an approximating Riemann sum can be interpreted as a definite integral.

We thoroughly reviewed using Riemann sums to approximate definite integrals in **EK 3.2B2**, and we reviewed the fact that the definite integral is *defined* as the limit of a Riemann sum in **EK 3.2A2**, so we will consider this EK covered.

LO 3.4B – Average Value of a Function

The Learning Objective for this section is as follows:

LO 3.4B: Apply definite integrals to problems involving the average value of a function.

In the lone EK for this LO, the average value of a function over a closed interval is defined.

EK 3.4B1: The average value of a function f over an interval $[a, b]$ is $\dfrac{1}{b-a}\displaystyle\int_a^b f(x)\,dx$.

The definition is straightforward enough, so let's use it in a couple of examples. We'll use the notation $\overline{f}$ to denote the average value of a function f.

Example 3.4.1: Find the average value of $f(x) = \sin(x)$ over the interval $[-\pi, \pi]$.

$$\overline{f} = \frac{1}{\pi - (-\pi)}\int_{-\pi}^{\pi} \sin(x)\,dx = \frac{1}{2\pi}\left(\cos(x)\right)\Big|_{-\pi}^{\pi} = \frac{1}{2\pi}\left(\cos(\pi) - \cos(-\pi)\right) = 0$$

The average value is zero. To understand why that is, look at the graph of $f(x) = \sin(x)$ on $[-\pi, \pi]$.

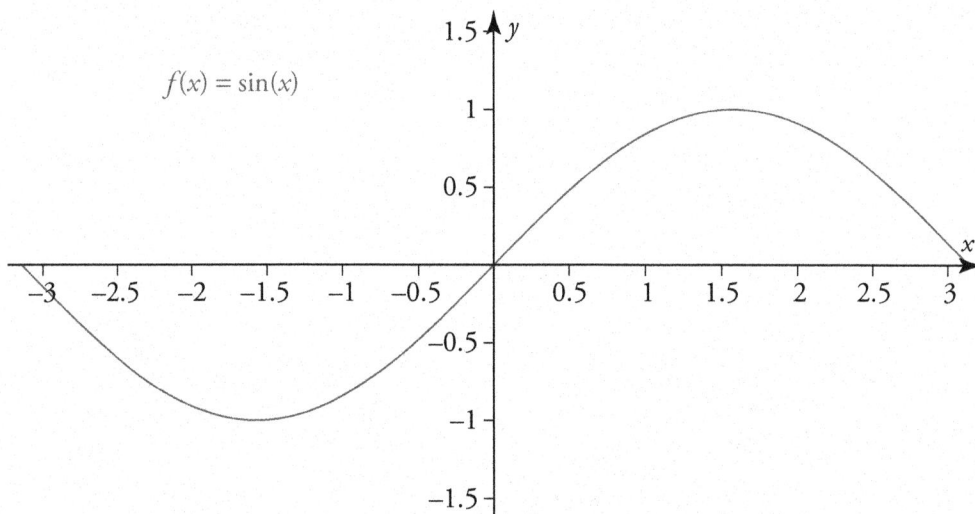

$f(x) = \sin(x)$

Since f is an odd function, its graph on $[-\pi, 0]$ is the *inverted mirror image* of its graph on $[0, \pi]$. This means that the contribution to $\overline{f}$ that comes from integrating f over $[-\pi, 0]$ exactly cancels out the contribution from the integral over $[0, \pi]$, giving us a total of zero.

Example 3.4.2: Find the average value of $f(x) = 4 - x^2$ over the interval $[-2, 2]$.

First we take a look at the graph of f on $[-2, 2]$.

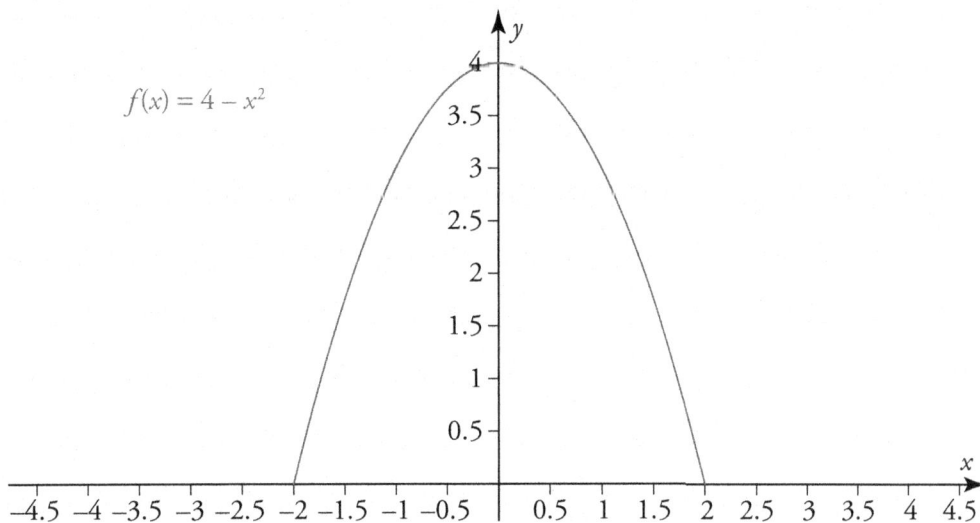

$f(x) = 4 - x^2$

This time, $f(x)$ is *never* negative, so we do not expect that the average value will be zero. To the contrary, since $0 \le f(x) \le 4$ for all x in $[-2, 2]$, we can be certain that $0 \le \overline{f} \le 4$ as well. Let's get to it.

$$\overline{f} = \frac{1}{2-(-2)} \int_{-2}^{2} \left(4-x^2\right) dx = \frac{1}{4}\left(4x - \frac{1}{3}x^3\right)\Big|_{-2}^{2} = \frac{1}{4}\left(\left(4(2) - \frac{1}{3}(2)^3\right) - \left(4(-2) - \frac{1}{3}(-2)^3\right)\right) = \frac{8}{3}$$

As expected, the average value is a positive number less than 4.

LO 3.4C – Motion of a Particle

The complete Learning Objective is given below.

LO 3.4C: Apply definite integrals to problems involving motion.

As we mentioned before: motion is the favorite application of the writers of the AP Calculus exams, both AB and BC. We start by reviewing the AB version.

EK 3.4C1: For a particle in rectilinear motion over an interval of time, the definite integral of velocity represents the particle's displacement over the interval of time, and the definite integral of speed represents the particle's total distance traveled over the interval of time.

Since we already reviewed the definitions displacement, velocity, and speed in **EK 2.3C1**, let's get right to an example.

Example 3.4.3: The velocity (in *m/s*) of a particle moving along the x-axis is given by $v(t) = t^2 - 4t + 3$ for time $t \ge 0\,s$. The particle is initially at $x = 0\,m$.

1. What is the displacement of the particle after $4\,s$?
2. What distance has the particle traveled after $4\,s$?

1. To find the displacement $x(4)$, we just integrate $v(t)$.

$$x(4) = \int_0^4 (t^2 - 4t + 3)\, dt = \left(\frac{1}{3}t^3 - 2t^2 + 3t\right)\Big|_0^4 = \left(\frac{1}{3}(4)^3 - 2(4)^2 + 3(4)\right) - 0 = \frac{4}{3} m$$

2. To find the distance $d(4)$, we integrate $|v(t)|$. We must take some care in doing this. A look at the graph of speed vs time will help here. The grid lines are 0.5 units apart.

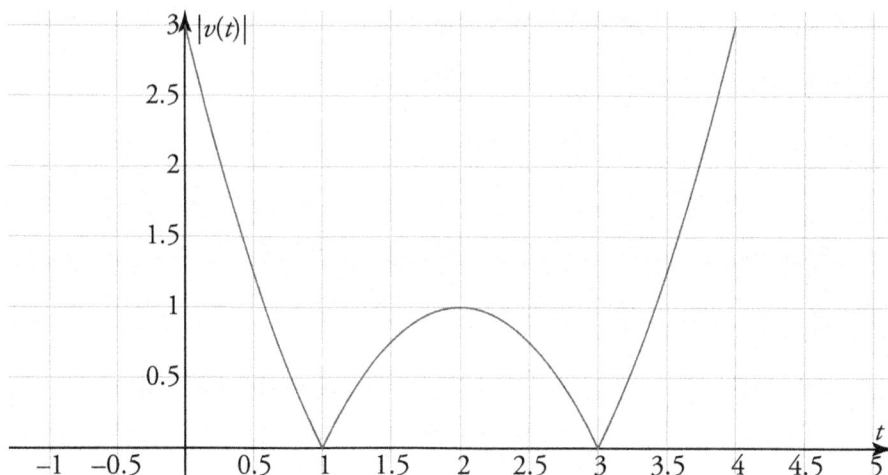

$$d(4) = \int_0^4 \left| t^2 - 4t + 3 \right| dt = \int_0^1 (t^2 - 4t + 3)\, dt - \int_1^3 (t^2 - 4t + 3)\, dt + \int_3^4 (t^2 - 4t + 3)\, dt$$

$$d(4) = \frac{4}{3} + \frac{4}{3} + \frac{4}{3} = 4\ m$$

Now we'll review the BC version.

EK 3.4C2: (BC) The definite integral can be used to determine displacement, distance, and position of a particle moving along a curve given by parametric or vector-valued functions.

Example 3.4.4: The velocity (in m/s) of a particle moving in the xy-plane is given by $\vec{v}(t) = (2t, -4t)$ for time $t \geq 0\,s$. The particle is initially at the origin.

1. What are the displacement and position of the particle after $2\,s$?
2. What distance has the particle traveled after $2\,s$?

1. Integrate to find the displacement $\vec{r}(2)$.

$$\vec{r}(2) = \int_0^2 (2t, -4t)\, dt = (t^2, -2t^2)\Big|_0^2 = (2^2, -2(2)^2) - (0,0) = (4, -8)\ m$$

The displacement of the particle is the *vector* $\vec{r}(2) = (4, -8)\, m$. The position of the particle is the *point* $(x, y) = (4, -8)\, m$.

2. The distance $d(2)$ traveled by the particle is given by the following integral:

$$d(2) = \int_0^2 \sqrt{(2t)^2 + (-4t)^2}\, dt = 2\sqrt{5} \int_0^2 t\, dt = 2\sqrt{5} \cdot \frac{1}{2}(t^2)\Big|_0^2 = \sqrt{5}(2^2 - 0^2) = 4\sqrt{5}\, m$$

LO 3.4D – Area, Volume, and Arc Length

The full Learning Objective for this section is as follows:

LO 3.4D: Apply definite integrals to problems involving area, volume, (**BC**) and length of a curve.

As you can see, we are focusing on geometrical applications of integration here. We'll start with area.

EK 3.4D1: Areas of certain regions in the plane can be calculated with definite integrals. (**BC**) Areas bounded by polar curves can be calculated with definite integrals.

We already reviewed the connection between definite integration and area in **EK 3.2C1**. In that EK, we only considered areas that could be computed using known geometric formulas. Here, we'll extend our review to regions whose areas have no basic formula.

We'll start with an example of a region bounded by the x-axis and the graph of a nonnegative function on an interval.

Example 3.4.5: Find the area of the region bounded by the graph of $f(x) = -x^2 + 4x - 1$ and the x-axis on the interval $[1, 3]$.

We sketch the region below, along with a rectangular differential area element.

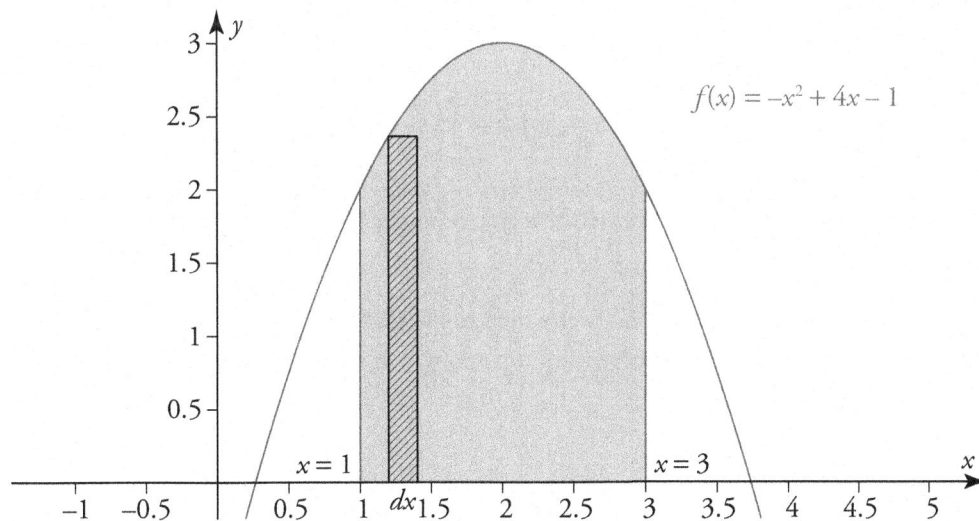

The area A is found by the following integral:

$$A = \int_1^3 \left(-x^2 + 4x - 1\right) dx = \left(-\frac{1}{3}x^3 + 2x^2 - x\right)\Big|_1^3$$

$$= \left(-\frac{1}{3}(3)^3 + 2(3)^2 - 3\right) - \left(-\frac{1}{3}(1)^3 + 2(1)^2 - 1\right) = \frac{16}{3}$$

There's no reason that the bottom boundary of a region has to be the x-axis. It can be the graph of another function instead. Consider the following figure:

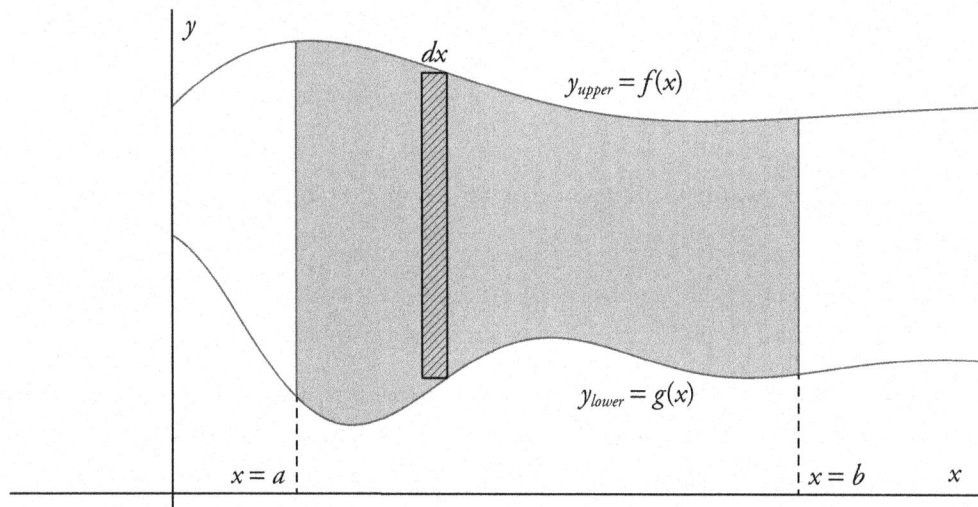

We have sketched two curves, one denoted $y_{upper} = f(x)$, and the other denoted $y_{lower} = g(x)$. The rectangular differential area element is given as follows:

$$dA = \left(y_{upper} - y_{lower}\right) dx = \left(f(x) - g(x)\right) dx$$

When subtracting the two functions in parentheses, it is essential that you get the order right. If you reverse y_{upper} and y_{lower}, then your area will come out *negative*. The area A of the shaded region is given as follows:

$$A = \int_a^b \left(y_{upper} - y_{lower}\right) dx = \int_a^b \left(f(x) - g(x)\right) dx$$

Example 3.4.6: Compute the area of the region bounded by the graphs of $f(x) = x^2 - 4x + 3$ and $g(x) = -x^2 + 2x + 3$.

We were not given an interval over which to integrate, so it's up to us to figure it out. Also, we need to know which curve is on top and which is on the bottom. A sketch of the graph will help on both counts.

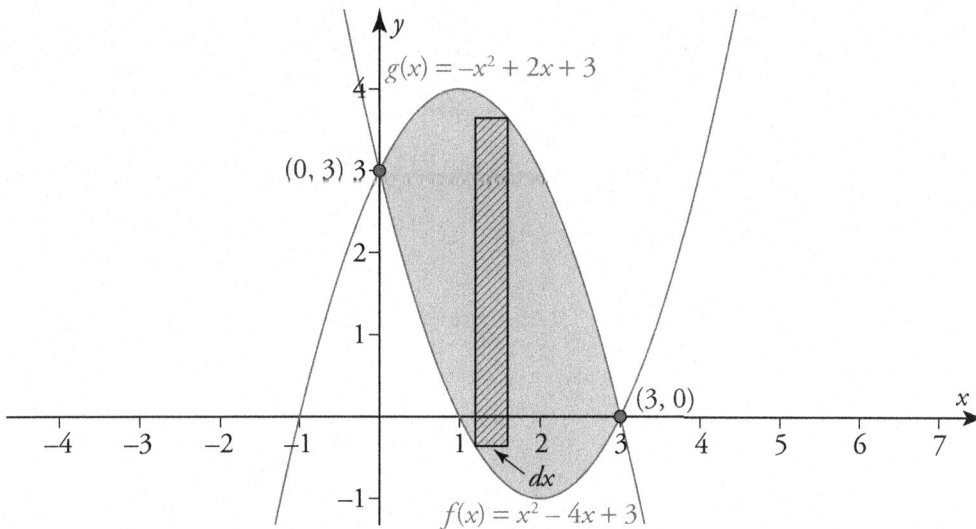

From the graph we see that the interval of integration is $[0, 3]$ and that on the entire interval, $g(x) \geq f(x)$. The area A of the region is thus found as follows:

$$A = \int_0^3 \left(\left(-x^2 + 2x + 3\right) - \left(x^2 - 4x + 3\right)\right) dx = \int_0^3 \left(-2x^2 + 6x\right) dx = \left(-\frac{2}{3}x^3 + 3x^2\right)\Big|_0^3$$

$$A = \left(-\frac{2}{3}(3)^3 + 3(3)^2\right) - 0 = 9$$

On the exam, you may also see a region that is bounded by functions of y, as in the following example:

Example 3.4.7: Compute the area of the region bounded by the graphs of $f(y) = 4 - y^2$ and $g(y) = y + 2$.

Again, it would be helpful to see a graph of the region.

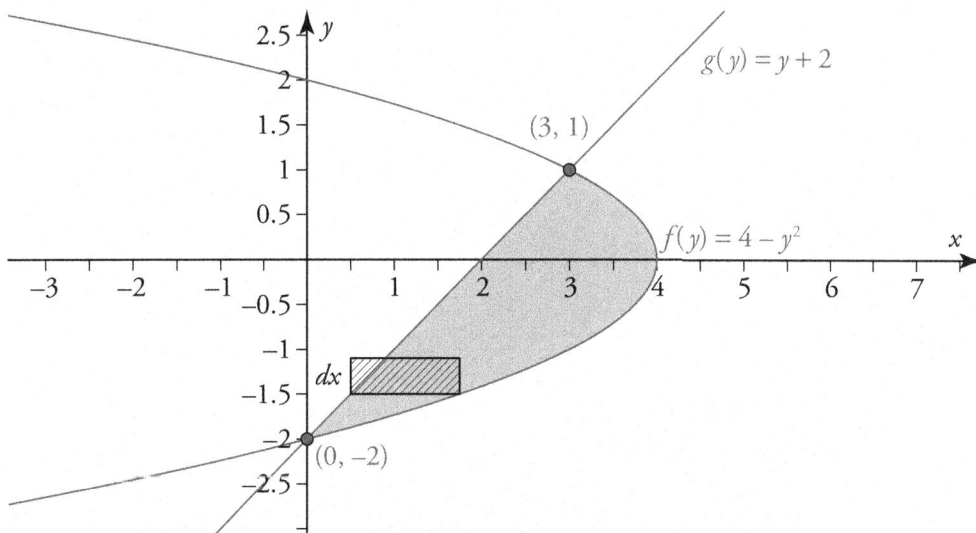

We have sketched the graphs of both functions, as well as a rectangular differential area element. Since both f and g are functions of y, we oriented the area element such that one of its dimensions is dy. In order to ensure that the differential area dA is positive, we compute it by subtracting the function whose graph is on the left from the function whose graph is on the right.

$$dA = \left(f(y) - g(y)\right)dy = \left(\left(4 - y^2\right) - \left(y + 2\right)\right)dy = \left(-y^2 - y + 2\right)dy$$

The area A of the region is then found by integrating dA over the interval $[-2, 1]$ on the y-axis.

$$A = \int_{-2}^{1} \left(-y^2 - y + 2\right)dy = \left(-\frac{1}{3}y^3 - \frac{1}{2}y^2 + 2y\right)\Bigg|_{-2}^{1}$$

$$A = \int_{-2}^{1} \left(-y^2 - y + 2\right)dy = \left(-\frac{1}{3}(1)^3 - \frac{1}{2}(1)^2 + 2(1)\right) - \left(-\frac{1}{3}(-2)^3 - \frac{1}{2}(-2)^2 + 2(-2)\right) = \frac{9}{2}$$

Next we'll move on to volumes.

EK 3.4D2: Volumes of solids with known cross sections, including discs and washers, can be calculated with definite integrals.

We'll begin our review with *solids of revolution*, which are solids that are obtained by revolving a plane region about an axis. You will need to know how to compute the volumes of such regions using the methods of discs and washers. The method of shells, while not explicitly required by the EK, may be useful to know for the exam, and so we will review that as well. Furthermore, you may be required to use definite integrals to calculate volumes of other solids whose cross sections are known, but are not obtained by revolving a plane area around an axis. We'll discuss discs, washers, shells, and other cross sections separately.

Disc Method

The method of discs is based on the formula for the volume of a right circular cylinder of radius R and height h: $V = \pi R^2 h$.

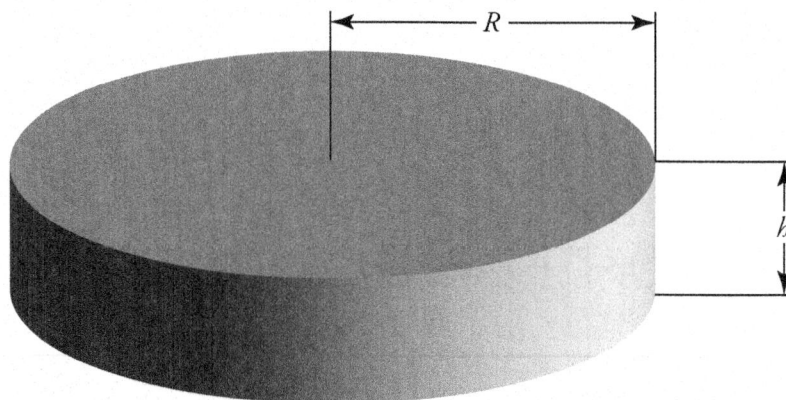

An easy way to remember this formula is that the volume is the area of the circular base (πR^2) times the height of the cylinder (h).

Now consider the plane region shown in the following figure:

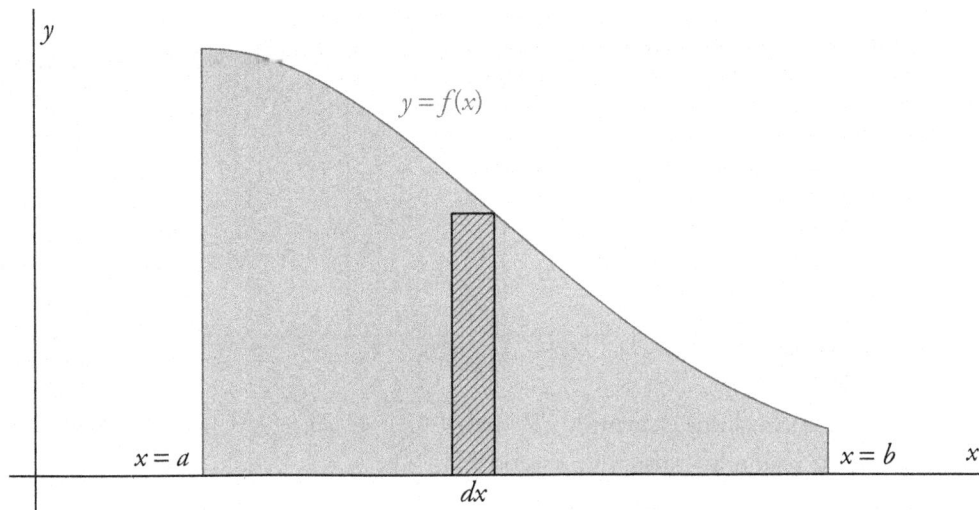

The region is bounded by the x-axis and the graph of $y = f(x)$ on the interval $[a, b]$. A solid of revolution is to be generated by revolving this region about the x-axis. When the rectangular differential *area* element with height $f(x)$ and base dx is revolved about the x-axis, a cylindrical differential *volume* element of radius $f(x)$ and height dx is obtained. This cylinder looks like a disc, hence the name of the method, and its volume dV is given as follows:

$$dV = \pi \left(f(x) \right)^2 dx$$

The total volume V of the solid is obtained by integrating this differential volume element over $[a, b]$.

$$V = \pi \int_a^b \left(f(x) \right)^2 dx$$

Example 3.4.8: Consider the region bounded by the graph of $f(x) = \sqrt{9 - x^2}$ and the x-axis on the interval $[-3, 3]$. Compute the volume of the solid of revolution obtained by revolving this region about the x-axis.

We'll sketch the region below. The grid lines are 0.5 units apart.

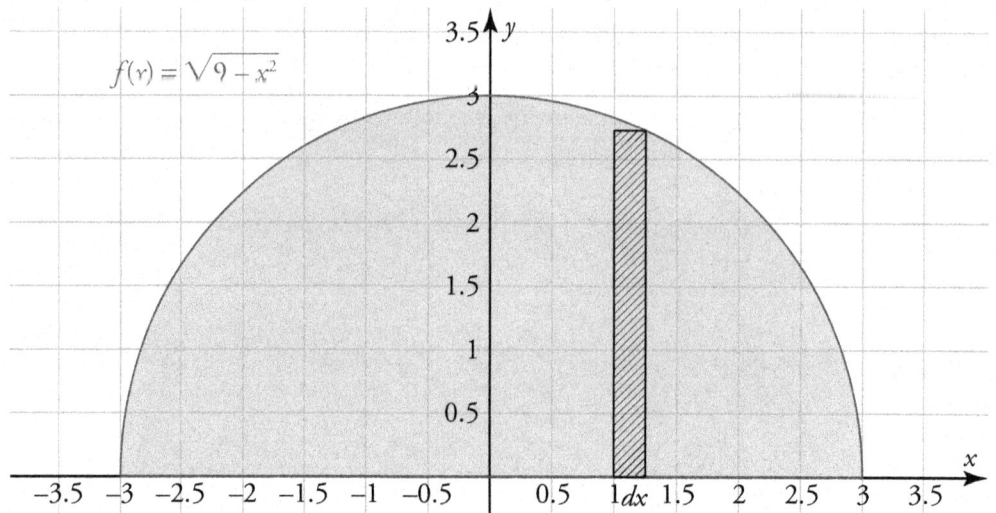

$$f(x) = \sqrt{9 - x^2}$$

The differential volume element is $dV = \pi\sqrt{9 - x^2}^2 \, dx = \pi\left(9 - x^2\right) dx$. We obtain the volume V of the region by integrating this over $[-3, 3]$.

$$V = \pi\int_{-3}^{3}\left(9 - x^2\right) dx = \left(9x - \frac{1}{3}x^3\right)\Bigg|_{-3}^{3} = \pi\left(\left(9(3) - \frac{1}{3}(3)^3\right) - \left(9(-3) - \frac{1}{3}(-3)^3\right)\right) = 36\pi$$

We can also obtain a solid of revolution by revolving a plane region about the y-axis. In this case, any functions whose graphs determine the region must be expressed as functions of y, not x.

Example 3.4.9: Consider the region bounded by the graph of $h(y) = -y^2 + 4$ and the y-axis on the interval $[-2, 2]$. Compute the volume of the solid of revolution obtained by revolving this region about the y-axis.

We'll sketch the region below. The grid lines are 0.5 units apart.

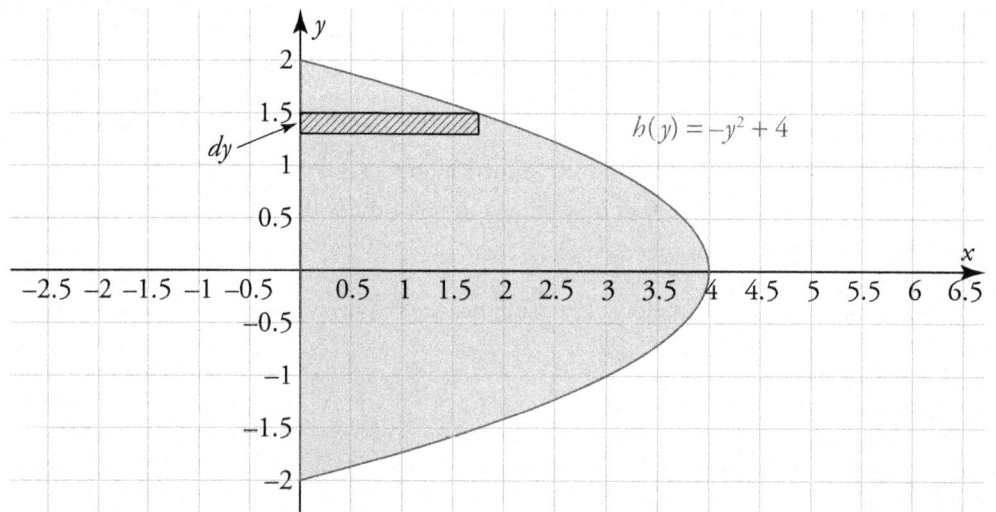

Since h is a function of y, we need to orient the differential volume element so that one of its dimensions is dy. Computing this element, we obtain $dV = \pi(-y^2 + 4)^2\, dy = \pi(y^4 - 8y^2 + 16)dy$. The volume V of the solid is found by integrating this over the interval $[-2, 2]$. Bear in mind that this interval is along the y-axis, not the x-axis.

$$V = \pi \int_{-2}^{2} \left(y^4 - 8y^2 + 16 \right) dy = \pi \left(\frac{1}{5} y^5 - \frac{8}{3} y^3 + 16 y \right) \Bigg|_{-2}^{2}$$

$$V = \pi \left(\left(\frac{1}{5}(2)^5 - \frac{8}{3}(2)^3 + 16(2) \right) - \left(\frac{1}{5}(-2)^5 - \frac{8}{3}(-2)^3 + 16(-2) \right) \right) = \frac{512\pi}{15}$$

Washer Method

A washer is a disk with a cylindrical hole punched out of its center, as shown in the following figure:

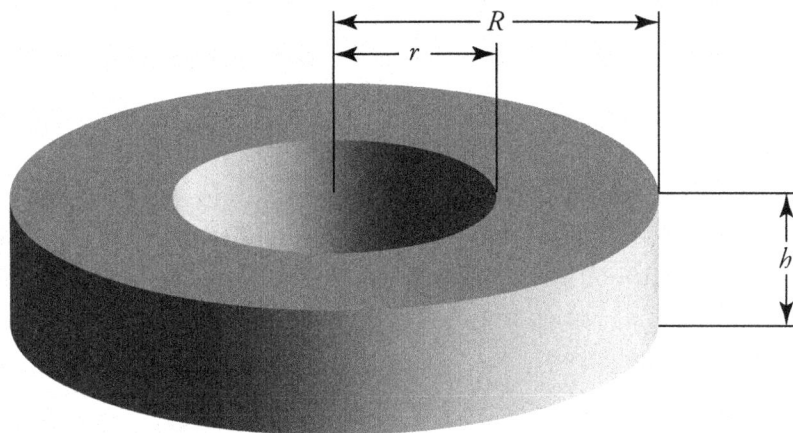

Just as with the disc, h is the height. But here there are *two* radii: an outer radius R and an inner radius r. The volume of the washer is the volume of the outer cylinder *minus* the volume of the inner cylinder: $V = \pi R^2 h - \pi r^2 h = \pi(R^2 - r^2)h$.

Now consider the plane region shown below.

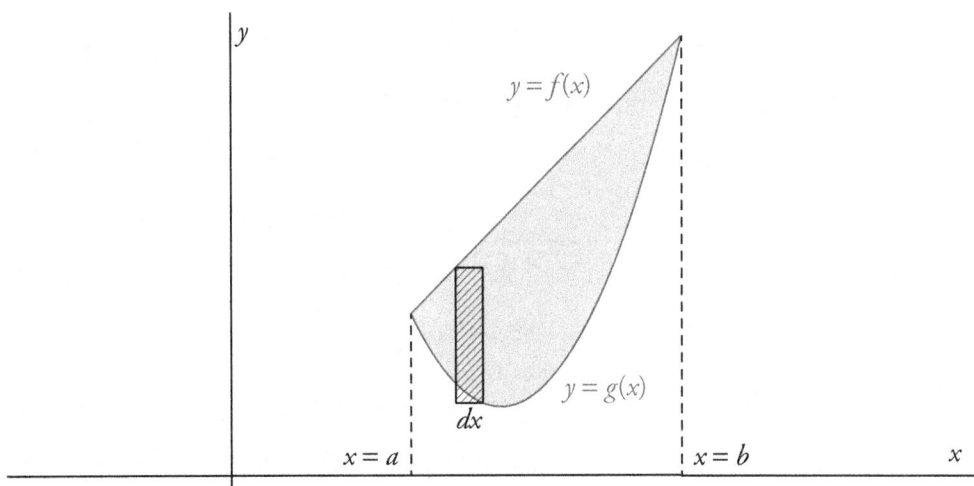

The region is bounded by the graphs of $y = f(x)$ and $y = g(x)$ on the interval $[a, b]$, and we will generate a solid of revolution by revolving this region about the x-axis. When the differential area element is revolved around the x-axis, we get not a disc, but a *washer* with height dx, outer radius $f(x)$, and inner radius $g(x)$. We obtain dV as follows:

$$dV = \pi\left(\left(f(x)\right)^2 - \left(g(x)\right)^2\right)dx$$

Just as in the disc method, we obtain V by integrating dV over $[a, b]$.

$$V = \pi\int_a^b\left(\left(f(x)\right)^2 - \left(g(x)\right)^2\right)dx$$

In our next example, we not only illustrate the washer method, but also we illustrate the fact that we need not revolve a plane region about a coordinate axis to obtain a solid of revolution.

Example 3.4.10: Consider the region bounded by the graphs of $f(x) = 4 - x^2$ and $g(x) = x^2 + 2$. Compute the volume of the solid of revolution obtained by revolving this region about the line $y = 1$.

This time we weren't given an interval over which to integrate, so we must find it ourselves by finding the points of intersection of the graphs of f and g. Equating $f(x)$ and $g(x)$, we obtain the equation $4 - x^2 = x^2 + 2$, which has solutions $x = \pm 1$. The interval of integration is then $[-1, 1]$, which you can clearly see in the following figure. Once again, the grid lines are 0.5 units apart.

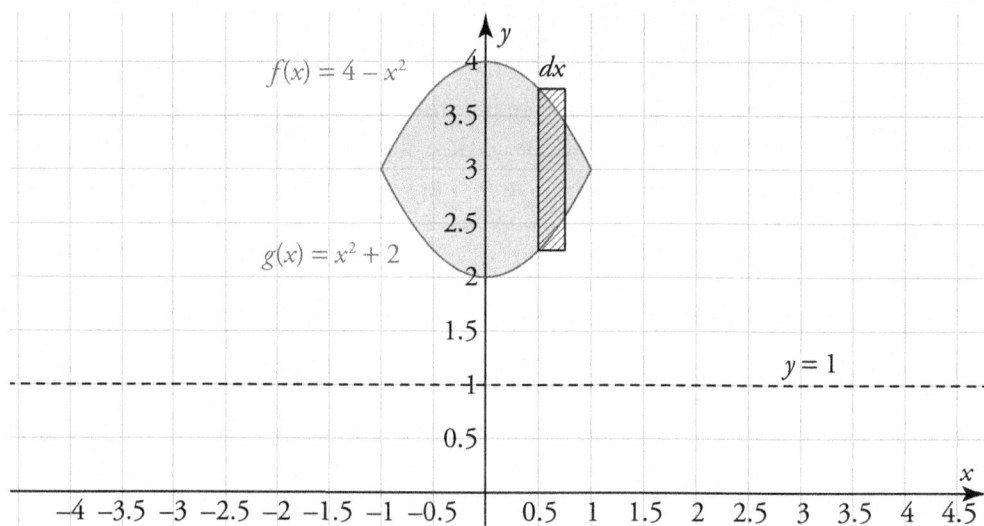

When the differential area element is revolved about the line $y = 1$, a *washer* is formed. The height of the washer is dx. The outer radius of the washer is the distance between $y = f(x)$ and $y = 1$, or $4 - x^2 - 1 = 3 - x^2$. The inner radius of the washer is the distance between $y = g(x)$ and $y = 1$, or $x^2 + 2 - 1 = x^2 + 1$. Our differential volume element dV is then found as follows:

$$dV = \pi\left(\left(3 - x^2\right)^2 - \left(x^2 + 1\right)^2\right)dx = \pi\left(-8x^2 + 8\right)dx$$

The volume V of the solid is then found by integrating dV over $[-1, 1]$.

$$V = \pi \int_{-1}^{1} \left(-8x^2 + 8\right)dx = \pi\left(-\frac{8}{3}x^3 + 8x\right)\Big|_{-1}^{1}$$

$$= \pi\left(\left(-\frac{8}{3}(1)^3 + 8(1)\right) - \left(-\frac{8}{3}(-1)^3 + 8(-1)\right)\right) = 16\pi$$

Shell Method

A shell is a (mostly) hollow right circular cylinder with a thin wall, as shown in the following figure:

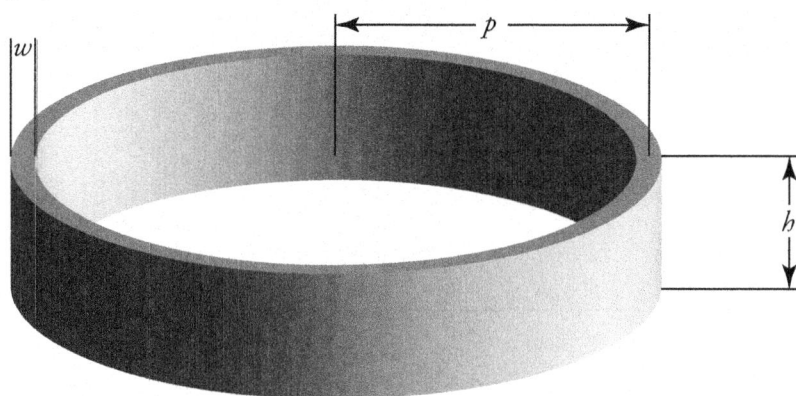

The relevant dimensions are the height h, the wall thickness w, and the radius p, which is the distance from the center of the cylinder to the *center* of the wall. The volume of the shell is $V = 2\pi phw$.

Now consider the following planar region:

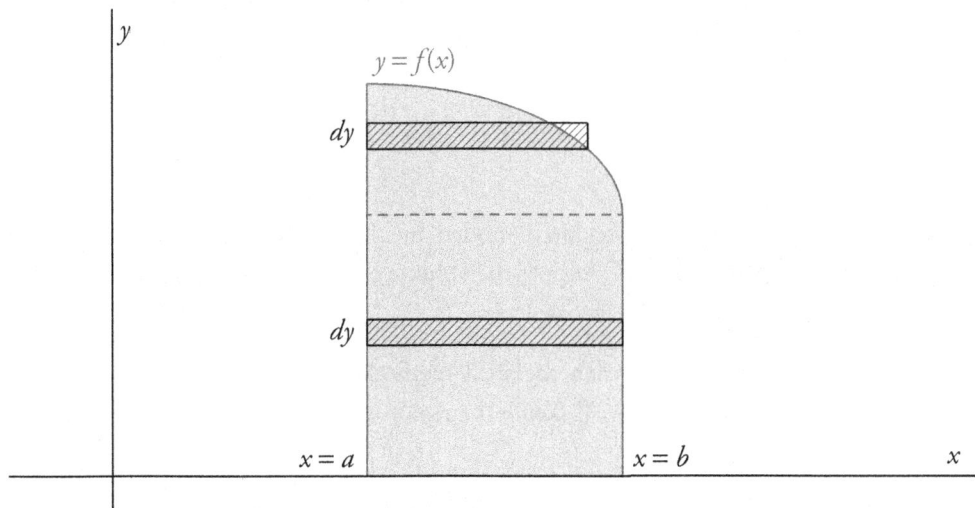

The region is bounded by the x-axis and the graph of $y = f(x)$ on the interval $[a, b]$. A solid of revolution is to be generated by revolving this region about the y-axis. Now, we *could* use the method of washers with horizontal rectangular strips. However, that would require us to do *two* integrals: one from $y = 0$ up to the blue dotted line, where the rectangular differential area is between $x = a$ and $x = b$, and another from the blue dotted line to the top of the graph, where the rectangular differential area is between $x = a$ and the graph of f. It turns out that we can get away with computing a *single* integral if we use the shell method. Let's take another look at this region.

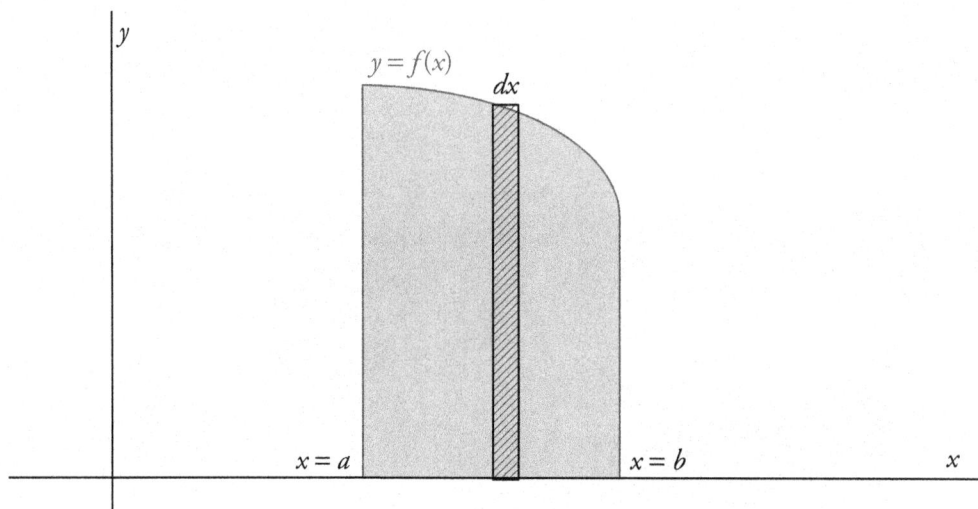

If the rectangular differential area shown above is revolved around the y-axis, then we would obtain a differential volume element that is a shell of height $f(x)$, radius x, and thickness dx. Since these dimensions do not change as we integrate over $[a, b]$, we can compute the volume using only one integral. The differential volume element is as follows:

$$dV = 2\pi x f(x)\, dx$$

As usual, the volume V of the solid is found by integrating dV over $[a, b]$.

$$V = 2\pi \int_a^b x f(x)\, dx$$

Example 3.4.11: Consider the region bounded by the graphs of $f(x) = x^2 + 1$ and $g(x) = 2x^2$ on the interval $[0, 1]$. Compute the volume of the solid of revolution obtained by revolving this region about the y-axis.

This is one of those times when the shell method is preferable to the disc/washer method. In the following figure, we'll sketch the region and a rectangular differential area.

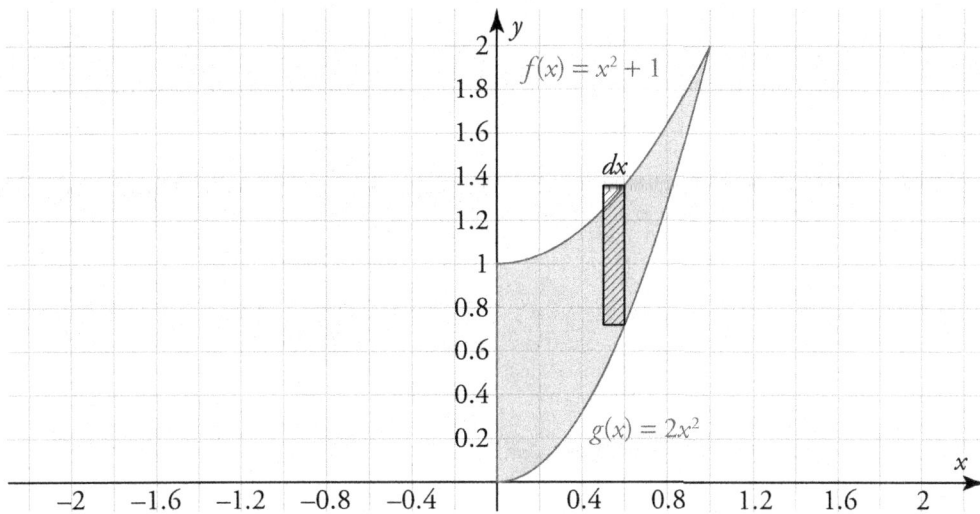

When the differential area is revolved around the y-axis, a differential volume in the shape of a shell is obtained. The shell has height $f(x) - g(x) = x^2 + 1 - 2x^2 = 1 - x^2$, radius x, and thickness dx. The differential volume element is $dV = 2\pi x(1 - x^2)\, dx$, and we integrate this over $[0, 1]$ to get V.

$$V = 2\pi \int_0^1 x\left(1 - x^2\right) dx = 2\pi \int_0^1 \left(x - x^3\right) dx = 2\pi \left(\frac{1}{2}x^2 - \frac{1}{4}x^4\right)\Bigg|_0^1$$

$$= 2\pi\left(\left(\frac{1}{2}(1)^2 - \frac{1}{4}(1)^4\right) - 0\right) = \frac{\pi}{2}$$

Other Cross Sections

We will show one example of finding the volume of a solid (*not* a solid of revolution) that has known cross sections. Coordinates of points will be given as ordered *triples* of the form (x, y, z).

Example 3.4.12: A pyramid has a rectangular base that has vertices at $(2, 1, 0)$, $(-2, 1, 0)$, $(-2, -1, 0)$, and $(2, -1, 0)$. The top of the pyramid is at the point $(0, 0, 2)$. Find the volume of the pyramid by integration.

We will give a sketch of the pyramid below.

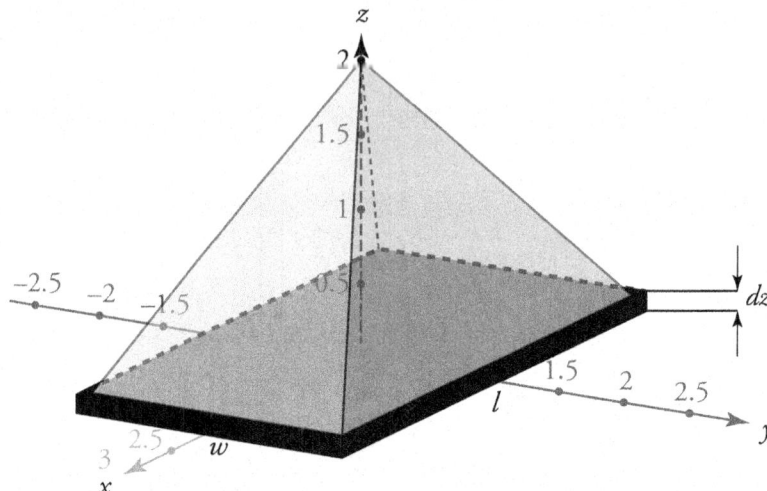

If we pass a plane through the pyramid parallel to its base, we see that the cross sections are all rectangles. We have sketched in a differential volume element, which is a rectangular prism with length l, width w, and height dz. The differential volume element is $dV = lw\,dz$. Before we can find the volume by integration, we must express both l and w in terms of z. We can do this by considering *similar pyramids*.

The differential volume element at $z = 0$ has $l = 4$ and $w = 2$. If we consider the element at a height z, then the pyramid that lies *above* that element, whose height is $2 - z$, is similar to the entire pyramid, and we can set up the following ratios to find its length and width:

$$\frac{l}{4} = \frac{w}{2} = \frac{2-z}{2}$$

From these ratios we find that $l = 4 - 2z$ and, $w = 2 - z$. So our differential volume element is $dV = (4 - 2z)(2 - z)\,dz = (2z^2 - 8z + 8)\,dz$. We compute the volume V by integrating over the interval $[0, 2]$, which is along the z-axis.

$$V = \int_0^2 \left(2z^2 - 8z + 8\right)dz = \left(\frac{2}{3}z^3 - 4z^2 + 8z\right)\Bigg|_0^2 = \left(\left(\frac{2}{3}(2)^3 - 4(2)^2 + 8(2)\right) - 0\right) = \frac{16}{3}$$

Now that we've thoroughly covered volume, we'll review arc length.

EK 3.4D3: (BC) The length of a planar curve defined by a function or by a parametrically defined curve can be calculated using a definite integral.

There are two components to this EK, and we'll cover them one at a time.

Planar Curve Defined by a Function

The formula for finding the length of a planar curve defined by a function is based on the Pythagorean Theorem. In the following figure, we'll show the graph of a function

$y = f(x)$, and we'll label a differential line element ds on the curve. Up and to the right, we will zoom on this line element.

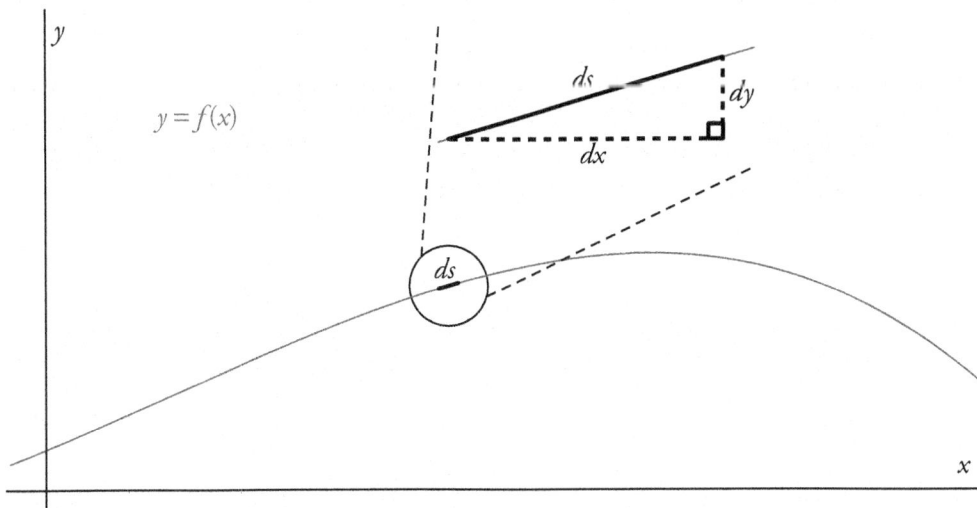

As you can see, we've constructed a right triangle with hypotenuse ds and legs dx and dy. As we mentioned earlier, we can apply the Pythagorean Theorem to get an expression for ds in terms of the other differentials.

$$ds = \sqrt{dx^2 + dy^2}$$

Since ds is on the graph of $y = f(x)$, dx and dy are not independent of one another. In particular, $dy = f'(x)\,dx$. Plugging this into the above expression for ds, we obtain the following:

$$ds = \sqrt{dx^2 + \left(f'(x)\right)^2 dx^2}$$

Next, we'll factor dx^2 out of the radicand and simplify to get our final result.

$$ds = \sqrt{1 + \left(f'(x)\right)^2}\,dx$$

To find the length of the graph of $y = f(x)$ over an interval $[a, b]$, we will integrate ds as follows:

$$s = \int_a^b \sqrt{1 + \left(f'(x)\right)^2}\,dx$$

Example 3.4.13: Find the arc length of the graph of $f(x) = \ln(\sin(x))$ over the interval $\left[\dfrac{\pi}{4}, \dfrac{\pi}{2}\right]$.

The formula for arc length calls for $f'(x)$, so we'll compute that first.

$$f'(x) = \frac{d}{dx}\left(\ln\left(\sin\left(x\right)\right)\right) = \frac{\cos\left(x\right)}{\sin\left(x\right)} = \cot\left(x\right)$$

The differential line element is found as follows:

$$ds = \sqrt{1 + \cot^2\left(x\right)}dx = \sqrt{\csc^2\left(x\right)}dx = \csc\left(x\right)dx$$

Now we determine the arc length by integration.

$$s = \int_{\pi/4}^{\pi/2} \csc\left(x\right)dx = -\left(\ln\left|\csc\left(x\right) + \cot\left(x\right)\right|\right)\Big|_{\pi/4}^{\pi/2}$$

$$= -\left(\left(\ln\left|\csc\left(\frac{\pi}{2}\right) + \cot\left(\frac{\pi}{2}\right)\right|\right) - \left(\ln\left|\csc\left(\frac{\pi}{4}\right) + \cot\left(\frac{\pi}{4}\right)\right|\right)\right)$$

$$s = \ln\left(\sqrt{2} + 1\right)$$

Planar Curve Defined by Parametric Equations

For a curve defined by parametric equations $x = f(t)$ and $y = g(t)$, we obtain the differential line element ds in terms of the parameter t as follows:

$$ds = \sqrt{dx^2 + dy^2} = \sqrt{\left(f'(t)\right)^2 dt^2 + \left(g'(t)\right)^2 dt^2}$$

Factoring dt^2 from the radicand and simplifying, we obtain our final result for ds.

$$ds = \sqrt{\left(f'(t)\right)^2 + \left(g'(t)\right)^2}\, dt$$

To find the arc length of the curve from $t = t_1$ to $t = t_2$, we integrate.

$$s = \int_{t_1}^{t_2} \sqrt{\left(f'(t)\right)^2 + \left(g'(t)\right)^2}\, dt$$

Example 3.4.14: Find the arc length of the curve defined by $x = t^2 + 1$ and $y = 4t^3 + 1$ on the interval $[0, 1]$.

First we'll work out the expression for ds.

$$ds = \sqrt{\left(2t\right)^2 + \left(12t^2\right)^2}\, dt = \sqrt{4t^2 + 144t^4}\, dt = 2t\sqrt{1 + 36t^2}\, dt$$

Now, we'll integrate to find s.

$$s = 2\int_0^1 t\sqrt{1 + 36t^2}\, dt$$

Let $u = 1 + 36t^2$, so then $du = 72t\,dt \Rightarrow t\,dt = \dfrac{1}{72}du$. Furthermore, $t = 0 \Rightarrow u = 1$, and $t = 1 \Rightarrow u = 37$.

$$s = 2 \cdot \frac{1}{72}\int_1^{37} u^{1/2}\,du = \frac{1}{36}\cdot\frac{2}{3}\left(u^{3/2}\right)\Big|_1^{37} = \frac{1}{54}\left(37^{3/2} - 1\right)$$

LO 3.4E – Miscellaneous Applications

This LO is a bit of a catch-all.

LO 3.4E: Use the definite integral to solve problems in various contexts.

The lone EK for this LO basically says the same thing.

EK 3.4E1: The definite integral can be used to express information about accumulation and net change in many applied contexts.

This EK pertains to the Net Change Theorem, which states that the definite integral of the rate of change of a function F over an interval $[a, b]$ gives the *net change* of F over $[a, b]$. We express this mathematically as follows:

$$\int_a^b F'(x)\,dx = F(b) - F(a)$$

As you can see, this is a direct consequence of the Fundamental Theorem of Calculus. We reviewed this in the context of motion in **LO 3.4A**. We'll present here an example in a different context.

Example 3.4.15: A water tank springs a leak at time $t = 0\,s$. At time t, the rate at which water is leaking out of the tank $10e^{-0.5t}\,\dfrac{m^3}{s}$. How much water leaks out of the tank from $t = 1\,s$ to $t = 5\,s$? Round your answer to the nearest $0.01\,m^3$.

Let $V(t)$ be the volume (in m^3) of water that has leaked out at time t. Then $V'(t) = 10e^{-0.5t}$, and the total amount of water that has leaked out of the tank during the indicated time interval is given by $V(5) - V(1)$. We can use the Net Change Theorem to find this quantity.

$$V(5) - V(1) = 10\int_1^5 e^{-0.5t}\,dt$$

Let $u = -0.5t$, so then $du = -0.5\,dt \Rightarrow dt = -2\,du$. Furthermore, $t = 1 \Rightarrow u = -0.5$, and $t = 5 \Rightarrow u = -2.5$.

$$V(5) - V(1) = -20\int_{-0.5}^{-2.5} e^u\,du = -20\left(e^u\right)\Big|_{-0.5}^{-2.5} = -20\left(e^{-2.5} - e^{-0.5}\right) \approx 10.49\,m^3$$

EU 3.5 – Differential Equations _____

LO 3.5A – General and Specific Solutions

Here, we'll review how to solve differential equations to obtain a general family of solutions, and we also review how to solve initial value problems to obtain specific solutions.

LO 3.5A: Analyze differential equations to obtain general and specific solutions.

Refer back to **EK 2.3E1**. We reviewed there that the *general* solution to the differential equation $\frac{dy}{dx} = 2x$ is the family of functions $y = x^2 + C$, and that the *specific* solution to that same differential equation, subject to the initial condition $y = 3$ when $x = 1$, is $y = x^2 + 2$. A differential equation, together with enough initial conditions to determine all constants of integration, is called an *initial value problem*. In **LO 2.3E**, we reviewed how to *verify* solutions to differential equations, and in **LO 2.3F**, we reviewed how to *approximate* solutions to differential equations. Here, we will review how to find *exact solutions* to differential equations.

EK 3.5A1: Antidifferentiation can be used to find specific solutions to differential equations with given initial conditions, including applications to motion along a line, exponential growth and decay, **(BC)** and logistic growth.

There are no new mathematical concepts to introduce here, so we'll get right to the examples.

Example 3.5.1: Find the specific solution to $\frac{dy}{dx} = xe^{-x^2}$ subject to the initial condition $y(0) = 3$.

First, we'll find the general solution by antidifferentiation.

$$y = \int xe^{-x^2}\,dx$$

Let $u = -x^2$, so then $du = -2x\,dx \Rightarrow x\,dx = -\frac{1}{2}\,du$.

$$y = -\frac{1}{2}\int e^u\,du = -\frac{1}{2}e^u + C = -\frac{1}{2}e^{-x^2} + C$$

Now apply the initial condition to find the constant of integration.

$$y(0) = -\frac{1}{2}e^{-0^2} + C$$

$$3 = -\frac{1}{2} + C \Rightarrow C = \frac{7}{2}$$

The specific solution is $y = -\dfrac{1}{2}e^{-x^2} + \dfrac{7}{2}$.

A differential equation in which the highest order derivative is a first derivative is called a **first order differential equation**. In general, a differential equation in which the highest order derivative is an nth derivative is called an **n^{th} order differential equation**. A differential equation of order n has n integration constants in its general solution, and so n initial conditions are needed to determine them all when finding a specific solution. In our next example, we will review how to find the specific solution to a second order differential equation.

Example 3.5.2: Find the specific solution to $y'' = 8\cos(2x)$ subject to the initial conditions $y'(0) = -2$ and $y(0) = 1$.

Integrate once to find y'.

$$y' = 8\int \cos(2x)\,dx$$

Let $u = 2x$, so $du = 2dx \Rightarrow dx = \dfrac{1}{2}du$.

$$y' = 8\left(\frac{1}{2}\right)\int \cos(u)\,du = 4\sin(u) + C_1 = 4\sin(2x) + C_1$$

Now use the initial condition on y' to find C_1.

$$y'(0) = 4\sin(2\cdot 0) + C_1$$

$$-2 = C_1$$

So thus far, we have $y' = 4\sin(2x) - 2$. Integrate once more to find y.

$$y = \int \left(4\sin(2x) - 2\right)dx$$

Once again, let $u = 2x$ so that $dx = \dfrac{1}{2}du$.

$$y = \frac{1}{2}\int \left(4\sin(u) - 2\right)du = \frac{1}{2}\left(-4\cos(u) - 2u\right) + C_2 = -2\cos(2x) - 2x + C_2$$

Finally, use the initial condition on y to find C_2.

$$y(0) = -2\cos(2\cdot 0) - 2(0) + C_2$$

$$1 = -2 + C_2 \Rightarrow C_2 = 3$$

The specific solution is thus $y = -2\cos(2x) - 2x + 3$.

The EK explicitly mentions some applications, so we'll review those with some examples now.

Motion Along a Line

We've already reviewed the connections between position, velocity, and acceleration, so we'll get right to an example.

Example 3.5.3: A particle is dropped from rest from a height of $19.6\,m$, and it falls straight down to the ground. The acceleration of gravity is $9.8\,\dfrac{m}{s^2}$ downward. With what velocity does the particle hit the ground? Ignore air resistance.

Let y be the position of the particle above the ground. Then the velocity and acceleration of the particle are $\dfrac{dy}{dt}$ and $\dfrac{d^2y}{dt^2}$, respectively, and the motion of the particle is described by the following initial value problem:

Solve $\dfrac{d^2y}{dt^2} = -9.8$, subject to $\dfrac{dy}{dt}\Big|_{t=0} = 0$ and $y(0) = 19.6$.

Integrate once to find $\dfrac{dy}{dt}$.

$$\frac{dy}{dt} = \int -9.8\,dt = -9.8t + C_1$$

Now use the initial condition on $\dfrac{dy}{dt}$ to find C_1.

$$\frac{dy}{dt}\Big|_{t=0} = -9.8(0) + C_1$$

$$0 = C_1$$

So we have $\dfrac{dy}{dt} = -9.8t$ Now integrate once more to find y.

$$y = -9.8\int t\,dt = -\frac{9.8}{2}t^2 + C_2 = -4.9t^2 + C_2$$

Next, use the initial condition on y to find C_2.

$$y(0) = -4.9(0)^2 + C_2$$

$$19.6 = C_2$$

So our specific solution is $y = -4.9t^2 + 19.6$. Now we can use this to answer the question. First, we'll find the time at which the particle hits the ground by setting $y = 0$.

$$y = -4.9t^2 + 19.6 = 0 \Rightarrow t = \pm\sqrt{\frac{19.6}{4.9}} = \pm 2\,s$$

Since the time $t = -2\,s$ is *before* the particle was dropped, we reject this solution and keep $t = 2\,s$. We will now plug this solution in to $\dfrac{dy}{dt}$ to find the velocity with which the particle hits the ground.

$$\left.\frac{dy}{dt}\right|_{t=2} = -9.8(2) = -19.6\,\frac{m}{s}$$

Exponential Growth

The initial value problem that gives rise to solutions that exhibit exponential growth is as follows:

Solve $\dfrac{dy}{dx} = ky$, subject to $y(0) = y_0$.

The positive constant k is called the **growth constant**. We can solve this differential equation by a method called **separation of variables**. What we do is divide both sides by y and multiply both sides by dx, as follows:

$$\frac{dx}{y} \cdot \frac{dy}{dx} = \frac{dx}{y} \cdot ky$$

$$\frac{dy}{y} = kdx$$

Now we'll integrate both sides.

$$\int \frac{dy}{y} = k \int dx$$

$$\ln|y| = kx + C$$

To solve for y, we'll rewrite this logarithmic equation in exponential form.

$$|y| = e^{kx+C} = e^C e^{kx}$$

$$y = \pm e^c e^{kx}$$

To simplify this, let $\pm e^C = A$

$$y = Ae^{kx}$$

Now use the initial condition on y to evaluate the constant A.

$$y(0) = Ae^{k(0)}$$

$$y_0 = A$$

So the specific solution is $y = y_0 e^{kx}$. A graph of this solution is sketched in the following figure:

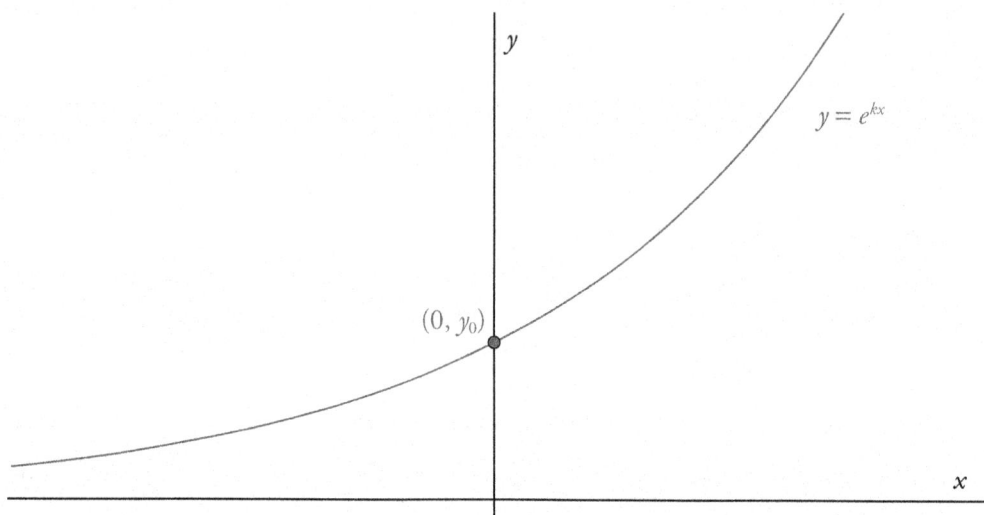

An application of exponential growth is *continuously compounded interest*. If an amount of money P, called the *principal*, is invested in an account that bears continuously compounded interest at an annual rate r, then the account balance at time t *yrs* is given by $A(t)$, and it grows exponentially. The growth constant is r, and the initial value of the account balance is $A(0) = P$. Thus, $A(t) = Pe^{rt}$.

Example 3.5.4: An investor deposits $P = \$10,000$ in an account that bears continuously compounded interest at an annual interest rate of $r = 6\%$. Assuming that the investor does not withdraw any money from the account, at what time t is his account balance equal to $\$12,000$? Round your answer to the nearest year.

The account balance at time t is $A(t) = 10,000e^{0.06t}$. To find the time at which the balance is equal to $\$12,000$, set $A(t) = 12,000$, and solve for t.

$$10,000e^{0.06t} = 12,000$$

$$e^{0.06t} = \frac{12,000}{10,000} = 1.2$$

Rewrite this exponential equation in logarithmic form, and solve for t.

$$0.06t = \ln(1.2) \Rightarrow t = \frac{\ln(1.2)}{0.06} \approx 3 \text{ yrs}$$

Exponential Decay

The initial value problem that gives rise to solutions that exhibit exponential decay is as follows:

Solve $\dfrac{dy}{dx} = -ky$, subject to $y(0) = y_0$.

The positive constant k is called the **decay constant**. The specific solution to this initial value problem is found by simply replacing k with $-k$ in the solution for exponential growth. This replacement gives us $y = y_0 e^{-kx}$. A graph of this solution is shown in the following figure:

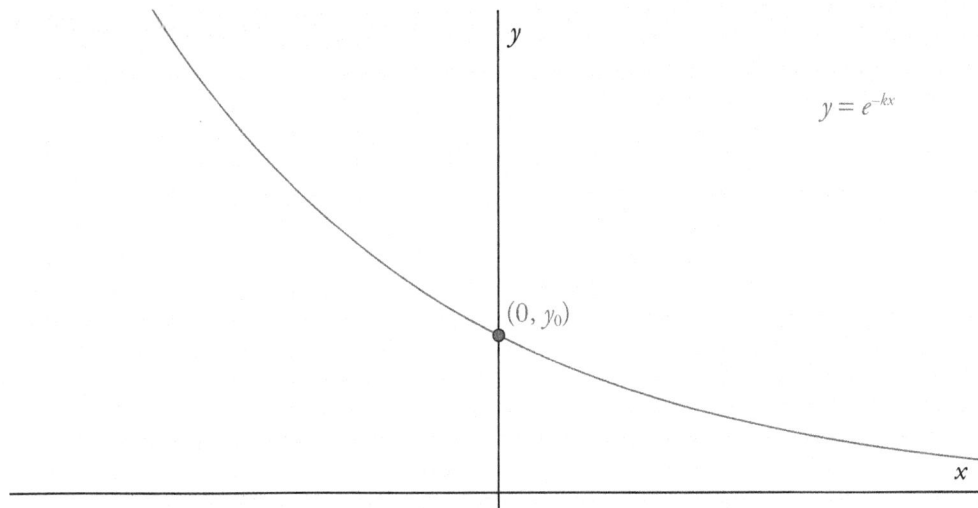

$y = e^{-kx}$

$(0, y_0)$

An application of exponential decay is *radioactive decay*. An amount N_0 (in *kg*) of a radioactive substance at time $t = 0\,yrs$ decays in such a way that the amount of the substance at time $t\,yrs$ is given by a function $N(t)$, which decays exponentially with decay constant k. Thus, $N(t) = N_0 e^{-kt}$.

The *half-life T* of a radioactive substance is the time required for an amount N_0 of the substance to decay to half of its initial value. The half-life is often reported in lieu of the decay constant, so we'll need to review how the two are related. We translate the verbal definition of half-life given here into mathematical symbols as follows:

$$N(T) = N_0 e^{-kT}$$

$$\frac{1}{2} N_0 = N_0 e^{-kT}$$

Note that we can divide both sides by N_0.

$$\frac{1}{2} = e^{-kT}$$

Now write this exponential equation in logarithmic form.

$$\ln\left(\frac{1}{2}\right) = -kT$$

$$-\ln(2) = -kT$$

This equation can be solved for k to yield $k = \dfrac{\ln(2)}{T}$, or for T to yield $T = \dfrac{\ln(2)}{k}$. Note that, since N_0 is not present in the expression for half-life, *the half-life of a sample of a radioactive substance does not depend on the initial amount in the sample.*

Example 3.5.5: Carbon-14 has a half-life of $5730\,yrs$. How long does it take for a $0.05\,kg$ sample to decay to $0.03\,kg$? Round your answer to the nearest year.

The amount of carbon-14 left in the sample at time t is $N(t) = 0.05e^{-kt}$, where $k = \dfrac{\ln(2)}{5730}$. We find the desired time t by setting $N(t) = 0.03$.

$$0.05e^{-\frac{\ln(2)t}{5730}} = 0.03$$

$$e^{-\frac{\ln(2)t}{5730}} = 0.6$$

Now we write this exponential equation in logarithmic form, and then solve for t.

$$-\frac{\ln(2)t}{5730} = \ln(0.6) \Rightarrow t = -\frac{5730\ln(0.6)}{\ln(2)} \approx 4223 \; yrs$$

Logistic Growth

The initial value problem that gives rise to solutions that exhibit logistic growth is as follows:

Solve $\dfrac{dy}{dx} = ky(a - y)$, subject to $y(0) = y_0$, where $0 < y_0 < a$.

Just as with exponential growth, the positive constant k is called the **growth constant**. The positive constant a is called the **carrying capacity**. We can solve this differential equation by separation of variables if we multiply both sides by dx and divide by $y(a - y)$.

$$\frac{dx}{y(a - y)} \cdot \frac{dy}{dx} = \frac{dx}{y(a - y)} \cdot ky(a - y)$$

$$\frac{dy}{y(a - y)} = kdx$$

Now we'll integrate both sides.

$$\int \frac{dy}{y(a - y)} = k \int dx$$

To compute the integral on the left side, we'll find the partial fraction decomposition of the integrand.

$$\frac{1}{y(u-y)} = \frac{A_1}{y} + \frac{A_2}{u-y}$$

$$1 = A_1(a-y) + A_2 y$$

Plug in $y = 0$ to obtain $A_1 = \dfrac{1}{a}$, and plug in $y = a$ to obtain $A_2 = \dfrac{1}{a}$. We will then rewrite the equation as follows:

$$\frac{1}{a}\int\left(\frac{1}{y} + \frac{1}{a-y}\right) = k\int dx$$

$$\frac{1}{a}\left(\ln|y| - \ln|a-y|\right) = kx + C$$

$$\frac{1}{a}\ln\left|\frac{y}{a-y}\right| = kx + C$$

$$\ln\left|\frac{y}{a-y}\right| = akx + aC$$

To solve this logarithmic equation for y, write it in exponential form.

$$\left|\frac{y}{a-y}\right| = e^{akx+aC} = e^{aC}e^{akx}$$

$$\frac{y}{a-y} = \pm e^{aC}e^{akx}$$

Let $A = \pm e^{aC}$, and multiply both sides of the equation by $a - y$.

$$\frac{y}{a-y} \cdot (a-y) = Ae^{akx}(a-y)$$

$$y = aAe^{akx} - yAe^{akx}$$

$$y + yAe^{akx} = aAe^{akx}$$

$$y\left(1 + Ae^{akx}\right) = aAe^{akx}$$

$$y = \frac{aAe^{akx}}{1 + Ae^{akx}}$$

We'll write this in its more common form by dividing the numerator and denominator by Ae^{akx}.

$$y = \dfrac{a}{1 + \dfrac{1}{A}e^{-akx}}$$

Now we'll apply the initial condition on y to evaluate A.

$$y(0) = \dfrac{a}{1 + \dfrac{1}{A}e^{-ak(0)}}$$

$$y_0 = \dfrac{a}{1 + \dfrac{1}{A}} \Rightarrow A = \dfrac{y_0}{a - y_0}$$

So, our specific solution is as follows:

$$y = \dfrac{a}{1 + \dfrac{a - y_0}{y_0}e^{-akx}}$$

We can make this look a little nicer by multiplying the top and bottom by y_0.

$$y = \dfrac{ay_0}{y_0 + (a - y_0)e^{-akx}}$$

The graph of this solution is shown in the following figure:

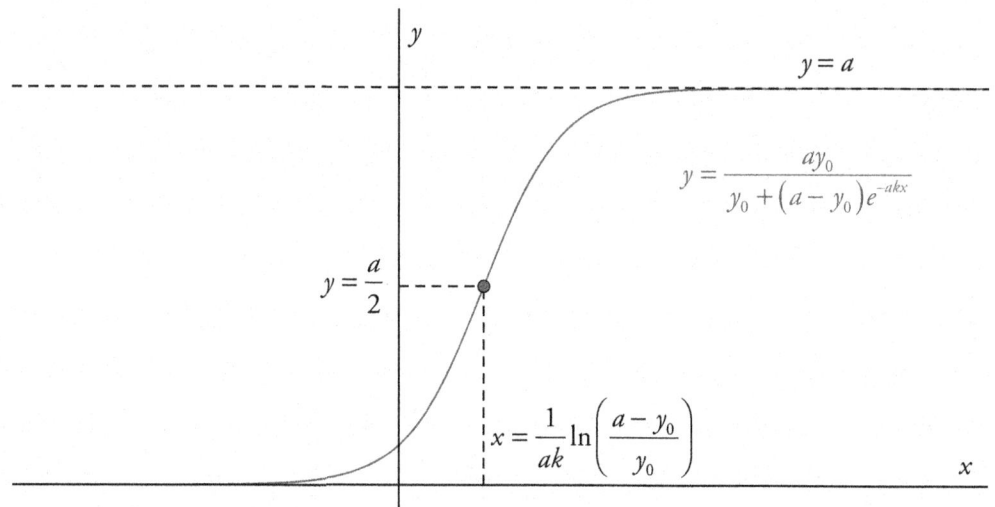

The logistic growth curve has two horizontal asymptotes: $y = 0$ and $y = a$. It also has an inflection point at $\left(\dfrac{1}{ak}\ln\left(\dfrac{a - y_0}{y_0}\right), \dfrac{a}{2} \right)$.

An application of logistic growth is population growth when limited resources are available. Suppose a population has a growth constant k, but the environment can only support a maximum population of P_{max}. Suppose further that the initial population is P_0, such that $0 < P_0 < P_{max}$. Then, if the population P undergoes logistic growth, it is given at time t by the function $P(t) = \dfrac{P_{max}P_0}{P_0 + (P_{max} - P_0)e^{-P_{max}kt}}$.

Example 3.5.6: A colony of bacteria is grown in a petri dish that can support a maximum population of $150\,mg$ of bacteria. The initial population is $30\,mg$ of bacteria, and the growth constant is $0.005\,\dfrac{mg}{h}$. What is the mass of the population at $t = 4\,h$? Round your answer to the nearest mg.

The population P (in mg) at time t (in h) is given by the following function:

$$P(t) = \frac{(150)(30)}{30 + (150 - 30)e^{-(150)(0.005)t}} = \frac{150}{1 + 4e^{-0.75t}}$$

We can answer the question by evaluating this function at $t = 4$.

$$P(4) = \frac{150}{1 + 4e^{-0.75(4)}} \approx 125\,mg$$

In the next EK, we will focus on a method of solution that we have already touched on here: separation of variables.

EK 3.5A2: Some differential equations can be solved by separation of variables.

Suppose that a first order differential equation can be written in the following form:

$$\frac{dy}{dx} = f(x)g(y)$$

Then we can separate the variables x and y in the equation by dividing both sides by $g(y)$ and multiplying both sides by dx.

$$\frac{dx}{g(y)} \cdot \frac{dy}{dx} = \frac{dx}{g(y)} \cdot f(x)g(y)$$

$$\frac{dy}{g(y)} = f(x)dx$$

We can then solve the equation by antidifferentiation.

$$\int \frac{dy}{g(y)} = \int f(x)dx$$

This is as far as we can go without knowing the functions f and g, so now we'll turn to a concrete example.

Example 3.5.7: Find the specific solution to the differential equation $\dfrac{dy}{dx} = \dfrac{2e^x}{y}$, subject to $y(0) = 4$

We begin by separating variables.

$$y\,dx \cdot \frac{dy}{dx} = y\,dx \cdot \frac{2e^x}{y}$$

$$y\,dy = 2e^x\,dx$$

Now we integrate both sides, and solve for y.

$$\int y\,dy = 2\int e^x\,dx$$

$$\frac{1}{2}y^2 = 2e^x + C$$

$$y^2 = 4e^x + 2C$$

$$y = \pm\sqrt{4e^x + 2C}$$

Since $y(0) > 0$, we reject the negative root. Now we apply the initial condition on y to find C.

$$y(0) = \sqrt{4e^0 + 2C}$$

$$4 = \sqrt{4 + 2C} \Rightarrow C = 6$$

So the specific solution is $y = \sqrt{4e^x + 12}$.

EK 3.5A3: Solutions to differential equations may be subject to domain restrictions.

Restrictions on the domain can be inferred from the equation itself. We illustrate with an example.

Example 3.5.8: What are the domain restrictions for the equation $\dfrac{dy}{dx} = \dfrac{y}{\sqrt[3]{x}}$?

The right side of this equation is not defined for $x = 0$. This means that we can define a solution to this equation on $(-\infty, 0)$, or $(0, \infty)$, or on any subinterval of these two intervals. We can also define a solution on a union of these intervals, such as $(-3, -1) \cup (5, 9)$, just as long as the domain does not include $x = 0$.

EK 3.5A4: The function F defined by $F(x) = C + \displaystyle\int_{x_a}^{x} f(t)\,dt$ is a general solution to the differential equation $\dfrac{dy}{dx} = f(x)$, and $F(x) = y_0 + \displaystyle\int_{a}^{x} f(t)\,dt$ is a particular solution to the differential equation $\dfrac{dy}{dx} = f(x)$ satisfying $F(a) = y_0$.

The statements of this EK are easy to verify. We begin by verifying the general solution.

$$\frac{d}{dx}\left(C+\int_a^x f(t)\,dt\right)=f(x)$$

Using the result of **EK 3.3A2** on the left side, we obtain the result immediately.

$$f(x)=f(x)$$

Now, we'll verify the statement regarding the initial condition.

$$F(a)=y_0$$

$$y_0+\int_a^a f(t)\,dt=y_0$$

Using the Zero Interval Rule, we obtain the desired result.

$$y_0=y_0$$

Example 3.5.9: Find the specific solution to the equation $\dfrac{dy}{dx}=e^{-x^2}$, subject to $y(0)=2$.

The specific solution is $y=2+\int_0^x e^{-t^2}\,dt$. Since we cannot express the antiderivative of e^{-t^2} in terms of elementary functions, we cannot take this any further.

LO 3.5B – Modeling with Differential Equations

The complete Learning Outcome statement is as follows:

LO 3.5B: Interpret, create, and solve differential equations from problems in context.

We focus here on two particular classes of differential equations: exponential growth/decay and logistic growth.

EK 3.5B1: The model for exponential growth and decay that arises from the statement "The rate of change of a quantity is proportional to the size of the quantity," and is $\dfrac{dy}{dt}=ky$.

Example 3.5.10: The intensity I of a signal (in W/m^2) from a faulty cellular tower decreases with time at a rate that is equal to $-0.15I\ \dfrac{W}{m^2\cdot h}$, starting at $t=0\,h$. After how many hours will the signal reach half of its initial strength? Round your answer to the nearest hundredth of an hour.

The differential equation that describes this phenomenon is $\dfrac{dI}{dt} = -0.15I$, which is an equation for exponential decay. The specific solution is $I(t) = I_0 e^{-0.15t}$, and we want to know the time at which $I(t) = \dfrac{1}{2} I_0$.

$$\frac{1}{2}I_0 = I_0 e^{-0.15t}$$

$$\frac{1}{2} = e^{-0.15t}$$

$$\ln\left(\frac{1}{2}\right) = -0.15t \Rightarrow t = \frac{\ln\left(\dfrac{1}{2}\right)}{-0.15} \approx 4.62 \; hrs$$

EK 3.5B2: (BC) The model for logistic growth that arises from the statement "The rate of change of a quantity is jointly proportional to the size of the quantity and the difference between the quantity and the carrying capacity," and is $\dfrac{dy}{dt} = ky(a - y)$.

Example 3.5.11: The rate of change of a population P of a herd of buffalo with respect to time t is jointly proportional to P and $40{,}000 - P$, where $40{,}000$ is the carrying capacity of the environment, and the constant of proportionality is $0.00005 \; day^{-1}$. If the initial population is $12{,}000$ buffalo, find the population after two weeks. Round your answer to the nearest unit.

The differential equation that models the population growth is $\dfrac{dP}{dt} = 0.00005P(40{,}000 - P)$, and the specific solution is as follows:

$$P(t) = \frac{(40{,}000)(12{,}000)}{12{,}000 + (40{,}000 - 12{,}000)e^{-(40{,}000)(0.00005)x}} = \frac{120{,}000}{3 + 7e^{-0.2t}}$$

The population at $t = 2 \; weeks = 14 \; days$ is $P(14)$.

$$P(14) = \frac{120{,}000}{3 + 7e^{-0.2(14)}} \approx 35{,}030 \text{ buffalo.}$$

Chapter 4: Series (BC Only)

Chapter 4 Overview

Chapter 4 is divided into the following two Enduring Understandings:

- **EU 4.1:** The sum of an infinite number of real numbers may converge.
- **EU 4.2:** A function can be represented by an associated power series over the interval of convergence for the power series.

EU 4.1 – Infinite Series

LO 4.1A – Convergence and Divergence of Series

When working with infinite series, the principal task is testing the series for convergence or divergence. That's what the first LO is all about.

LO 4.1A: Determine whether a series converges or diverges.

Before we get to the notion of convergence, we need to review the notion of a *partial sum*.

EK 4.1A1: The nth partial sum is defined as the sum of the first n terms of a sequence.

Suppose $\sum_{i=1}^{\infty} a_i$ is an infinite series of real numbers. Its first few partial sums are as follows:

- First partial sum: $S_1 = \sum_{i=1}^{1} a_i = a_1$

- Second partial sum: $S_2 = \sum_{i=1}^{2} a_i = a_1 + a_2$

- Third partial sum: $S_3 = \sum_{i=1}^{3} a_i = a_1 + a_2 + a_3$

- n^{th} partial sum: $S_n = \sum_{i=1}^{n} a_i = a_1 + a_2 + a_3 + \cdots + a_n$

We note here that the function a_i in the infinite series is called the **summand** of the series.

Example 4.1.1: Find the first, second, third, and n^{th} partial sums for $\sum_{i=1}^{\infty} \left(\frac{1}{n} - \frac{1}{n+1} \right)$.

- First partial sum: $S_1 = \sum_{i=1}^{1} \left(\dfrac{1}{n} - \dfrac{1}{n+1} \right) = \dfrac{1}{1} - \dfrac{1}{2} = \dfrac{1}{2}$

- Second partial sum: $S_2 = \sum_{i=1}^{2} \left(\dfrac{1}{n} - \dfrac{1}{n+1} \right) = \left(\dfrac{1}{1} - \dfrac{1}{2} \right) + \left(\dfrac{1}{2} - \dfrac{1}{3} \right) = 1 - \dfrac{1}{3} = \dfrac{2}{3}$

- Third partial sum: $S_3 = \sum_{i=1}^{3} \left(\dfrac{1}{n} - \dfrac{1}{n+1} \right) = \left(\dfrac{1}{1} - \dfrac{1}{2} \right) + \left(\dfrac{1}{2} - \dfrac{1}{3} \right) + \left(\dfrac{1}{3} - \dfrac{1}{4} \right) = 1 - \dfrac{1}{4} = \dfrac{3}{4}$

- n^{th} partial sum: $S_n = \sum_{i=1}^{n} \left(\dfrac{1}{n} - \dfrac{1}{n+1} \right) = \left(\dfrac{1}{1} - \dfrac{1}{2} \right) + \left(\dfrac{1}{2} - \dfrac{1}{3} \right) + \cdots + \left(\dfrac{1}{n} - \dfrac{1}{n+1} \right) = 1 - \dfrac{1}{n+1}$

Notice how, in each partial sum, all of the terms in between $\dfrac{1}{1}$ and the last term cancel out. The nth partial sum collapses down to $1 - \dfrac{1}{n+1}$, with all the terms in between cancelling each other out. A series whose sequence of partial sums exhibits this behavior is called a **telescoping series**.

EK 4.1A2: An infinite series of numbers converges to a real number S (or has a sum S), if and only if the limit of its sequence of partial sums exists and equals S.

If $\displaystyle\sum_{i=1}^{\infty} a_i$ is an infinite series of numbers, then its partial sums $S_1, S_2, S_3, \ldots = \left\{ S_n \right\}_{n=1}^{\infty}$ forms a sequence. If the sequence of partial sums converges to a real number S, then we call S the **sum** of the series, and vice versa. We express this symbolically as follows:

$$\lim_{n \to \infty} S_n = S \Leftrightarrow \sum_{i=1}^{\infty} a_n = S$$

We use the double arrow because the statement in the EK is an "if and only if" statement.

Example 4.1.2: Find the sum of the series $\displaystyle\sum_{i=1}^{\infty} \left(\dfrac{1}{n} - \dfrac{1}{n+1} \right)$.

We found in Example 4.1.1 that the nth partial sum of this series is $S_n = 1 - \dfrac{1}{n+1}$. We'll compute the limit of the sequence of partial sums.

$$\lim_{n \to \infty} S_n = \lim_{n \to \infty} \left(1 - \dfrac{1}{n+1} \right) = 1$$

Since the limit of the sequence of partial sums is 1, the sum of the series is 1 as well.

Example 4.1.3: Show that the infinite series $\displaystyle\sum_{i=1}^{\infty} (-1)^n$ does not converge to a sum.

Let's look at the first few partial sums.

- First partial sum: $S_1 = \displaystyle\sum_{i=1}^{1} (-1)^n = -1$

- Second partial sum: $S_2 = \displaystyle\sum_{i=1}^{2} (-1)^n = -1 + 1 = 0$

- Third partial sum: $S_3 = \sum_{i=1}^{3} (-1)^n = -1 + 1 - 1 = -1$

- Fourth partial sum: $S_4 = \sum_{i=1}^{4} (-1)^n = -1 + 1 - 1 + 1 = 0$

The partial sums keep alternating between -1 and 0, so the nth partial sum does not approach a limit as n increases without bound. That is, $\lim_{n \to \infty} S_n$ does not exist, so the infinite series $\sum_{i=1}^{\infty} (-1)^n$ does not converge to a sum. When a series does not converge, we say that it **diverges**.

Next, we will look at some types of series that you are likely to see on the exam.

EK 4.1A3: Common series of numbers include geometric series, the harmonic series, and p-series.

We'll review each of these one at a time.

Geometric Series

A geometric series is an infinite series in which successive terms have a common ratio. The form of a geometric series is as follows:

$$\sum_{n=0}^{\infty} ar^n = a + ar + ar^2 + ar^3 + \cdots$$

Here, a and r are real numbers. a is the first term in the series (corresponding to $n = 0$), and r is the common ratio of successive terms in the series: $\dfrac{ar^{n+1}}{ar^n_\infty} = r$.

Example 4.1.4: Write the infinite series $\sum_{n=0}^{\infty} \dfrac{2^{n+2}}{3^n}$ in the form $\sum_{n=0}^{\infty} ar^n$.

We'll manipulate the summand as follows:

$$\frac{2^{n+2}}{3^n} = \frac{2^n \cdot 2^2}{3^n} = 2^2 \cdot \frac{2^n}{3^n} = 4\left(\frac{2}{3}\right)^n$$

So, we can write the series as $\sum_{n=0}^{\infty} 4\left(\dfrac{2}{3}\right)^n$, which is a geometric series with $a = 4$ and $r = \dfrac{2}{3}$.

We wish to note that a geometric series needn't start at $n = 0$. For instance, $\sum_{n=2}^{\infty} 4\left(\dfrac{2}{3}\right)^n$ is still an infinite series whose successive terms have a common ratio of $r = \dfrac{2}{3}$. However, the first term in the series (which corresponds to $n = 2$) is not 4, but rather $4\left(\dfrac{2}{3}\right)^2 = \dfrac{16}{9}$.

Example 4.1.5: Find the sum of the infinite series $\displaystyle\sum_{n=0}^{\infty} \frac{2^{n+2}}{3^n}$.

As we showed in Example 4.1.4, this series can be written as $\displaystyle\sum_{n=0}^{\infty} 4\left(\frac{2}{3}\right)^n$. We'll find an expression for the nth partial sum S_n, and then take its limit. We'll find S_n using a clever trick.

- n^{th} partial sum: $S_n = 4 + 4\left(\frac{2}{3}\right) + 4\left(\frac{2}{3}\right)^2 + \cdots + 4\left(\frac{2}{3}\right)^{n-1}$

- $\dfrac{2}{3}$ times n^{th} partial sum: $\dfrac{2}{3}S_n = 4\left(\frac{2}{3}\right) + 4\left(\frac{2}{3}\right)^2 + \cdots + 4\left(\frac{2}{3}\right)^{n-1} + 4\left(\frac{2}{3}\right)^n$

Now, we'll *subtract* these results, and then solve for S_n. Notice how most of the terms cancel out upon subtraction.

$$S_n - \frac{2}{3}S_n = 4 - 4\left(\frac{2}{3}\right)^n$$

$$\frac{1}{3}S_n = 4 - 4\left(\frac{2}{3}\right)^n$$

$$S_n = 12 - 12\left(\frac{2}{3}\right)^n$$

Now, we will compute the limit of the nth partial sum.

$$\lim_{n\to\infty} S_n = \lim_{n\to\infty}\left(12 - 12\left(\frac{2}{3}\right)^n \right) = 12 - 12(0) = 12$$

The sum of the series is therefore 12.

Harmonic Series

A harmonic series is of the following form:

$$\sum_{n=1}^{\infty} \frac{1}{n} = 1 + \frac{1}{2} + \frac{1}{3} + \cdots$$

As with geometric series, a harmonic series can start at any positive value of n. For instance, $\displaystyle\sum_{n=4}^{\infty} \frac{1}{n}$ is a harmonic series as well.

p-Series

A p-series is any series of the following form.

$$\sum_{n=1}^{\infty} \frac{1}{n^p} = \frac{1}{1^p} + \frac{1}{2^p} + \frac{1}{3^p} + \cdots$$

Here, p is any real number. Note that a harmonic series is just a p-series with $p = 1$. As with harmonic series, a p-series can start at any positive value of n.

We'll discuss the convergence of p-series when we get to the convergence tests in **EK 4.1A6**.

EK 4.1A4: A series may be absolutely convergent, conditionally convergent, or divergent.

We defined in **EK 4.1A2** what it means for a series to converge. A series that does not converge is said to **diverge**. You saw a divergent series in Example 4.1. We need to define absolute and conditional convergence.

Let $\displaystyle\sum_{n=0}^{\infty} a_n$ be a convergent infinite series.

- $\displaystyle\sum_{n=0}^{\infty} a_n$ **converges absolutely** if $\displaystyle\sum_{n=0}^{\infty} |a_n|$ also converges.

- $\displaystyle\sum_{n=0}^{\infty} a_n$ **converges conditionally** if $\displaystyle\sum_{n=0}^{\infty} |a_n|$ diverges.

We'll get to some concrete examples after we cover the alternating series test in **EK 4.1A6**.

EK 4.1A5: If a series converges absolutely, then it converges.

This is just an obvious fact. If you look back at the definition of absolute convergence, you can see right at the start that the convergence of the series is assumed.

Testing series for convergence or divergence by constructing the nth partial sum and taking its limit is not always convenient or even possible. For instance, there is no easy way to find S_n for a p-series. We need an arsenal of convergence and divergence tests for infinite series, and in the next EK, we cover them.

EK 4.1A6: In addition to examining the limit of the sequence of partial sums, methods for determining whether a series of numbers converges or diverges are the nth term test, the comparison test, the limit comparison test, the integral test, the ratio test, and the alternating series test.

In your AP Calculus course, you probably learned about other tests for infinite series (e.g., the root test). For the exam, you only need to worry about the tests that are explicitly mentioned in the syllabus. We'll review the tests listed here one at a time.

The nth Term Test for Divergence

Let $\{a_n\}_{n=1}^{\infty}$ be a sequence of real numbers. If $\displaystyle\lim_{n \to \infty} a_n \neq 0$, then the infinite series $\displaystyle\sum_{n=0}^{\infty} a_n$ diverges.

In the EK, this test is listed simply as "the nth term test". We add the words "*for divergence*" to emphasize the fact that *the nth term test cannot tell you if a series converges*. It can *only* be used to test for *divergence*.

Example 4.1.6: Show that the series $\sum_{n=0}^{\infty} \dfrac{2n+1}{3n+2}$ diverges.

Since we only have one test in our arsenal so far, it's pretty clear what we should do here: take the limit of the summand.

$$\lim_{n \to \infty} \frac{2n+1}{3n+2} = \frac{2}{3} \neq 0$$

Since the limit of the summand is not zero, the series diverges.

We wish to note here that if $\lim_{n \to \infty} a_n = 0$, then the nth term test yields no information at all about $\sum_{n=0}^{\infty} a_n$. For instance, with the harmonic series $\sum_{n=1}^{\infty} \dfrac{1}{n}$ we have $\lim_{n \to \infty} \dfrac{1}{n} = 0$, but that does not mean that the harmonic series converges (in fact, it *diverges*, as we'll see later).

We're going to review the tests out of the order in which they're listed in the EK, and in the order in which they're covered in a typical calculus book.

The Integral Test

If f is positive, continuous, and decreasing for $x \geq 1$, and $a_n = f(n)$, then $\sum_{n=1}^{\infty} a_n$ and $\int_1^{\infty} f(x)\,dx$ either both converge or both diverge.

Example 4.1.7: Determine whether the series $\sum_{n=2}^{\infty} \dfrac{1}{n \ln(n)}$ converges or diverges.

First, we'll check to see whether the integral test applies. Certainly, $f(x) = \dfrac{1}{x \ln(x)}$ is both positive and continuous for $x \geq 2$. We need to check to see if f is decreasing, which we do by taking the derivative of f.

$$f'(x) = \frac{d}{dx}\left(x \ln(x)\right)^{-1} = -\left(x\ln(x)\right)^{-2} \cdot \frac{d}{dx}\left(x\ln(x)\right) = -\left(x\ln(x)\right)^{-2} \cdot \left(\ln(x)+1\right)$$

$$f'(x) = -\frac{\ln(x)+1}{\left(x\ln(x)\right)^2}$$

Since $f'(x) \leq 0$ for $x \geq 2$, f is decreasing for $x \geq 2$, and so we can apply the integral test.

$$\int_2^{\infty} \frac{1}{x\ln(x)}\,dx = \lim_{b \to \infty} \int_2^{\infty} \frac{1}{x\ln(x)}\,dx$$

Let $u = \ln(x)$, so $du = \dfrac{1}{x}\,dx$. Furthermore, $x = 2 \Rightarrow u = \ln(2)$, and $x = b \Rightarrow u = \ln(b)$.

$$\int_2^{\infty} \frac{1}{x\ln(x)}\,dx = \lim_{b \to \infty} \int_{\ln(2)}^{\ln(b)} \frac{1}{u}\,du = \lim_{b \to \infty} \left(\ln|u|\right)\Big|_{\ln(2)}^{\ln(b)} = \lim_{b \to \infty} \left(\ln\left|\ln(b)\right| - \ln\left|\ln(2)\right|\right) = \infty$$

Since the integral $\int_2^{\infty} \dfrac{1}{x\ln(x)}\,dx$ diverges, the series $\sum_{n=2}^{\infty} \dfrac{1}{n\ln(n)}$ diverges also.

Example 4.1.8: Let p be a real number. Show that $\int_1^\infty \frac{1}{x^p}\,dx = \frac{1}{p-1}$ if $p > 1$, and that $\int_1^\infty \frac{1}{x^p}\,dx$ diverges if $p \le 1$.

We'll prove the result by cases: $p > 1$, $p = 1$, and $p < 1$.

Case 1: $p > 1$

$$\int_1^\infty \frac{1}{x^p}\,dx = \lim_{b\to\infty}\int_1^b x^{-p}\,dx = \frac{1}{1-p}\lim_{b\to\infty}\left(x^{1-p}\right)\Big|_1^b = \frac{1}{1-p}\lim_{b\to\infty}\left(b^{1-p} - 1^{1-p}\right)$$

$p > 1 \Rightarrow 1 - p < 0$, so $\lim_{b\to\infty} b^{1-p} = 0$, and we obtain the following:

$$\int_1^\infty \frac{1}{x^p}\,dx = \lim_{b\to\infty}\int_1^b x^{-p}\,dx = \frac{1}{1-p}(0-1) = \frac{1}{p-1}$$

Case 2: $p = 1$

$$\int_1^\infty \frac{1}{x}\,dx = \lim_{b\to\infty}\int_1^b \frac{1}{x}\,dx = \lim_{b\to\infty}\left(\ln|x|\right)\Big|_1^b = \lim_{b\to\infty}\left(\ln|b| - \ln|1|\right) = \infty$$

Case 3: $p < 1$

$$\int_1^\infty \frac{1}{x^p}\,dx = \lim_{b\to\infty}\int_1^b x^{-p}\,dx = \frac{1}{1-p}\lim_{b\to\infty}\left(x^{1-p}\right)\Big|_1^b = \frac{1}{1-p}\lim_{b\to\infty}\left(b^{1-p} - 1^{1-p}\right)$$

$p < 1 \Rightarrow 1 - p > 0$, so $\lim_{b\to\infty} b^{1-p} = \infty$, and we obtain the following:

$$\int_1^\infty \frac{1}{x^p}\,dx = \infty$$

This completes the proof.

The integral in Example 4.1.8, combined with the integral test, gives rise to a corollary to the integral test, namely **the p-series test**. Let p be a real number. Then the series $\sum_{n=1}^\infty \frac{1}{n^p}$ converges if $p > 1$, and diverges if $p \le 1$.

This allows us to tell at a glance if a p-series converges or diverges.

Example 4.1.9: For each series given below, determine whether it converges or diverges.

1. $\displaystyle\sum_{n=1}^\infty \frac{1}{n}$

2. $\displaystyle\sum_{n=1}^\infty \frac{1}{n^3}$

3. $\displaystyle\sum_{n=1}^\infty \frac{1}{\sqrt[3]{n^2}}$

Each one of these is a *p*-series.

1. This is the harmonic series, which is a *p*-series with $p = 1$, so it diverges. This is a result that you should remember: *the harmonic series diverges.*

2. This is a *p*-series with $p = 3$, so it converges.

3. This may not look like a *p*-series, but it will after we rewrite it.

$$\sum_{n=1}^{\infty} \frac{1}{\sqrt[3]{n^2}} = \sum_{n=1}^{\infty} \frac{1}{n^{2/3}}$$

4. This is a *p*-series with $p = \dfrac{2}{3}$, so it diverges.

The Comparison Test

Let $0 < a_n \le b_n$ for all *n*. Then the following results hold.

1. $\displaystyle\sum_{n=1}^{\infty} b_n$ converges $\Rightarrow \displaystyle\sum_{n=1}^{\infty} a_n$ converges.

2. $\displaystyle\sum_{n=1}^{\infty} a_n$ diverges $\Rightarrow \displaystyle\sum_{n=1}^{\infty} b_n$ diverges.

The first statement is telling us that if a series of positive terms converges, then a second series of positive terms whose terms are *less than* those of the first series also converges. The second statement is telling us that if a series of positive terms diverges, then a second series of positive terms whose terms are *greater than* those of the first series also diverges.

In order to use the comparison test effectively, you will need to have a large store of convergent and divergent series in your memory. We will illustrate with two examples.

Example 4.1.10: Determine whether the series $\displaystyle\sum_{n=1}^{\infty} \frac{1}{n^2 + 2}$ converges or diverges.

This series reminds us of the *p*-series $\displaystyle\sum_{n=1}^{\infty} \frac{1}{n^2}$, which is known to converge. Since $\dfrac{1}{n^2 + 2} \le \dfrac{1}{n^2}$ for all *n*, it follows from the comparison test that $\displaystyle\sum_{n=1}^{\infty} \frac{1}{n^2 + 2}$ converges also.

Example 4.1.11: Determine whether the series $\displaystyle\sum_{n=2}^{\infty} \frac{1}{\ln(n)}$ converges or diverges.

Refer back to Example 4.1.7. The series $\displaystyle\sum_{n=2}^{\infty} \frac{1}{n\ln(n)}$ is known to diverge. Furthermore, for all *n*, $\dfrac{1}{n\ln(n)} \le \dfrac{1}{\ln(n)}$. Therefore, it follows from the comparison test that $\displaystyle\sum_{n=2}^{\infty} \frac{1}{\ln(n)}$ diverges also.

The Limit Comparison Test

Let $a_n, b_n > 0$ for all *n*, and let $\displaystyle\lim_{n \to \infty} \left(\frac{a_n}{b_n} \right) = R$, where $0 < R < \infty$. Then the two series $\displaystyle\sum_{n=1}^{\infty} a_n$ and $\displaystyle\sum_{n=1}^{\infty} b_n$ either both converge or both diverge.

To see why a second comparison test is needed, consider the series $\displaystyle\sum_{n=1}^{\infty} \frac{1}{\sqrt{n} + 1}$. This sort of looks like the divergent *p*-series $\displaystyle\sum_{n=1}^{\infty} \frac{1}{\sqrt{n}}$, except that we can't use the comparison test, because $\dfrac{1}{\sqrt{n}+1} \le \dfrac{1}{\sqrt{n}}$ for all $n \ge 1$, which tells us nothing. It turns out that the limit comparison test will help us out here, as we'll show in the next example.

Example 4.1.12: Determine whether the series $\displaystyle\sum_{n-1}^{\infty} \dfrac{1}{\sqrt{n}+1}$ converges or diverges.

As we mentioned right before the example, we'll use the limit comparison test with the known divergent series $\displaystyle\sum_{n=1}^{\infty} \dfrac{1}{\sqrt{n}}$.

$$\lim_{n\to\infty}\left(\dfrac{1/\sqrt{n}}{1/\left(\sqrt{n}+1\right)} \right) = \lim_{n\to\infty}\left(\dfrac{\sqrt{n}+1}{\sqrt{n}} \right) = \lim_{n\to\infty}\left(1 + \dfrac{1}{\sqrt{n}} \right) = 1$$

Since this limit is positive and finite, it follows from the limit comparison test that $\displaystyle\sum_{n=1}^{\infty} \dfrac{1}{\sqrt{n}+1}$ diverges, just as $\displaystyle\sum_{n=1}^{\infty} \dfrac{1}{\sqrt{n}}$ does.

With the integral test, the p-series test, the comparison test, and the limit comparison test, the terms in the series were all required to be *positive*. In the next test, the terms in the series are required to have *alternating signs*. When the signs of the terms in a series alternate, we call such a series an **alternating series**.

The Alternating Series Test

Let $a_n > 0$ for all n. The alternating series $\displaystyle\sum_{n=1}^{\infty} (-1)^n\, a_n$ and $\displaystyle\sum_{n=1}^{\infty} (-1)^{n+1}\, a_n$ converge if the following two conditions are satisfied.

1. $\displaystyle\lim_{n\to\infty} a_n = 0$
2. a_n is non-increasing for all n (i.e., $a_{n+1} \le a_n$ for all n).

Note that the alternating series test has no condition for *divergence*. This test can only tell you if a series *converges*. Also note that, since the terms of these series are not all positive, if an alternating series converges, then we will need to test for *absolute vs conditional* convergence, as defined in **EK 4.1A4**.

Example 4.1.13: For each of the following series, determine if it converges or diverges. If it converges, determine if it converges absolutely or conditionally.

1. $\displaystyle\sum_{n=1}^{\infty} \dfrac{(-1)^n}{n}$

2. $\displaystyle\sum_{n=1}^{\infty} \dfrac{(-1)^{n+1}}{n^2}$

3. $\displaystyle\sum_{n=1}^{\infty} \cos(\pi n)$

Each of these series is an alternating series.

1. In this series, $a_n = \dfrac{1}{n}$. Since $\lim\limits_{n\to\infty} \dfrac{1}{n} = 0$, and $\dfrac{1}{n+1} \le \dfrac{1}{n}$, the series converges. To check for absolute vs conditional convergence, look at the series $\displaystyle\sum_{n=1}^{\infty} \left| \dfrac{(-1)^n}{n} \right| = \sum_{n=1}^{\infty} \dfrac{1}{n}$. This is the harmonic series, which diverges. Since $\displaystyle\sum_{n=1}^{\infty} \dfrac{(-1)^n}{n}$ converges and $\displaystyle\sum_{n=1}^{\infty} \left| \dfrac{(-1)^n}{n} \right|$ diverges, it follows that $\displaystyle\sum_{n=1}^{\infty} \dfrac{(-1)^n}{n}$ converges conditionally.

   The series $\displaystyle\sum_{n=1}^{\infty} \dfrac{(-1)^n}{n}$ is called the **alternating harmonic series**, and you should make it a point to remember that it converges conditionally.

2. In this series, $a_n = \dfrac{1}{n^2}$. Since $\lim\limits_{n\to\infty} \dfrac{1}{n^2} = 0$, and $\dfrac{1}{(n+1)^2} \le \dfrac{1}{n^2}$, the series converges. To check for absolute vs conditional convergence, look at the series $\displaystyle\sum_{n=1}^{\infty} \left| \dfrac{(-1)^{n+1}}{n^2} \right| = \sum_{n=1}^{\infty} \dfrac{1}{n^2}$. This is a p-series with $p = 2$, and it converges. Since $\displaystyle\sum_{n=1}^{\infty} \dfrac{(-1)^{n+1}}{n^2}$ and $\displaystyle\sum_{n=1}^{\infty} \left| \dfrac{(-1)^{n+1}}{n^2} \right|$ both converge, it follows that $\displaystyle\sum_{n=1}^{\infty} \dfrac{(-1)^{n+1}}{n^2}$ converges absolutely.

3. This series doesn't look like an alternating series until we note that

   $$\{\cos(\pi n)\}_{n=1}^{\infty} = \{-1, 1, -1, 1, -1, 1, \ldots\}, \text{ so } \sum_{n=1}^{\infty} (-1)^n. \text{ For this series, } a_n = 1. \text{ Since}$$

   $\lim\limits_{n\to\infty} 1 = 1 \ne 0$, the alternating series test yields no information. We turn instead to the nth term test for divergence: $\lim\limits_{n\to\infty} \cos(\pi n) = \lim\limits_{n\to\infty} (-1)^n$. Since this limit does not exist (and hence does not equal 0), it follows that the series diverges.

The final test is important not only for infinite series, but also for power series.

The Ratio Test

Let $\displaystyle\sum_{n=1}^{\infty} a_n$ be a series such that, for all n, $a_n \ne 0$.

1. $\lim\limits_{n\to\infty}\left|\dfrac{a_{n+1}}{a_n}\right|<1\Rightarrow\sum\limits_{n=1}^{\infty}a_n$ converges absolutely.

2. $\lim\limits_{n\to\infty}\left|\dfrac{a_{n+1}}{a_n}\right|>1\Rightarrow\sum\limits_{n=1}^{\infty}a_n$ diverges.

3. $\lim\limits_{n\to\infty}\left|\dfrac{a_{n+1}}{a_n}\right|=1\Rightarrow$ The ratio test yields no information about $\sum\limits_{n=1}^{\infty}a_n$.

If $\lim\limits_{n\to\infty}\left|\dfrac{a_{n+1}}{a_n}\right|=1$, it means that you have to use one of the other tests.

Example 4.1.14: Determine whether the series $\sum\limits_{n=1}^{\infty}\dfrac{(-2)^n(n+1)}{n!}$ converges or diverges.

We'll use the ratio test. We have $a_n\,\dfrac{(-2)^n(n+1)}{n!}$, and $a_{n+1}\,\dfrac{(-2)^{n+1}(n+2)}{(n+1)!}$.

$$\lim_{n\to\infty}\left|\frac{a_{n+1}}{a_n}\right|=\lim_{n\to\infty}\left|\frac{(-2)^{n+1}(n+2)}{(n+1)!}\cdot\frac{n!}{(-2)^n(n+1)}\right|=\lim_{n\to\infty}\left|\frac{-2(n+2)}{(n+1)^2}\right|=2\lim_{n\to\infty}\frac{n+2}{n^2+2n+1}=0$$

Since $\lim\limits_{n\to\infty}\left|\dfrac{a_{n+1}}{a_n}\right|<1$, the series converges absolutely.

LO 4.1B – The Sum of a Series

In Example 4.1.2, we determined the sum of a telescoping series. Here in this LO, we will focus on finding either an exact or an approximate value of the sum of a series.

LO 4.1B: Determine or estimate the sum of a series.

You may have noticed that the geometric series test was absent from the list of tests given in **EK 4.1A6**. We'll remedy that omission in the next EK.

EK 4.1B1: If a is a real number and r is a real number such that $|r|<1$, then the geometric series $\sum\limits_{n=0}^{\infty}ar^n=\dfrac{a}{1-r}$.

The **geometric series test** says that a series of the form $\sum\limits_{n=0}^{\infty}ar^n$ diverges if $|r|\geq1$, and converges absolutely to the sum $\dfrac{a}{1-r}$ if $|r|<1$.

Example 4.1.15: For each of the following series, determine whether converges or diverges. If it converges, find its sum.

1. $\sum\limits_{n=0}^{\infty}\dfrac{3^{n+2}}{5^{n+1}}$

2. $\sum\limits_{n=0}^{\infty}\dfrac{\pi^n}{e^n}$

Each one of these series is a geometric series.

1. $$\sum_{n=0}^{\infty} \frac{3^{n+2}}{5^{n+1}} = \sum_{n=0}^{\infty} \frac{3^n \cdot 3^2}{5^n \cdot 5^1} = \sum_{n=0}^{\infty} \frac{9}{5}\left(\frac{3}{5}\right)^n$$

For this series, $a = \frac{9}{5}$, and $r = \frac{3}{5}$. Since $|r| < 1$, this geometric series converges absolutely. The sum is as follows:

$$\frac{a}{1-r} = \frac{9/5}{1-3/5} = \frac{9/5}{2/5} = \frac{9}{2}$$

2. $$\sum_{n=0}^{\infty} \frac{\pi^n}{e^n} = \sum_{n=0}^{\infty} \left(\frac{\pi}{e}\right)^n$$

For this series, $a = 1$, and $r = \frac{\pi}{e}$. Since $0 < e < \pi$, $\frac{\pi}{e} > 1$, and so $|r| > 1$. Hence, this series diverges.

Next we'll look at estimating the sums of alternating series.

EK 4.1B2: If an alternating series converges by the alternating series test, then the alternating series error bound can be used to estimate how close a partial sum is to the value of the infinite series.

Suppose we have a convergent alternating series whose sum is S. When S is approximated by the Nth partial sum S_N, the error $R_N = S - S_N$ of the approximation satisfies the following condition.

$$|R_N| \leq a_{N+1}$$

That is, the absolute value of the error is less than or equal to the absolute value of the first term that was excluded from the partial sum. The inequality given above is called the **alternating series error bound**.

Example 4.1.16: Consider the absolutely convergent alternating series $\sum_{n=1}^{\infty} \frac{(-1)^{n+1}}{n^2}$ from Example 4.1.13b. Compute S_5, and use the alternating series error bound to place upper and lower bounds on the sum S of the series.

First, we compute the 5th partial sum.

$$S_5 = \sum_{n=1}^{\infty} \frac{(-1)^{n+1}}{n^2} = 1 - \frac{1}{4} + \frac{1}{9} - \frac{1}{16} + \frac{1}{25} = \frac{3019}{3600}$$

Next, we find the error bound.

$$|R_5| \leq \frac{1}{6^2} = \frac{1}{36}$$

Finally, we place upper and lower bounds on the sum S of the series as follows:

$$|R_5| = |S - S_5| \leq \frac{1}{36}$$

$$-\frac{1}{36} \leq S - S_5 \leq \frac{1}{36}$$

$$S_5 - \frac{1}{36} \leq S \leq S_5 + \frac{1}{36}$$

$$\frac{973}{1200} \leq S \leq \frac{3119}{3600}$$

$$0.8108\overline{3} \leq S \leq 0.86638$$

EK 4.1B3: If a series converges absolutely, then any series obtained from it by regrouping or rearranging the terms has the same value.

All this says is that, if a series converges *absolutely*, then its sum does not depend on the association or order of its terms. This probably doesn't seem surprising. The astonishing fact is that, if a series converges *conditionally*, then the sum of the series *does* depend on these things. Let's revisit the alternating harmonic series, which converges *conditionally*. With the aid of power series, the following can be shown:

$$\sum_{n=1}^{\infty} \frac{(-1)^n}{n} = -1 + \frac{1}{2} - \frac{1}{3} + \frac{1}{4} - \frac{1}{5} + \frac{1}{6} - \frac{1}{7} + \frac{1}{8} - \frac{1}{9} + \frac{1}{10} - \cdots = \ln(2)$$

Now observe what happens when we rearrange and regroup the terms.

$$\sum_{n=1}^{\infty} \frac{(-1)^n}{n} = \left(-1 + \frac{1}{2}\right) + \frac{1}{4} + \left(-\frac{1}{3} + \frac{1}{6}\right) + \frac{1}{8} + \left(-\frac{1}{5} + \frac{1}{10}\right) + \frac{1}{12} + \cdots$$

$$\sum_{n=1}^{\infty} \frac{(-1)^n}{n} = -\frac{1}{2} + \frac{1}{4} - \frac{1}{6} + \frac{1}{8} - \frac{1}{10} + \frac{1}{12} + \cdots = \frac{1}{2}\left(-1 + \frac{1}{2} - \frac{1}{3} + \frac{1}{4} - \frac{1}{5} + \frac{1}{6} + \cdots\right)$$

$$\sum_{n=1}^{\infty} \frac{(-1)^n}{n} = -\frac{1}{2}\ln(2)$$

What happened here is that the sum of the alternating harmonic series was $-\ln(2)$ with the original arrangement of the terms, and then the sum became $-\frac{1}{2}\ln(2)$ after the terms were rearranged and regrouped. It turns out that a conditionally convergent series can be made to converge to *any* real number by appropriately rearranging and regrouping the terms. That is not the case with absolutely convergent series.

EU 4.2 – Power Series

LO 4.2A – Taylor Polynomials

Here, we will generalize the concept of the tangent line approximation that we discussed in **EK 2.3B2**. The tangent line approximation is also known as a *first degree Taylor polynomial*.

LO 4.2A: Construct and use Taylor polynomials.

EK 4.2A1: The coefficient of the nth degree term in a Taylor polynomial centered at $x = a$ for the function f is $\dfrac{f^{(n)}(a)}{n!}$.

We'll introduce a little terminology here. If a Taylor polynomial is centered at $x = 0$, then we usually call it a **Maclaurin polynomial**. In our examples, we'll use the notation $P_n(x)$ to denote a Taylor polynomial of degree n.

Example 4.2.1: Compute $P_1(x)$, $P_2(x)$, $P_3(x)$, and $P_4(x)$, all centered at $x = 0$, for $f(x) = e^{-2x}$.

We'll work out the coefficients in a table.

k	$f^{(k)}(x)$	$f^{(k)}(0)$	$\dfrac{f^{(k)}(0)}{k!}$
0	e^{-2x}	$e^{-2(0)} = 1$	$\dfrac{1}{0!} = 1$
1	$-2e^{-2x}$	$-2e^{-2(0)} = -2$	$\dfrac{-2}{1!} = -2$
2	$4e^{-2x}$	$4e^{-2(0)} = 4$	$\dfrac{4}{2!} = 2$
3	$-8e^{-2x}$	$-8e^{-2(0)} = -8$	$\dfrac{-8}{3!} = -\dfrac{4}{3}$
4	$16e^{-2x}$	$16e^{-2(0)} = 16$	$\dfrac{16}{4!} = \dfrac{2}{3}$

The Taylor polynomials are as follows:

$$P_1(x) = 1 - 2x$$

$$P_2(x) = 1 - 2x + 2x^2$$

$$P_3(x) = 1 - 2x + 2x^2 - \frac{4}{3}x^3$$

$$P_4(x) = 1 - 2x + 2x^2 - \frac{4}{3}x^3 + \frac{2}{3}x^4$$

Example 4.2.2: Compute $P_1(x)$, $P_2(x)$, $P_3(x)$, and $P_4(x)$, all centered at $x = \dfrac{\pi}{4}$, for $f(x) = \sin(x)$.

Again, we'll use a table to work out the coefficients.

k	$f^{(k)}(x)$	$f^{(k)}(0)$	$\dfrac{f^{(k)}(0)}{k!}$
0	$\sin(x)$	$\sin\left(\dfrac{\pi}{4}\right) = \dfrac{\sqrt{2}}{2}$	$\dfrac{\sqrt{2}}{2 \cdot 0!} = \dfrac{\sqrt{2}}{2}$
1	$\cos(x)$	$\cos\left(\dfrac{\pi}{4}\right) = \dfrac{\sqrt{2}}{2}$	$\dfrac{\sqrt{2}}{2 \cdot 1!} = \dfrac{\sqrt{2}}{2}$
2	$-\sin(x)$	$-\sin\left(\dfrac{\pi}{4}\right) = -\dfrac{\sqrt{2}}{2}$	$-\dfrac{\sqrt{2}}{2 \cdot 2!} = -\dfrac{\sqrt{2}}{4}$
3	$-\cos(x)$	$-\cos\left(\dfrac{\pi}{4}\right) = -\dfrac{\sqrt{2}}{2}$	$-\dfrac{\sqrt{2}}{2 \cdot 3!} = -\dfrac{\sqrt{2}}{12}$
4	$\sin(x)$	$\sin\left(\dfrac{\pi}{4}\right) = \dfrac{\sqrt{2}}{2}$	$\dfrac{\sqrt{2}}{2 \cdot 4!} = \dfrac{\sqrt{2}}{48}$

Now we can write down the Taylor polynomials.

$$P_1(x) = \frac{\sqrt{2}}{2} + \frac{\sqrt{2}}{2}\left(x - \frac{\pi}{4}\right)$$

$$P_2(x) = \frac{\sqrt{2}}{2} + \frac{\sqrt{2}}{2}\left(x - \frac{\pi}{4}\right) - \frac{\sqrt{2}}{4}\left(x - \frac{\pi}{4}\right)^2$$

$$P_3(x) = \frac{\sqrt{2}}{2} + \frac{\sqrt{2}}{2}\left(x - \frac{\pi}{4}\right) - \frac{\sqrt{2}}{4}\left(x - \frac{\pi}{4}\right)^2 - \frac{\sqrt{2}}{12}\left(x - \frac{\pi}{4}\right)^3$$

$$P_4(x) = \frac{\sqrt{2}}{2} + \frac{\sqrt{2}}{2}\left(x - \frac{\pi}{4}\right) - \frac{\sqrt{2}}{4}\left(x - \frac{\pi}{4}\right)^2 - \frac{\sqrt{2}}{12}\left(x - \frac{\pi}{4}\right)^3 + \frac{\sqrt{2}}{48}\left(x - \frac{\pi}{4}\right)^4$$

In the next EK, we'll see what these Taylor polynomials are used for.

EK 4.2A2: Taylor polynomials for a function f centered at $x = a$ can be used to approximate function values of f near $x = a$.

We'll illustrate this visually.

Example 4.2.3: Refer to Example 4.2.1. Sketch the graphs of $f(x) = e^{-2x}$, $P_1(x)$, $P_2(x)$, $P_3(x)$, and $P_4(x)$ all on the same grid, and observe how the Taylor polynomials approximate f near $x = 0$.

The graph is sketched below.

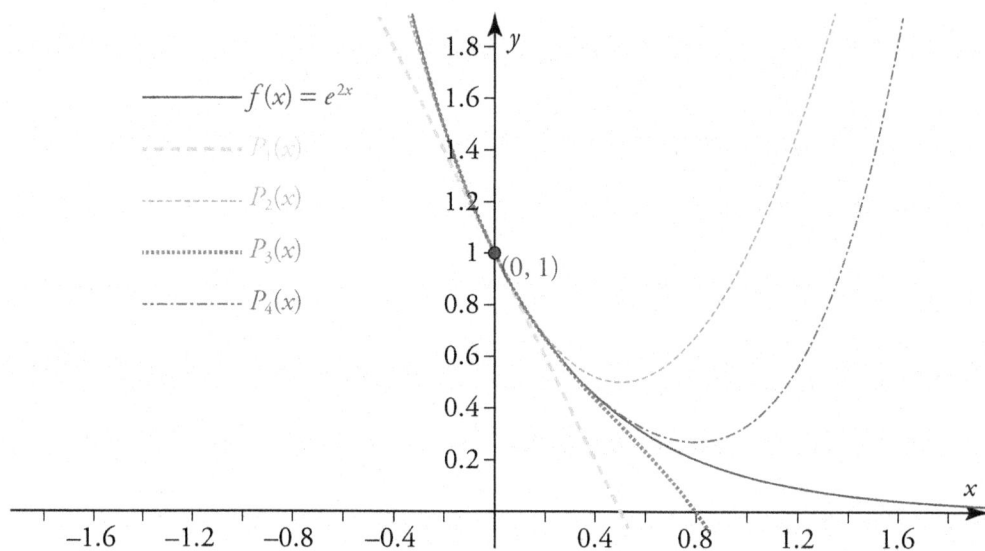

Example 4.2.4: Refer to Example 4.2.2. Sketch the graphs of $f(x) = \sin(x)$, $P_1(x)$, $P_2(x)$, $P_3(x)$, and $P_4(x)$ all on the same grid, and observe how the Taylor polynomials approximate f near $x = \dfrac{\pi}{4}$.

The graph is sketched below.

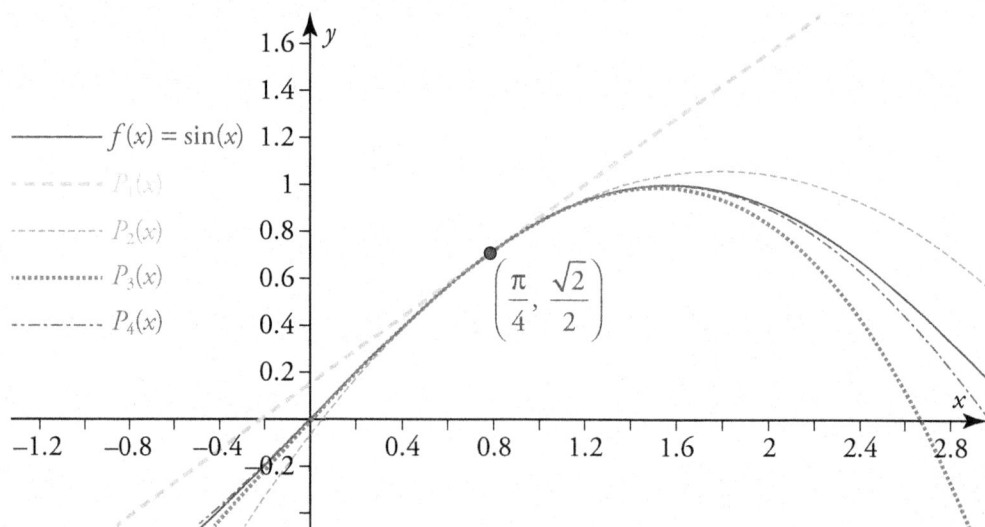

Note how in each of the previous two examples, the Taylor polynomial approximates f better and over a longer interval as the degree of the polynomial increases. This leads us to our next EK.

EK 4.2A3: In many cases, as the degree of a Taylor polynomial increases, the nth degree polynomial will converge to the original function over some interval.

We can write down a formula for the Taylor polynomial centered at $x = a$ for a function f as follows:

$$P_n(x) = \sum_{k=0}^{n} \frac{f^{(k)}(a)}{k!}(x-a)^k$$

What this EK is saying is that, in many cases, $\displaystyle\lim_{n \to \infty} P_n(x) = \sum_{k=0}^{\infty} \frac{f^{(k)}(a)}{k!}(x-a)^k$ converges to $f(x)$ over some interval I. That interval *could* be the entire x-axis. Some examples that you should know for the exam are given in the following table:

$f(x)$	$x = a$	$\displaystyle\lim_{n \to \infty} P_n(x)$	I
$\dfrac{1}{x}$	1	$\displaystyle\sum_{k=0}^{\infty}(-1)^k(x-1)^k$	$(0,2)$
$\dfrac{1}{1+x}$	0	$\displaystyle\sum_{k=0}^{\infty}(-1)^k x^k$	$(-1,1)$
$\ln(x)$	1	$\displaystyle\sum_{k=0}^{\infty}\frac{(-1)^{k+1}}{k}(x-1)^k$	$(0,2]$
e^x	0	$\displaystyle\sum_{k=0}^{\infty}\frac{1}{k!}x^k$	$(-\infty,\infty)$
$\sin(x)$	0	$\displaystyle\sum_{k=0}^{\infty}\frac{(-1)^k}{(2k+1)!}x^{2k+1}$	$(-\infty,\infty)$
$\cos(x)$	0	$\displaystyle\sum_{k=0}^{\infty}\frac{(-1)^k}{(2k)!}x^{2k}$	$(-\infty,\infty)$
$\arctan(x)$	0	$\displaystyle\sum_{k=0}^{\infty}\frac{(-1)^k}{2k+1}x^{2k+1}$	$[-1,1]$
$\arcsin(x)$	0	$\displaystyle\sum_{k=0}^{\infty}\frac{(2k)!}{\left(2^k k!\right)^2(2k+1)}x^{2k+1}$	$[-1,1]$

The next two items of Essential Knowledge are all about placing bounds on the error of a Taylor polynomial approximation to a function.

EK 4.2A4: The Lagrange error bound can be used to bound the error of a Taylor polynomial approximation to a function.

Let $P_n(x)$ be the nth degree Taylor polynomial centered at $x = a$ for a function f. When $P_n(x)$ used to approximate $f(x)$ near $x = a$, there will be a **remainder**, or **error**, $R_n(x)$ in the approximation. We express this error symbolically as follows:

$$R_n(x) = f(x) - P_n(x)$$

It can be shown that $R_n(x) = \dfrac{f^{(n+1)}(z)}{(n+1)!}(x - a)^{n+1}$, where z is a number between x and a. We would be able to evaluate the error exactly if we knew the exact value of z, but we don't. So we have to settle for approximating the error, and this is where the **Lagrange error bound** comes in. The Lagrange error bound is an upper bound on the absolute value of the error, and it is stated as follows:

$$\left| R_n(x) \right| \le \frac{|x - a|^{n+1}}{(n+1)!} \max \left| f^{(n+1)}(z) \right|$$

Now this we *can* evaluate, because we have the tools to locate the maximum value of $f^{(n+1)}$ on a closed interval. Let's do it in an example.

Example 4.2.5: Let $f(x) = e^{-2x}$.

1. Use $P_4(x) = 1 - 2x + 2x^2 - \dfrac{4}{3}x^3 + \dfrac{2}{3}x^4$ to approximate $f(0.3) = e^{-0.6}$.

2. Use the Lagrange error bound to bound the error of the approximation.

3. Use the result of part b to place lower and upper bounds on the approximation itself.

1. $P_4(0.3) = 1 - 2(0.3) + 2(0.3)^2 - \dfrac{4}{3}(0.3)^3 + \dfrac{2}{3}(0.3)^4 = 0.5494$

2. In order to determine the Lagrange error bound, we need to know $\max \left| f^{(5)}(z) \right|$ for z in $[0, 0.3]$. $\left| f^{(5)}(z) \right| = \left| -32e^{-2z} \right| = 32e^{-2z}$, and since this is function is decreasing on $[0, 0.3]$, its maximum value occurs at the left endpoint, which is $x = 0$. That is, $\max \left| f^{(5)}(z) \right| = 32e^{-2(0)} = 32$. So the Lagrange error bound is as follows:

$$\left| R_4(0.3) \right| \le \frac{(0.3)^5}{5!} \cdot 32 = 0.000648$$

3. Recalling that $R_4(0.3) = f(0.3) - P_4(0.3) = e^{-0.6} - 0.5494$, we use the Lagrange error bound to place bounds on the approximation as follows.

$$\left| e^{-0.6} - 0.5494 \right| \le 0.000648$$

$$-0.000648 \le e^{-0.6} - 0.5494 \le 0.000648$$

$$0.548752 \le e^{-0.6} \le 0.550048$$

EK 4.2A5: In some situations where the signs of a Taylor polynomial are alternating, the alternating series error bound can be used to bound the error of a Taylor polynomial approximation to the function.

Refer back to **EK 4.1B2**, where we discussed the alternating series error bound. We'll use this to place another bound on the error of the approximation in Example 4.2.5.

Example 4.2.6: Let $f(x) = e^{-2x}$. In Example 4.2.5, we used $P_4(x) = 1 - 2x + 2x^2 - \frac{4}{3}x^3 + \frac{2}{3}x^4$ to approximate $f(0.3) = e^{-0.6}$.

Use the alternating series error bound to bound the error of the approximation. When we compute the approximation $P_4(0.3)$, we get the following:

$$P_4(0.3) = 1 - 2(0.3) + 2(0.3)^2 - \frac{4}{3}(0.3)^3 + \frac{2}{3}(0.3)^4$$

This can be viewed as a partial sum of the alternating series $\sum_{k=0}^{\infty} \frac{(-1)^k 2^k}{k!}(0.3)^k$, so $a_k = \frac{2^k}{k!}(0.3)^k$. The first excluded term corresponds to $k = 5$, and so $a_5 = \frac{2^5}{5!}(0.3)^5 = 0.000648$. This is the alternating series error bound. We write this symbolically as follows:

$$|R_4(0.3)| \le \frac{(0.3)^5}{5!} \cdot 32 = 0.000648$$

As you can see, the alternating series error bound gave us the same bound on the error that the Lagrange error bound did.

LO 4.2B – Power Series Representations of Functions

Here, we will move on to power series, which are an extension of Taylor polynomials.

LO 4.2B: Write a power series representing a given function.

The first EK of this LO formally introduces us to the notion of a power series.

EK 4.2B1: A power series is a series of the form $\sum_{n=0}^{\infty} a_n(x - r)^n$ where n is a non-negative integer, $\{a_n\}$ is a sequence of real numbers, and r is a real number.

As you can see, the power series in the EK is expressed in terms of the variable x, which suggests that it is a *function* of x. The number r is called the **center** of the series. If the coefficients of the power series are of the form $a_n = \frac{f^{(n)}(r)}{n!}$, then the power series is called a **Taylor series**. A Taylor series centered at $r = 0$ is called a **Maclaurin series**.

EK 4.2B2: The Maclaurin series for $\sin(x)$, $\cos(x)$, and e^x provide the foundation for constructing the Maclaurin series for other functions.

The Maclaurin series mentioned in the EK are given as follows:

$$\sin(x) = x - \frac{x^3}{3!} + \frac{x^5}{5!} - \frac{x^7}{7!} + \cdots = \sum_{k=0}^{\infty} \frac{(-1)^k}{(2k+1)!} x^{2k+1}$$

$$\cos(x) = 1 - \frac{x^2}{2!} + \frac{x^4}{4!} - \frac{x^6}{6!} + \cdots = \sum_{k=0}^{\infty} \frac{(-1)^k}{(2k)!} x^{2k}$$

$$e^x = 1 + x + \frac{x^2}{2!} + \frac{x^3}{3!} + \cdots = \sum_{k=0}^{\infty} \frac{1}{k!} x^k$$

We will illustrate how to use these to construct Maclaurin series for other functions in the examples.

Example 4.2.7: Find the Maclaurin series for $f(x) = xe^x$.

We *could* determine this series from scratch by computing the Maclaurin coefficients: $\frac{f^{(n)}(0)}{n!}$. However, we will instead simply multiply the Maclaurin series for e^x by x, which is a lot less work.

$$f(x) = xe^x = x \sum_{n=0}^{\infty} \frac{1}{n!} x^n = \sum_{n=0}^{\infty} \frac{1}{n!} x^{n+1}$$

Example 4.2.8: Find the Maclaurin series for $f(x) = \sin(x^2)$.

Again, we will not determine this series from scratch. Instead, we will substitute x^2 into the Maclaurin series for $\sin(x)$.

$$f(x) = \sin(x^2) = \sum_{n=0}^{\infty} \frac{(-1)^n}{(2n+1)!} (x^2)^{2n+1} = \sum_{n=0}^{\infty} \frac{(-1)^n}{(2n+1)!} x^{4n+2}$$

EK 4.2B3: The Maclaurin series for $\frac{1}{1-x}$ is a geometric series.

Refer back to **EK 4.1B1**. If $|r| < 1$, then $\sum_{n=0}^{\infty} ar^n = \frac{a}{1-r}$. Now compare that result with the function $f(x) = \frac{1}{1-x}$. We *could* compute the Maclaurin series for this function from scratch, but we don't have to. Notice that this function looks exactly like the sum of a geometric series with $a = 1$ and $r = x$. If we make these substitutions in the geometric series, we immediately get the Maclaurin series for f.

$$f(x) = \frac{1}{1-x} = \sum_{n=0}^{\infty} x^n, \ |x| < 1$$

EK 4.2B4: A Taylor polynomial for $f(x)$ is a partial sum of the Taylor series for $f(x)$.

We already alluded to this **EK 4.2A3**. The nth degree Taylor *polynomial* $P_n(x) = \sum_{k=0}^{n} \frac{f^{(k)}(a)}{k!}(x-a)^k$ is a partial sum of $\lim_{n \to \infty} P_n(x) = \sum_{k=0}^{\infty} \frac{f^{(k)}(a)}{k!}(x-a)^k$, which is a Taylor *series*.

A power series for a given function can either be computed from scratch by computing the Taylor coefficients, or it can be *derived* from a power series of a related function. We review the latter option in the next EK.

EK 4.2B5: A power series for a given function can be derived by various methods (e.g., algebraic processes, substitutions, using properties of geometric series, and operations on known series such as term-by-term integration or term-by-term differentiation).

We already illustrated how to derive a power series for a given function by algebraic processes and substitutions in Examples 4.2.7 and 4.2.8, so we'll move on to the other methods.

In our next example, we will use properties of geometric series to derive a power series.

Example 4.2.9: Find a power series for $f(x) = \frac{4}{5+x}$ with center $a = -3$. For what interval does the series converge?

In order to have the series centered at $a = -3$, we need to see the combination $(x + 3)$ in $f(x)$. We can make that happen by adding and subtracting 3, as follows.

$$f(x) = \frac{4}{5+(x+3-3)} = \frac{4}{2+(x+3)}$$

In order to use the properties of geometric series, we need to make $f(x)$ look like $\frac{a}{1-r}$. First, we'll divide the numerator and denominator by 2 to get something of the form $1 + r$ in the denominator.

$$f(x) = \frac{4/2}{(2+(x+3))/2} = \frac{2}{1+\dfrac{x+3}{2}}$$

Next, we need a *difference* of two terms in the denominator so that we get something of the form $1 - r$. We'll make that happen by introducing a double negative.

$$f(x) = \frac{2}{1-\left(-\dfrac{x+3}{2}\right)}$$

Now, $f(x)$ looks like $\frac{a}{1-r}$, with $a = 2$ and $r = -\dfrac{x+3}{2}$. We are ready to write down the power series.

$$f(x) = \frac{2}{1-\left(-\dfrac{x+3}{2}\right)} = \sum_{n=0}^{\infty} 2\left(-\frac{x+3}{2}\right)^n$$

This series converges for $\left|-\dfrac{x+3}{2}\right| < 1$. To find the interval on which the series converges, we need to solve this inequality for x.

$$\left|-\frac{x+3}{2}\right| < 1 \Rightarrow |x+3| < 2$$

$$-2 < x+3 < 2 \Rightarrow -5 < x < -1$$

This series converges on the interval $(-5, -1)$, which we call the **interval of convergence** of the series. We will have more to say about the interval of convergence in **EK 4.2C1**.

In the next example, we will review how to derive a power series for a function using term-by-term differentiation and term-by-term integration.

Example 4.2.10:

1. Use the result from Example 4.2.8 to obtain a Maclaurin series for $g(x) = 2x\cos(x^2)$.
   Hint: $\dfrac{d}{dx}\left(\sin\left(x^2\right)\right) = 2x\cos\left(x^2\right)$.

2. Use the result from Example 4.2.7 to obtain a Maclaurin series for $g(x) = xe^x - e^x$.
   Hint: $\displaystyle\int xe^x\,dx = xe^x - e^x + C$.

1. Differentiate the Maclaurin series for $f(x) = \sin(x^2)$ term-by-term.

$$\frac{d}{dx}\left(\sin\left(x^2\right)\right) = \frac{d}{dx}\left(\sum_{n=0}^{\infty} \frac{(-1)^n}{(2n+1)!} x^{4n+2}\right) =$$

$$\sum_{n=0}^{\infty} \frac{(-1)^n}{(2n+1)!} \frac{d}{dx}\left(x^{4n+2}\right) = \sum_{n=0}^{\infty} \frac{(-1)^n}{(2n+1)!}(4n+2)x^{4n+1}$$

$$2x\cos\left(x^2\right) = \sum_{n=0}^{\infty} \frac{(-1)^n}{(2n+1)!}2(2n+1)x^{4n+1}$$

Simplifying, we get $g(x) = 2x\cos\left(x^2\right) = \displaystyle\sum_{n=0}^{\infty} \frac{2(-1)^n}{(2n)!}x^{4n+1}$.

2. Integrate the Maclaurin series for $f(x) = xe^x$ term-by-term.

$$\int xe^x\,dx = \int\left(\sum_{n=0}^{\infty} \frac{1}{n!}x^{n+1}\right)dx = \sum_{n=0}^{\infty} \frac{1}{n!}\int x^{n+1}dx$$

$$xe^x - e^x = C + \sum_{n=0}^{\infty} \frac{1}{n!(n+2)} x^{n+2}$$

Plug in $x = 0$ to determine the constant C.

$$0e^0 - e^0 = \sum_{n=0}^{\infty} \frac{1}{n!(n+2)}(0)^{n+2} + C \Rightarrow C = -1$$

So we get $g(x) = xe^x - e^x = -1 + \sum_{n=0}^{\infty} \frac{1}{n!(n+2)} x^{n+2}$.

LO 4.2C – Radius and Interval of Convergence

Welcome to the final Learning Objective of the book!

LO 4.2C: Determine the radius and interval of convergence of a power series.

We briefly mentioned the term interval of convergence in Example 4.2.9. We have more to say about it here.

EK 4.2C1: If a power series converges, it either converges at a single point or has an interval of convergence.

Every power series converges at its center. It's easy to see why if you consider the following calculation:

$$\left(\sum_{n=0}^{\infty} a_n(x-r)^n \right)\bigg|_{x=r} = a_0 + a_1(r-r) + a_2(r-r)^2 + \cdots = a_0$$

Some power series do not converge at any point beyond its center, but others do. For any power series $\sum_{n=0}^{\infty} a_n(x-r)^n$, precisely one of the following statements is true:

1. $\sum_{n=0}^{\infty} a_n(x-r)^n$ converges at $x = r$ only.

2. $\sum_{n=0}^{\infty} a_n(x-r)^n$ converges absolutely on an interval of the form $(r-R, r+R)$, where

   R is finite and nonzero, and diverges on the intervals $(-\infty, r-R)$ and $(r+R, \infty)$.

   The number R is called the **radius of convergence** of the power series.

3. $\sum_{n=0}^{\infty} a_n(x-r)^n$ converges absolutely for all real x.

In the next EK, we will review a method for *finding* the interval and radius of convergence for a power series.

EK 4.2C2: The ratio test can be used to determine the radius of convergence of a power series.

Recall from **EK 4.1A6** that the infinite series $\sum_{n=0}^{\infty} a_n$ $(a_n \neq 0)$ converges absolutely if $\lim_{n \to \infty} \left| \frac{a_{n+1}}{a_n} \right| < 1$. Let the nth term of a power series be denoted $u_n(x)$. When we apply this test to $\sum_{n=0}^{\infty} u_n(x)$, we impose the following convergence condition:

$$\lim_{n \to \infty} \left| \frac{u_{n+1}(x)}{u_n(x)} \right| < 1$$

We'll then find the radius of convergence as follows:

1. If $\lim_{n \to \infty} \left| \frac{u_{n+1}(x)}{u_n(x)} \right| = \infty$, then the radius of convergence is 0. The series only converges at its center $x = r$.

2. If $\lim_{n \to \infty} \left| \frac{u_{n+1}(x)}{u_n(x)} \right|$ is finite and nonzero, then the radius of convergence is found by rearranging the inequality $\lim_{n \to \infty} \left| \frac{u_{n+1}(x)}{u_n(x)} \right| < 1$ into the form $|x - r| < R$.

3. If $\lim_{n \to \infty} \left| \frac{u_{n+1}(x)}{u_n(x)} \right| = 0$, then the radius of convergence is infinite. The power series converges absolutely for all real x.

In case 2, the interval of convergence is either $(r - R, r + R)$, $(r - R, r + R]$, $[r - R, r + R)$, or $[r - R, r + R]$. The ratio test cannot tell us what happens at the endpoints, and in this case, we must test separately for endpoint convergence.

Example 4.2.11: Find the radius of convergence of each of the following power series. If the power series has an interval of convergence, find that, too.

1. $\sum_{n=0}^{\infty} \frac{(2n)! x^n}{(n+1)!}$

2. $\sum_{n=0}^{\infty} \frac{2^{n+1} (x+1)^n}{n!}$

3. $\sum_{n=1}^{\infty} \frac{(x-2)^n}{n}$

1. Here $u_n = \frac{(2n)! x^n}{(n+1)!}$, and $u_{n+1} = \frac{(2n+2)! x^{n+1}}{(n+2)!}$.

$$\lim_{n \to \infty} \left| \frac{u_{n+1}}{u_n} \right| = \lim_{n \to \infty} \left| \frac{(2n+2)! x^{n+1}}{(n+2)!} \cdot \frac{(n+1)!}{(2n)! x^n} \right| = |x| \lim_{n \to \infty} \left(\frac{(2n+2)(2n+1)}{n+2} \right) = \infty$$

The radius of convergence is $R = 0$. The power series has no interval of convergence. It converges only at its center, $x = 0$.

2. Here $u_n = \dfrac{2^{n+1} (x+1)^n}{n!}$, and $u_{n+1} = \dfrac{2^{n+2} (x+1)^{n+1}}{(n+1)!}$.

$$\lim_{n \to \infty} \left| \frac{u_{n+1}}{u_n} \right| = \lim_{n \to \infty} \left| \frac{2^{n+2} (x+1)^{n+1}}{(n+1)!} \cdot \frac{n!}{2^{n+1} (x+1)^n} \right| = 2|x+1| \lim_{n \to \infty} \left(\frac{1}{n+1} \right) = 0$$

The radius of convergence is $R = \infty$. The power series converges for all real x, so the interval of convergence is $(-\infty, \infty)$.

3. Here $u_n = \dfrac{(x-2)^n}{n}$, and $u_{n+1} = \dfrac{(x-2)^{n+1}}{n+1}$.

$$\lim_{n \to \infty} \left| \frac{u_{n+1}}{u_n} \right| = \lim_{n \to \infty} \left| \frac{(x-2)^{n+1}}{n+1} \cdot \frac{n}{(x-2)^n} \right| = |x-2| \lim_{n \to \infty} \left(\frac{n}{n+1} \right) = |x-2|$$

Imposing the convergence condition from the ratio test, we obtain $|x-2| < 1$, so the radius of convergence is $R = 1$. The series converges absolutely on the open interval $(2-1, 2+1) = (1,3)$, but the ratio test cannot tell us what happens at the endpoints. We'll test for endpoint convergence by plugging $x = 1$ and $x = 3$ into the power series.

$x = 1$:

$$\sum_{n=1}^{\infty} \frac{(1-2)^n}{n} = \sum_{n=1}^{\infty} \frac{(-1)^n}{n}$$

This is the alternating harmonic series, which converges conditionally. The power series thus converges at $x = 1$.

$x = 3$:

$$\sum_{n=1}^{\infty} \frac{(3-2)^n}{n} = \sum_{n=1}^{\infty} \frac{1}{n}$$

This is the harmonic series, which diverges. The power series thus diverges at $x = 3$.

The interval of convergence of the power series is thus $[1,3)$.

EK 4.2C3: If a power series has a positive radius of convergence, then the power series is the Taylor series of the function to which it converges over the open interval.

This is true whether the radius of convergence is positive and finite or $+\infty$. For instance, in Example 4.2.11b, the power series $\displaystyle\sum_{n=0}^{\infty} \frac{2^{n+1} (x+1)^n}{n!}$ is the Taylor series of the function to which it converges over the open interval $(-\infty, \infty)$.

EK 4.2C4: The radius of convergence of a power series obtained by term-by-term differentiation or term-by-term integration is the same as the radius of convergence of the original power series.

Example 4.2.12: Let $f(x) = \displaystyle\sum_{n=1}^{\infty} \frac{(-1)^n x^n}{n}$. Show that $f(x)$, $f'(x)$, and $\int f(x)\, dx$ all have the same radius of convergence.

First, we apply the ratio test to $f(x)$.

$$\lim_{n \to \infty} \left| \frac{(-1)^{n+1} x^{n+1}}{n+1} \cdot \frac{n}{(-1)^n x^n} \right| < 1$$

$$|x| \lim_{n \to \infty} \left(\frac{n}{n+1} \right) < 1$$

$$|x| < 1$$

The radius of convergence of $f(x)$ is 1. We need to show that this is the case for $f'(x)$ and $\int f(x)\, dx$ also.

$$f'(x) = \frac{d}{dx} \left(\sum_{n=1}^{\infty} \frac{(-1)^n x^n}{n} \right) = \sum_{n=2}^{\infty} \frac{(-1)^n}{n} \cdot \frac{d}{dx} (x^n)$$

Note that the lower limit on the index changed from $n = 1$ to $n = 2$ in the last step. That's because the $n = 1$ term is constant, and we lose it when we differentiate term-by-term.

$$f'(x) = \sum_{n=2}^{\infty} \frac{(-1)^n}{n} \cdot n x^{n-1} = \sum_{n=2}^{\infty} (-1)^n x^{n-1}$$

Next, we'll apply the ratio test to $f'(x)$.

$$\lim_{n \to \infty} \left| \frac{(-1)^{n+1} x^n}{(-1)^n x^{n-1}} \right| < 1$$

$$|x| \lim_{n \to \infty} (1) < 1$$

$$|x| < 1$$

The radius of convergence of $f'(x)$ is also 1. We have one more radius to check.

$$\int f(x)\, dx = \int \left(\sum_{n=1}^{\infty} \frac{(-1)^n x^n}{n} \right) dx = C + \sum_{n=1}^{\infty} \frac{(-1)^n}{n} \int x^n dx = C + \sum_{n=1}^{\infty} \frac{(-1)^n x^{n+1}}{n(n+1)}$$

The constant of integration has no bearing on the convergence of the power series, and so we may ignore it when finding the radius of convergence. We'll now apply the ratio test.

$$\lim_{n \to \infty} \left| \frac{(-1)^{n+1} x^{n+2}}{(n+1)(n+2)} \cdot \frac{n(n+1)}{(-1)^n x^{n+1}} \right| < 1$$

$$|x| \lim_{n \to \infty} \left(\frac{n}{n+2} \right) < 1$$

$$|x| < 1$$

The radius of convergence of $\int f(x)\,dx$ is 1. This completes the exercise.

We wish to note that, while $f(x)$, $f'(x)$, and $\int f(x)\,dx$ do all have the same *radius* of convergence, they do not all have the same *interval* of convergence. We will leave it as an exercise for you to show that the intervals of convergence of these three power series are, respectively, $(-1, 1]$, $(-1, 1)$, and $[-1, 1]$.

PART IV:
Practice Exams

AP Calculus AB Practice Examination 1

CALCULUS AB

A CALCULATOR **CANNOT BE USED ON PART A OF SECTION I OR ON PART B OF SECTION II**. A GRAPHING CALCULATOR FROM THE APPROVED LIST **IS REQUIRED ON PART B OF SECTION I AND ON PART A OF SECTION II** OF THE EXAMINATION. CALCULATOR MEMORIES NEED NOT BE CLEARED. COMPUTERS, NONGRAPHING SCIENTIFIC CALCULATORS, CALCULATORS WITH QWERTY KEYBOARDS, AND ELECTRONIC WRITING PADS ARE NOT ALLOWED. CALCULATORS MAY NOT BE SHARED AND COMMUNICATION BETWEEN CALCULATORS IS PROHIBITED DURING THE EXAMINATION. ATTEMPTS TO REMOVE TEST MATERIALS FROM THE ROOM BY ANY METHOD WILL RESULT IN INVALIDATION OF TEST SCORES.

CALCULUS AB – SECTION I

Time – 1 hour and 45 minutes
All questions are given equal weight.
Percent of total grade – 50

Part A: 55 minutes, 28 multiple-choice questions
A calculator is NOT allowed.

Part B: 50 minutes, 17 multiple-choice questions
A graphing calculator is required.

Parts A and B of Section I are printed in this examination booklet. Section II, which consists of longer problems, is printed in a separate booklet.

GENERAL INSTRUCTIONS

DO NOT OPEN THIS BOOKLET UNTIL YOU ARE TOLD TO DO SO.

INDICATE YOUR ANSWERS TO QUESTIONS IN PART A ON PAGE 2 OF THE SEPARATE ANSWER SHEET. THE ANSWERS TO QUESTIONS IN PART B SHOULD BE INDICATED ON PAGE 3 OF THE ANSWER SHEET. No credit will be given for anything written in this examination booklet, but you may use the booklet for notes or scratchwork. After you have decided which of the suggested answers is best, COMPLETELY fill In the corresponding oval on the answer sheet. Give only one answer to each question. If you change an answer, be sure that the previous mark is erased completely.

Example: Sample Answer: Ⓐ Ⓑ Ⓒ ⬤ Ⓔ

1. What is the value of the numerical expression $4 - 2(3 - 5)$?

    A. -12

    B. -8

    C. 0

    D. 8

    E. 12

Many candidates wonder whether or not to guess the answers to questions about which they are not certain. It is improbable, that mere guessing will improve your score significantly; it may even lower your score, and it does take time. If, however, you are not sure of the best answer but have some knowledge of the question and are able to eliminate one or more of the answer choices as wrong, your chance of answering correctly is improved, and it may be to your advantage to answer such a question.

Use your time effectively, working as rapidly as you can without losing accuracy. Do not spend too much time on questions that are too difficult. Go on to other questions and come back to the difficult ones later if you have time. It is not expected that you will be able to answer all of the multiple choice questions.

CALCULUS AB

SECTION I, Part A

Time – 55 Minutes

Number of questions – 28

A CALCULATOR MAY NOT BE USED ON THIS PART OF THE EXAMINATION.

Directions: Solve each of the following problems, using the available space for scratchwork. After examining the form of the choices, decide which is the best of the choices given and fill in the corresponding oval on the answer sheet. No credit will be given for anything written in the test book. Do not spend too much time on any one problem.

In this test: Unless otherwise specified, the domain of a function *f* is assumed to be the set of all real numbers *x* for which $f(x)$ is a real number.

1. If $f(x) = \dfrac{x}{x^2 + 2}$, then $f'(x)$ is

    A. $\dfrac{1}{2x}$

    B. $\dfrac{1}{x^2 + 2}$

    C. $\dfrac{-x^2 + 2}{\left(x^2 + 2\right)^2}$

    D. $\dfrac{x^2 - 2}{\left(x^2 + 2\right)^2}$

    E. $\dfrac{3x^2 + 2}{\left(x^2 + 2\right)^2}$

2. If $f(x) = 2x^{\frac{2}{3}}$, then $f'(-8) =$

    A. $-\dfrac{8}{3}$

    B. $-\dfrac{2}{3}$

    C. 0

    D. $\dfrac{2}{3}$

    E. $\dfrac{8}{3}$

GO ON TO THE NEXT PAGE.

3. $\lim\limits_{x \to \infty} \dfrac{2 - 3x^2}{6x^2 + 5} =$

   A. $-\dfrac{3}{5}$

   B. $\dfrac{1}{2}$

   C. $\dfrac{1}{3}$

   D. $\dfrac{2}{5}$

   E. The limit does not exist.

4. Let the function f be defined as follows:

$$f(x) = \begin{cases} \dfrac{x^2 - 5x + 6}{x - 3} & x \neq 3 \\ k & x = 3 \end{cases}$$

   If f is continuous at $x = 3$, then $k =$

   A. -1

   B. $-\dfrac{2}{3}$

   C. $\dfrac{2}{3}$

   D. 1

   E. There is no value of k for which f is continuous at $x = 3$.

5. If $x^3 - y^3 = 2xy$, then $\dfrac{dy}{dx} =$

   A. $\dfrac{3x^2 - 2y}{2x + 3y^2}$

   B. $\dfrac{2x + 3y^2}{3x^2 - 2y}$

   C. $\dfrac{3x^2}{2x + 3y^2}$

   D. $\dfrac{3x^2 - 3y^2}{2x}$

   E. $\dfrac{3x^2 - 3y^2}{2}$

GO ON TO THE NEXT PAGE.

6. If $y = \sin(x)\cos(x)$, then $\dfrac{dy}{dx} =$

   A. 1

   B. $\cos^2(x) - \sin^2(x)$

   C. $-2\sin(x)\cos(x)$

   D. $-\sin(x)\cos(x)$

   E. $\sin(x)\cos(x)$

7. If $f(x) = 2^{-x}$ and $g(x) = x - 1$, then $\lim\limits_{x \to 0} f(g(x)) =$

   A. -2

   B. -1

   C. $\dfrac{1}{2}$

   D. 1

   E. 2

8. If $f(x) = \sec(x)$, then $f''\left(\dfrac{\pi}{4}\right) =$

   A. 0

   B. $\sqrt{2}$

   C. $2\sqrt{2}$

   D. $3\sqrt{2}$

   E. Undefined

9. The slope of the line tangent to the graph of $x^2 y + 3y - 4 = 0$ at $(-1, 1)$ is

   A. $-\dfrac{3}{2}$

   B. $-\dfrac{1}{2}$

   C. 0

   D. $\dfrac{1}{2}$

   E. $\dfrac{3}{2}$

GO ON TO THE NEXT PAGE.

10. An equation of the graph of the line tangent to the graph of $y = (6x+2)^{1/3}$ at $(1,2)$ is

A. $x - 2y - 3 = 0$

B. $x - 2y + 3 = 0$

C. $x + 2y - 3 = 0$

D. $x + 2y + 3 = 0$

E. The graph has no tangent at the point $(1,2)$.

11. $\int x^2 \sin(x^3)\,dx =$

A. $-3\cos(x^3) + C$

B. $-\dfrac{1}{3}\cos(x^3) + C$

C. $-\cos(x^3) + C$

D. $\dfrac{1}{3}\cos(x^3) + C$

E. $3\cos(x^3) + C$

12. $\displaystyle\int_{1/2}^{\sqrt{3}/2} \dfrac{2}{\sqrt{1-x^2}}\,dx =$

A. $-\dfrac{\pi}{3}$

B. $-\dfrac{\pi}{6}$

C. 0

D. $\dfrac{\pi}{6}$

E. $\dfrac{\pi}{3}$

13. The function $f(x) = 3\tan\left(\pi x - \dfrac{\pi}{4}\right) + 2$ has a fundamental period of

A. π

B. 3

C. 2

D. 1

E. $\dfrac{\pi}{2}$

GO ON TO THE NEXT PAGE.

14. If $f(x) = \begin{cases} -2x^2 & x < 0 \\ 3x^2 & x \geq 0 \end{cases}$, which of the following must be true?

I. $f'(x)$ is continuous everywhere.

II. $f(x)$ is differentiable everywhere.

III. $f(x)$ is integrable everywhere.

A. I only

B. I and II only

C. I and III only

D. II and III only

E. I, II, and III

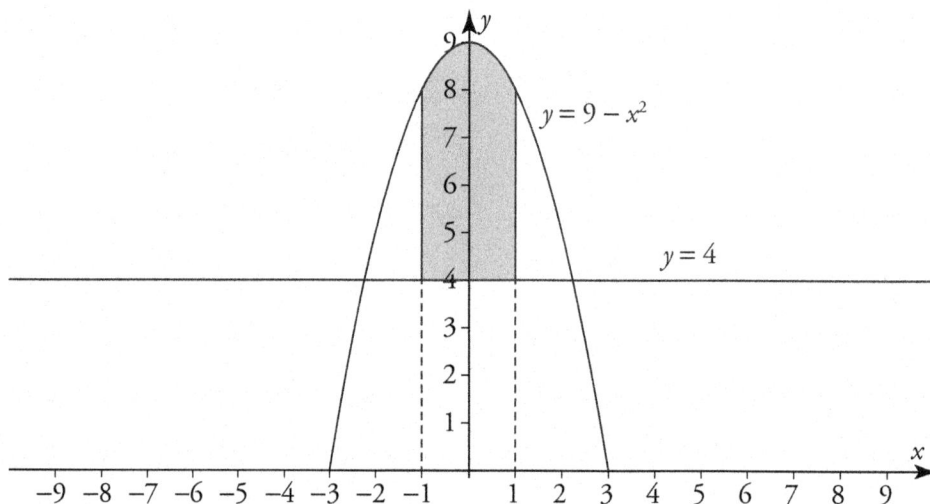

15.

Which of the following integrals correctly yields the area of the shaded region in the above figure?

A. $\int_1^9 (x^2 - 5)\, dx$

B. $\int_1^9 (5 - x^2)\, dx$

C. $\int_{-1}^1 (x^2 - 5)\, dx$

D. $\int_{-1}^1 (5 - x^2)\, dx$

E. $\int_{-1}^1 (13 - x^2)\, dx$

GO ON TO THE NEXT PAGE.

16. If $y = 2^{-x^2}$, then $\dfrac{dy}{dx} =$

  A. $-\dfrac{x}{\ln(2)} 2^{-x^2+1}$

  B. $\dfrac{1}{\ln(2)} 2^{-x^2}$

  C. $-x^2 2^{-x^2-1}$

  D. $\ln(2) 2^{-x^2}$

  E. $-x\ln(2) 2^{-x^2+1}$

17. The average value of the function $f(x) = \dfrac{1}{x^2 - 2x + 1}$ on the interval $[2,5]$ is

  A. $\dfrac{3}{4}$

  B. $\dfrac{1}{4}$

  C. $-\dfrac{1}{4}$

  D. $-\dfrac{3}{4}$

  E. $-\dfrac{1}{3}\ln(4)$

18. For what value of x does the function $f(x) = \dfrac{1}{3}x^3 + \dfrac{1}{2}x^2 - 6x + 8$ have a relative maximum?

  A. -3

  B. -2

  C. 0

  D. 2

  E. 3

19. $\displaystyle\lim_{x \to 0} \dfrac{\tan(2x) - \sin(x)}{x} =$

  A. 0

  B. 1

  C. 2

  D. ∞

  E. The limit does not exist.

GO ON TO THE NEXT PAGE.

20. $\int \dfrac{e^{2x} - e^x}{e^x}\, dx =$

A. $2e^{2x} - e^x + C$

B. $2e^{2x} - x + C$

C. $e^x - x + C$

D. $\ln\left(e^{2x} - e^x\right) + C$

E. $\dfrac{1}{2}\left(e^{2x} - e^x\right)^2 + C$

21. $\displaystyle\int_{-\sqrt{6}}^{\sqrt{6}} \dfrac{1}{x^2 + 2}\, dx =$

A. $\dfrac{\pi\sqrt{2}}{3}$

B. $\dfrac{\pi}{3\sqrt{2}}$

C. 0

D. $\sqrt{2}\tan^{-1}\left(\sqrt{6}\right)$

E. $\dfrac{1}{\sqrt{2}}\tan^{-1}\left(\sqrt{6}\right)$

22. Let R be the region bounded by the graph of $f(x) = e^{x/2}$ and the lines $y = 0$, $x = 0$, and $x = 2$. The volume of the solid obtained when R is revolved about the x-axis is given by which of the following integrals?

A. $\pi\displaystyle\int_0^{\ln(2)} e^{x/2}\, dx$

B. $\pi\displaystyle\int_0^{\ln(2)} e^x\, dx$

C. $\pi\displaystyle\int_0^2 e^{x/2}\, dx$

D. $\pi\displaystyle\int_0^2 e^x\, dx$

E. $\pi\displaystyle\int_0^2 e^{x^2}\, dx$

GO ON TO THE NEXT PAGE.

23. If $f(x) = (x^2 + 3x)\sqrt{x^3 + 3}$, then $f'(0) =$

A. 0

B. $\sqrt{3}$

C. $2\sqrt{3}$

D. $3\sqrt{3}$

E. $4\sqrt{3}$

24. If $\dfrac{dy}{dx} = 2xe^{-y}$ and $y = 0$ when $x = 2$, then when $x = 4$, $y =$

A. 0

B. $\ln(13)$

C. $\ln(17)$

D. e^{13}

E. e^{17}

25. If $f(x) = \sin(\cos(2x))$, then $f'(x) =$

A. $-2\sin(2x)\cos(\cos(2x))$

B. $-\sin(2x)\cos(\cos(2x))$

C. $2\sin(2x)\cos(\cos(2x))$

D. $\cos^2(2x) - \sin^2(2x)$

E. $2\cos^2(2x) - 2\sin^2(2x)$

26. If $f(x) = (\ln(x))^5$, then $f'(e^2) =$

A. $\dfrac{5}{e^2}$

B. $\dfrac{80}{e^2}$

C. $5e^2$

D. 80

E. $80e^2$

GO ON TO THE NEXT PAGE.

27. The displacement s of a particle moving along the x-axis at time $t \geq 0$ is given by $s(t) = t^3 - 9t^2 + 24t$. For which time intervals is the particle traveling in the **positive** x-direction?

A. $(0,2)$ only

B. $(2,4)$ only

C. $(3,\infty)$ only

D. $(4,\infty)$ only

E. $(0,2)$ or $(4,\infty)$

28. $\displaystyle\int_{1}^{2} \frac{x}{x^2+1}\,dx =$

A. $\dfrac{1}{2}\ln(2)$

B. $\dfrac{1}{2}\ln\left(\dfrac{5}{2}\right)$

C. $\ln(2)$

D. $\ln\left(\dfrac{5}{2}\right)$

E. $\ln(5)$

STOP

END OF PART A, SECTION I
IF YOU FINISH BEFORE TIME IS CALLED, YOU MAY CHECK YOUR WORK ON PART A ONLY.

DO NOT GO ON TO PART B UNTIL YOU ARE TOLD TO DO SO.

CALCULUS AB

SECTION I, Part B

Time – 50 Minutes

Number of questions – 17

A GRAPHING CALCULATOR IS REQUIRED FOR SOME
QUESTIONS ON THIS PART OF THE EXAMINATION.

Directions: Solve each of the following problems, using the available space for scratchwork. After examining the form of the choices, decide which is the best of the choices given and fill in the corresponding oval on the answer sheet. No credit will be given for anything written in the test book. Do not spend too much time on any one problem.

In this test:

1. The **exact** numerical value of the correct answer does not always appear among the choices given. When this happens, select from among the choices the number that best approximates the exact numerical value.

2. Unless otherwise specified, the domain of a function f is assumed to be the set of all real numbers x for which $f(x)$ is a real number.

29. $\lim\limits_{h\to 0} \dfrac{\sec\left(h+\dfrac{\pi}{3}\right) - \sec\left(\dfrac{\pi}{3}\right)}{h} =$

  A. 0

  B. 0.289

  C. 0.866

  D. 1.155

  E. 3.464

30. The graph of $y = x^3 - 5x^2 + 4x + 2$ has a local minimum at

  A. (0.46,–4.06)

  B. (0.46,2.88)

  C. (1.53,0)

  D. (2.87,–4.06)

  E. (2.87,2.88)

GO ON TO THE NEXT PAGE.

31. Two particles leave the origin of the xy-plane at the same time. Particle 1 travels to the right on the x-axis at a constant rate of $2\frac{units}{s}$, and Particle 2 travels upwards on the y-axis at a constant rate of $3\frac{units}{s}$. After 15s, at what rate is the distance between the two particles changing (in $\frac{units}{s}$)?

A. 0.0667

B. 0.291

C. 1.81

D. 3.61

E. 7.22

32. If $f(x) = x^3 + 2x^2 + 1$, then $\left(f^{-1}\right)'(-3) =$

A. -8

B. $-\dfrac{1}{8}$

C. 0

D. $\dfrac{1}{8}$

E. 8

33. $\displaystyle\int_{-\pi/3}^{0} \tan(x)\,dx + \int_{0}^{\pi/3} \sec(x)\,dx =$

A. $-\ln(2) - \ln\left(2 + \sqrt{3}\right)$

B. $-\ln(2) + \ln\left(2 + \sqrt{3}\right)$

C. $\ln\left(\sqrt{3}\right)$

D. $\ln(2) - \ln\left(2 + \sqrt{3}\right)$

E. $\ln(2) + \ln\left(2 + \sqrt{3}\right)$

GO ON TO THE NEXT PAGE.

34. $\dfrac{d}{dx}\displaystyle\int_{0}^{x^2}\dfrac{1}{t+1}dt =$

   A. $\dfrac{2x}{x^2+1}$

   B. $\dfrac{1}{x^2+1}$

   C. $x^2\ln\left(x^2+1\right)$

   D. $2x\ln\left(x^2+1\right)$

   E. $\ln\left(x^2+1\right)$

35. If the definite integral $\displaystyle\int_{0}^{2}\left(9-x^2\right)dx$ is approximated by using an **upper sum** with $n=4$, the error is

   A. $\dfrac{11}{12}$

   B. $\dfrac{13}{12}$

   C. $\dfrac{57}{4}$

   D. $\dfrac{46}{3}$

   E. $\dfrac{65}{4}$

36. If $\displaystyle\int_{3}^{7}f\left(x\right)dx=-2$ and $\displaystyle\int_{5}^{7}f\left(x\right)dx=3$, then $\displaystyle\int_{3}^{5}f\left(x\right)dx=$

   A. Cannot be determined.

   B. -5

   C. -1

   D. 1

   E. 5

37. The average value of the function $f\left(x\right)=\dfrac{1}{x^4+1}$ on the interval $\left[1,3\right]$ is

   A. 0.12

   B. 0.20

   C. 0.23

   D. 0.41

   E. 0.99

GO ON TO THE NEXT PAGE.

38. $\int \dfrac{\ln(x)}{x} dx =$

A. $\dfrac{\ln(x)}{x^2} + C$

B. $\dfrac{1}{2}\left(\ln(x)\right)^2 + C$

C. $\left(\ln(x)\right)^2 + C$

D. $\dfrac{1}{2}\ln\left(\ln(x)\right) + C$

E. $\ln\left(\ln(x)\right) + C$

39. The radius of a circle is increasing at a rate of $4\dfrac{in}{s}$. At what rate (in $\dfrac{in^2}{s}$) is the area of the circle increasing when its radius is 21 *in*?

A. 65.973

B. 131.947

C. 527.788

D. 1385.442

E. 5541.769

40. If the function $f(x)$ is differentiable at $x = 2$ and $f(x) = \begin{cases} ax^3 + 2 & x \le 2 \\ 4x^2 + b & x > 2 \end{cases}$, then $a =$

A. $-\dfrac{10}{3}$

B. $\dfrac{3}{4}$

C. $\dfrac{4}{3}$

D. 3

E. 4

41. The length of each side of a cube is *s*. If *s* is increased from 2 *cm* to 2.03 *cm*, then use differentials to estimate (in *cm*) the change in volume of the cube.

A. 8

B. 0.36

C. 0

D. –0.36

E. –8

GO ON TO THE NEXT PAGE.

42. $\int \sin(x)\sec^{3}(x)\,dx =$

A. $-\dfrac{1}{2}\sec^{2}(x)+C$

B. $-\dfrac{1}{4}\sec^{4}(x)+C$

C. $\dfrac{1}{4}\sec^{4}(x)+C$

D. $\dfrac{1}{2}\sec^{2}(x)+C$

E. $\ln\left(\cos^{3}(x)\right)+C$

43. A point travels along the graph of $f(x) = \sin(x)$ such that the x-coordinate of the particle increases at a rate of $5\dfrac{units}{s}$. At what rate (in $\dfrac{units}{s}$) is the point moving away from the origin when it is at the point $\left(\dfrac{\pi}{6}, \dfrac{1}{2}\right)$?

A. 2.05

B. 4.10

C. 6.15

D. 8.20

E. 10.25

44. Let R be the region bounded by the graphs of $f(x) = 4 - x^2$ and $g(x) = -x + 2$. The volume generated by revolving R about the x-axis is

A. 21.6

B. 33.9

C. 43.2

D. 67.9

E. 135.7

45. A particle travels in a straight line, and its velocity v (in $\dfrac{cm}{s}$) is given as a function of time t (in s) by $v(t) = 2t\sin(t)$. The distance (in cm) traveled by the particle from $t = 0s$ to $t = 2s$ is

A. 1.742

B. 3.483

C. 5.225

D. 6.966

E. 8.708

STOP

END OF PART B, SECTION I
IF YOU FINISH BEFORE TIME IS CALLED, YOU MAY CHECK YOUR WORK ON PART B ONLY.

DO NOT GO ON TO SECTION II UNTIL YOU ARE TOLD TO DO SO.

CALCULUS AB – SECTION II

Time – 1 hour and 30 minutes
Number of problems – 6
Percent of total grade – 50

GENERAL INSTRUCTIONS

You may wish to look over the problems before starting to work on them, since it is not expected that everyone will be able to complete all parts of all problems. All problems are given equal weight, but the parts of a particular problem are not necessarily given equal weight.

A GRAPHING CALCULATOR IS REQUIRED FOR SOME PROBLEMS OR PARTS OF PROBLEMS ON THIS SECTION OF THE EXAMINATION.

- You should write all work for each part of each problem in the space provided for that part in the booklet. Be sure to write clearly and legibly. If you make an error, you may save time by crossing it out rather than trying to erase it. Erased or crossed-out work will not be graded.

- Show all your work. You will be graded on the correctness and completeness of your methods as well as your answers. Correct answers without supporting work may not receive credit.

- Justifications require that you give mathematical (noncalculator) reasons and that you clearly identify functions, graphs, tables, or other objects you use.

- You are permitted to use your calculator to solve an equation, find the derivative of a function at a point, or calculate the value of a definite integral. However, you must clearly indicate the setup of your problem, namely the equation, function, or integral you are using. If you use other built-in features or programs, you must show the mathematical steps necessary to produce your results.

- Your work must be expressed in standard mathematical notation rather than calculator syntax. For example, $\int_1^5 x^2 dx$ may not be written as $fnInt(X^2,X,1,5)$.

- Unless otherwise specified, answers (numeric or algebraic) need not be simplified. If your answer is given as a decimal approximation, it should be correct to three places after the decimal point.

- Unless otherwise specified, the domain of a function f is assumed to be the set of all real numbers x for which $f(x)$ is a real number.

SECTION II, PART A

Time—30 Minutes

Number of problems—2

A graphing calculator is required for some problems or parts of problems.

During the timed portion for Part A, you may work only on the problems in Part A.

On Part A, you are permitted to use your calculator to solve an equation, find the derivative of a function at a point, or calculate the value of a definite integral. However, you must clearly indicate the setup of your problem, namely the equation, function, or integral you are using. If you use other built-in features or programs, you must show the mathematical steps necessary to produce your results.

1. Let R be the region enclosed by the graphs of $f(x) = \dfrac{1}{x^3+1}$, $g(x) = x^3 + 1$, and the lines $x = 1$ and $x = 2$. For all parts of this problem, round your answers to the nearest thousandth.

   A. Find the area of R.

   B. Find the volume of the solid generated by revolving R about the x-axis.

   C. Find the volume of the solid generated by revolving R about the y-axis.

GO ON TO THE NEXT PAGE.

2. A colony of bacteria grows in a petri dish. Over a certain time interval, the population N of the colony has a rate of growth with respect to time t (in hrs) of $\dfrac{dN}{dt} = kN$, where k is a constant.

    A. Given that the population of the colony is 10,000 initially and 22,500 at $t = 6$ hrs, find an expression for N in terms of t.

    B. Assuming that the growth of the population of the colony continues according to the expression you found in part A, determine the time (in hrs) required for the population of the colony to reach 40,000. Round your answer to the nearest hundredth.

    C. Assuming that the growth of the population of the colony continues according to the expression you found in part A, determine the rate at which the *rate of growth* (in $\dfrac{bacteria}{hr}$) is increasing when the population of the colony reaches 40,000. Round your answer to the nearest unit.

STOP

END OF PART A, SECTION II
IF YOU FINISH BEFORE TIME IS CALLED, YOU MAY CHECK YOUR WORK ON PART A ONLY.

DO NOT GO ON TO PART B UNTIL YOU ARE TOLD TO DO SO.

SECTION II, PART B

Time—1 hour
Number of problems—4

NO CALCULATOR IS ALLOWED FOR THESE PROBLEMS.

During the timed portion for Part B, you may continue to work on the
problems in Part A without the use of any calculator.

3. Consider the equation $x^3 - 3xy + y^3 = 1$.

   A. Find $\dfrac{dy}{dx}$ in terms of x and y.

   B. Find $\dfrac{d^2 y}{dx^2}$ in terms of x and y. Express your answer with no compound fractions.

   C. Find the equations of the lines tangent to the graph of the equation at each of the points on the graph whose x-coordinate is 1.

GO ON TO THE NEXT PAGE.

4. Consider the function $f(x) = x^3 - 3x^2 - 9x - 1$.

A. Find and identify the open intervals on which f is increasing or decreasing.

B. Find the coordinates of the absolute extrema of f on the interval $[1, 5]$.

C. Find the coordinates of any points of inflection of the graph of f.

GO ON TO THE NEXT PAGE.

5. Let $F(x) = \int_0^x \left(2^t + t^2\right) dt$ on the closed interval $[0,4]$.

A. Approximate $F(4)$ with a **lower sum** using $n = 4$.

B. Find the exact values of $F(0)$ and $F(4)$.

C. Find $F'(x)$.

D. Find the average value of $F'(x)$ on $[0,4]$.

GO ON TO THE NEXT PAGE.

6. A particle moves in a straight line. Its velocity v (in $\frac{m}{s}$) at time $t \geq 0$ (in s) is given by $v(t) = 2t^2 - 5t + 3$. The initial position of the particle is $x = 3m$

A. Write a function for the position x (in m) of the particle for all $t \geq 0$.

B. Write a function for the acceleration a (in $\frac{m}{s^2}$) of the particle for all $t \geq 0$.

C. At what time(s) is the direction of motion of the particle changing?

STOP

END OF EXAM

Answer Key – Section I

1. C	10. B	19. B	28. B	37. A
2. B	11. B	20. C	29. E	38. B
3. B	12. E	21. A	30. D	39. C
4. D	13. D	22. D	31. D	40. C
5. A	14. E	23. D	32. D	41. B
6. B	15. D	24. B	33. B	42. D
7. E	16. E	25. A	34. A	43. B
8. D	17. B	26. B	35. A	44. D
9. D	18. A	27. E	36. B	45. B

Solutions – Section I, Part A

1. **Answer: C**

$$f'(x) = \frac{d}{dx}\left(\frac{x}{x^2 + 2}\right)$$

Use the Quotient Rule.

$$f'(x) = \frac{\frac{d}{dx}(x) \cdot (x^2 + 2) - x \cdot \frac{d}{dx}(x^2 + 2)}{(x^2 + 2)^2}$$

$$f'(x) = \frac{(1)(x^2 + 2) - x(2x)}{(x^2 + 2)^2}$$

$$f'(x) = \frac{-x^2 + 2}{(x^2 + 2)^2}$$

2. **Answer: B**

$$f'(x) = \frac{d}{dx}\left(2x^{\frac{2}{3}}\right)$$

Use the Power Rule.

$$f'(x) = \frac{2}{3} \cdot 2x^{-\frac{1}{3}} = \frac{4}{3x^{\frac{1}{3}}}$$

$$f'(-8) = \frac{4}{3(-8)^{\frac{1}{3}}} = \frac{4}{3(-2)} = -\frac{2}{3}$$

3. **Answer: B**

Since the limit yields the indeterminate form $\frac{-\infty}{\infty}$, use L'Hôpital's rule.

$$\lim_{x \to \infty} \frac{2 - 3x^2}{6x^2 + 5} = \lim_{x \to \infty} \frac{\frac{d}{dx}(2 - 3x^2)}{\frac{d}{dx}(6x^2 + 5)}$$

$$\lim_{x \to \infty} \frac{2 - 3x^2}{6x^2 + 5} = \lim_{x \to \infty} \frac{-6x}{12x} = \lim_{x \to \infty} \frac{-6}{12} = \frac{-6}{12} = -\frac{1}{2}$$

4. **Answer: D**

In order for f to be continuous at $x = 3$, we must have $\lim_{x \to 3} f(x) = f(3)$. Since $f(3) = k$, we require the following:

$$\lim_{x \to 3} \frac{x^2 - 5x + 6}{x - 3} = k$$

$$\lim_{x \to 3} \frac{(x - 2)(x - 3)}{(x - 3)} = k$$

$$\lim_{x \to 3} (x - 2) = k$$

$$1 = k$$

5. **Answer: A**

Use Implicit Differentiation.

$$\frac{d}{dx}\left(x^3 - y^3\right) = \frac{d}{dx}(2xy)$$

$$3x^2 - 3y^2 y' = 2y + 2xy'$$

In the last step, we used the Product Rule on the right-hand side. Now algebraically solve for y'.

$$\left(-2x - 3y^2\right)y' = -3x^2 + 2y$$

$$y' = \frac{-3x^2 + 2y}{-\left(2x + 3y^2\right)}$$

$$y' = \frac{3x^2 - 2y}{2x + 3y^2}$$

6. **Answer: B**

$$\frac{dy}{dx} = \frac{d}{dx}\big(\sin(x)\cos(x)\big)$$

Use the Product Rule.

$$\frac{dy}{dx} = \frac{d}{dx}\big(\sin(x)\big) \cdot \cos(x) + \sin(x) \cdot \frac{d}{dx}\big(\cos(x)\big)$$

$$\frac{dy}{dx} = \cos(x)\cos(x) + \sin(x)\big(-\sin(x)\big)$$

$$\frac{dy}{dx} = \cos^2(x) - \sin^2(x)$$

7. Answer: E

$$\lim_{x \to 0} f\left(g(x)\right) = \lim_{x \to 0} 2^{-(x-1)}$$

$$\lim_{x \to 0} f\left(g(x)\right) = 2^{-(0-1)} = 2$$

8. Answer: D

$$f'(x) = \frac{d}{dx}\left(\sec(x)\right) = \sec(x)\tan(x)$$

$$f''(x) = \frac{d}{dx}\left(\sec(x)\tan(x)\right)$$

Use the Product Rule.

$$f''(x) = \frac{d}{dx}\left(\sec(x)\right) \cdot \tan(x) + \sec(x) \cdot \frac{d}{dx}\left(\tan(x)\right)$$

$$f''(x) = \sec(x)\tan(x) \cdot \tan(x) + \sec(x) \cdot \sec^2(x)$$

$$f''(x) = \sec(x)\tan^2(x) + \sec^3(x)$$

$$f''\left(\frac{\pi}{4}\right) = \sec\left(\frac{\pi}{4}\right)\tan^2\left(\frac{\pi}{4}\right) + \sec^3\left(\frac{\pi}{4}\right)$$

$$f''\left(\frac{\pi}{4}\right) = \sqrt{2} \cdot 1^2 + \sqrt{2}^3 = 3\sqrt{2}$$

9. Answer: D

First, find y' using Implicit Differentiation. Use the Product Rule when differentiating the first term on the left-hand side.

$$\frac{d}{dx}\left(x^2 y + 3y - 4\right) = \frac{d}{dx}(0)$$

$$2xy + x^2 y' + 3y' = 0$$

Next, algebraically solve for y'.

$$\left(x^2 + 3\right)y' = -2xy$$

$$y' = \frac{-2xy}{x^2 + 3}$$

Finally, find the slope m of the tangent line by evaluating y' at the point $(-1,1)$.

$$m = y'\big|_{(-1,1)}$$

$$m = \frac{-2(-1)(1)}{(-1)^2 + 3} = \frac{1}{2}$$

10. Answer: B

First, find y' using the Chain Rule.

$$y' = \frac{d}{dx}\left((6x+2)^{1/3}\right)$$

$$y' = \frac{1}{3}(6x+2)^{-2/3} \cdot \frac{d}{dx}(6x+2)$$

$$y' = \frac{1}{3}(6x+2)^{-2/3} \cdot 6$$

$$y' = \frac{2}{(6x+2)^{2/3}}$$

Next, find the slope of the tangent line by evaluating y' at $x = 1$.

$$m = y'\big|_1$$

$$m = \frac{2}{(6(1)+2)^{2/3}} = \frac{1}{2}$$

Finally, find the equation of the tangent line. Since the answer choices are all in general form, we will put our answer in general form as well.

$$y - 2 = \frac{1}{2}(x-1)$$

$$y - 2 = \frac{1}{2}x - \frac{1}{2}$$

$$-\frac{1}{2}x + y - \frac{3}{2} = 0$$
$$x - 2y + 3 = 0$$

11. Answer: B

Let $u = x^3$. Then $du = 3x^2 dx$, or $x^2 dx = \frac{1}{3}du$.

$$\int x^2 \sin(x^3)\,dx = \frac{1}{3}\int \sin(u)\,du$$

$$\int x^2 \sin(x^3)\,dx = -\frac{1}{3}\cos(u) + C$$

$$\int x^2 \sin(x^3)\,dx = -\frac{1}{3}\cos(x^3) + C$$

12. Answer: E

$$\int_{1/2}^{\sqrt{3}/2} \frac{2}{\sqrt{1-x^2}}\,dx = 2\left(\sin^{-1}(x)\right)\Big|_{1/2}^{\sqrt{3}/2}$$

$$\int_{1/2}^{\sqrt{3}/2} \frac{2}{\sqrt{1-x^2}}\,dx = 2\left(\sin^{-1}\left(\frac{\sqrt{3}}{2}\right) - \sin^{-1}\left(\frac{1}{2}\right)\right)$$

$$\int_{1/2}^{\sqrt{3}/2} \frac{2}{\sqrt{1-x^2}}\,dx = 2\left(\frac{\pi}{3} - \frac{\pi}{6}\right) = \frac{\pi}{3}$$

13. Answer: D

In general, the fundamental period of $y = a\tan(bx+c)+d$ is equal to $\dfrac{\pi}{a}$. Since $a = \pi$ in $f(x)$, the fundamental period is equal to $\dfrac{\pi}{\pi} = 1$.

14. Answer: E

Since differentiability implies continuity, and since continuity implies integrability, we will check statement II first, as it is the most restrictive condition on f. Since f is defined piecewise by polynomial functions, it is piecewise differentiable on either $(-\infty, 0)$ or $(0, \infty)$. It remains only to check whether f is differentiable at $x = 0$. Since $f'(0) = \lim\limits_{x \to 0} \dfrac{f(x) - f(0)}{x - 0}$, we check to see if this limit exists by checking the right- and left-handed limits.

$$\lim_{x \to 0^-} \frac{f(x) - f(0)}{x - 0} = \lim_{x \to 0^-} \frac{-2x^2 - 0}{x - 0} = \lim_{x \to 0^-} (-2x) = 0$$

$$\lim_{x \to 0^+} \frac{f(x) - f(0)}{x - 0} = \lim_{x \to 0^+} \frac{3x^2 - 0}{x - 0} = \lim_{x \to 0^+} (3x) = 0$$

Since $\lim\limits_{x \to 0^-} \dfrac{f(x) - f(0)}{x - 0} = \lim\limits_{x \to 0^+} \dfrac{f(x) - f(0)}{x - 0}$, f is differentiable at $x = 0$ as well as on $(-\infty, 0) \cup (0, \infty)$. Hence, f is differentiable (and therefore continuous and integrable) everywhere.

15. Answer: D

The area A is given by the following integral:

$$A = \int_{-1}^{1} \left((9 - x^2) - 4\right)dx$$

$$A = \int_{-1}^{1} \left(-x^2 + 5\right)dx$$

16. Answer: E

$$\frac{dy}{dx} = \frac{d}{dx}\left(2^{-x^2}\right)$$

Use the Chain Rule.

$$\frac{dy}{dx} = \ln(2)2^{-x^2} \cdot \frac{d}{dx}\left(-x^2\right)$$

$$\frac{dy}{dx} = \ln(2)2^{-x^2} \cdot \left(-2x\right)$$

$$\frac{dy}{dx} = -x\ln(2)2^{-x^2+1}$$

17. Answer: B

Let $\bar{f}$ denote the desired average value of f.

$$\bar{f} = \frac{1}{5-2}\int_2^5 \frac{1}{x^2-2x+1}\,dx$$

$$\bar{f} = \frac{1}{3}\int_2^5 \frac{1}{(x-1)^2}\,dx$$

Let $u = x-1$. Then $\dfrac{1}{(x-1)^2} = u^{-2}, du = dx, x = 2 \Rightarrow u = 1$, and $x = 5 \Rightarrow u = 4$.

$$\bar{f} = \frac{1}{3}\int_1^4 u^{-2}\,du$$

$$\bar{f} = \frac{1}{3}\left(-u^{-1}\right)\Big|_1^4$$

$$\bar{f} = -\frac{1}{3}\left(\frac{1}{u}\right)\Big|_1^4$$

$$\bar{f} = -\frac{1}{3}\left(\frac{1}{4}-\frac{1}{1}\right) = \frac{1}{4}$$

18. Answer: A

First, find $f'(x)$.

$$f'(x) = \frac{d}{dx}\left(\frac{1}{3}x^3 + \frac{1}{2}x^2 - 6x + 8\right)$$

$$f'(x) = x^2 + x - 6$$

Next, find the critical numbers of f. Since f is differentiable everywhere, all of its critical numbers occur when $f'(x) = 0$.

$$f'(x) = 0$$

$$x^2 + x - 6 = 0$$

$$(x+3)(x-2) = 0$$

$$x = -3 \text{ or } x = 2$$

We will use the following table to apply the First Derivative Test to each critical number.

Interval	Test Value x	$f'(x)$	Behavior of f
$(-\infty, -3)$	$x = -4$	$f'(-4) = 6 > 0$	Increasing
$(-3, 2)$	$x = 0$	$f'(0) = -6 < 0$	Decreasing
$(2, \infty)$	$x = 3$	$f'(3) = 6 > 0$	Increasing

Since f changes from increasing to decreasing at $x = -3$, that is value of x that corresponds to the relative maximum of f.

19. Answer: B

Since the limit yields the indeterminate form $\dfrac{0}{0}$, apply L'Hôpital's rule.

$$\lim_{x \to 0} \frac{\tan(2x) - \sin(x)}{x} = \lim_{x \to 0} \frac{\dfrac{d}{dx}\left(\tan(2x) - \sin(x)\right)}{\dfrac{d}{dx}(x)}$$

$$\lim_{x \to 0} \frac{\tan(2x) - \sin(x)}{x} = \lim_{x \to 0} \frac{2\sec^2(2x) - \cos(x)}{1}$$

$$\lim_{x \to 0} \frac{\tan(2x) - \sin(x)}{x} = \lim_{x \to 0} \left(2\sec^2(2x) - \cos(x)\right)$$

$$\lim_{x \to 0} \frac{\tan(2x) - \sin(x)}{x} = 2\sec^2(2 \cdot 0) - \cos(0) = 2 - 1 = 1$$

20. Answer: C

$$\int \frac{e^{2x} - e^x}{e^x} \, dx = \int \left(e^x - 1\right) dx$$

$$\int \frac{e^{2x} - e^x}{e^x} \, dx = e^x - x + C$$

21. Answer: A

Since the integrand is an even function, and the interval over which we are integrating is symmetric about $x = 0$, we can rewrite the integral as follows:

$$\int_{-\sqrt{6}}^{\sqrt{6}} \frac{1}{x^2 + 2} \, dx = 2 \int_{0}^{\sqrt{6}} \frac{1}{x^2 + 2} \, dx$$

$$\int_{-\sqrt{6}}^{\sqrt{6}} \frac{1}{x^2 + 2} \, dx = 2 \cdot \frac{1}{\sqrt{2}} \left(\tan^{-1}\left(\frac{x}{\sqrt{2}}\right)\right)\Bigg|_{0}^{\sqrt{6}}$$

$$\int_{-\sqrt{6}}^{\sqrt{6}} \frac{1}{x^2+2} dx = \sqrt{2}\left(\tan^{-1}\left(\frac{\sqrt{6}}{\sqrt{2}}\right) - \tan^{-1}\left(\frac{0}{\sqrt{2}}\right)\right)$$

$$\int_{-\sqrt{6}}^{\sqrt{6}} \frac{1}{x^2+2} dx = \sqrt{2}\left(\tan^{-1}\left(\sqrt{3}\right) - \tan^{-1}(0)\right)$$

$$\int_{-\sqrt{6}}^{\sqrt{6}} \frac{1}{x^2+2} dx = \sqrt{2}\left(\frac{\pi}{3} - 0\right) = \frac{\pi\sqrt{2}}{3}$$

22. Answer: D

We begin by sketching the region. Since the curves that form the boundaries of the region are all elementary functions, this is not difficult to do without a graphing calculator.

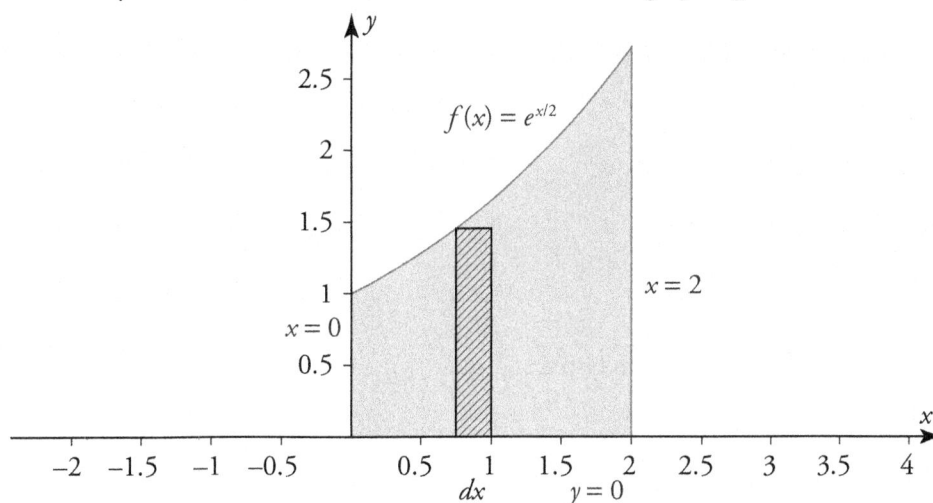

The differential volume element is $dV = \pi\left(f(x)\right)^2 dx = \pi\left(e^{x/2}\right)^2 dx = \pi e^x dx$. The total volume V of the solid of revolution is given by the following integral:

$$V = \int dV = \pi\int_0^2 e^x dx$$

23. Answer: D

First, find $f'(x)$ using both the Product Rule and the Chain Rule.

$$f'(x) = \frac{d}{dx}\left(\left(x^2+3x\right)\sqrt{x^3+3}\right)$$

$$f'(x) = \frac{d}{dx}\left(\left(x^2+3x\right)\left(x^3+3\right)^{1/2}\right)$$

$$f'(x) = \frac{d}{dx}\left(x^2+3x\right)\cdot\left(x^3+3\right)^{1/2} + \left(x^2+3x\right)\cdot\frac{d}{dx}\left(\left(x^3+3\right)^{1/2}\right)$$

$$f'(x) = (2x+3) \cdot (x^3+3)^{1/2} + (x^2+3x) \cdot \frac{1}{2}(x^3+3)^{-1/2} \cdot \frac{d}{dx}(x^3)$$

$$f'(x) = (2x+3) \cdot (x^3+3)^{1/2} + (x^2+3x) \cdot \frac{1}{2}(x^3+3)^{-1/2} \cdot 3x^2$$

$$f'(x) = (2x+3)\sqrt{x^3+3} + \frac{3x^2(x^2+3x)}{2\sqrt{x^3+3}}$$

Next, evaluate $f'(0)$.

$$f'(0) = (2(0)+3)\sqrt{0^3+3} + \frac{3 \cdot 0^2(0^2+3(0))}{2\sqrt{0^3+3}} = 3\sqrt{3}$$

24. Answer: B

First, find the general solution of the differential equation by separation of variables and integration.

$$\frac{dy}{dx} = 2xe^{-y}$$

$$e^y \, dy = 2x \, dx$$

$$\int e^y \, dy = \int 2x \, dx$$

$$e^y = x^2 + C$$

Next, use the solution point $(x, y) = (2,0)$ to find C.

$$e^0 = 2^2 + C$$

$$1 = 4 + C$$

$$-3 = C$$

So the particular solution is given by $e^y = x^2 - 3$. Finally, find y when $x = 4$.

$$e^y = 4^2 - 3 = 13$$

$$y = \ln(13)$$

25. Answer: A

Use the Chain Rule twice, as this is a doubly composite function.

$$f'(x) = \frac{d}{dx}\left(\sin(\cos(2x))\right)$$

$$f'(x) = \cos(\cos(2x)) \cdot \frac{d}{dx}(\cos(2x))$$

$$f'(x) = \cos(\cos(2x)) \cdot (-\sin(2x)) \cdot \frac{d}{dx}(2x)$$

$$f'(x) = \cos(\cos(2x)) \cdot (-\sin(2x)) \cdot 2$$

$$f'(x) = -2\sin(2x)\cos\big(\cos(2x)\big)$$

26. Answer: B

First, find $f'(x)$ using the Chain Rule.

$$f'(x) = \frac{d}{dx}\Big(\big(\ln(x)\big)^5\Big)$$

$$f'(x) = 5\big(\ln(x)\big)^4 \cdot \frac{d}{dx}\big(\ln(x)\big)$$

$$f'(x) = 5\big(\ln(x)\big)^4 \cdot \frac{1}{x}$$

$$f'(x) = \frac{5\big(\ln(x)\big)^4}{x}$$

Next, evaluate $f'\big(e^2\big)$.

$$f'\big(e^2\big) = \frac{5\big(\ln\big(e^2\big)\big)^4}{e^2} = \frac{5 \cdot 2^4}{e^2} = \frac{80}{e^2}$$

27. Answer: E

The particle is moving in the positive x-direction whenever the velocity $s'(t)$ is positive.

$$s'(t) = \frac{d}{dt}\big(t^3 - 9t^2 + 24t\big)$$

$$s'(t) = 3t^2 - 18t + 24$$

$$s'(t) = 3(t - 2)(t - 4)$$

The velocity function has zeros at $t = 2$ and $t = 4$, so its sign is constant on $(0,2)$, $(2,4)$, and $(4,\infty)$. We determine the sign of $s'(t)$ on each of these intervals by using test values, as follows:

Interval	Test Value t	$s'(t)$
$(0,2)$	$t = 1$	$s'(1) = 9 > 0$
$(2,4)$	$t = 3$	$s'(3) = -3 < 0$
$(4,\infty)$	$t = 5$	$s'(5) = 9 > 0$

So $s'(t) > 0$ for all t in $(0,2)$ or $(4,\infty)$.

28. Answer: B

Let $u = x^2 + 1$. Then $du = 2x\,dx$, or $x\,dx = \dfrac{1}{2}du$. Also $x = 1 \Rightarrow u = 2$, and $x = 2 \Rightarrow u = 5$.

$$\int_1^2 \frac{x}{x^2+1}\,dx = \frac{1}{2}\int_2^5 \frac{du}{u}$$

$$\int_1^2 \frac{x}{x^2+1}\,dx = \frac{1}{2}\left(\ln|u|\right)\Big|_2^5$$

$$\int_1^2 \frac{x}{x^2+1}\,dx = \frac{1}{2}\left(\ln(5) - \ln(2)\right) = \frac{1}{2}\ln\left(\frac{5}{2}\right)$$

Solutions – Section I, Part B

29. Answer: E

Let $f(x) = \sec(x)$, and recognize that this limit is equal to $f'\left(\dfrac{\pi}{3}\right)$.

$$f'(x) = \frac{d}{dx}\left(\sec(x)\right)$$

$$f'(x) = \sec(x)\tan(x)$$

$$\lim_{h \to 0} \frac{\sec\left(h + \dfrac{\pi}{3}\right) - \sec\left(\dfrac{\pi}{3}\right)}{h} = \sec\left(\frac{\pi}{3}\right)\tan\left(\frac{\pi}{3}\right) = 2\sqrt{3} \approx 3.464$$

30. Answer: D

See the following graph.

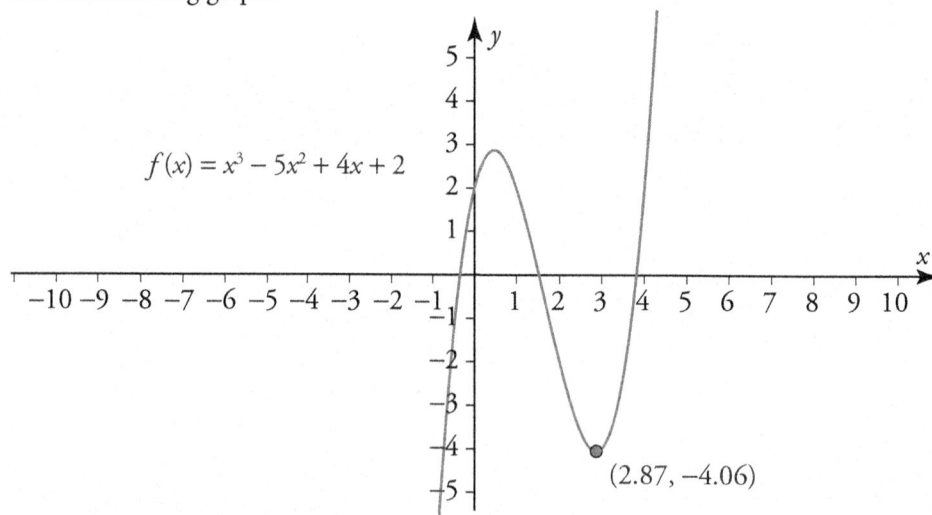

$f(x) = x^3 - 5x^2 + 4x + 2$

$(2.87, -4.06)$

31. Answer: D

The setup is given in the following figure. The distance between the particles is denoted by r.

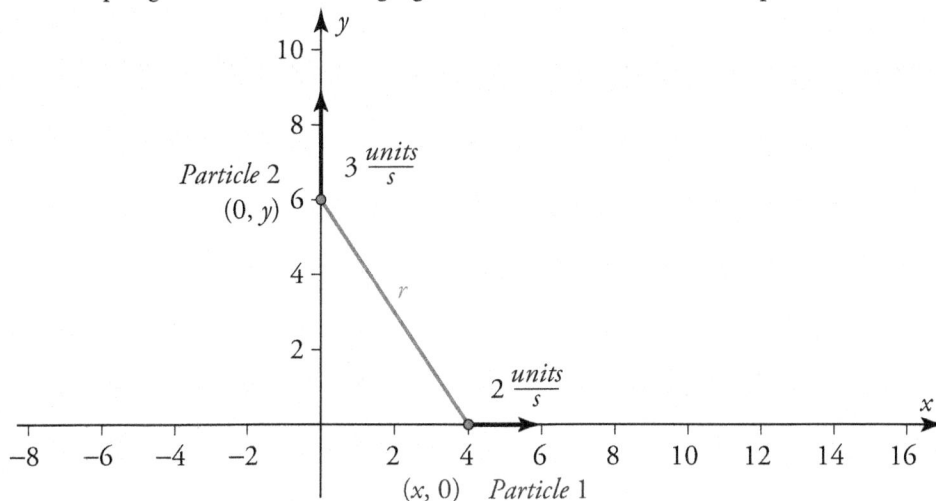

The distance r between the two particles is given in terms of x and y as follows:

$r^2 = x^2 + y^2$

Differentiating implicitly with respect to time t, we obtain the following:

$$2r\frac{dr}{dt} = 2x\frac{dx}{dt} + 2y\frac{dy}{dt}$$

$$\frac{dr}{dt} = \frac{x\dfrac{dx}{dt} + y\dfrac{dy}{dt}}{r}$$

At $t = 15s$, $x = \left(2\dfrac{units}{s}\right)(15s) = 30\,units$, $y = \left(3\dfrac{units}{s}\right)(15s) = 45\,units$, and

$r = \sqrt{(30\,units)^2 + (45\,units)^2} = 15\sqrt{13}\,units$.

$$\frac{dr}{dt} = \frac{(30)(2) + (45)(3)}{15\sqrt{13}} \approx 3.61$$

32. Answer: D

The graph of f is given below.

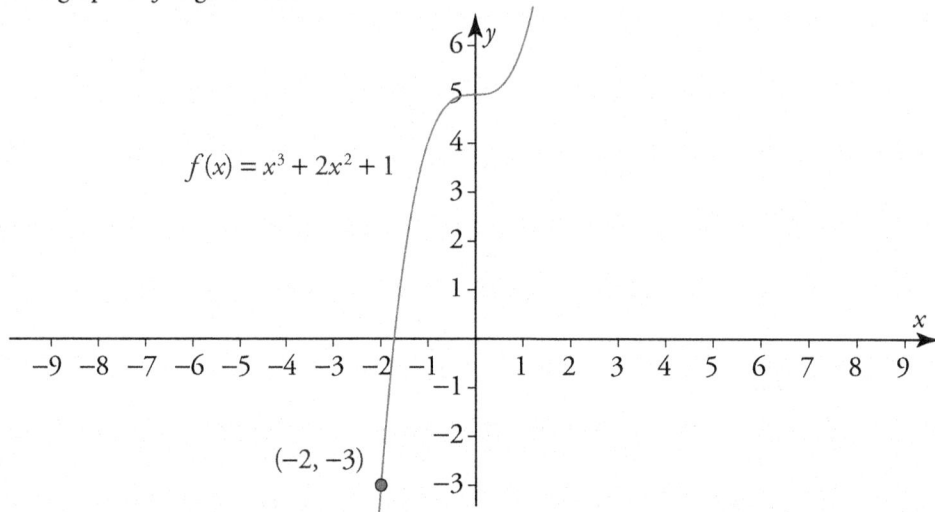

$$f(x) = x^3 + 2x^2 + 1$$

$$(-2, -3)$$

As you can see from the graph, f is one-to-one, so it has an inverse function f^{-1}. Furthermore, the point $(-2, -3)$ is on the graph of f, which means that $f(-2) = -3$, and hence $f^{-1}(-3) = -2$.

Using the formula $\left(f^{-1}\right)'(x) = \dfrac{1}{f'\left(f^{-1}(x)\right)}$ together with $f'(x) = 3x^2 + 2x$, we obtain $\left(f^{-1}\right)'(-3)$.

$$\left(f^{-1}\right)'(-1) = \frac{1}{f'(-2)}$$

$$\left(f^{-1}\right)'(-1) = \frac{1}{3(-2)^2 + 2(-2)} = \frac{1}{8}$$

33. Answer: B

First, evaluate the integral of the tangent function.

$$\int_{-\pi/3}^{0} \tan(x)\,dx = -\ln|\cos(x)|\Big|_{-\pi/3}^{0}$$

$$\int_{-\pi/3}^{0} \tan(x)\,dx = -\ln|\cos(0)| - \left(-\ln\left|\cos\left(-\frac{\pi}{3}\right)\right|\right)$$

$$\int_{-\pi/3}^{0} \tan(x)\,dx = -\ln(2)$$

Second, evaluate the integral of the secant function.

$$\int_{0}^{\pi/3} \sec(x)\,dx = \ln|\sec(x) + \tan(x)|\Big|_{0}^{\pi/3}$$

$$\int_{0}^{\pi/3} \sec(x)\,dx = \ln\left|\sec\left(\frac{\pi}{3}\right) + \tan\left(\frac{\pi}{3}\right)\right| - \ln|\sec(0) + \tan(0)|$$

$$\int_0^{\pi/3} \sec(x)\,dx = \ln\left(2+\sqrt{3}\right)$$

Finally, add the two integrals to get $-\ln(2)+\ln\left(2+\sqrt{3}\right)$. Note that you can also use the calculator to compute the integrals.

34. Answer: A

The calculator will not be helpful on this problem, so solve it using the Second Fundamental Theorem of Calculus, combined with the Chain Rule.

$$\frac{d}{dx}\int_0^{x^2} \frac{1}{t+1}\,dt = \frac{1}{x^2+1}\cdot\frac{d}{dx}\left(x^2\right)$$

$$\frac{d}{dx}\int_0^{x^2} \frac{1}{t+1}\,dt = \frac{2x}{x^2+1}$$

35. Answer: A

First, use the calculator to obtain $\int_0^2 \left(9-x^2\right)dx = \dfrac{46}{3}$.

Second, set up the upper sum U. In the following figure, the rectangles whose areas comprise the upper sum are drawn in, and their heights are labeled. The width of each rectangle is $\Delta x = \dfrac{2-0}{4} = \dfrac{1}{2}$.

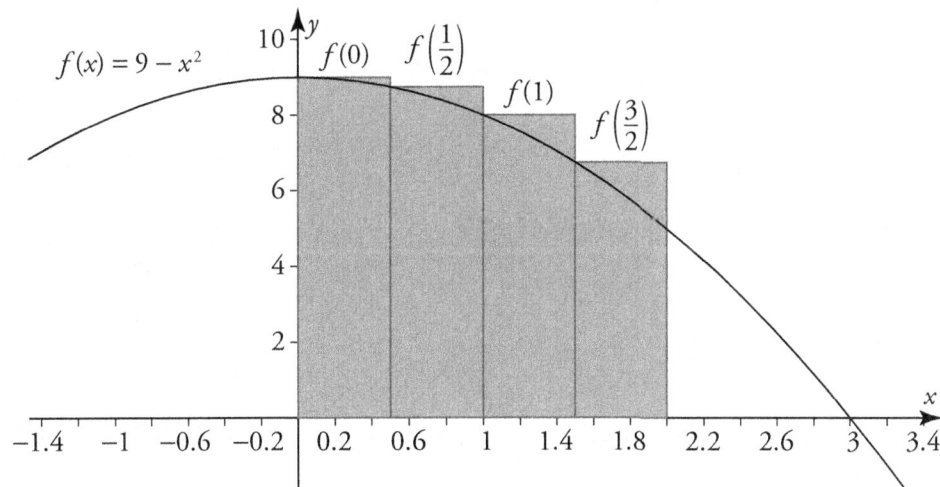

$$U = f(0)\Delta x + f\left(\frac{1}{2}\right)\Delta x + f(1)\Delta x + f\left(\frac{3}{2}\right)\Delta x$$

$$U = \left[f(0)+f\left(\frac{1}{2}\right)+f(1)+f\left(\frac{3}{2}\right)\right]\Delta x$$

$$U = \left[9+\frac{35}{4}+8+\frac{27}{4}\right]\cdot\frac{1}{2} = \frac{65}{4}$$

The error E is given as follows:

$$E = U - \int_0^2 \left(9 - x^2\right) dx$$

$$E = \frac{65}{4} - \frac{46}{3} = \frac{11}{12}$$

36. Answer: B

Use the Additive Interval Property for Integrals

$$\int_3^7 f(x)\,dx = \int_3^5 f(x)\,dx + \int_5^7 f(x)\,dx$$

$$-2 = \int_3^5 f(x)\,dx + 3$$

$$-5 = \int_3^5 f(x)\,dx$$

37. Answer: A

Use the calculator to find $\int_1^3 \frac{1}{x^4+1}\,dx \approx 0.231$. The average value $\overline{f}$ of f on the interval $[1,3]$ is given as follows:

$$\overline{f} = \frac{1}{3-1} \int_1^3 \frac{1}{x^4+1}\,dx$$

$$\overline{f} \approx 0.12$$

38. Answer: B

Let $u = \ln(x)$. Then $du = \dfrac{dx}{x}$.

$$\int \frac{\ln(x)}{x}\,dx = \int u\,du$$

$$\int \frac{\ln(x)}{x}\,dx = \frac{1}{2}u^2 + C$$

$$\int \frac{\ln(x)}{x}\,dx = \frac{1}{2}\left(\ln(x)\right)^2 + C$$

39. Answer: C

The area A of the circle is given in terms of its radius r by the relationship $A = \pi r^2$. Differentiating both sides with respect to time t yields the following:

$$\frac{dA}{dt} = 2\pi r \frac{dr}{dt}$$

Given that $r = 21\,in$ and $\frac{dr}{dt} = 4\frac{in}{s}$, we obtain the following for $\frac{dA}{dt}$:

$$\frac{dA}{dt} = 2\pi(4)(21) \approx 527.788$$

40. Answer: C

In order for f to be differentiable at $x = 2$, it must first be continuous at $x = 2$. This means that we must have $\lim\limits_{x \to 2} f(x) = f(2)$. This limit is clearly satisfied from the left. In order for it to be satisfied from the right, we must have the following:

$$\lim_{x \to 2^+} f(x) = f(2)$$

$$\lim_{x \to 2^+} \left(4x^2 + b\right) = 8a + 2$$

$$16 + b = 8a + 2$$

$$b = 8a - 14$$

By definition $f'(2) = \lim\limits_{x \to 2} \dfrac{f(x) - f(2)}{x - 2}$, so this limit must exist if f is to be differentiable at $x = 2$. We will evaluate the left- and right-handed limits separately.

Left-handed limit:

$$\lim_{x \to 2^-} \frac{f(x) - f(2)}{x - 2} = \lim_{x \to 2^-} \frac{ax^3 + 2 - (8a + 2)}{x - 2}$$

$$\lim_{x \to 2^-} \frac{f(x) - f(2)}{x - 2} = \lim_{x \to 2^-} \frac{ax^3 - 8a}{x - 2}$$

$$\lim_{x \to 2^-} \frac{f(x) - f(2)}{x - 2} = \lim_{x \to 2^-} \frac{a(x - 2)(x^2 + 2x + 4)}{x - 2}$$

$$\lim_{x \to 2^-} \frac{f(x) - f(2)}{x - 2} = \lim_{x \to 2^-} a(x^2 + 2x + 4)$$

$$\lim_{x \to 2^-} \frac{f(x) - f(2)}{x - 2} = 12a$$

Right-handed limit:

$$\lim_{x \to 2^+} \frac{f(x) - f(2)}{x - 2} = \lim_{x \to 2^+} \frac{4x^2 + b - (8a + 2)}{x - 2}$$

Using the result $b = 8a - 14$, we obtain the following:

$$\lim_{x \to 2^+} \frac{f(x) - f(2)}{x - 2} = \lim_{x \to 2^+} \frac{4x^2 + 8a - 14 - (8a + 2)}{x - 2}$$

$$\lim_{x \to 2^+} \frac{f(x) - f(2)}{x - 2} = \lim_{x \to 2^+} \frac{4x^2 - 16}{x - 2}$$

$$\lim_{x \to 2^+} \frac{f(x) - f(2)}{x - 2} = \lim_{x \to 2^+} \frac{4(x - 2)(x + 2)}{x - 2}$$

$$\lim_{x \to 2^+} \frac{f(x) - f(2)}{x - 2} = \lim_{x \to 2^+} 4(x + 2)$$

$$\lim_{x \to 2^+} \frac{f(x) - f(2)}{x - 2} = 16$$

Equating the left- and right-handed limits yields $12a = 16 \Rightarrow a = \dfrac{4}{3}$.

41. Answer: B

The volume V of the cube is given in terms of the side length s by the relation $V = s^3$. The change in volume ΔV of the cube is approximated as follows:

$$\Delta V \approx dV$$

$$\Delta V \approx 3s^2 ds$$

Evaluating ΔV for $s = 2\,cm$ and $ds = 0.03\,cm$, we obtain $\Delta V \approx 3(2\,cm)^2(0.03\,cm) = 0.36\,cm^3$.

42. Answer: D

$$\int \sin(x)\sec^3(x)\,dx = \int \frac{\sin(x)}{\cos^3(x)}\,dx$$

Let $u = \cos(x)$. Then $du = -\sin(x)\,dx$, or $\sin(x)\,dx = -du$.

$$\int \sin(x)\sec^3(x)\,dx = -\int \frac{1}{u^3}\,du$$

$$\int \sin(x)\sec^3(x)\,dx = -\int u^{-3}\,du$$

$$\int \sin(x)\sec^3(x)\,dx = -\frac{1}{-2}u^{-2} + C$$

$$\int \sin(x)\sec^3(x)\,dx = \frac{1}{2}(\cos(x))^{-2} + C$$

$$\int \sin(x)\sec^3(x)\,dx = \frac{1}{2}\sec^2(x) + C$$

43. Answer: B

Refer to the following graph. The distance between the particle and the origin is denoted by r.

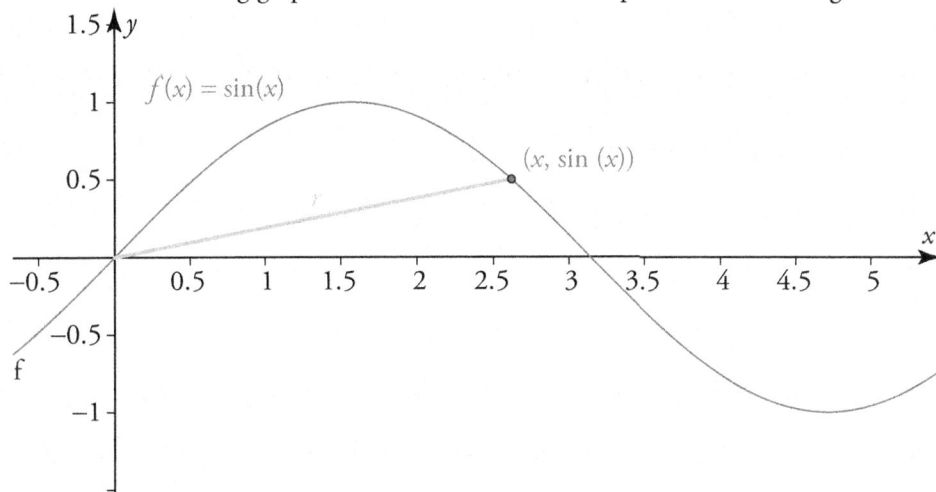

r is given in terms of x and y by the relation $r^2 = x^2 + \sin^2(x)$. Differentiating implicitly with respect to time t, we obtain the following:

$$2r\frac{dr}{dt} = 2x\frac{dx}{dt} + 2\sin(x)\cos(x)\frac{dx}{dt}$$

$$\frac{dr}{dt} = \frac{\left(x + \sin(x)\cos(x)\right)\frac{dx}{dt}}{r}$$

At $(x, y) = \left(\dfrac{5\pi}{6}, \dfrac{1}{2}\right)$, $r = \sqrt{\left(\dfrac{5\pi}{6}\right)^2 + \left(\dfrac{1}{2}\right)^2} = \sqrt{\dfrac{25\pi^2}{36} + \dfrac{1}{4}}$.

$$\frac{dr}{dt} = \frac{\left(\dfrac{5\pi}{6} + \sin\left(\dfrac{5\pi}{6}\right)\cos\left(\dfrac{5\pi}{6}\right)\right)(5)}{\sqrt{\dfrac{25\pi^2}{36} + \dfrac{1}{4}}} \approx 4.10\,\frac{units}{s}$$

44. Answer: D

We begin by sketching the region.

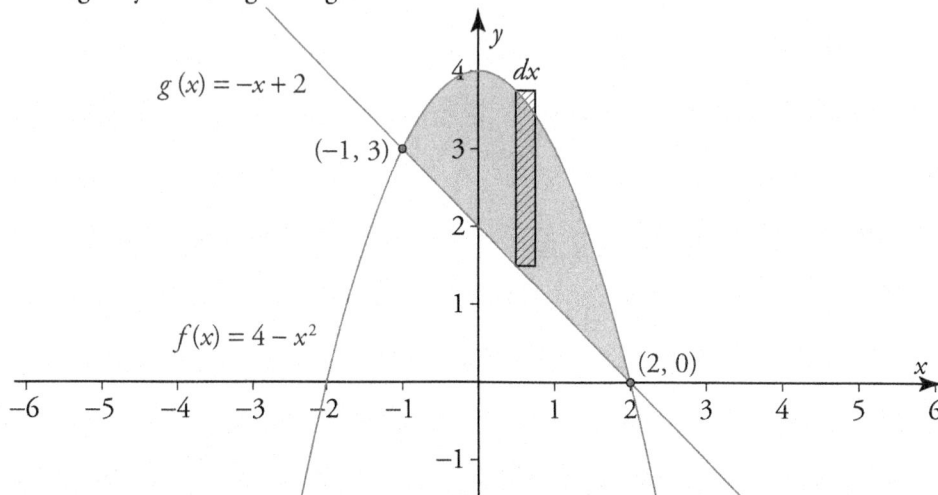

The differential volume element dV is given as follows:

$$dV = \pi\left(\left(f(x)\right)^2 - \left(g(x)\right)^2\right)dx$$

$$dV = \pi\left(\left(4-x^2\right)^2 - \left(-x+2\right)^2\right)dx$$

The volume of the solid of revolution is found by integrating dV on the interval $[-1,2]$. Using the calculator to do the integration, we find that $V \approx 67.9$.

45. Answer: B

We begin by sketching the function $v(t)$ on the interval $[0,2]$.

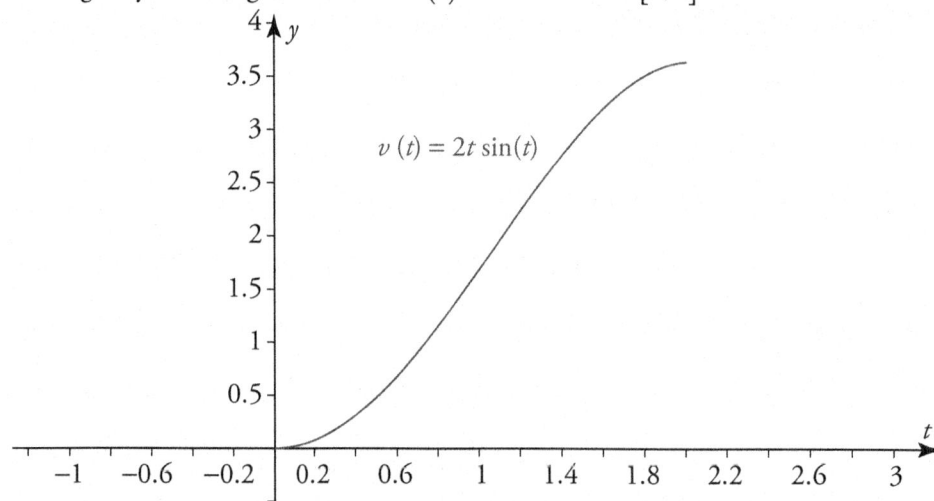

Since $v(t) \geq 0$ for all t in $[0,2]$, the distance traveled by the particle is found by integrating $v(t)$ on $[0,2]$. Using the calculator, we find that $\int_0^2 2t\sin(t)\,dt \approx 3.483$.

1.

A. We begin by sketching the region.

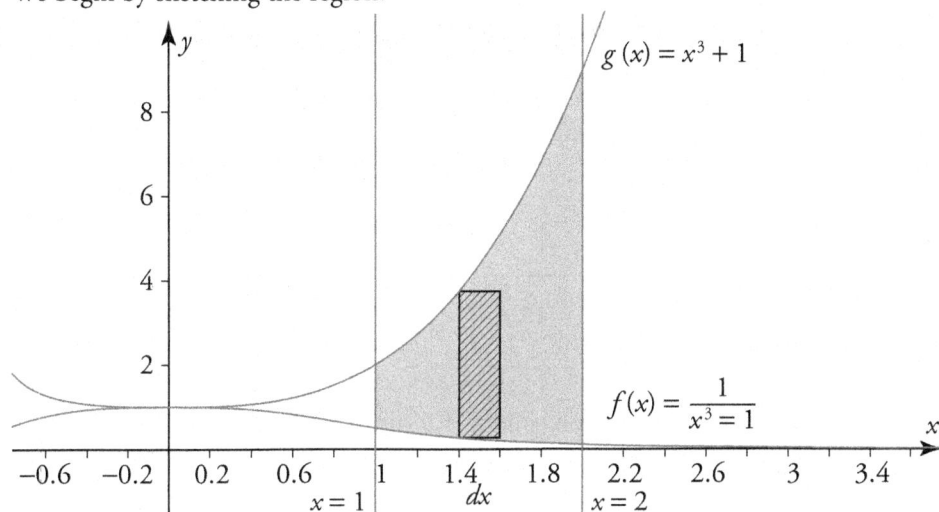

The differential area element is $dA = \big(g(x) - f(x)\big)dx$. The area A of the region is found by integrating dA on the interval $[1,2]$. Using the calculator, we find that $A \approx 4.496$.

B. The figure from part A can be used to determine that the differential volume element dV is given as follows:

$$dV = \pi\left(\left(x^3 + 1\right)^2 - \left(\frac{1}{x^3 + 1}\right)^2\right)dx$$

The volume V of the solid of revolution is found by integrating dV on the interval $[1,2]$. Using the calculator, we find that $V \approx 83.459$.

C. We will need to express the functions f and g as functions of y. With a little algebra, the following can be obtained:

$$y = x^3 + 1 \Rightarrow x = \sqrt[3]{y - 1} = h(y)$$

$$y = \frac{1}{x^3 + 1} \Rightarrow x = \sqrt[3]{\frac{1}{y} - 1} = i(y)$$

We will need three different expressions for the differential volume element: one on the interval $\left[\frac{1}{9}, \frac{1}{2}\right]$, one on the interval $\left[\frac{1}{2}, 2\right]$, and one on the interval $[2, 9]$. We will look at these intervals one at a time.

First, we will look at $\left[\frac{1}{9}, \frac{1}{2}\right]$.

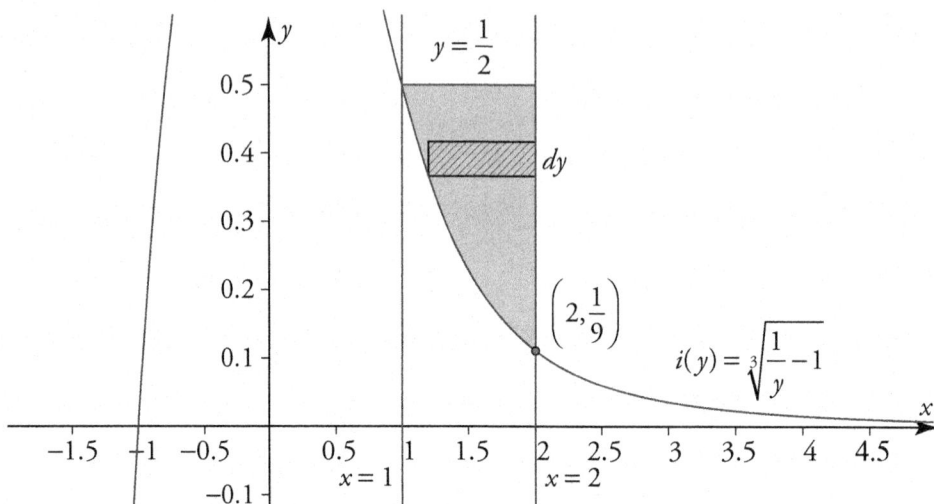

On this interval, the differential volume element $dV_1 = \pi\left(2^2 - \left(i(y)\right)^2\right)dy$. We integrate dV_1 on $\left[\dfrac{1}{9}, \dfrac{1}{2}\right]$ to obtain V_1. Using the calculator, we find that $V_1 \approx 2.512$.

Next, we will look at $\left[\dfrac{1}{2}, 2\right]$.

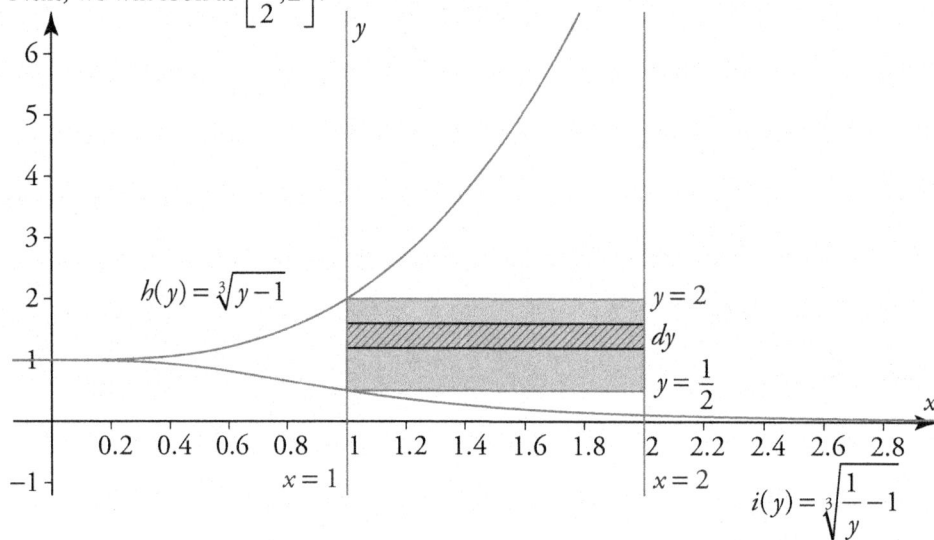

On this interval, the differential volume element $dV_2 = \pi\left(2^2 - 1^2\right)dy$. We integrate dV_2 on $\left[\dfrac{1}{2}, 2\right]$ to obtain V_2. Using the calculator, we find that $V_2 \approx 14.137$.

Finally, we will look at $\left[2, 9\right]$.

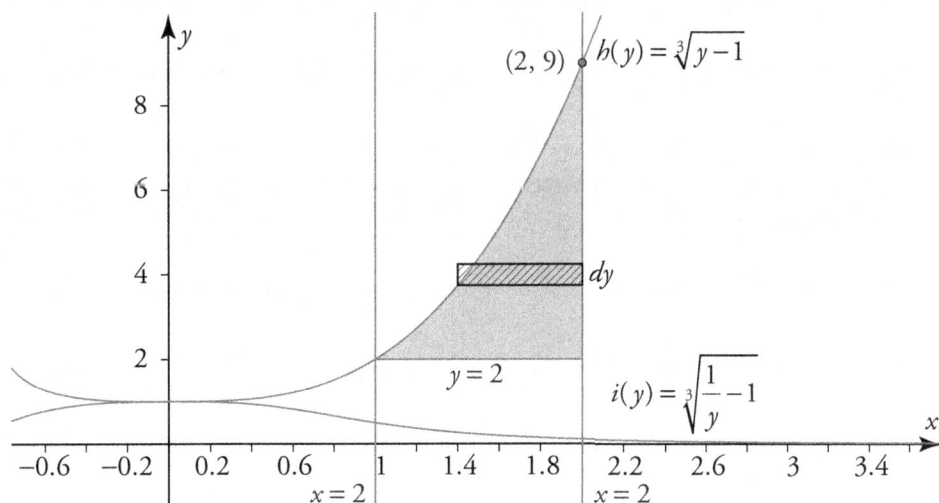

On this interval, the differential volume element $dV_3 = \pi\left(2^2 - \left(h(y)\right)^2\right)dy$. We integrate dV_3 on $[2,9]$ to obtain V_3. Using the calculator, we find that $V_3 \approx 29.531$.

The total volume of the solid of revolution is $V = V_1 + V_2 + V_3 \approx 46.180$.

2.

A. First solve the differential equation for $N(t)$.

$$\frac{dN}{dt} = kN$$

$$\frac{dN}{N} = k\, dt$$

$$\int \frac{dN}{N} = \int k\, dt$$

$$\ln|N| = kt + C$$

$$e^{\ln|N|} = e^{kt+C}$$

$$|N| = e^C e^{kt}$$

$$N = \pm e^C e^{kt}$$

Letting $A = \pm e^C$, we obtain $N(t) = Ae^{kt}$. Next, use the initial data to find A.

$$N(0) = Ae^{0k}$$

$$10{,}000 = A$$

So now we have $N(t) = 10{,}000e^{kt}$. Finally, use the data at $t = 6$ *hrs* to find k.

$$N(6) = 10{,}000e^{6k}$$

$$22{,}500 = 10{,}000e^{6k}$$

$$e^{6k} = 22.5$$

$$\ln\left(e^{6k}\right) = \ln\left(22.5\right)$$

$$6k = \ln\left(22.5\right)$$

$$k = \frac{\ln\left(22.5\right)}{6}$$

So our final result is $N\left(t\right) = 10{,}000e^{\frac{\ln(22.5)}{6}t}$.

B. Now we find t such that $N\left(t\right) = 40{,}000$.

$$10{,}000e^{\frac{\ln(22.5)}{6}t} = 40{,}000$$

$$e^{\frac{\ln(22.5)}{6}t} = 4$$

$$\ln\left(e^{\frac{\ln(22.5)}{6}t}\right) = \ln\left(4\right)$$

$$\frac{\ln\left(22.5\right)}{6}t = \ln\left(4\right)$$

$$t = \frac{6\ln\left(4\right)}{\ln\left(22.5\right)} \approx 2.67\,hrs$$

C. The rate of growth of the population is $N'\left(t\right)$, which is equal to $\frac{\ln\left(22.5\right)}{6}N\left(t\right)$ via the differential equation. To find the rate of growth of the population when the population reaches 40,000, we evaluate $N'\left(t\right)$ at $t = \frac{6\ln\left(4\right)}{\ln\left(22.5\right)}$.

$$N'\left(\frac{6\ln\left(4\right)}{\ln\left(22.5\right)}\right) = \frac{\ln\left(22.5\right)}{6}N\left(\frac{6\ln\left(4\right)}{\ln\left(22.5\right)}\right)$$

$$N'\left(\frac{6\ln\left(4\right)}{\ln\left(22.5\right)}\right) = \frac{\ln\left(22.5\right)}{6}\cdot 40{,}000 \approx 20{,}757\frac{bacteria}{hr}$$

3.

A. Use Implicit Differentiation. Be sure to use the Product Rule when differentiating $-3xy$.

$$\frac{d}{dx}\left(x^3 - 3xy + y^3\right) = \frac{d}{dx}(1)$$

$$3x^2 - 3y - 3xy' + 3y^2 y' = 0$$

Now, algebraically solve for y'.

$$-3\left(x - y^2\right)y' = -3\left(x^2 - y\right)$$

$$y' = \frac{x^2 - y}{x - y^2}$$

B. Use Implicit Differentiation combined with the Quotient Rule.

$$y'' = \frac{d}{dx}\left(\frac{x^2 - y}{x - y^2}\right)$$

$$y'' = \frac{\frac{d}{dx}\left(x^2 - y\right) \cdot \left(x - y^2\right) - \left(x^2 - y\right) \cdot \frac{d}{dx}\left(x - y^2\right)}{\left(x - y^2\right)^2}$$

$$y'' = \frac{\left(2x - y'\right)\left(x - y^2\right) - \left(x^2 - y\right)\left(1 - 2yy'\right)}{\left(x - y^2\right)^2}$$

$$y'' = \frac{2x^2 - 2xy^2 - xy' + y^2 y' - x^2 + 2x^2 yy' + y - 2y^2 y'}{\left(x - y^2\right)^2}$$

$$y'' = \frac{\left(x^2 - 2xy^2 + y\right) + \left(2x^2 y - x - y^2\right)y'}{\left(x - y^2\right)^2}$$

Next we must make the substitution $y' = \frac{x^2 - y}{x - y^2}$.

$$y'' = \frac{\left(x^2 - 2xy^2 + y\right) + \left(2x^2 y - x - y^2\right)\left(\dfrac{x^2 - y}{x - y^2}\right)}{\left(x - y^2\right)^2}$$

$$y'' = \frac{\left(x^2 - 2xy^2 + y\right) + \left(2x^2 y - x - y^2\right)\left(\dfrac{x^2 - y}{x - y^2}\right)}{\left(x - y^2\right)^2} \cdot \frac{x - y^2}{x - y^2}$$

$$y'' = \frac{\left(x^2 - 2xy^2 + y\right)\left(x - y^2\right) + \left(2x^2y - x - y^2\right)\left(x^2 - y\right)}{\left(x - y^2\right)^3}$$

C. First, we find the y-coordinates of the points of tangency by substituting $x = 1$ and solving the equation for y.

$$1^3 - 3(1)y + y^3 = 1$$

$$1 - 3y + y^3 = 1$$

$$y\left(y^2 - 3\right) = 0$$

$$y = -\sqrt{3}, 0, \text{ or } \sqrt{3}$$

We will now find the equations of the three tangent lines.

Point of tangency: $\left(1, -\sqrt{3}\right)$:

$$m = y'\big|_{\left(1, -\sqrt{3}\right)} = \frac{1^2 - \left(-\sqrt{3}\right)}{1 - \left(-\sqrt{3}\right)^2} = -\frac{1 + \sqrt{3}}{2}$$

$$y - \left(-\sqrt{3}\right) = -\frac{1 + \sqrt{3}}{2}(x - 1)$$

$$y = -\frac{1 + \sqrt{3}}{2}(x - 1) - \sqrt{3}$$

Point of tangency: $\left(1, 0\right)$:

$$m = y'\big|_{\left(1,0\right)} = \frac{1^2 - 0}{1 - 0} = 1$$

$$y - 0 = 1(x - 1)$$

$$y = x - 1$$

Point of tangency: $\left(1, \sqrt{3}\right)$:

$$m = y'\big|_{\left(1, \sqrt{3}\right)} = \frac{1^2 - \sqrt{3}}{1 - \sqrt{3}^2} = \frac{\sqrt{3} - 1}{2}$$

$$y - \sqrt{3} = \frac{\sqrt{3} - 1}{2}(x - 1)$$

$$y = \frac{\sqrt{3} - 1}{2}(x - 1) + \sqrt{3}$$

4.

A. First find the critical numbers of f. Since f is differentiable everywhere, the only critical numbers occur when $f'(x) = 0$.

$$f'(x) = 3x^2 - 6x - 9 = 0$$

$$3(x+1)(x-3) = 0 \Rightarrow x = -1 \text{ or } x = 3$$

The intervals of interest are $(-\infty, -1), (-1, 3)$, and $(3, \infty)$. We will choose a test value in each interval to determine the behavior of f on that interval.

Interval	Test Value x	$f'(x)$	Behavior of f
$(-\infty, -1)$	$x = -2$	$f'(-2) = 15 > 0$	Increasing
$(-1, 3)$	$x = 0$	$f'(0) = -9 < 0$	Decreasing
$(3, \infty)$	$x = 4$	$f'(4) = 15 > 0$	Increasing

So f is increasing when x is in $(-\infty, -1)$ or $(3, \infty)$, and f is decreasing when x is in $(-1, 3)$.

B. The absolute extrema can only occur either at an endpoint of $[1, 5]$ or at a critical number of f in $(1, 5)$. Since f only has one critical number in $(1, 5)$, we have only three points to test: $x = 1, x = 3$, and $x = 5$.

$$f(1) = -12$$

$$f(3) = -28$$

$$f(5) = 4$$

So the absolute minimum of f on $[1, 5]$ occurs at the point $(3, -28)$, and the absolute maximum of f on $[1, 5]$ occurs at the point $(5, 4)$.

C. First find the critical numbers of f'. Since f' is differentiable everywhere, the only critical numbers occur when $f''(x) = 0$.

$$f''(x) = 6x - 6 = 0$$

$$6(x - 1) = 0 \Rightarrow x = 1$$

There is a *possible* point of inflection corresponding to $x = 1$. We must check to see whether the concavity of the graph of f changes at $x = 1$.

Interval	Test Value x	$f''(x)$	Behavior of f
$(-\infty, 1)$	$x = 0$	$f''(0) = -6 < 0$	Concave Down
$(1, \infty)$	$x = 2$	$f''(2) = 6 > 0$	Concave Up

Since the graph of f does change concavity at $x = 1$, it has a point of inflection at $(1, -12)$.

5.

A. $F(4) = \int_0^4 \left(2^t + t^2\right) dt$, and $\Delta t = \dfrac{4-0}{4} = 1$. Let $f(t) = 2^t + t^2$. Then $f'(t) = \ln(2)2^t + 2t$. Since $f'(t) > 0$ on all of $[0,4]$, it follows that f is a strictly increasing function on all of $[0,4]$. Thus, the lower sum is obtained by evaluating $f(t)$ at the *left* endpoint of each subinterval. Those endpoints are $x_1 = 0, x_2 = 1, x_3 = 2$, and $x_4 = 3$.

$$F(4) \approx f(0)\Delta t + f(1)\Delta t + f(2)\Delta t + f(3)\Delta t$$

$$F(4) \approx \left[f(0) + f(1) + f(2) + f(3) \right]\Delta t$$

$$F(4) \approx \left[1 + 3 + 8 + 17 \right] \cdot 1 = 29$$

B. $F(0) = \int_0^0 \left(2^t + t^2\right) dt = 0$, because the interval on which we are integrating contains only one point. We will evaluate $F(4)$ using the Fundamental Theorem of Calculus.

$$F(4) = \int_0^4 \left(2^t + t^2\right) dt$$

$$F(4) = \left(\ln(2)2^t + \frac{1}{3}t^3 \right)\Bigg|_0^4$$

$$F(4) = \left(\ln(2)2^4 + \frac{1}{3} \cdot 4^3 \right) - \left(\ln(2)0^4 + \frac{1}{3} \cdot 0^3 \right)$$

$$F(4) = 16\ln(2) + \frac{64}{3}$$

C. Use the Second Fundamental Theorem of Calculus.

$$F'(x) = \frac{d}{dx} \int_0^x \left(2^t + t^2\right) dt$$

$$F'(x) = 2^x + x^2$$

D. We will denote the average value of $F'(x)$ on $[0,4]$ by $\overline{F'}$.

$$\overline{F'} = \frac{1}{4-0} \int_0^4 F'(x)\, dx$$

$$\overline{F'} = \frac{1}{4} \int_0^4 \left(2^x + x^2\right) dx$$

$$\overline{F'} = \frac{1}{4}\left(F(4) - F(0)\right)$$

Using our results from part B, we obtain the following:

$$\overline{F'} = \frac{1}{4}\left(16\ln(2) + \frac{64}{3} - 0 \right)$$

$$\overline{F'} = 4\ln(2) + \frac{16}{3}$$

6.

A. We find $x(t)$ by integrating $v(t)$.

$$x(t) = \int \left(2t^2 - 5t + 3\right) dt$$

$$x(t) = \frac{2}{3}t^3 - \frac{5}{2}t^2 + 3t + C$$

Now we use the fact that $x = 3m$ at $t = 0s$ to find C.

$$x(0) = \frac{2}{3} \cdot 0^3 - \frac{5}{2} \cdot 0^2 + 3 \cdot 0 + C$$

$$3 = C$$

So our final result is $x(t) = \frac{2}{3}t^3 - \frac{5}{2}t^2 + 3t + 3$.

B. We find $a(t)$ by differentiating $v(t)$.

$$a(t) = \frac{d}{dt}\left(2t^2 - 5t + 3\right)$$

$$a(t) = 4t - 5$$

C. The direction of motion of the particle changes whenever $v(t)$ changes signs. We begin by finding the zeros of v.

$$v(t) = 2t^2 - 5t + 3 = 0$$

$$(t - 1)(2t - 3) = 0 \Rightarrow t = 1 \text{ or } t = \frac{3}{2}.$$

We will now check the sign of $v(t)$ on the intervals between the zeros of v.

Interval	Test Value t	$v(t)$
$(-\infty, 1)$	$t = 0$	$v(0) = 3 > 0$
$(1, 3/2)$	$t = 5/4$	$v(5/4) = -1/8 < 0$
$(3/2, \infty)$	$t = 2$	$v(2) = 1 > 0$

So the direction of the motion of the particle changes at $t = 1s$ and at $t = \frac{3}{2}s$.

AP Calculus AB Practice Examination 2

CALCULUS AB

A CALCULATOR **CANNOT BE USED ON PART A OF SECTION I OR ON PART B OF SECTION II**. A GRAPHING CALCULATOR FROM THE APPROVED LIST **IS REQUIRED ON PART B OF SECTION I AND ON PART A OF SECTION II** OF THE EXAMINATION. CALCULATOR MEMORIES NEED NOT BE CLEARED. COMPUTERS, NONGRAPHING SCIENTIFIC CALCULATORS, CALCULATORS WITH QWERTY KEYBOARDS, AND ELECTRONIC WRITING PADS ARE NOT ALLOWED. CALCULATORS MAY NOT BE SHARED AND COMMUNICATION BETWEEN CALCULATORS IS PROHIBITED DURING THE EXAMINATION. ATTEMPTS TO REMOVE TEST MATERIALS FROM THE ROOM BY ANY METHOD WILL RESULT IN INVALIDATION OF TEST SCORES.

CALCULUS AB – SECTION I

Time – 1 hour and 45 minutes
All questions are given equal weight.
Percent of total grade – 50

Part A: 55 minutes, 28 multiple-choice questions
A calculator is NOT allowed

Part B: 50 minutes, 17 multiple-choice questions
A graphing calculator is required.

Parts A and B of Section I are printed in this examination booklet. Section II, which consists of longer problems, is printed in a separate booklet.

GENERAL INSTRUCTIONS

DO NOT OPEN THIS BOOKLET UNTIL YOU ARE TOLD TO DO SO.

INDICATE YOUR ANSWERS TO QUESTIONS IN PART A ON PAGE 2 OF THE SEPARATE ANSWER SHEET. THE ANSWERS TO QUESTIONS IN PART B SHOULD BE INDICATED ON PAGE 3 OF THE ANSWER SHEET. No credit will be given for anything written in this examination booklet, but you may use the booklet for notes or scratchwork. After you have decided which of the suggested answers is best, COMPLETELY fill in the corresponding oval on the answer sheet. Give only one answer to each question. If you change an answer, be sure that the previous mark is erased completely.

Example: Sample Answer: (A) (B) (C) ● (E)

1. What is the value of the numerical expression $4 - 2(3 - 5)$?
   A. -12
   B. -8
   C. 0
   D. 8
   E. 12

Many candidates wonder whether or not to guess the answers to questions about which they are not certain. It is improbable, that mere guessing will improve your score significantly; it may even lower your score, and it does take time. If, however, you are not sure of the best answer but have some knowledge of the question and are able to eliminate one or more of the answer choices as wrong, your chance of answering correctly is improved, and it may be to your advantage to answer such a question.

Use your time effectively, working as rapidly as you can without losing accuracy. Do not spend too much time on questions that are too difficult. Go on to other questions and come back to the difficult ones later if you have time. It is not expected that you will be able to answer all of the multiple choice questions.

Directions: Solve each of the following problems, using the available space for scratchwork. After examining the form of the choices, decide which is the best of the choices given and fill in the corresponding oval on the answer sheet. No credit will be given for anything written in the test book. Do not spend too much time on any one problem.

In this test: Unless otherwise specified, the domain of a function *f* is assumed to be the set of all real numbers *x* for which $f(x)$ is a real number.

1. What are the coordinates of the point of inflection of the graph of $y = x^3 - 12x^2 + 15x + 45$?

   A. $(0, 45)$

   B. $(1, 49)$

   C. $(4, -23)$

   D. $(4, 23)$

   E. $(6, -81)$

2. If $x^2 y + 2x - 3y = -6$, then when $x = 3$, $\dfrac{dy}{dx} =$

   A. -2

   B. $-\dfrac{5}{3}$

   C. 0

   D. $\dfrac{5}{3}$

   E. 2

3. If *f* is continuous on the interval $[a, b]$ and differentiable on the interval (a, b), then each of the following must be true *except*

   A. For some c in (a, b), $f(c) = \dfrac{f(b) - f(a)}{b - a}$.

   B. For some c in (a, b), $f'(c) = \dfrac{f(b) - f(a)}{b - a}$.

   C. For all c in (a, b), $\lim\limits_{x \to c} f(x) = f(c)$

   D. *f* has an absolute maximum on $[a, b]$.

   E. *f* has an absolute minimum on $[a, b]$.

GO ON TO THE NEXT PAGE.

4. If $f(x) = \dfrac{\sqrt{x^2+4}}{x}$, then $f'(x) =$

   A. $\sqrt{x^2+4}$

   B. $\dfrac{x}{\sqrt{x^2+4}}$

   C. $\dfrac{-4}{x^2\sqrt{x^2+4}}$

   D. $\dfrac{1}{2\sqrt{x^2+4}}$

   E. $\dfrac{-2x^2+x-8}{x^2\sqrt{x^2+4}}$

5. What is the instantaneous rate of change of $f(x) = x\sin(x)$ at $x = \pi$?

   A. $-\pi$

   B. -1

   C. 0

   D. 1

   E. π

6. $\dfrac{d}{dx}\left[\sec^2\left(x^2\right)\right] =$

   A. $\sec(2x)$

   B. $\sec^2(2x)\tan(2x)$

   C. $4x\sec\left(x^2\right)$

   D. $\sec^2\left(x^2\right)\tan\left(x^2\right)$

   E. $4x\sec^2\left(x^2\right)\tan\left(x^2\right)$

GO ON TO THE NEXT PAGE.

7. A particle moves in a straight line. The velocity v (in $\frac{m}{s}$) of the particle as a function of time t (in s) is shown in the graph below. How far (in m) does the particle travel from $t = 0\ s$ to $t = 20\ s$?

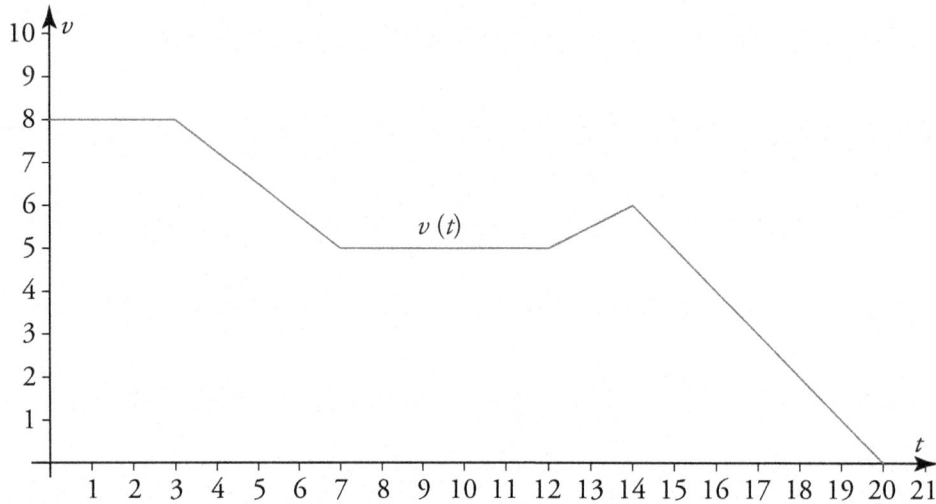

A. 100

B. 102

C. 104

D. 106

E. 108

8. $\int_0^x 2te^{-t^2}\,dt$

A. $1 - e^{-x^2}$

B. $-e^{-x^2}$

C. $2xe^{-x^2}$

D. $\left(2 - 4x^4\right)e^{-x^2}$

E. $x^4 - 2x^2e^{-x^2}$

GO ON TO THE NEXT PAGE.

9. $\int_{\pi/6}^{\pi/3} \tan(x)\,dx$

   A. $\ln\sqrt{3}$

   B. $-\ln\left(\sqrt{3}\right)$

   C. $\dfrac{2\sqrt{3}}{3}$

   D. $-\dfrac{2\sqrt{3}}{3}$

   E. 0

10. If $f(x) = \int_{0}^{x} \dfrac{t}{t^2+1}\,dt$, then $f'(1) =$

   A. 0

   B. $\dfrac{1}{2}$

   C. $\dfrac{1}{2x}$

   D. $\dfrac{1}{2}x$

   E. $\dfrac{1}{2}\ln\left(x^2+1\right)$

11. $\int \sin(2x)\,e^{\cos^2(x)}\,dx =$

   A. $-e^{\cos^2(x)}+C$

   B. $e^{\cos^2(x)}+C$

   C. $-\dfrac{1}{2}\cos(2x)\,e^{\cos^2(x)}+C$

   D. $\dfrac{1}{2}\cos(2x)\,e^{\cos^2(x)}+C$

   E. $\ln\left|\sin(2x)\,e^{\cos^2(x)}\right|+C$

GO ON TO THE NEXT PAGE.

12. The graph of a piecewise linear function f is shown below. What is the average value of f on the interval $[0,18]$?

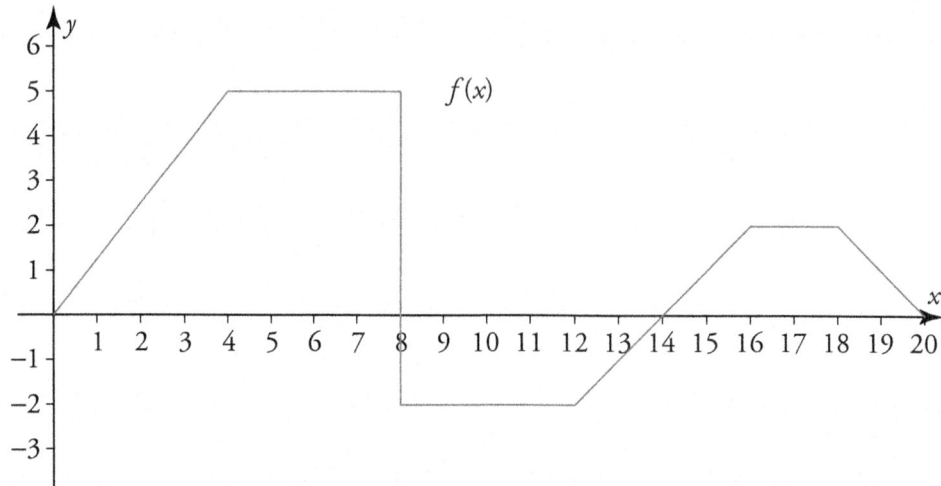

A. $\dfrac{5}{9}$

B. $\dfrac{14}{9}$

C. $\dfrac{19}{9}$

D. 28

E. 38

13. $\displaystyle\int_{3}^{e+2} \frac{1}{x-2}\,dx$

A. $-e$

B. -1

C. 0

D. 1

E. e

GO ON TO THE NEXT PAGE.

14. $\int \dfrac{\sin\left(\sqrt{x}\right)}{\sqrt{x}}\,dx =$

A. $-2\cos\left(\sqrt{x}\right)+C$

B. $\cos\left(\sqrt{x}\right)+C$

C. $\cos\left(\sqrt{x}\right)+C$

D. $2\cos\left(\sqrt{x}\right)+C$

E. $\dfrac{2\cos\left(\sqrt{x}\right)}{\sqrt{x}}+C$

15. $\lim\limits_{x\to-3}\dfrac{x+5}{x^2+8x+15}$

A. $-\dfrac{1}{6}$

B. 0

C. $\dfrac{1}{6}$

D. $\dfrac{1}{5}$

E. The limit does not exist.

16. What is the equation of the line tangent to the graph of $f(x)=\tan\left(\dfrac{x}{4}\right)$ at $x=\pi$?

A. $y-1=\dfrac{1}{2}(x-\pi)$

B. $y-1=\dfrac{1}{2}(x+\pi)$

C. $y+1=\dfrac{1}{2}(x-\pi)$

D. $y-1=2(x-\pi)$

E. $y+1=-2(x-\pi)$

GO ON TO THE NEXT PAGE.

17. The function $f(x) = x^3 - 3x^2 - 9x$ is increasing for

A. $x < -1$ only

B. $x < -1$ or $x > 3$

C. $x > 3$ only

D. $-1 < x < 3$

E. None of the above

18. $\lim\limits_{x \to 0} \dfrac{e^{2x} - 1}{3x}$

A. 0

B. $\dfrac{1}{3}$

C. $\dfrac{2}{3}$

D. $\dfrac{e^2}{3}$

E. $\dfrac{2e^2}{3}$

19. The graph of $f(x) = x^4 + 4x^3 - 48x^2 + 8x - 12$ is concave downwards for

A. $x < -4$ only

B. $x < -4$ or $x > 2$

C. $x > 2$ only

D. $-4 < x < 2$

E. None of the above

20. If $f(x) = \left(\tan^{-1}(3x)\right)^2$, then $f'(x) =$

A. $\dfrac{6\tan^{-1}(3x)}{1 + 9x^2}$

B. $\dfrac{3\tan^{-1}(3x)}{1 + 9x^2}$

C. $\dfrac{2\tan^{-1}(3x)}{1 + 9x^2}$

D. $-6\cot(3x)\csc^2(3x)$

E. $6\cot(3x)\csc^2(3x)$

GO ON TO THE NEXT PAGE.

21. $\lim\limits_{h \to 0} \dfrac{\ln\left(e^2 + h\right) - 2}{h}$

   A. -2

   B. 2

   C. e^2

   D. $\dfrac{1}{e^2}$

   E. $2e$

22. If $f(x) = \cot(\pi x)$, then $f'\left(\dfrac{1}{3}\right) =$

   A. $-\dfrac{4\pi}{3}$

   B. $-\dfrac{3\pi}{4}$

   C. $-\dfrac{4}{3}$

   D. $\dfrac{4}{3}$

   E. $\dfrac{4\pi}{3}$

23. Let $f(x) = x^3 - 2x$ on $[1,3]$. The number c in $(1,3)$ which is guaranteed by the Mean Value Theorem for derivatives is

   A. $-\sqrt{\dfrac{13}{3}}$

   B. $-\sqrt{\dfrac{2}{3}}$

   C. 0

   D. $\sqrt{\dfrac{2}{3}}$

   E. $\sqrt{\dfrac{13}{3}}$

GO ON TO THE NEXT PAGE.

24. A particle moves along the *x*-axis. Its displacement *x* at time *t* is given by $x(t) = t^3 - 9t^2 + 15t + 3$. At what time(s) is the acceleration of the particle zero?

A. 1

B. 3

C. 5

D. 3 or 5

E. 1 or 3 or 5

25. The sum of two positive numbers is 20. If the product of the square of one number and the cube of the other is a maximum, then the two numbers are

A. 10 and 10

B. 0 and 20

C. 8 and 12

D. 2 and 18

E. 5 and 15

26. The area of the region bounded by the graphs of $f(x) = x^2$ and $g(x) = \sqrt{x}$ is

A. $\dfrac{1}{3}$

B. $\dfrac{2}{3}$

C. 1

D. $\dfrac{4}{3}$

E. $\dfrac{5}{3}$

GO ON TO THE NEXT PAGE.

27. The graph of a differentiable function $f(x)$ is shown in the figure below.

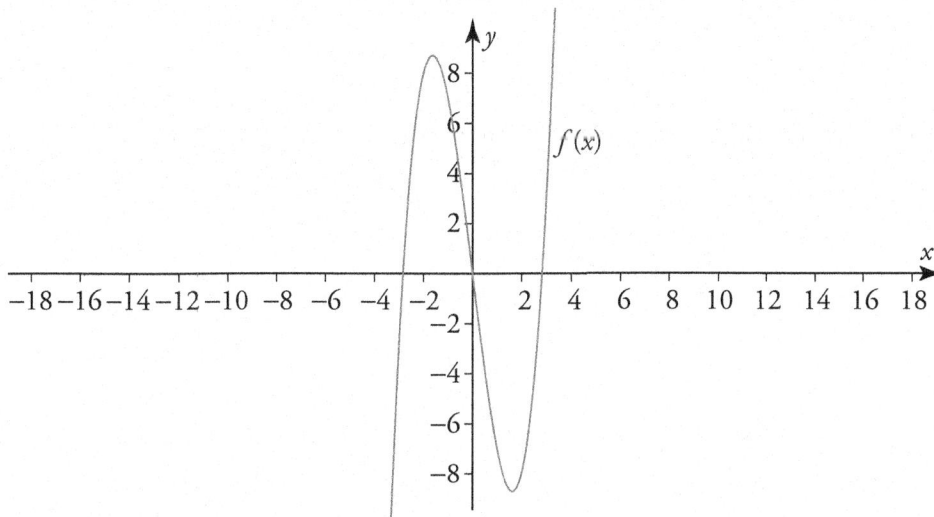

Which of the following is the graph of $f'(x)$?

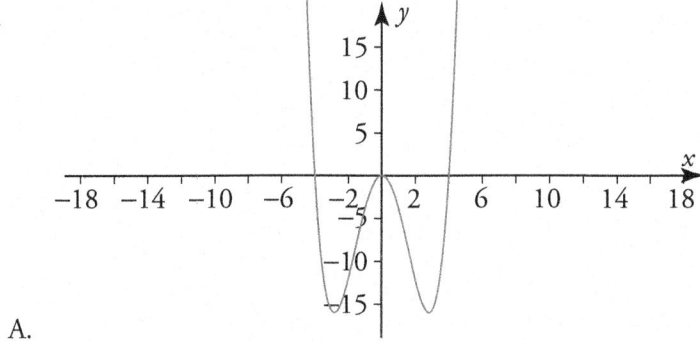

A.

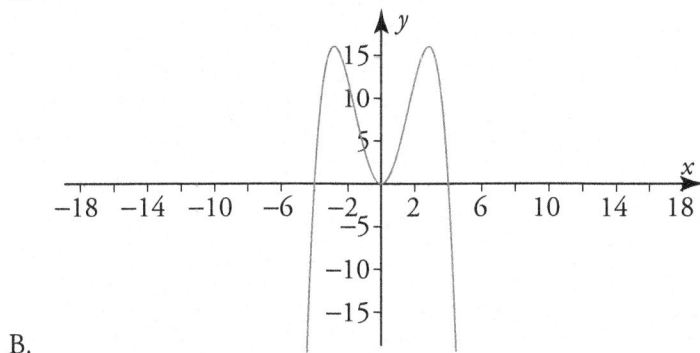

B.

GO ON TO THE NEXT PAGE.

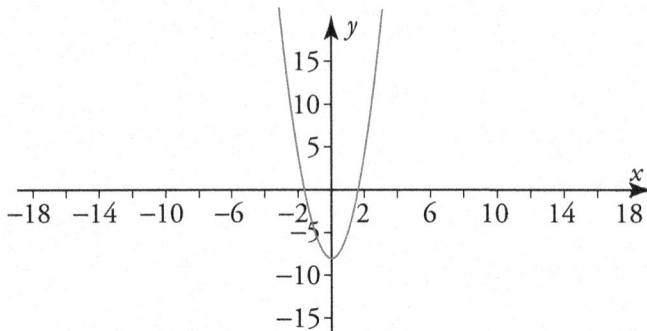

C.

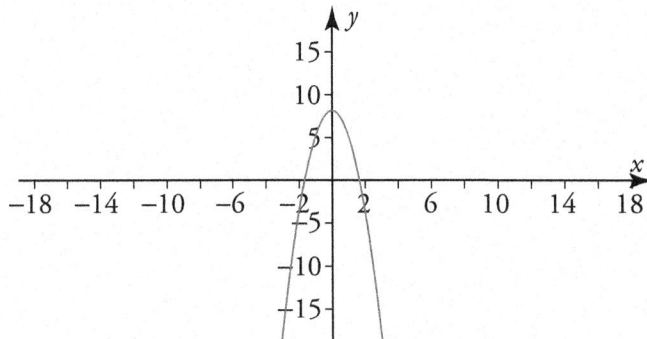

D.

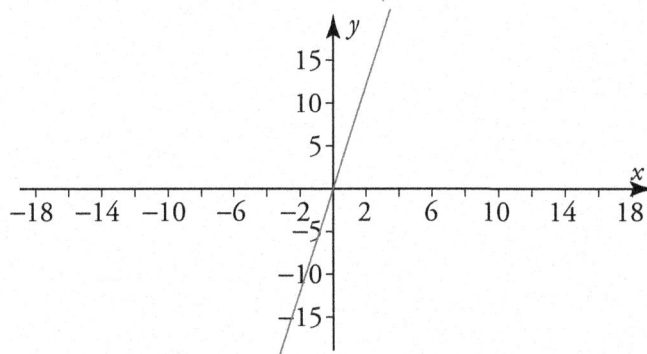

E.

28. Let R be the region bounded by the graphs of $f(x) = 2x, x = 6,$ and $y = 0$. The volume of the solid generated by revolving R about the y-axis is

A. 144

B. 288

C. 144π

D. 288π

E. 432π

STOP

END OF PART A, SECTION I
IF YOU FINISH BEFORE TIME IS CALLED, YOU MAY CHECK YOUR WORK ON PART A ONLY.

DO NOT GO ON TO PART B UNTIL YOU ARE TOLD TO DO SO.

CALCULUS AB

SECTION I, Part B

Time – 50 Minutes

Number of questions – 17

A GRAPHING CALCULATOR IS REQUIRED FOR SOME
QUESTIONS ON THIS PART OF THE EXAMINATION

Directions: Solve each of the following problems, using the available space for scratchwork. After examining the form of the choices, decide which is the best of the choices given and fill in the corresponding oval on the answer sheet. No credit will be given for anything written in the test book. Do not spend too much time on any one problem.

In this test:

1. The **exact** numerical value of the correct answer does not always appear among the choices given. When this happens, select from among the choices the number that best approximates the exact numerical value.

2. Unless otherwise specified, the domain of a function *f* is assumed to be the set of all real numbers *x* for which $f(x)$ is a real number.

29. A particle moves in a straight line. It's displacement *s* (in *m*) at time *t* (in *s*) is given by
$s(t) = 3t^3 - 18t^2 + 10t + 28$ for $0 \le t \le 4$. The maximum *speed* (in *m/s*) of the particle is

A. 28

B. 26

C. 10

D. –26

E. –28

30. The average value of $f(x) = \sin(x^2)$ on the interval $\left[0, \sqrt{\dfrac{\pi}{2}}\right]$ is

A. 0

B. 0.438

C. 0.549

D. 0.798

E. 1

GO ON TO THE NEXT PAGE.

31. $\dfrac{d}{dx}\displaystyle\int_1^{3x}\dfrac{1}{t^3+1}\,dt =$

   A. $\dfrac{3}{27x^3+1}$

   B. $\dfrac{1}{27x^3+1}$

   C. $3\tan^{-1}(3x)$

   D. $\tan^{-1}(3x)$

   E. $-\dfrac{3}{\left(27x^3+1\right)^2}$

32. $\displaystyle\int_2^4 e^{x/2}\,dx =$

   A. 4.67

   B. 9.34

   C. 12.70

   D. 25.39

   E. 94.42

33. Find $\displaystyle\int_0^1 \cos^2(x)\,dx$ to the nearest thousandth using the Trapezoidal Rule with n = 4.

   A. 0.634

   B. 0.723

   C. 0.727

   D. 0.781

   E. 0.811

34. The volume of a sphere is increasing at a rate proportional to its volume at any time t. The volume of the sphere is $0.5m^3$ at $t = 0$ s and $1.2m^3$ at $t = 5$ s. To the nearest hundredth, its volume (in m^3) at $t = 10$ s will be

   A. 1.70

   B. 1.90

   C. 2.40

   D. 2.88

   E. 3.38

GO ON TO THE NEXT PAGE.

35. The slope of the normal line to the graph of $x^2 + 2y^2 = 2$ in the fourth quadrant at $x = 1$ is

   A. -1.414

   B. $-0,707$

   C. 0

   D. 0.707

   E. 1.414

36. The graph of $f(x) = x^3 + 5x^2 - 3x + 1$ has a local minimum at

   A. $(-3.61, 0.57)$

   B. $(-3.61, 29.9)$

   C. $(0.28, 0)$

   D. $(0.28, 0.57)$

   E. $(0.28, 29.9)$

37. The radius of a sphere is increasing at a constant rate of $0.2 \frac{m}{s}$. When the radius of the sphere is 2 m, its volume is increasing at what rate (in $\frac{m^3}{s}$)?

   A. 10.1

   B. 50.3

   C. 53.6

   D. 251.3

   E. 268.1

38. If $f(x) = \dfrac{x}{x^2 + 9}$, then $f''(2) =$

   A. -0.042

   B. -0.030

   C. 0

   D. 0.030

   E. 0.042

GO ON TO THE NEXT PAGE.

39. If $f(x)$ is differentiable at $x = 1$, and $f(x) = \begin{cases} ax^3 - 2x & x \leq 1 \\ bx^2 + x & x > 1 \end{cases}$, then $b =$

A. 6

B. 3

C. 0

D. –3

E. –6

40. A projectile is thrown straight upwards, and its height y (in m) at time $t \geq 0$ (in s) is given by $y(t) = -4.9t^2 + 50t$. What is the velocity (in $\frac{m}{s}$) of the projectile at $t = 8.2\ s$?

A. –80.52

B. –30.36

C. –9.80

D. 30.36

E. 80.52

41. $\lim\limits_{x \to \infty} \dfrac{8 - 3x^2}{5x^2 + 2x - 1}$

A. –8

B. $-\dfrac{3}{5}$

C. $\dfrac{8}{5}$

D. 3

E. ∞

42. $\int \sec^5(x)\tan(x)\,dx =$

A. $\dfrac{1}{6}\sec^6(x) + C$

B. $\dfrac{1}{6}\sec^6(x) \cdot \ln|\sec(x)| + C$

C. $\dfrac{1}{5}\sec^5(x) + C$

D. $\dfrac{1}{5}\sec^5(x) \cdot \ln|\sec(x)| + C$

E. $5\sec^5(x)\tan^2(x) + \sec^7(x) + C$

GO ON TO THE NEXT PAGE.

43. $\int \dfrac{\sin\left(\dfrac{1}{x^2}\right)}{x^3}\,dx =$

A. $-\dfrac{1}{2}\cos\left(\dfrac{1}{x^2}\right)+C$

B. $-\cos\left(\dfrac{1}{x^2}\right)+C$

C. $\dfrac{1}{2}\cos\left(\dfrac{1}{x^2}\right)+C$

D. $\cos\left(\dfrac{1}{x^2}\right)+C$

E. $\dfrac{4\sin\left(\dfrac{-2}{x^3}\right)}{x^4}+C$

44. Find the area of the region bounded by the graphs of $f(x)=\sin(x)$ and $g(x)=\dfrac{2\sqrt{2}}{3\pi}x$.

A. 0

B. 0.874

C. 1.748

D. 2.622

E. 3.496

45. A particle travels in a straight line, and its velocity v (in $\frac{m}{s}$) is given as a function of time t (in s) by $v(t)=te^{-t^2/2}$ (for $t \geq 0$ s). The distance (in m) traveled by the particle from $t = 2$ s to $t = 4$ s is

A. 0.067

B. 0.135

C. 0.202

D. 0.270

E. 0.337

STOP

END OF PART B, SECTION I
IF YOU FINISH BEFORE TIME IS CALLED, YOU MAY CHECK YOUR WORK ON PART B ONLY.

DO NOT GO ON TO SECTION II UNTIL YOU ARE TOLD TO DO SO.

CALCULUS AB – SECTION II

Time – 1 hour and 30 minutes
Number of problems – 6
Percent of total grade – 50

GENERAL INSTRUCTIONS

You may wish to look over the problems before starting to work on them, since it is not expected that everyone will be able to complete all parts of all problems. All problems are given equal weight, but the parts of a particular problem are not necessarily given equal weight.

A GRAPHING CALCULATOR IS REQUIRED FOR SOME PROBLEMS OR
PARTS OF PROBLEMS ON THIS SECTION OF THE EXAMINATION.

- You should write all work for each part of each problem in the space provided for that part in the booklet. Be sure to write clearly and legibly. If you make an error, you may save time by crossing it out rather than trying to erase it. Erased or crossed-out work will not be graded.

- Show all your work. You will be graded on the correctness and completeness of your methods as well as your answers. Correct answers without supporting work may not receive credit.

- Justifications require that you give mathematical (noncalculator) reasons and that you clearly identify functions, graphs, tables, or other objects you use.

- You are permitted to use your calculator to solve an equation, find the derivative of a function at a point, or calculate the value of a definite integral. However, you must clearly indicate the setup of your problem, namely the equation, function, or integral you are using. If you use other built-in features or programs, you must show the mathematical steps necessary to produce your results.

- Your work must be expressed in standard mathematical notation rather than calculator syntax. For example, $\int_{1}^{5} x^2 dx$ may not be written as *fnInt*(X^2,X,1,5).

- Unless otherwise specified, answers (numeric or algebraic) need not be simplified. If your answer is given as a decimal approximation, it should be correct to three places after the decimal point.

- Unless otherwise specified, the domain of a function *f* is assumed to be the set of all real numbers *x* for which *f* (*x*) is a real number.

A graphing calculator is required for some problems or parts of problems.

During the timed portion for Part A, you may work only on the problems in Part A.

On Part A, you are permitted to use your calculator to solve an equation, find the derivative of a function at a point, or calculate the value of a definite integral. However, you must clearly indicate the setup of your problem, namely the equation, function, or integral you are using. If you use other built-in features or programs, you must show the mathematical steps necessary to produce your results.

1. Consider the integral $I = \int_0^\pi \dfrac{\sin(x)}{x}\,dx$. For all parts, round your answer to the nearest thousandth.

    A. Use a lower sum with $n = 4$ to approximate I.

    B. Use an upper sum with $n = 4$ to approximate I.

    C. Use the trapezoidal rule with $n = 4$ to approximate I.

GO ON TO THE NEXT PAGE.

2. A particle moves along the x-axis such that its acceleration a (in $\frac{m}{s^2}$) at time $t > 0$ s is given by $a(t) = 2t - 7$. At time $t = 2$ s the velocity v of the particle is $1\frac{m}{s}$, and the position x of the particle is 8 m.

A. Find $v(t)$, the velocity of the particle as a function of time.

B. At what time(s) does the particle change direction? Round your answer(s) to the nearest thousandth of a second.

C. Find $x(t)$, the position of the particle as a function of time.

D. Find the distance that the particle travels from $t = 2$ s to $t = 4$ s. Round your answer to the nearest thousandth of a meter.

STOP

END OF PART A, SECTION II
IF YOU FINISH BEFORE TIME IS CALLED, YOU MAY CHECK YOUR WORK ON PART A ONLY.
DO NOT GO ON TO PART B UNTIL YOU ARE TOLD TO DO SO.

During the timed portion for Part B, you may continue to work on the problems in Part A without the use of any calculator.

3. Consider the function $f(x) = -2x^4 + 16x^2$

    A. Find the equation of the line normal to the graph of f at $x = 1$.

    B. Find the coordinates of any relative maxima of f.

    C. Find the coordinates of any relative minima of f.

GO ON TO THE NEXT PAGE.

4. Let $F(x) = \int_0^x f(t)\,dt$ be defined on the interval $[0,16]$, where $f(t)$ is graphed below.

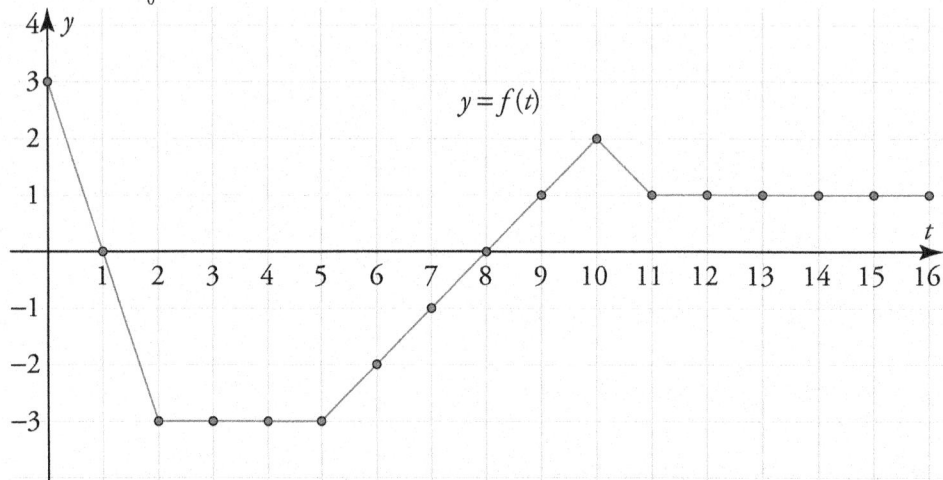

$y = f(t)$

A. Find and identify the open intervals on which F is increasing.

B. Find and identify the open intervals on which F is decreasing.

C. Find the coordinates of the absolute extrema of F on the interval $[0,16]$.

GO ON TO THE NEXT PAGE.

5. Let R be the region bounded by the graphs of $f(x) = 1 - x^2$ and $g(x) = x^2 - 2x + 1$.

A. Find the area of R.

B. Find the volume of the solid generated when R is revolved about the x-axis.

C. Find the volume of the solid generated when R is revolved about the line $y = -1$.

GO ON TO THE NEXT PAGE.

6. Consider the function $f(x) = \dfrac{2}{x-2} + 3$.

A. Find the average rate of change of f on the interval $[3,5]$.

B. Find $f'(4)$.

C. Find all values of c that satisfy the Mean Value Theorem for Derivatives for f on the interval $[3,5]$.

STOP

END OF EXAM

Answer Key – Section I

1. C	10. B	19. D	28. D	37. A
2. D	11. A	20. A	29. B	38. D
3. A	12. B	21. D	30. B	39. D
4. C	13. D	22. B	31. A	40. B
5. A	14. A	23. E	32. B	41. B
6. E	15. E	24. B	33. B	42. C
7. C	16. A	25. C	34. D	43. C
8. A	17. B	26. A	35. A	44. C
9. A	18. C	27. D	36. D	45. B

Solutions

Solutions – Section I, Part A

1. **Answer: C**

   First find y''.

   $$y' = \frac{dy}{dx} = \frac{d}{dx}\left(x^3 - 12x^2 + 15x + 45\right) = 3x^2 - 24x + 15$$

   $$y'' = \frac{dy'}{dx} = \frac{d}{dx}\left(3x^2 - 24x + 15\right) = 6x - 24$$

   Since y is twice differentiable everywhere, all critical points correspond to values of x for which $y'' = 0$.
   $6x - 24 = 0 \Rightarrow x = 4$

   We will test this value of x by checking the second derivative on either side of it.

Interval	Test Value x	$y''(x)$	Behavior of Graph of $y(x)$
$(-\infty, 4)$	$x = 3$	$y''(3) = -6$	Concave Down
$(4, \infty)$	$x = 5$	$y''(5) = 6$	Concave Up

   Since the graph of $y(x)$ changes concavity at $x = 4$, the graph has a point of inflection corresponding to this value of x. Since $y(4) = -23$, the point of inflection is at the point $(4, -23)$.

2. **Answer: D**

   First, find y when $x = 3$.
   $$3^2 y + 2(3) - 3y = -6$$
   $$9y + 6 - 3y = -6$$
   $$6y = -12 \Rightarrow y = -2$$

   So the point on the graph at which we will evaluate y' is $(3, -2)$.

   Next, use Implicit Differentiation to find y'. Be sure to use the Product Rule when differentiating $x^2 y$.

   $$\frac{d}{dx}\left(x^2 y + 2x - 3y\right) = \frac{d}{dx}(-6)$$
   $$2xy + x^2 y' + 2 - 3y' = 0$$
   $$\left(x^2 - 3\right)y' = -2xy - 2$$
   $$y' = \frac{-2xy - 2}{x^2 - 3}$$

Finally, evaluate y' at $(3,-2)$.

$$y'\big|_{(3,-2)} = \frac{-2(3)(-2)-2}{3^2-3} = \frac{5}{3}$$

3. Answer: A

The Mean Value Theorem for Derivatives implies that B is true, the definition of continuity implies that C is true, and the Extreme Value Theorem implies that both D and E are true.

4. Answer: C

$$f'(x) = \frac{d}{dx}\left(\frac{\sqrt{x^2+4}}{x}\right)$$

$$f'(x) = \frac{d}{dx}\left(\frac{(x^2+4)^{1/2}}{x}\right)$$

Use the Quotient Rule and the Chain Rule.

$$f'(x) = \frac{\frac{d}{dx}\left((x^2+4)^{1/2}\right) \cdot x - (x^2+4)^{1/2} \cdot \frac{d}{dx}(x)}{x^2}$$

$$f'(x) = \frac{\frac{1}{2}(x^2+4)^{-1/2} \cdot \frac{d}{dx}(x^2+4) \cdot x - (x^2+4)^{1/2} \cdot 1}{x^2}$$

$$f'(x) = \frac{\frac{1}{2}(x^2+4)^{-1/2} \cdot 2x \cdot x - (x^2+4)^{1/2}}{x^2}$$

$$f'(x) = \frac{\dfrac{x^2}{\sqrt{x^2+4}} - \sqrt{x^2+4}}{x^2}$$

$$f'(x) = \frac{\dfrac{x^2}{\sqrt{x^2+4}} - \sqrt{x^2+4}}{x^2} \cdot \frac{\sqrt{x^2+4}}{\sqrt{x^2+4}}$$

$$f'(x) = \frac{x^2 - (x^2+4)}{x^2\sqrt{x^2+4}}$$

$$f'(x) = \frac{-4}{x^2\sqrt{x^2+4}}$$

5. Answer: A

First, find $f'(x)$. Use the Product Rule.

$$f'(x) = \frac{d}{dx}\big(x\sin(x)\big)$$

$$f'(x) = \frac{d}{dx}(x)\cdot\sin(x) + x\cdot\frac{d}{dx}\big(\sin(x)\big)$$

$$f'(x) = \sin(x) + x\cos(x)$$

Second, compute $f'(\pi)$.
$$f'(\pi) = \sin(\pi) + \pi\cos(\pi) = -\pi$$

6. Answer: E

$$f'(x) = \frac{d}{dx}\Big(\sec^2\big(x^2\big)\Big)$$

$$f'(x) = \frac{d}{dx}\Big(\big(\sec\big(x^2\big)\big)^2\Big)$$

Use the Chain Rule twice, as this is a doubly composite function.

$$f'(x) = 2\sec\big(x^2\big)\cdot\frac{d}{dx}\big(\sec\big(x^2\big)\big)$$

$$f'(x) = 2\sec\big(x^2\big)\cdot\sec\big(x^2\big)\tan\big(x^2\big)\cdot\frac{d}{dx}\big(x^2\big)$$

$$f'(x) = 2\sec\big(x^2\big)\cdot\sec\big(x^2\big)\tan\big(x^2\big)\cdot 2x$$

$$f'(x) = 4x\sec^2\big(x^2\big)\tan\big(x^2\big)$$

7. Answer: C

Since $v(t) \geq 0$ for $0 \leq t \leq 20$, the distance traveled by the particle is equal to $\int_0^{20} v(t)\,dt$, which is equal to the area of the region bounded by the graph of $v(t)$, the x-axis, and the lines $t = 0$ and $t = 20$. This area is most easily found by partitioning the region into rectangles and triangles, as shown below.

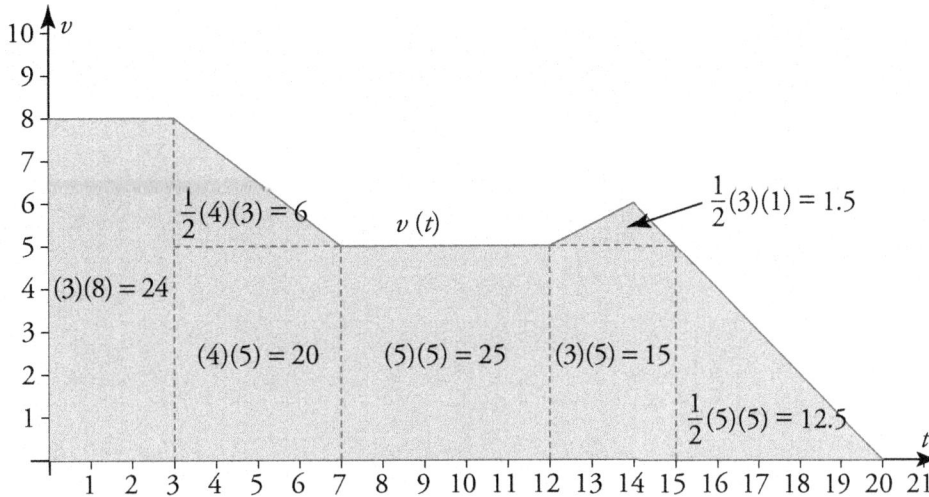

The area of each sub-region has been labeled in the figure. Summing the areas, we get the following:

$$\int_0^{20} v(t)\,dt = 24 + 6 + 20 + 25 + 1.5 + 15 + 12.5 = 104\,m$$

8. **Answer: A**

Let $u = -t^2$. Then $du = -2t\,dt$, or $2t\,dt = -du$. Also, $t = 0 \Rightarrow u = 0$, and $t = x \Rightarrow u = -x^2$.

$$\int_0^x 2te^{-t^2}\,dt = -\int_0^{-x^2} e^u\,du$$

Now use the Fundamental Theorem of Calculus.

$$\int_0^x 2te^{-t^2}\,dt = -\left(e^u\right)\Big|_0^{-x^2}$$

$$\int_0^x 2te^{-t^2}\,dt = -\left(e^{-x^2} - e^0\right)$$

$$\int_0^x 2te^{-t^2}\,dt = 1 - e^{-x^2}$$

9. Answer: A

Use the Fundamental Theorem of Calculus.

$$\int_{\pi/6}^{\pi/3} \tan(x)\, dx = -\left(\ln|\cos(x)|\right)\Big|_{\pi/6}^{\pi/3}$$

$$\int_{\pi/6}^{\pi/3} \tan(x)\, dx = -\left(\ln\left(\cos\left(\frac{\pi}{3}\right)\right) - \ln\left(\cos\left(\frac{\pi}{6}\right)\right)\right)$$

$$\int_{\pi/6}^{\pi/3} \tan(x)\, dx = -\left(\ln\left(\frac{1}{2}\right) - \ln\left(\frac{\sqrt{3}}{2}\right)\right)$$

$$\int_{\pi/6}^{\pi/3} \tan(x)\, dx = -\ln\left(\frac{1}{\sqrt{3}}\right)$$

$$\int_{\pi/6}^{\pi/3} \tan(x)\, dx = \ln\left(\sqrt{3}\right)$$

10. Answer: B

First, find $f'(x)$. Use the Second Fundamental Theorem of Calculus.

$$f'(x) = \frac{d}{dx}\int_0^x \frac{t}{t^2+1}\, dt$$

$$f'(x) = \frac{x}{x^2+1}$$

Next, compute $f'(1)$.

$$f'(1) = \frac{1}{1^2+1} = \frac{1}{2}$$

11. Answer: A

First, note that we can use a double-angle formula to write $\sin(2x) = 2\sin(x)\cos(x)$.

$$\int \sin(2x) e^{\cos^2(x)}\, dx = \int 2\sin(x)\cos(x) e^{\cos^2(x)}\, dx$$

Next, change variables and integrate. Let $u = \cos^2(x)$. Then $du = -2\sin(x)\cos(x)\,dx$, or $2\sin(x)\cos(x)\,dx = -du$.

$$\int \sin(2x) e^{\cos^2(x)}\, dx = -\int e^u\, du$$

$$\int \sin(2x) e^{\cos^2(x)}\, dx = -e^u + C$$

Finally, undo the change of variable.

$$\int \sin(2x) e^{\cos^2(x)}\, dx = -e^{\cos^2(x)} + C$$

12. Answer: B

Denote the average value of $f(x)$ on $[0,18]$ by $\overline{f}$. By definition, $\overline{f} = \frac{1}{18}\int_0^{18} f(x)\,dx$. This integral is best evaluated by partitioning the region bounded by the graph of f and the x-axis into polygons, as shown below.

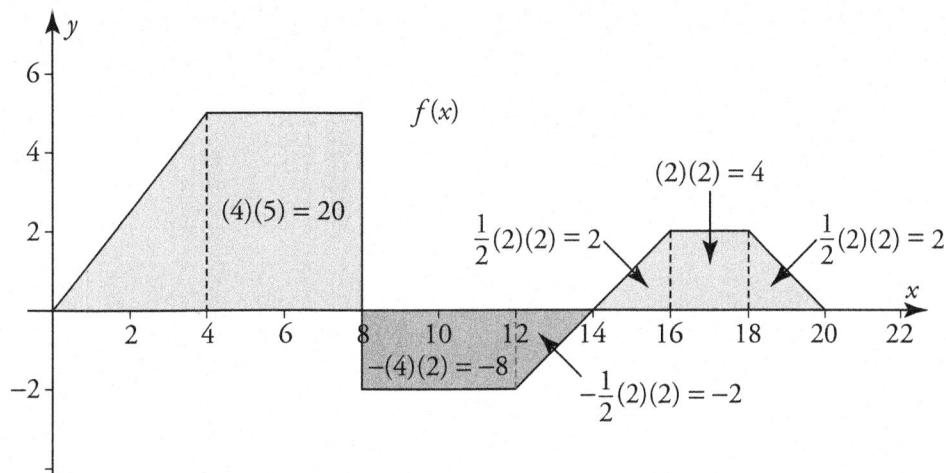

The numerical value of the integral of f on each subinterval is given in the figure. Integrals on subintervals for which $f(x) \le 0$ give a negative contribution to the total integral. Summing all of the contributions, we get the following:

$$\overline{f} = \frac{1}{18}(10 + 20 - 8 - 2 + 2 + 4 + 2) = \frac{14}{9}$$

13. Answer: D

Let $u = x - 2$. Then $du = dx, x = e + 2 \Rightarrow u = e,$ and $x = 3 \Rightarrow u = 1$.

$$\int_3^{e+2} \frac{1}{x-2}\,dx = \int_1^e \frac{1}{u}\,du$$

Now use the Fundamental Theorem of Calculus.

$$\int_3^{e+2} \frac{1}{x-2}\,dx = \ln(u)\Big|_1^e$$

$$\int_3^{e+2} \frac{1}{x-2}\,dx = \ln(e) - \ln(1) = 1$$

14. Answer: A

First, do a change of variable. Let $u = \sqrt{x} = x^{1/2}$. Then $du = \frac{1}{2}x^{-1/2}dx$, or $\frac{1}{\sqrt{x}}dx = 2\,du$.

$$\int \frac{\sin\left(\sqrt{x}\right)}{\sqrt{x}}dx = 2\int \sin\left(u\right)du$$

Next, integrate.

$$\int \frac{\sin\left(\sqrt{x}\right)}{\sqrt{x}}dx = -2\cos\left(u\right)+C$$

Finally, undo the change of variable.

$$\int \frac{\sin\left(\sqrt{x}\right)}{\sqrt{x}}dx = -2\cos\left(\sqrt{x}\right)+C$$

15. Answer: E

Direct substitution yields $\frac{2}{0}$, which is undefined. We will investigate further by reducing $\frac{x+5}{x^2+8x+15}$ to lowest terms.

$$\lim_{x \to -3} \frac{x+5}{x^2+8x+15} = \lim_{x \to -3} \frac{x+5}{(x+3)(x+5)}$$

$$\lim_{x \to -3} \frac{x+5}{x^2+8x+15} = \lim_{x \to -3} \frac{1}{x+3}$$

Looking at the right- and left-handed limits, we see the following:

$$\lim_{x \to -3^-} \frac{1}{x+3} = -\infty$$

$$\lim_{x \to -3^+} \frac{1}{x+3} = \infty$$

Since the right- and left-handed limits do not agree, the limit does not exist.

16. Answer: A

First, note that the y-coordinate of the point of tangency is $f\left(\pi\right) = \tan\left(\frac{\pi}{4}\right) = 1$. So the point of tangency is $(\pi,1)$

Next, find the slope m of the line, where $m = f'(\pi)$. Use the Chain Rule when calculating $f'(x)$.

$$f'\left(x\right) = \frac{d}{dx}\left(\tan\left(\frac{x}{4}\right)\right)$$

$$f'\left(x\right) = \sec^2\left(\frac{x}{4}\right) \cdot \frac{d}{dx}\left(\frac{x}{4}\right)$$

$$f'(x) = \frac{1}{4}\sec^2\left(\frac{x}{4}\right)$$

$$m = \frac{1}{4}\sec^2\left(\frac{\pi}{4}\right) = \frac{1}{2}$$

Finally, obtain the equation of the tangent line. Since the line passes through $(\pi, 1)$ and has slope $\frac{1}{2}$, we can immediately write down its equation in point-slope form as follows:

$$y - 1 = \frac{1}{2}(x - \pi)$$

17. Answer: B

First, find the critical numbers of f. Since f is differentiable everywhere, the critical numbers occur only when $f'(x) = 0$.

$$f'(x) = \frac{d}{dx}\left(x^3 - 3x^2 - 9x\right)$$

$$f'(x) = 3x^2 - 6x - 9 = 0$$

$$3(x+1)(x-3) = 0 \Rightarrow x = -1 \, or \, x = 3$$

Second, check the sign of $f'(x)$ in between the critical numbers.

Interval	Test Value x	$f'(x)$	Behavior of f
$(-\infty, -1)$	$x = -2$	$f'(-2) = 15 > 0$	Increasing
$(-1, 3)$	$x = 0$	$f'(0) = -9 < 0$	Decreasing
$(3, \infty)$	$x = 4$	$f'(4) = 15 > 0$	Increasing

So f is increasing for $x < -1$ or $x > 3$.

18. Answer: C

This limit yields the indeterminate form $\frac{0}{0}$. Apply L'Hôpital's Rule.

$$\lim_{x \to 0}\frac{e^{2x} - 1}{3x} = \lim_{x \to 0}\frac{\frac{d}{dx}\left(e^{2x} - 1\right)}{\frac{d}{dx}(3x)}$$

$$\lim_{x \to 0}\frac{e^{2x} - 1}{3x} = \lim_{x \to 0}\frac{2e^{2x}}{3}$$

$$\lim_{x \to 0}\frac{e^{2x} - 1}{3x} = \frac{2}{3}$$

19. Answer: D

First, find the critical numbers of f'. Since f is twice differentiable everywhere, these critical numbers only occur when $f''(x) = 0$.

$$f'(x) = \frac{d}{dx}\left(x^4 + 4x^3 - 48x^2 + 8x - 12\right)$$

$$f'(x) = 4x^3 + 12x^2 - 96x + 8$$

$$f''(x) = \frac{d}{dx}\left(4x^3 + 12x^2 - 96x + 8\right)$$

$$f''(x) = 12x^2 + 24x - 96 = 0$$

$$12(x+4)(x-2) = 0 \Rightarrow x = -4 \text{ or } x = 2$$

Second, check the sign of f'' in between the critical numbers.

Interval	Test Value x	$f''(x)$	Behavior of Graph of f
$(-\infty, -4)$	$x = -5$	$f''(-5) = 84 > 0$	Concave Up
$(-4, 2)$	$x = 0$	$f''(0) = -96 < 0$	Concave Down
$(2, \infty)$	$x = 3$	$f''(3) = 84 > 0$	Concave Up

So the graph of f is concave downwards for $-4 < x < 2$.

20. Answer: A

Use the Chain Rule twice, as this is a doubly composite function.

$$f'(x) = \frac{d}{dx}\left(\left(\tan^{-1}(3x)\right)^2\right)$$

$$f'(x) = 2\tan^{-1}(3x) \cdot \frac{d}{dx}\left(\tan^{-1}(3x)\right)$$

$$f'(x) = 2\tan^{-1}(3x) \cdot \frac{1}{1+9x^2} \cdot \frac{d}{dx}(3x)$$

$$f'(x) = 2\tan^{-1}(3x) \cdot \frac{1}{1+9x^2} \cdot 3$$

$$f'(x) = \frac{6\tan^{-1}(3x)}{1+9x^2}$$

21. Answer: D

Let $f(x) = \ln(x)$, and recognize that $\displaystyle\lim_{h \to 0} \frac{\ln(e^2 + h) - 2}{h} = f'(e^2)$.

$$f'(x) = \frac{d}{dx}(\ln(x))$$

$$f'(x) = \frac{1}{x}$$

$$\lim_{h \to 0} \frac{\ln(e^2 + h) - 2}{h} = \frac{1}{e^2}$$

22. Answer: B

First, use the Chain Rule to find $f'(x)$.

$$f'(x) = \frac{d}{dx}(\cot(\pi x))$$

$$f'(x) = -\csc^2(\pi x) \cdot \frac{d}{dx}(\pi x)$$

$$f'(x) = -\pi \csc^2(\pi x)$$

Now, compute $f'\left(\dfrac{1}{3}\right)$.

$$f'\left(\frac{1}{3}\right) = -\pi \csc^2\left(\frac{\pi}{3}\right)$$

$$f'\left(\frac{1}{3}\right) = -\frac{3\pi}{4}$$

23. Answer: E

First, find $f'(x)$.

$$f'(x) = \frac{d}{dx}(x^3 - 2x)$$

$$f'(x) = 3x^2 - 2$$

Second, find the mean rate of change of f on $[1, 3]$.

$$\frac{f(3) - f(1)}{3 - 1} = \frac{(3^3 - 2(3)) - (1^3 - 2(1))}{3 - 1}$$

$$\frac{f(3) - f(1)}{3 - 1} = 11$$

Finally, find c in $(1,3)$ such that the instantaneous rate of change of f at c equals the mean rate of change of f on $[1,3]$.

$$f'(c) = \frac{f(3) - f(1)}{3 - 1}$$

$$3c^2 - 2 = 11$$

$$c^2 = \frac{13}{3} \Rightarrow c = \pm\sqrt{\frac{13}{3}}$$

Since $-\sqrt{\frac{13}{3}}$ is not in the interval $(1,3)$, the only admissible value is $c = \sqrt{\frac{13}{3}}$.

24. Answer: B

First, find the velocity v of the particle: $v(t) = s'(t)$.

$$v(t) = \frac{d}{dt}\left(t^3 - 9t^2 + 15t + 3\right)$$

$$v(t) = 3t^2 - 18t + 15$$

Second, find the acceleration a of the particle: $a(t) = v'(t)$.

$$a(t) = \frac{d}{dt}\left(3t^2 - 18t + 15\right)$$

$$a(t) = 6t - 18$$

Finally, find the time at which the acceleration is zero.
$$6t - 18 = 0 \Rightarrow t = 3$$

25. Answer: C

Let x be one of the numbers, and let y be the other number. The first sentence of the problem statement gives us the following *constraint equation*:
$$x + y = 20$$

We are to maximize the product $P = x^2 y^3$. Solving the constraint equation for x, we obtain $x = 20 - y$. Substituting this for x in P, we obtain the following:

$$P = (20 - y)^2 y^3$$

$$P = (400 - 40y + y^2)y^3$$

$$P = y^5 - 40y^4 + 400y^3$$

Find the critical numbers of P. Since P is differentiable everywhere, the critical numbers will only occur when $\frac{dP}{dy} = 0$.

$$\frac{dP}{dy} = \frac{d}{dy}\left(y^5 - 40y^4 + 400y^3\right)$$

$$\frac{dP}{dy} = 5y^4 - 160y^3 + 1200y^2 = 0$$

$$5y^2(y-12)(y-20) = 0 \Rightarrow y = 0 \; or \; y = 12 \; or \; y = 20$$

We may disregard $y = 0$ and $y = 20$ because for these values of y, $P = 0$. For $y = 12$, we will get a *positive* value of P, so 0 is certainly not the maximum value. The maximum of P therefore occurs when $y = 12$ and $x = 20 - 12 = 8$.

26. Answer: A

These are elementary functions, and they are easy to graph without a calculator. A sketch of the region is given below.

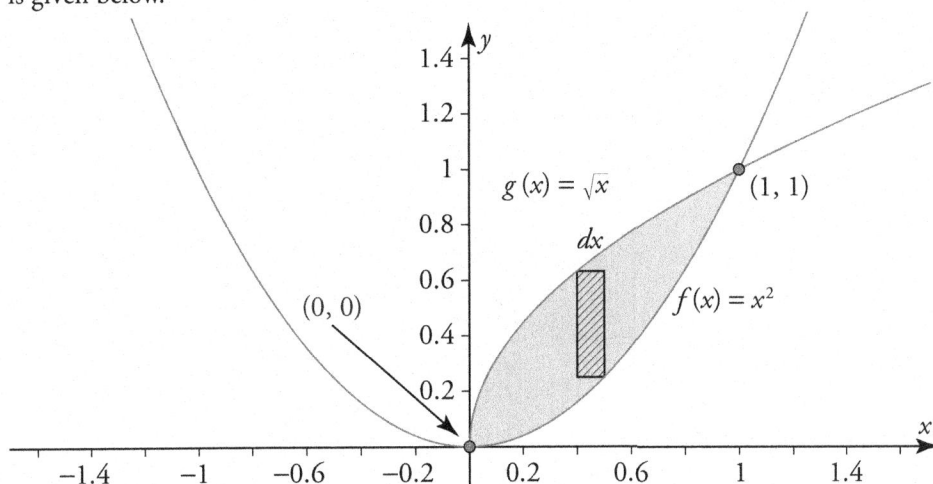

The differential area element is $dA = \left(\sqrt{x} - x^2\right)dx = \left(x^{1/2} - x^2\right)dx$. The area A is found as follows:

$$A = \int_0^1 \left(x^{1/2} - x^2\right)dx$$

$$A = \left(\frac{2}{3}x^{3/2} - \frac{1}{3}x^3\right)\Bigg|_0^1$$

$$A = \left(\frac{2}{3} - \frac{1}{3}\right) - 0 = \frac{1}{3}$$

27. Answer: D

f has a relative maximum for some x in the interval $(-2,-1)$, and it has a relative minimum for some x in the interval $(1,2)$. Since f is differentiable, this implies that, for some x in $(-2,-1)$ *and* for some x in $(1,2)$, $f'(x) = 0$. Since answer choices A, B, and E do not have zeros in these intervals, eliminate them. Note also that f is *decreasing* at $x = 0$, which means that $f'(0) < 0$. Eliminate C, and choose D.

28. Answer: D

Since R is to be revolved about the y-axis, we will need to express the left and right boundaries of the region as functions of y.

$$f(x) = y = 2x \Rightarrow x = \frac{y}{2} = g(y)$$

$$x = 6 = h(y)$$

A sketch of the region is given below.

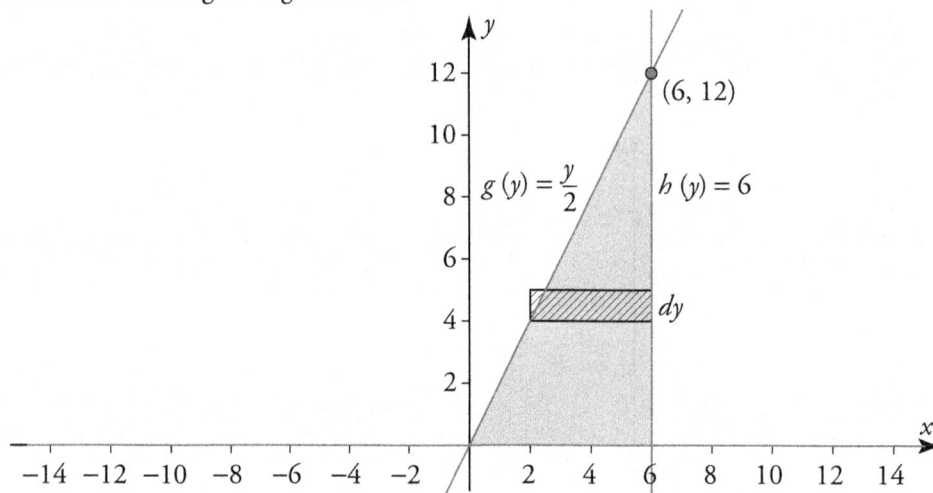

The differential volume element is given as follows:

$$dV = \pi\left(\left(h(y)\right)^2 - \left(g(y)\right)^2\right)dy$$

$$dV = \pi\left(6^2 - \left(\frac{y}{2}\right)^2\right)dy$$

$$dV = \pi\left(36 - \frac{y^2}{4}\right)dy$$

The total volume V of the solid is given by the following integral:

$$V = \pi\int_0^{12}\left(36 - \frac{y^2}{4}\right)dy$$

$$V = \pi\left(36y - \frac{y^3}{12}\right)\Big|_0^{12}$$

$$V = \pi\left((432 - 144) - 0\right) = 288\pi$$

29. Answer: B

The velocity v of the particle is given by $v(t) = s'(t)$.

$$v(t) = \frac{d}{dt}\left(3t^3 - 18t^2 + 10t + 28\right)$$

$$v(t) = 9t^2 - 36t + 10$$

The speed of the particle is given by the absolute value of the velocity: $|v(t)|$.

$$|v(t)| = |9t^2 - 36t + 10|$$

We will find the maximum value of this function on $[0, 4]$ graphically.

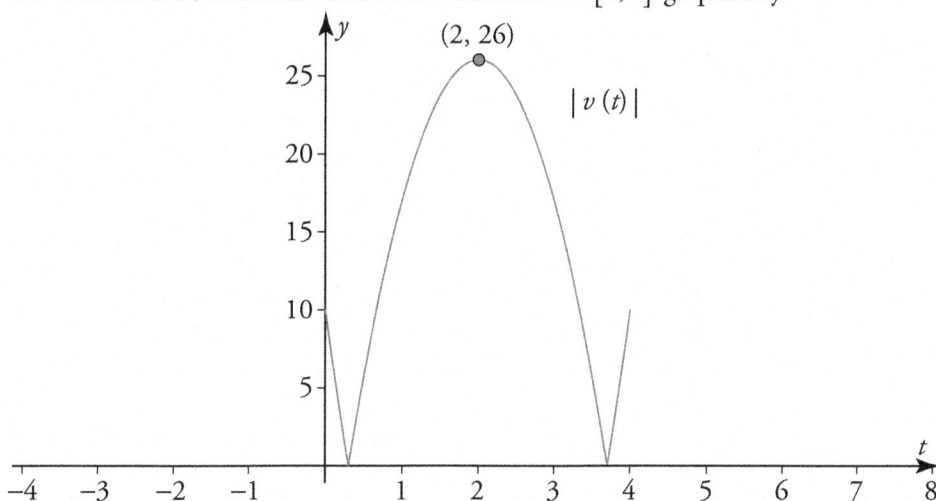

The maximum speed of the particle for $0 \le t \le 4$ is $26\frac{m}{s}$.

30. Answer: B

Denote the average value of f on $\left[0, \sqrt{\frac{\pi}{2}}\right]$ by $\overline{f}$.

$$\overline{f} = \frac{1}{\sqrt{\frac{\pi}{2}} - 0} \int_0^{\sqrt{\frac{\pi}{2}}} \sin\left(x^2\right) dx$$

$$\overline{f} = \sqrt{\frac{2}{\pi}} \int_0^{\sqrt{\frac{\pi}{2}}} \sin\left(x^2\right) dx$$

Using the calculator to compute the integral, we find that, to the nearest thousandth, $f = 0.438$.

31. Answer: A

Use the Second Fundamental Theorem of Calculus and the Chain Rule.

$$\frac{d}{dx}\int_0^{3x}\frac{1}{t^3+1}dt = \frac{1}{(3x)^3+1}\cdot\frac{d}{dx}(3x)$$

$$\frac{d}{dx}\int_0^{3x}\frac{1}{t^3+1}dt = \frac{3}{27x^3+1}$$

32. Answer: B

Let $u=\dfrac{x}{2}$. Then $du=\dfrac{dx}{2}$, or $dx=2\,du$. Also $x=2\Rightarrow u=1$, and $x=4\Rightarrow u=2$.

$$\int_2^4 e^{x/2}dx = 2\int_1^2 e^u\,du$$

$$\int_2^4 e^{x/2}dx = 2\left(e^u\right)\Big|_1^2$$

$$\int_2^4 e^{x/2}dx = 2\left(e^2-e^1\right)\approx 9.34$$

33. Answer: B

We sketch the region of integration with the trapezoids below.

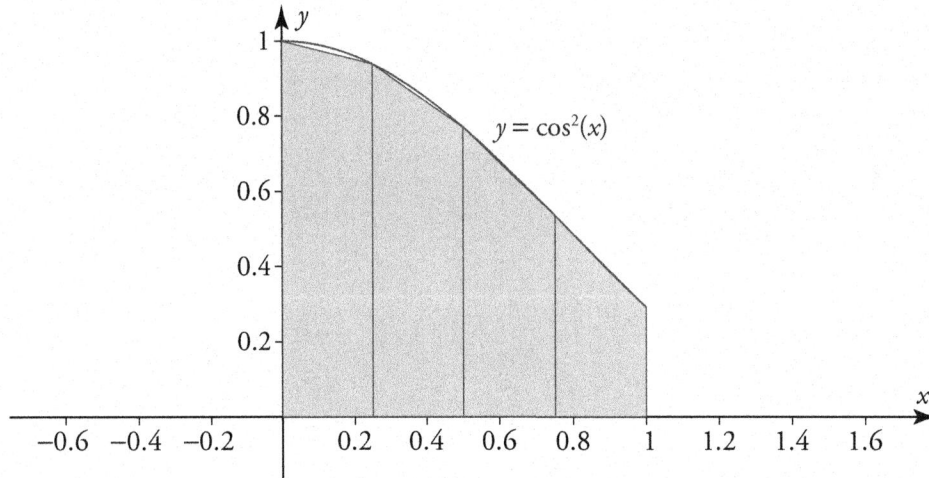

Let $f(x)=\cos^2(x)$. Applying the Trapezoidal Rule with $n=4$, we get the following:

$$\int_0^1 \cos^2(x)\,dx \approx \frac{1-0}{2(4)}\big(f(0)+2f(0.25)+2f(0.75)+f(1)\big)$$

$$\int_0^1 \cos^2(x) \approx 0.723$$

34. Answer: D

The rate of change of the volume V of the sphere with respect to time t is modeled by the following differential equation:

$$\frac{dV}{dt} = kV$$

In the above equation, k is a positive constant. To solve the equation for V, we separate variables and integrate.

$$\frac{dV}{V} = k\,dt$$

$$\int \frac{dV}{V} = k \int dt$$

$$\ln|V| = kt + C$$

Exponentiating both sides, we obtain the following:

$$e^{\ln|V|} = e^{kt+C}$$

$$|V| = e^C e^{kt}$$

$$V(t) = \pm e^C e^{kt}$$

Let $A = \pm e^C$. Then we obtain the following:

$$V(t) = Ae^{kt}$$

Use the fact that $V = 0.5\,m^3$ when $t = 0\,s$ to find A.

$$V(0) = Ae^{k(0)}$$

$$0.5 = A$$

So thus far, we have $V(t) = 0.5e^{kt}$. Next, use the fact that $V = 1.2\,m^3$ when $t = 5\,s$ to find k.

$$V(5) = 0.5e^{k(5)}$$

$$1.2 = 0.5e^{5k}$$

$$e^{5k} = 2.4$$

$$5k = \ln(2.4)$$

$$k = \frac{\ln(2.4)}{5}$$

So our particular solution is $V(t) = 0.5e^{\frac{\ln(2.4)}{5}t}$. Finally, find V when $t = 10\,s$.

$$V(10) = 0.5e^{\frac{\ln(2.4)}{5}(10)} = 2.88\,m^3$$

35. Answer: A

Solving the equation for y, we obtain $y = \pm\sqrt{1 - \dfrac{x^2}{2}}$. Since the point at which the normal line

intersects the graph is in the fourth quadrant, we need only concern ourselves with the negative root.

Thus, $y = -\sqrt{1 - \dfrac{x^2}{2}}$. Use the calculator to get the slope m_t of the line tangent to the graph of this

equation when $x = 1$.

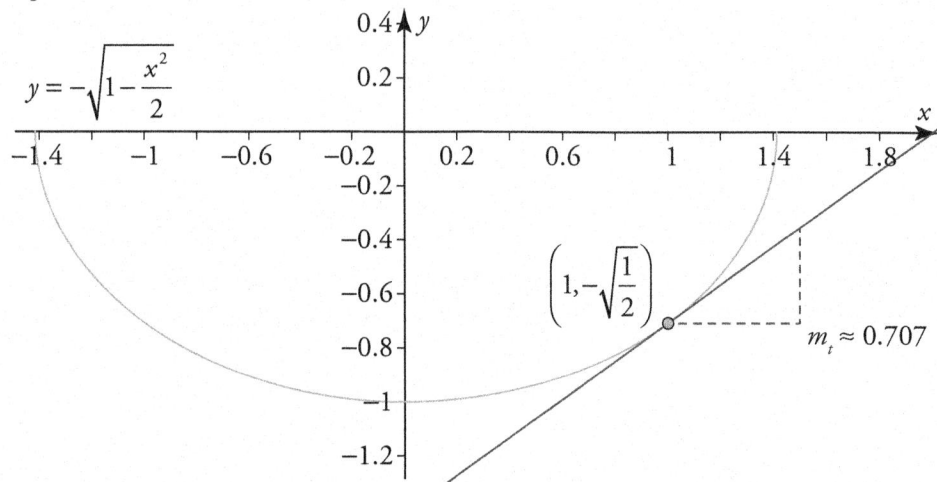

The slope m_n of the normal line is the *negative reciprocal* of m_t. Thus, we obtain the following:

$$m_n = -\frac{1}{m_t}$$

$$m_n \approx -\frac{1}{0.707} \approx -1.414$$

36. Answer: D

Use the calculator to graph the function and locate the local minimum.

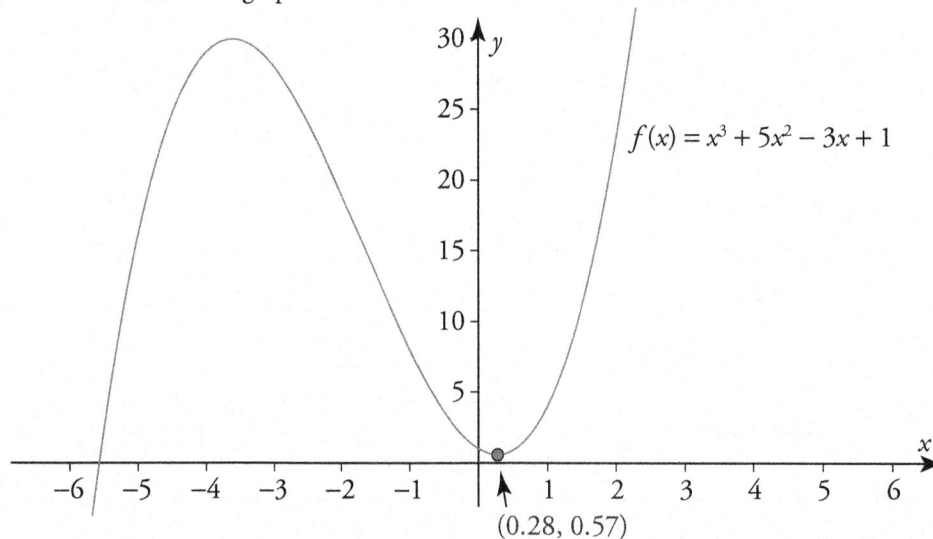

37. Answer: A

The volume V of a sphere is given in terms of its radius r by the relation $V = \dfrac{4}{3}\pi r^3$. Differentiating this with respect to time t yields the following:

$$\frac{dV}{dt} = 4\pi r^2 \frac{dr}{dt}$$

Substituting in the given data, we obtain the following:

$$\frac{dV}{dt} = 4\pi(2)^2(0.2) \approx 10.1\frac{m^3}{s}$$

38. Answer: D

First, use the Quotient Rule to find $f'(x)$.

$$f'(x) = \frac{d}{dx}\left(\frac{x}{x^2+9}\right)$$

$$f'(x) = \frac{\dfrac{d}{dx}(x)\cdot(x^2+9) - x\cdot\dfrac{d}{dx}(x^2+9)}{(x^2+9)^2}$$

$$f'(x) = \frac{x^2+9-2x^2}{(x^2+9)^2}$$

$$f'(x) = \frac{-x^2+9}{(x^2+9)^2}$$

Next, compute $f'(2)$.

$$f'(2) = \frac{-2^2+9}{(2^2+9)^2} \approx 0.030$$

39. Answer: D

In order for $f(x)$ to be differentiable at $x = 1$, it must first be continuous at $x = 1$. This implies the following:

$$a(1)^3 - 2(1) = b(1)^2 + 1$$

$$a - 2 = b + 1$$

$$b = a - 3$$

The requirement that f be differentiable at $x = 1$ means that the limit $\lim\limits_{x\to 1}\dfrac{f(x)-f(1)}{x-1}$ must exist. The value of this limit is equal to $f'(1)$. First, we will look at the left-handed limit.

$$\lim_{x\to 1^-}\frac{f(x)-f(1)}{x-1} = \lim_{x\to 1^-}\frac{(ax^3-2x)-(a-2)}{x-1}$$

Since this limit yields the indeterminate form $\frac{0}{0}$, apply L'Hôpital's Rule.

$$\lim_{x \to 1^-} \frac{f(x) - f(1)}{x - 1} = \lim_{x \to 1^-} \frac{3ax^2 - 2}{1}$$

$$\lim_{x \to 1^-} \frac{f(x) - f(1)}{x - 1} = 3a - 2$$

Next, we will look at the right-handed limit.

$$\lim_{x \to 1^+} \frac{f(x) - f(1)}{x - 1} = \lim_{x \to 1^+} \frac{\left(bx^2 + x\right) - (a - 2)}{x - 1}$$

Substituting $b = a - 3$, we obtain the following:

$$\lim_{x \to 1^+} \frac{f(x) - f(1)}{x - 1} = \lim_{x \to 1^+} \frac{\left((a - 3)x^2 + x\right) - (a - 2)}{x - 1}$$

Since this limit yields the indeterminate form $\frac{0}{0}$, apply L'Hôpital's Rule.

$$\lim_{x \to 1^+} \frac{f(x) - f(1)}{x - 1} = \lim_{x \to 1^+} \frac{2(a - 3)x + 1}{1}$$

$$\lim_{x \to 1^+} \frac{f(x) - f(1)}{x - 1} = 2a - 5$$

Equating the left- and right-handed limits yields the following:

$$\lim_{x \to 1^-} \frac{f(x) - f(1)}{x - 1} = \lim_{x \to 1^+} \frac{f(x) - f(1)}{x - 1}$$

$$3a - 2 = 2a - 5 \Rightarrow a = -3$$

40. Answer: B

The velocity v of the projectile is given as a function of time t by $v(t) = y'(t)$.

$$v(t) = \frac{d}{dt}\left(-4.9t^2 + 50t\right)$$

$$v(t) = -9.8t + 50$$

Now, find the velocity when $t = 8.2\,s$.

$$v(8.2) = -9.8(8.2) + 50 = -30.36\frac{m}{s}$$

41. Answer: B

Divide the numerator and denominator by x^2 inside the limit.

$$\lim_{x \to \infty} \frac{8 - 3x^2}{5x^2 + 2x - 1} = \lim_{x \to \infty} \frac{\dfrac{8}{x^2} - \dfrac{3x^2}{x^2}}{\dfrac{5x^2}{x^2} + \dfrac{2x}{x^2} - \dfrac{1}{x^2}}$$

$$\lim_{x \to \infty} \frac{8 - 3x^2}{5x^2 + 2x - 1} = \lim_{x \to \infty} \frac{\dfrac{8}{x^2} - 3}{5 + \dfrac{2}{x} - \dfrac{1}{x^2}}$$

$$\lim_{x \to \infty} \frac{8 - 3x^2}{5x^2 + 2x - 1} = \frac{0 - 3}{5 + 0 - 0} = -\frac{3}{5}$$

42. Answer: C

$$\int \sec^5(x)\tan(x)\,dx = \int \sec^4(x) \cdot \sec(x)\tan(x)\,dx$$

Do a change of variable and integrate. Let $u = \sec(x)$. Then $du = \sec(x)\tan(x)\,dx$.

$$\int \sec^5(x)\tan(x)\,dx = \int u^4\,du$$

$$\int \sec^5(x)\tan(x)\,dx = \frac{1}{5}u^5 + C$$

Finally, undo the change of variable.

$$\int \sec^5(x)\tan(x)\,dx = \frac{1}{5}\sec^5(x) + C$$

43. Answer: C

Do a change of variable and integrate. Let $u = \dfrac{1}{x^2} = x^{-2}$. Then $du = -2x^{-3}\,dx = -\dfrac{2}{x^3}\,dx$, or $\dfrac{1}{x^3}\,dx = -\dfrac{1}{2}\,du$.

$$\int \frac{\sin\left(\dfrac{1}{x^2}\right)}{x^3}\,dx = -\frac{1}{2}\int \sin(u)\,du$$

$$\int \frac{\sin\left(\dfrac{1}{x^2}\right)}{x^3}\,dx = \frac{1}{2}\cos(u) + C$$

Finally, undo the change of variable.

$$\int \frac{\sin\left(\dfrac{1}{x^2}\right)}{x^3}\,dx = \frac{1}{2}\cos\left(\frac{1}{x^2}\right) + C$$

44. Answer: C

The region is sketched below.

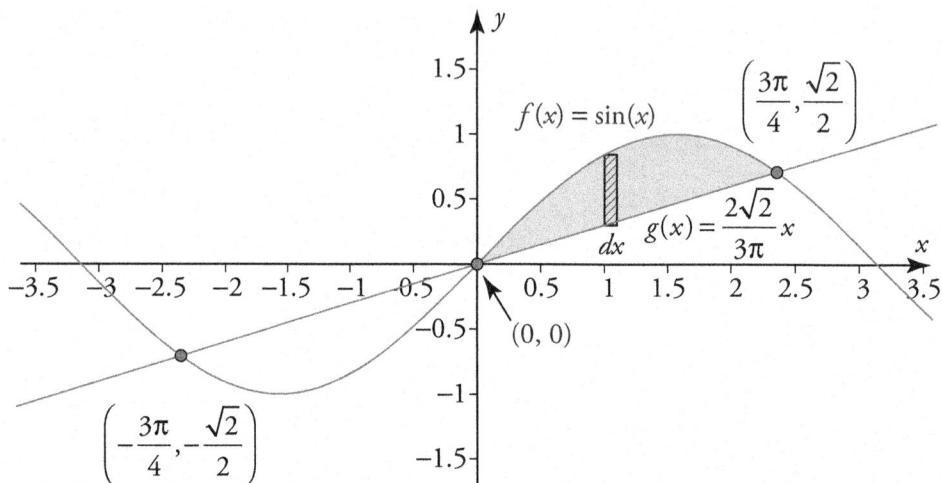

The differential area element on the interval $\left[-\dfrac{3\pi}{4}, 0\right]$ is $dA = \left[g(x) - f(x)\right]dx$, and the differential area element on the interval $\left[0, \dfrac{3\pi}{4}\right]$ is $dA = \left[f(x) - g(x)\right]dx$. However, since both f and g have odd symmetry, we can simply integrate on $\left[0, \dfrac{3\pi}{4}\right]$ and double the result. Do this on the calculator.

$$A = 2\int_0^{3\pi/4} \left(f(x) - g(x)\right)dx$$

$$A = 2\int_0^{3\pi/4} \left(\sin(x) - \frac{2\sqrt{2}}{3\pi}x\right)dx \approx 1.748$$

45. Answer: B

The graph of $v(t)$ is given below.

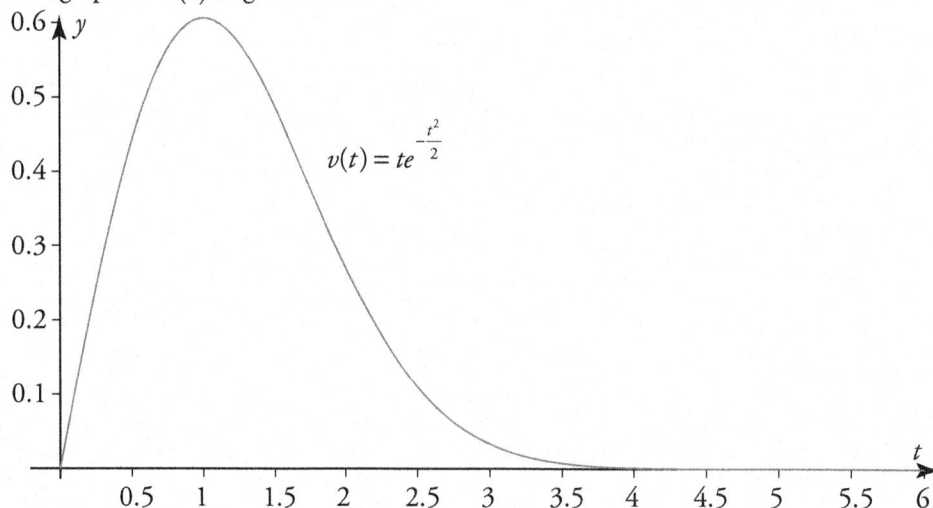

Since $v(t) \geq 0$ for all t in the interval $[2,4]$, the distance d traveled by the particle in this time interval is the integral of $v(t)$ on the interval. Use the calculator to compute this.

$$d = \int_{2}^{4} te^{-\frac{t^2}{2}} dt \approx 0.135\, m$$

1. Let $f(x) = \dfrac{\sin(x)}{x}$. For all three approximations, $\Delta x = \dfrac{\pi - 0}{4} = \dfrac{\pi}{4}$, so $x_0 = 0$, $x_1 = \dfrac{\pi}{4}$, $x_2 = \dfrac{\pi}{2}$, $x_4 = \dfrac{3\pi}{4}$, and $x_5 = \pi$.

A. The graph of f and the four inscribed rectangles are shown below.

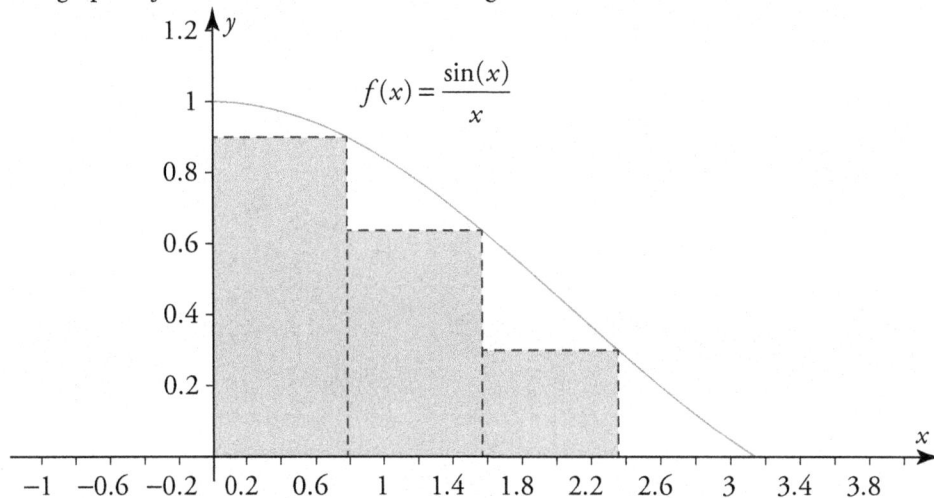

$$I \approx \left(f\left(\dfrac{\pi}{4}\right) + f\left(\dfrac{\pi}{2}\right) + f\left(\dfrac{3\pi}{4}\right) + f(\pi) \right)\Delta x$$

$$I \approx \left(\dfrac{\sin(\pi/4)}{\pi/4} + \dfrac{\sin(\pi/2)}{\pi/2} + \dfrac{\sin(3\pi/4)}{3\pi/4} + \dfrac{\sin(\pi)}{\pi} \right)\dfrac{\pi}{4}$$

$$I \approx 1.443$$

B. The graph of f and the four circumscribed rectangles are shown below.

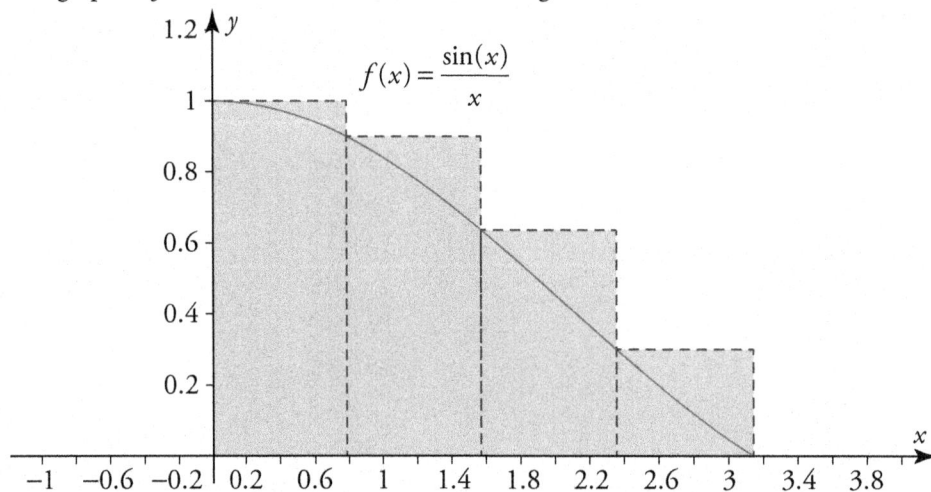

Since f is not defined at $x = 0$, we will use $f(0) = \lim_{x \to 0} \dfrac{\sin(x)}{x} = 1$.

$$I \approx \left(f(0) + f\left(\frac{\pi}{4}\right) + f\left(\frac{\pi}{2}\right) + f\left(\frac{3\pi}{4}\right) \right) \Delta x$$

$$I \approx \left(1 + \frac{\sin(\pi/4)}{\pi/4} + \frac{\sin(\pi/2)}{\pi/2} + \frac{\sin(3\pi/4)}{3\pi/4} \right) \frac{\pi}{4}$$

$$I \approx 2.228$$

C. The graph of f and the four trapezoids are shown below.

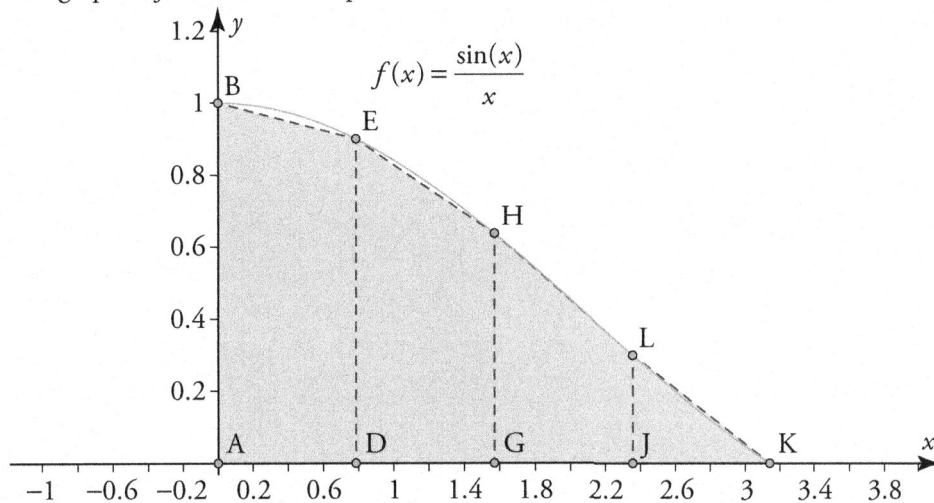

Again, we will use $f(0) = \lim\limits_{x \to 0} \dfrac{\sin(x)}{x} = 1$.

$$I \approx \left(f(0) + 2f\left(\frac{\pi}{4}\right) + 2f\left(\frac{\pi}{2}\right) + 2f\left(\frac{3\pi}{4}\right) + f(\pi) \right) \frac{\Delta x}{2}$$

$$I \approx \left(1 + 2 \cdot \frac{\sin(\pi/4)}{\pi/4} + 2 \cdot \frac{\sin(\pi/2)}{\pi/2} + 2 \cdot \frac{\sin(3\pi/4)}{3\pi/4} + \frac{\sin(\pi)}{\pi} \right) \frac{\pi}{8}$$

$$I \approx 1.836$$

2.

A. We find $v(t)$ by integrating $a(t)$.

$$v(t) = \int (2t - 7)\, dt$$

$$v(t) = t^2 - 7t + C_1$$

Use the fact that the velocity of the particle is $1\dfrac{m}{s}$ at $t = 2\,s$ to find C_1.

$$v(2) = 2^2 - 7(2) + C_1 = 1 \Rightarrow C_1 = 11$$

So $v(t) = t^2 - 7t + 11$.

B. The particle changes direction whenever $v(t)$ changes sign. Use the calculator to graph $v(t)$ and find its zeros.

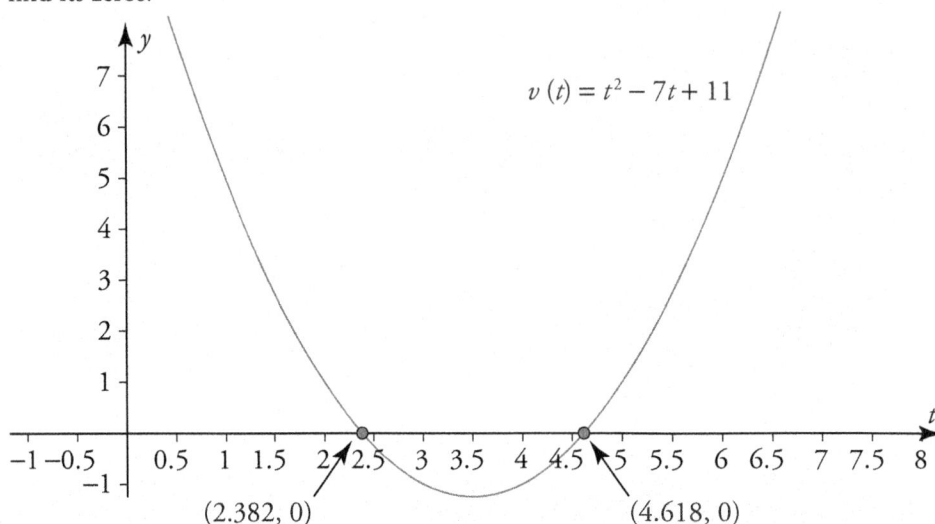

(2.382, 0)   (4.618, 0)

Since the sign of $v(t)$ changes at $t \approx 2.382\,s$ and $t \approx 4.618\,s$, the particle changes directions at those times as well.

C. We find $x(t)$ by integrating $v(t)$.

$$x(t) = \int (t^2 - 7t + 11)\,dt$$

$$x(t) = \frac{1}{3}t^3 - \frac{7}{2}t^2 + 11t + C_2$$

Use the fact that the velocity of the particle is $8m$ at $t = 2s$ to find C_2.

$$x(2) = \frac{1}{3}(2)^3 - \frac{7}{2}(2)^2 + 11(2) + C_2 = 8 \Rightarrow C_2 = -\frac{8}{3}$$

So $x(t) = \frac{1}{3}t^3 - \frac{7}{2}t^2 + 11t - \frac{8}{3}$.

D. Refer back to the figure for Part B. Since $v(t) \geq 0$ on $[2, 2.382]$ and $v(t) \leq 0$ on $[2.382, 4]$, the distance d traveled by the particle during the interval $[2, 4]$ is found as follows:

$$d = \int_{2}^{2.382} (t^2 - 7t + 11)\,dt - \int_{2.382}^{4} (t^2 - 7t + 11)\,dt$$

Do the integrations on the calculator to obtain $d \approx 1.697\,m$.

3. For all three parts, we will need the derivative of f.

$$f'(x) = \frac{d}{dx}\left(-2x^4 + 16x^2\right)$$

$$f'(x) = -8x^3 + 32x$$

A. The slope m_t of the line *tangent* to the graph of f at $x = 1$ is found as follows:

$$m_t = f'(1)$$

$$m_t = -8(1)^3 + 32(1) = 24$$

The slope m_n of the line *normal* to the graph of f at $x = 1$ is found as follows:

$$m_n = -\frac{1}{m_t}$$

$$m_n = -\frac{1}{24}$$

The y-coordinate of the point of intersection of the normal line with the graph of f is $f(1) = -2(1)^2 + 16(1)^2 = 14$. The equation of the normal line in point-slope form is $y - 14 = -\frac{1}{24}(x-1)$.

B. Since f is differentiable everywhere, relative extrema will only occur at critical numbers of f. We find the critical numbers by setting $f'(x) = 0$.

$$f'(x) = -8x^3 + 32x = 0$$

$$-8x(x+2)(x-2) = 0 \Rightarrow x = -2 \, or \, x = 0 \, or \, x = 2$$

We will use the First Derivative Test to classify the critical numbers.

Interval	Test Value x	$f'(x)$	Behavior of $f(x)$
$(-\infty, -2)$	$x = -3$	$f'(-3) = 120 > 0$	Increasing
$(-2, 0)$	$x = -1$	$f'(-1) = -24 < 0$	Decreasing
$(0, 2)$	$x = 1$	$f'(1) = 24 > 0$	Increasing
$(2, \infty)$	$x = 3$	$f'(3) = -120 < 0$	Decreasing

Since f changes from increasing to decreasing at $x = -2$ and $x = 2$, these critical numbers correspond to relative maxima. The coordinates of these maxima are as follows:

$$\left(-2, f(-2)\right) = \left(-2, 32\right)$$

$$\left(2, f(2)\right) = \left(2, 32\right)$$

C. Refer to the table in part B. Since f changes from decreasing to increasing at $x = 0$, this critical number corresponds to a relative minimum. The coordinates of this minimum are as follows:

$$\big(0, f(0)\big) = (0, 0)$$

4. For all three parts, we will need to know that $F'(x) = \dfrac{d}{dx} \displaystyle\int_0^x f(t)\,dt = f(x)$. This is the same function as $f(t)$, with t replaced by x.

   A. $F(x)$ is increasing whenever $F'(x) = f(x) > 0$. From the graph, we can see that this occurs whenever x is in the interval $(0,1)$ or $(8,16)$.

   B. $F(x)$ is decreasing whenever $F'(x) = f(x) < 0$. From the graph, we can see that this occurs whenever x is in the interval $(1,8)$.

   C. The absolute extrema of F on $[0,16]$ occur either at an endpoint of $[0,16]$ or at a critical number of F in the open interval $(0,16)$. Since $F'(x) = f(x)$ is defined everywhere in $(0,16)$, F is differentiable everywhere in $(0,16)$. If follows then that the only critical numbers of F in this interval are solutions of the equation $f(x) = 0$. We can see from the graph that these critical numbers are $x = 1$ and $x = 8$. We will now evaluate F at each critical number and at each endpoint.

   $$F(0) = \int_0^0 f(t)\,dt = 0$$

   $$F(1) = \int_0^1 f(t)\,dt = \frac{3}{2}$$

   This integral is equal to the shaded area shown in the figure to the right.

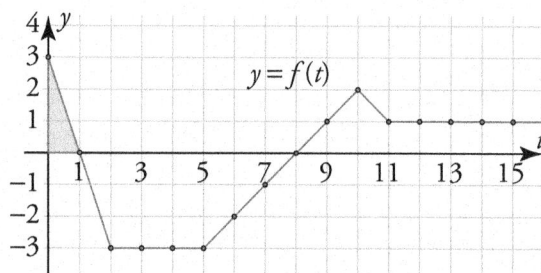

   $$F(8) = \int_0^8 f(t)\,dt$$

   $$F(8) = F(1) + \int_1^8 f(t)\,dt = -\frac{27}{2}$$

   The integral in the last step is equal to the *negative* of the shaded area shown in the figure to the right.

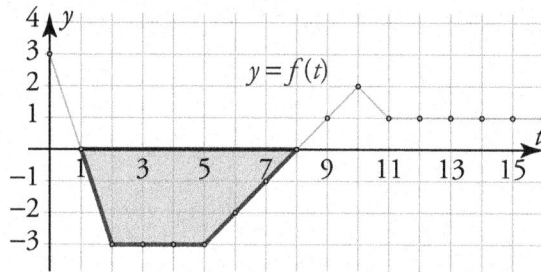

   $$F(16) = \int_0^{16} f(t)\,dt$$

   $$F(16) = F(8) + \int_8^{16} f(t)\,dt = -5$$

   The integral in the last step is equal to the shaded area shown in the figure to the right.

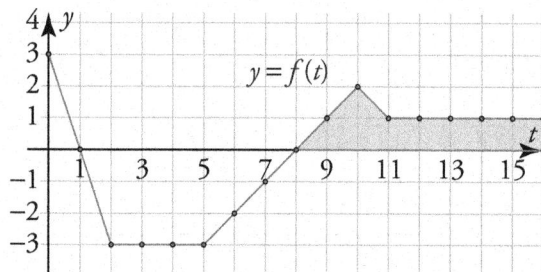

On $[0,16]$, the absolute maximum of F occurs at the point $\left(1, \dfrac{3}{2}\right)$ and the absolute minimum occurs at the point $\left(8, -\dfrac{27}{2}\right)$.

5. For all three parts, we will need a sketch of the region R. Since both f and g are quadratic functions, they are easy to graph without a calculator.

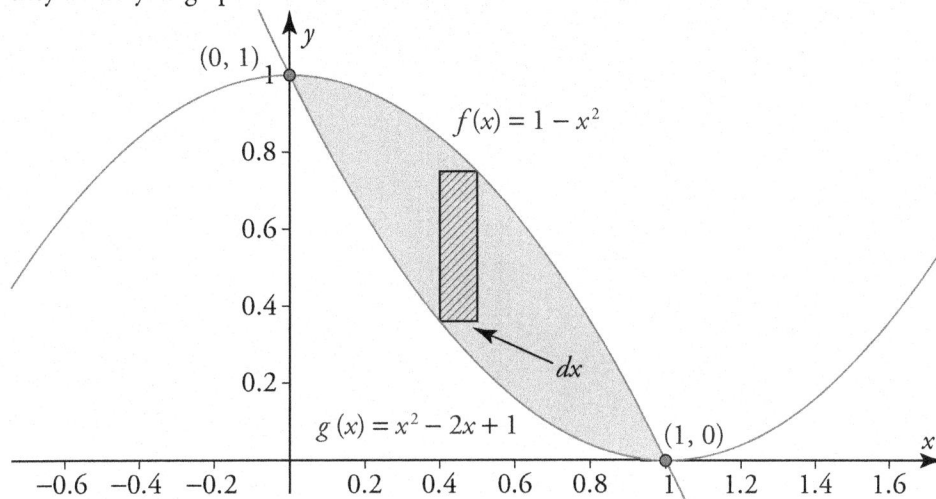

A. The differential area element is as follows:

$$dA = \left(f(x) - g(x) \right) dx$$

$$dA = \left(\left(1 - x^2\right) - \left(x^2 - 2x + 1\right) \right) dx$$

$$dA = \left(-2x^2 + 2x \right) dx$$

We obtain the area A of the region by integrating this on $[0,1]$.

$$A = \int_0^1 \left(-2x^2 + 2x \right) dx$$

$$A = \left(-\frac{2}{3}x^3 + x^2 \right) \bigg|_0^1$$

$$A = \left(-\frac{2}{3}(1)^3 + 1^2 \right) - \left(-\frac{2}{3}(0)^3 + 0^2 \right) = \frac{1}{3}$$

B. The differential volume element is as follows:

$$dV = \pi \left(\left(f(x) \right)^2 - \left(g(x) \right)^2 \right) dx$$

$$dV = \pi \left(\left(1 - x^2\right)^2 - \left(x^2 - 2x + 1\right)^2 \right) dx$$

$$dV = \pi \left(\left(1 - 2x^2 + x^4\right) - \left(x^4 - 4x^3 + 6x^2 - 4x + 1\right) \right) dx$$

$$dV = \pi \left(4x^3 - 8x^2 + 4x \right) dx$$

We obtain the volume V of the solid of revolution by integrating this on $[0,1]$.

$$V = \pi \int_0^1 \left(4x^3 - 8x^2 + 4x\right) dx$$

$$V = \pi \left(x^4 - \frac{8}{3}x^3 + 2x^2 \right)\Big|_0^1$$

$$V = \pi \left(\left(1^4 - \frac{8}{3}(1)^3 + 2(1)^2\right) - \left(0^4 - \frac{8}{3}(0)^3 + 2(0)^2\right) \right) = \frac{\pi}{3}$$

C. The differential volume element is as follows:

$$dV = \pi \left(\left(f(x) - (-1)\right)^2 - \left(g(x) - (-1)\right)^2 \right) dx$$

$$dV = \pi \left(\left(-x^2 + 2\right)^2 - \left(x^2 - 2x + 2\right)^2 \right) dx$$

$$dV = \pi \left(\left(4 - 2x^2 + x^4\right) - \left(x^4 - 4x^3 + 8x^2 - 8x + 4\right) \right) dx$$

$$dV = \pi \left(4x^3 - 10x^2 + 8x\right) dx$$

We obtain the volume V of the solid of revolution by integrating this on $[0,1]$.

$$V = \pi \int_0^1 \left(4x^3 - 10x^2 + 8x\right) dx$$

$$V = \pi \left(x^4 - \frac{10}{3}x^3 + 4x^2 \right)\Big|_0^1$$

$$V = \pi \left(\left(1^4 - \frac{10}{3}(1)^3 + 4(1)^2\right) - \left(0^4 - \frac{10}{3}(0)^3 + 4(0)^2\right) \right) = \frac{5\pi}{3}$$

6.

A. First, evaluate f at each endpoint of the interval $[3,5]$.

$$f(3) = \frac{2}{3-2} + 3 = 5$$

$$f(5) = \frac{2}{5-2} + 3 = \frac{11}{3}$$

Now, compute the average rate of change of f on $[3,5]$.

$$\frac{f(5) - f(3)}{5 - 3} = \frac{\frac{11}{3} - 5}{5 - 3} = -\frac{2}{3}$$

B. First, compute $f'(x)$.

$$f'(x) = \frac{d}{dx}\left(\frac{2}{x-2} + 3\right)$$

$$f'(x) = \frac{d}{dx}\left(2(x-2)^{-1} + 3\right)$$

$$f'(x) = -2(x-2)^{-2} \cdot \frac{d}{dx}(x-2)$$

$$f'(x) = -2(x-2)^{-2} \cdot 1$$

$$f'(x) = -\frac{2}{(x-2)^2}$$

Now, evaluate f' at $x = 4$.

$$f'(4) = -\frac{2}{(4-2)^2} = -\frac{1}{2}$$

C. Apply the Mean Value Theorem for Derivatives.

$$f'(c) = \frac{f(5) - f(3)}{5 - 3}$$

$$-\frac{2}{(c-2)^2} = -\frac{2}{3}$$

Cross-multiplication and simplification yields the following:

$$(c-2)^2 = 3 \Rightarrow c = 2 \pm \sqrt{3}$$

Reject $2 - \sqrt{3}$, since it is not in the interval [3,5]. The only solution is $c = 2 + \sqrt{3}$.

AP Calculus BC Practice Examination 1

CALCULUS BC

A CALCULATOR **CANNOT BE USED ON PART A OF SECTION I OR ON PART B OF SECTION II.** A GRAPHING CALCULATOR FROM THE APPROVED LIST **IS REQUIRED ON PART B OF SECTION I AND ON PART A OF SECTION II** OF THE EXAMINATION. CALCULATOR MEMORIES NEED NOT BE CLEARED. COMPUTERS, NONGRAPHING SCIENTIFIC CALCULATORS, CALCULATORS WITH QWERTY KEYBOARDS, AND ELECTRONIC WRITING PADS ARE NOT ALLOWED. CALCULATORS MAY NOT BE SHARED AND COMMUNICATION BETWEEN CALCULATORS IS PROHIBITED DURING THE EXAMINATION. ATTEMPTS TO REMOVE TEST MATERIALS FROM THE ROOM BY ANY METHOD WILL RESULT IN INVALIDATION OF TEST SCORES.

CALCULUS BC – SECTION I

Time – 1 hour and 45 minutes
All questions are given equal weight.
Percent of total grade – 50

Part A: 55 minutes, 28 multiple-choice questions
A calculator is NOT allowed

Part B: 50 minutes, 17 multiple-choice questions
A graphing calculator is required.

Parts A and B of Section I are printed in this examination booklet. Section II, which consists of longer problems, is printed in a separate booklet.

GENERAL INSTRUCTIONS

DO NOT OPEN THIS BOOKLET UNTIL YOU ARE TOLD TO DO SO.

INDICATE YOUR ANSWERS TO QUESTIONS IN PART A ON PAGE 2 OF THE SEPARATE ANSWER SHEET. THE ANSWERS TO QUESTIONS IN PART B SHOULD BE INDICATED ON PAGE 3 OF THE ANSWER SHEET. No credit will be given for anything written in this examination booklet, but you may use the booklet for notes or scratchwork. After you have decided which of the suggested answers is best, COMPLETELY fill in the corresponding oval on the answer sheet. Give only one answer to each question. If you change an answer, be sure that the previous mark is erased completely.

Example: Sample Answer: Ⓐ Ⓑ Ⓒ ● Ⓔ

1. What is the value of the numerical expression $4 - 2(3 - 5)$?

    A. -12

    B. -8

    C. 0

    D. 8

    E. 12

Many candidates wonder whether or not to guess the answers to questions about which they are not certain. It is improbable, therefore, that mere guessing will improve your score significantly; it may even lower your score, and it does take time. If, however, you are not sure of the best answer but have some knowledge of the question and are able to eliminate one or more of the answer choices as wrong, your chance of answering correctly is improved, and it may be to your advantage to answer such a question.

Use your time effectively, working as rapidly as you can without losing accuracy. Do not spend too much time on questions that are too difficult. Go on to other questions and come back to the difficult ones later if you have time. It is not expected that you will be able to answer all of the multiple choice questions.

A CALCULATOR MAY NOT BE USED ON THIS PART OF THE EXAMINATION.

<u>Directions</u>: Solve each of the following problems, using the available space for scratchwork. After examining the form of the choices, decide which is the best of the choices given and fill in the corresponding oval on the answer sheet. No credit will be given for anything written in the test book. Do not spend too much time on any one problem.

<u>In this test</u>: Unless otherwise specified, the domain of a function f is assumed to be the set of all real numbers x for which $f(x)$ is a real number.

1. If $f(x) = \dfrac{5x - 3}{2x + 7}$, then $f'(x) =$

    A. $\dfrac{5}{(2x + 7)^2}$

    B. $\dfrac{5}{2x + 7}$

    C. $\dfrac{3}{2}$

    D. $\dfrac{41}{(2x + 7)^2}$

    E. $\dfrac{20x + 29}{(2x + 7)^2}$

2. A particle moves along the x-axis so that its velocity is given by $v(t) = \dfrac{2t}{t^2 + 1}$ for all $t \geq 0$. If the displacement of the particle at time $t = 0$ is $x = 5$, then what is the displacement of the particle when $t = \sqrt{e^4 - 1}$?

    A. 4

    B. 9

    C. $\ln(4)$

    D. $\ln(9)$

    E. $5 + \ln(4)$

GO ON TO THE NEXT PAGE.

3. If possible, find the sum of the series $\sum_{n=0}^{\infty} \frac{2^{n+1}}{3^n}$.

A. $\frac{2}{3}$

B. 1

C. 3

D. 6

E. The series diverges.

4. $\displaystyle\lim_{h \to 0} \frac{\sec(x+h) - \sec(x)}{h} =$

A. $\sec(x)$

B. $\cos(x)$

C. $\sec(x)\tan(x)$

D. $\tan(x)$

E. $\sec^2(x)$

5. $\displaystyle\int_1^5 \frac{e^{5/x}}{x^2} =$

A. $\frac{1}{5}\left(e - e^5\right)$

B. $\frac{1}{5}\left(e + e^5\right)$

C. $\frac{1}{5}e^4$

D. $\frac{1}{5}e^5$

E. $\frac{1}{5}\left(e^5 - e\right)$

6. The length of the graph of the curve given by $x(t) = 4t^3$ and $y(t) = 3t^2$ for $0 \le t \le 2$ is given by

A. $\displaystyle\int_0^2 6t\sqrt{4t^2 + 1}\,dt$

B. $\displaystyle\int_0^2 \left(12t^2 + 6t\right)dt$

C. $\displaystyle\int_0^2 t^3\sqrt{t^2 + 1}\,dt$

D. $\displaystyle\int_0^2 \left(t^4 + t^6\right)dt$

E. $\displaystyle\int_0^2 \sqrt{1 + 36t^2}\,dt$

GO ON TO THE NEXT PAGE.

7. Consider the following direction field:

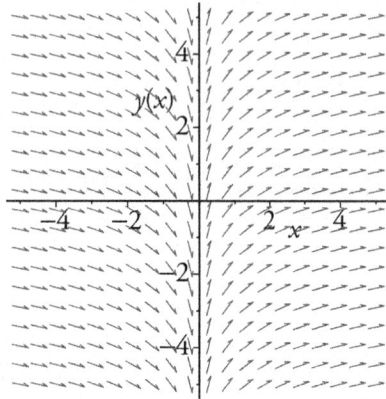

The differential equation that corresponds to this direction field is

A. $\dfrac{dy}{dx} = \dfrac{1}{x^2}$

B. $\dfrac{dy}{dx} = \dfrac{1}{x}$

C. $\dfrac{dy}{dx} = -\dfrac{1}{x^2}$

D. $\dfrac{dy}{dx} = -\dfrac{1}{x}$

E. $\dfrac{dy}{dx} = \ln(x)$

8. Below are listed some values for a function f that is continuous on the interval $[0,3]$.

x	0	1	2	3
$f(x)$	1	4	k	15

The left Riemann sum approximation for $\displaystyle\int_0^3 f(x)\,dx$ with a uniform partition of $[0,3]$ into three subintervals is 14. Find the value of k.

A. -9

B. -5

C. 5

D. 9

E. 14

GO ON TO THE NEXT PAGE.

9. If $f(x)$ and $g(x)$ are both continuous on the interval $[a,b]$, then each of the following functions must be continuous on $[a,b]$ except

A. $f(x)+g(x)$

B. $f(x)-g(x)$

C. $\dfrac{f(x)}{g(x)}$

D. $(f(x))^2$

E. $e^{g(x)}$

10. $x+\dfrac{x^3}{3!}+\dfrac{x^5}{5!}+\cdots+\dfrac{x^{2n+1}}{(2n+1)!}+\cdots$ is the Maclaurin series for

A. $f(x)=\sin(x)$

B. $f(x)=\cos(x)$

C. $f(x)=e^{x^2}$

D. $f(x)=\dfrac{1}{2}(e^x+e^{-x})$

E. $f(x)=\dfrac{1}{2}(e^x-e^{-x})$

11. The sides and altitude of the isosceles triangle below are increasing with time.

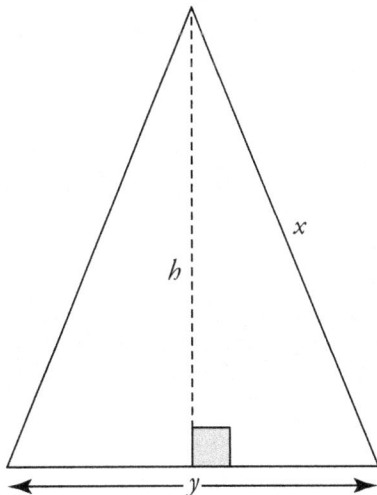

GO ON TO THE NEXT PAGE.

At the instant when $x = 13$ and $y = 10$, $\dfrac{dy}{dt} = 4\dfrac{dx}{dt}$ and $\dfrac{dh}{dt} = a\dfrac{dx}{dt}$. At that instant, $a =$

A. $\dfrac{1}{13}$

B. $\dfrac{1}{12}$

C. $\dfrac{1}{10}$

D. $\dfrac{1}{4}$

E. $\dfrac{1}{3}$

12. If $\dfrac{dy}{dx} = 4x\sqrt{y+1}$ and $y = 8$ when $x = 1$, then $y =$

A. $y = \left(x^2 + 2\right)^2 - 1$

B. $y = \left(x^2 + 1\right)^2 - 1$

C. $y = \left(x^2 - 1\right)^2 - 1$

D. $y = \left(x^2 - 4\right)^2 - 1$

E. $y = \left(x^2 + 3\right)^2 - 8$

13. If $x = e^{-3t}$ and $y = \ln(t)$, then $\dfrac{dy}{dx} =$

A. $3te^{3t}$

B. $\dfrac{e^{3t}}{3t}$

C. $-\dfrac{e^{3t}}{3t}$

D. $-\dfrac{3t}{e^{3t}}$

E. $-3te^{3t}$

14. $\int \left(\sin(x) + \cos(x) \right)^2 dx$

A. $x + \dfrac{1}{2}\cos(2x) + C$

B. $x - \dfrac{1}{2}\cos(2x) + C$

C. $\cos(2x) + C$

D. $x + \dfrac{1}{2}\sin(2x) + C$

E. $x - \dfrac{1}{2}\sin(2x) + C$

15. Consider the integral $\int_0^2 \left(4 - x^2 \right) dx$. Let there be a uniform partition of [0,2] with four subintervals. For this partition, let L be the left Riemann approximation of the integral, let R be the right Riemann approximation of the integral, and let T be the trapezoidal approximation of the integral. Which of the following inequalities is true?

A. $R < T < L < \int_0^2 \left(4 - x^2 \right) dx$

B. $R < T < \int_0^2 \left(4 - x^2 \right) dx < L$

C. $R < \int_0^2 \left(4 - x^2 \right) dx < T < L$

D. $L < T < \int_0^2 \left(4 - x^2 \right) dx < R$

E. $L < T < R < \int_0^2 \left(4 - x^2 \right) dx$

16. The graph of $y = f(x)$ is shown below.

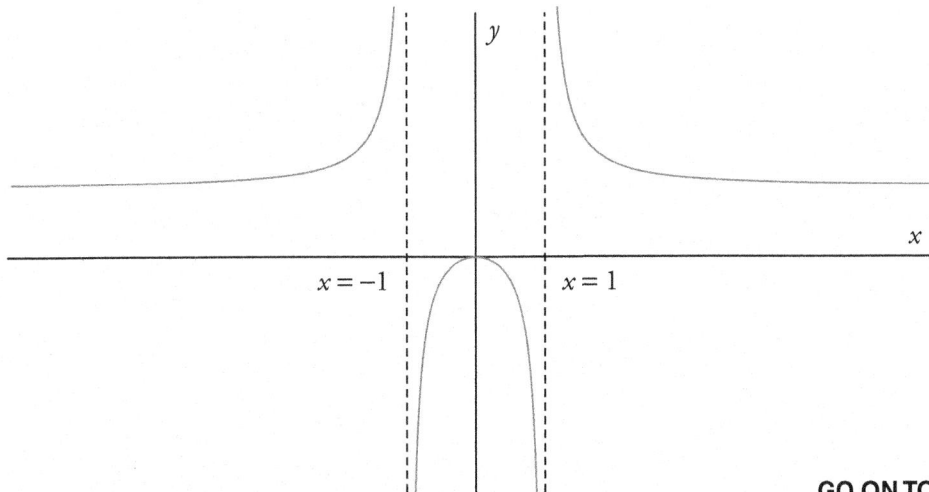

GO ON TO THE NEXT PAGE.

Which of the following could be the graph of $y = f'(x)$?

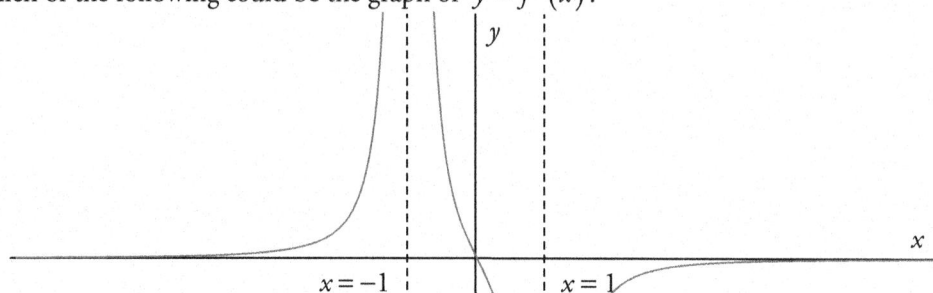

A.

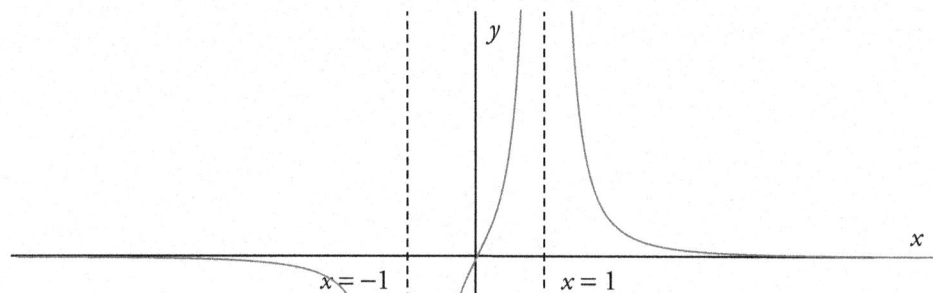

B.

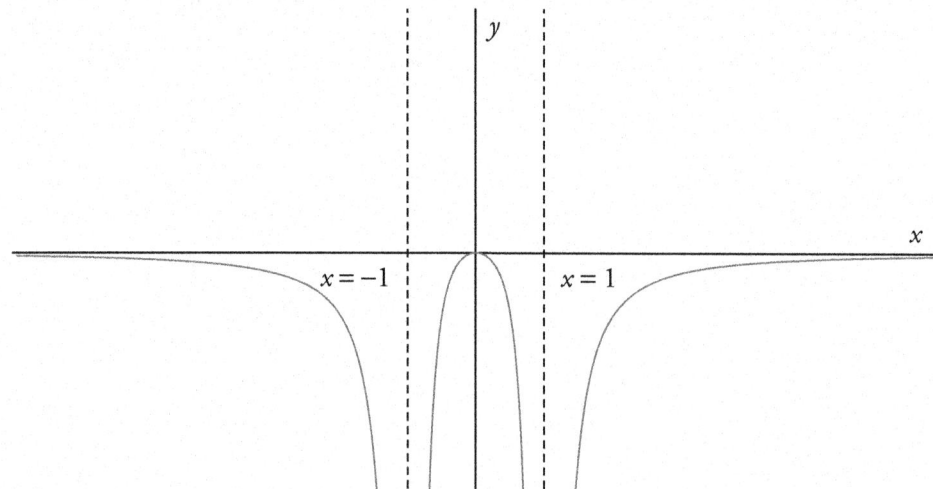

C.

GO ON TO THE NEXT PAGE.

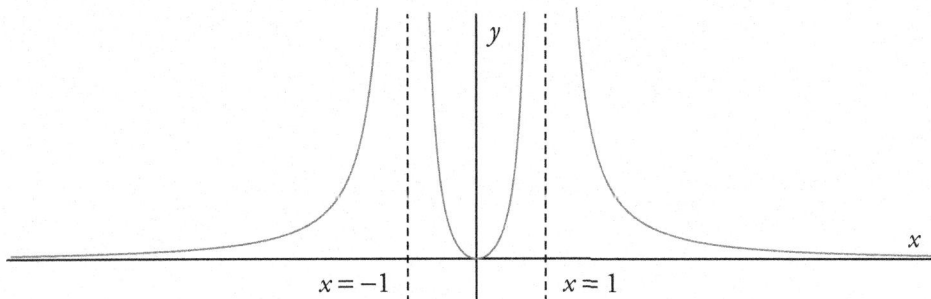

D.

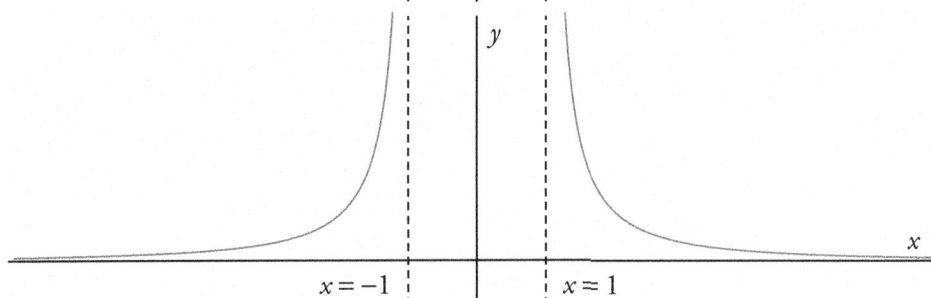

E.

17. The interval of convergence of $\sum_{n=1}^{\infty} \dfrac{x^n}{n(n+1)}$ is $[-1,1]$. The interval of convergence of $\sum_{n=1}^{\infty} \dfrac{x^{n-1}}{(n+1)}$ is

A. $[-1,1]$

B. $[-1,1)$

C. $(-1,1]$

D. $(-1,1)$

E. $(-\infty,\infty)$

GO ON TO THE NEXT PAGE.

18. If $f(x) = \sin^3(x^2)$, then $f'(x) =$

A. $3\sin^2(x^2)$

B. $3\sin^2(x^2)\cos(x^2)$

C. $3x\sin^2(x^2)\cos(x^2)$

D. $6x\sin^2(x^2)\cos(x^2)$

E. $6\sin^2(2x)\cos(2x)$

19. Which of the following sequences converges?

I. $a_n = \dfrac{(-1)^n}{n}$ II. $b_n = (-1)^n$ III. $c_n = (-1)^n n$

A. I only

B. II only

C. III only

D. I and II only

E. II and III only

20. Find the slope at the point $(2,1)$ of the graph of $x^2 - y^2 - x = 1$.

A. $-\dfrac{3}{2}$

B. -1

C. 0

D. 1

E. $\dfrac{3}{2}$

21. Let $f(x)$ be twice differentiable on $[0,1]$. If $f(1) = 2, f'(1) = -3,$ and $\int_0^1 f(x)\,dx = 4$, then $\int_0^1 x^2 f''(x)\,dx =$

A. -7

B. -4

C. 1

D. 4

E. 7

GO ON TO THE NEXT PAGE.

22. Let f and g be differentiable functions on the whole real line, and let $h(x) = f\big(g(x)\big)$. Use the table of values below to find $h'(1)$.

x	$f(x)$	$f'(x)$	$g(x)$	$g'(x)$
-2	-1	2	-3	0
1	4	5	-2	3

A. -15

B. -4

C. 0

D. 6

E. 15

23. Find the slope of the polar curve $r = 1 - \cos(\theta)$ at $\theta = \dfrac{\pi}{6}$.

A. $1 - \dfrac{\sqrt{3}}{2}$

B. $\dfrac{1}{2}$

C. 1

D. $\dfrac{\sqrt{3}}{2}$

E. $1 + \dfrac{\sqrt{3}}{2}$

24. $\displaystyle\int \frac{1}{x^2 - 1}\,dx =$

A. $\dfrac{1}{2}\ln\left|\dfrac{x-1}{x+1}\right| + C$

B. $\dfrac{1}{2}\ln\left|\dfrac{x+1}{x-1}\right| + C$

C. $\dfrac{1}{2}\ln\left|x^2 - 1\right| + C$

D. $\ln\left|x^2 - 1\right| + C$

E. $\tan^{-1}(x) + C$

GO ON TO THE NEXT PAGE.

25. $\displaystyle \lim_{x \to -\infty} \frac{e^{(x+1)^2}}{x^2 + 2x} =$

A. $-\infty$

B. 0

C. 1

D. e

E. ∞

26. $\displaystyle \int_{-1}^{1} \frac{3}{x^4} \, dx =$

A. -2

B. -1

C. 0

D. 2

E. The integral diverges.

27. The Riemann sum $\displaystyle \sum_{i=1}^{360} \sin\left(\frac{\pi i}{360}\right) \cdot \frac{\pi}{360}$ is an approximation for which of the following integrals?

A. $\displaystyle \int_{1}^{360} \sin(x) \, dx$

B. $\displaystyle \int_{1}^{360} \sin(\pi x) \, dx$

C. $\displaystyle \int_{0}^{\pi} \sin(x) \, dx$

D. $\displaystyle \int_{0}^{\pi} \sin(\pi x) \, dx$

E. $\displaystyle \int_{0}^{\pi} \sin(360 x) \, dx$

GO ON TO THE NEXT PAGE.

28. The graph of $f(t)$ is given below, and a, b, c, d, and e are real numbers.

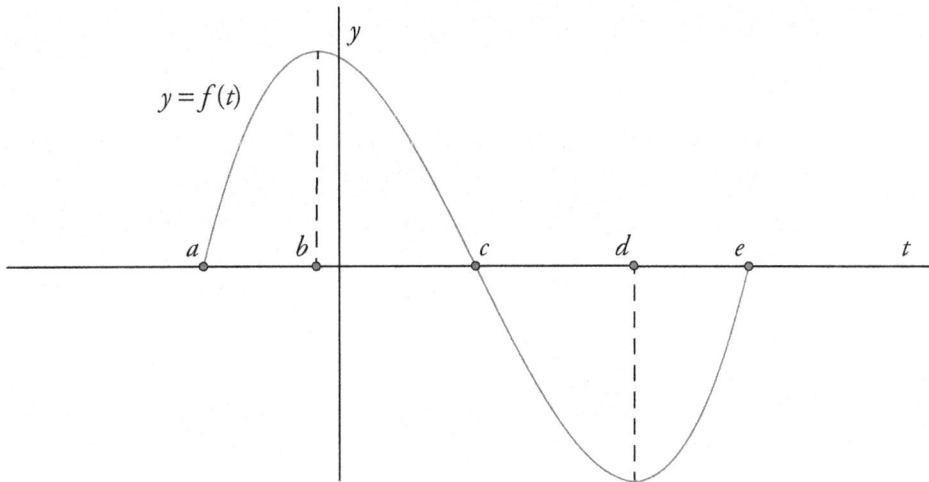

Let $g(x) = \int_a^x f(t)\,dt$. On which of the following intervals is the graph of g concave up?

A. (a,b) only

B. (a,c) only

C. (a,b) and (c,d)

D. (a,b) and (d,e)

E. (b,d) only

CALCULUS BC

SECTION I, Part B

Time – 50 Minutes

Number of questions – 17

A GRAPHING CALCULATOR IS REQUIRED FOR SOME
QUESTIONS ON THIS PART OF THE EXAMINATION.

Directions: Solve each of the following problems, using the available space for scratchwork. After examining the form of the choices, decide which is the best of the choices given and fill in the corresponding oval on the answer sheet. No credit will be given for anything written in the test book. Do not spend too much time on any one problem.

In this test:

1. The **exact** numerical value of the correct answer does not always appear among the choices given. When this happens, select from among the choices the number that best approximates the exact numerical value.

2. Unless otherwise specified, the domain of a function f is assumed to be the set of all real numbers x for which $f(x)$ is a real number.

29. The polynomial function $P_4(x) = 1 - x + \dfrac{x^2}{2!} - \dfrac{x^3}{3!} + \dfrac{x^4}{4!}$ is used to approximate the function $f(x) = e^{-x}$. Find the Lagrange bound on the error when approximating $e^{0.1}$ with $P_4(0.1)$.

    A. 4.17×10^{-3}

    B. 8.33×10^{-4}

    C. 4.17×10^{-6}

    D. 8.33×10^{-8}

    E. 1.39×10^{-9}

30. If $y = \left(\cos(x)\right)^{x^2+1}$, then $y' =$

    A. $\left(x^2 + 1\right)\left(\cos(x)\right)^{x^2}$

    B. $\left(x^2 + 1\right)\left(\cos(x)\right)^{x^2} \sin(x)$

    C. $2x \ln\left(\cos(x)\right) - \left(x^2 + 1\right)\tan(x)$

    D. $\left(2x \ln\left(\cos(x)\right) + \left(x^2 + 1\right)\tan(x)\right)\left(\cos(x)\right)^{x^2+1}$

    E. $\left(2x \ln\left(\cos(x)\right) - \left(x^2 + 1\right)\tan(x)\right)\left(\cos(x)\right)^{x^2+1}$

GO ON TO THE NEXT PAGE.

31. If f is a continuous function on $(-\infty, \infty)$, and $f(x) = \dfrac{\sin(2x)}{5xe^{3x}}$ when $x \neq 0$, then $f(0) =$

A. 0

B. $\dfrac{2}{5}$

C. $\dfrac{2}{3}$

D. 1

E. 2

32. If f is a vector-valued function given by $f(t) = \left(t\sin(t), \cos(t^2)\right)$, then $f'\left(\dfrac{\pi}{3}\right) =$

A. $(1.39, -1.86)$

B. $(0.5, 1.86)$

C. $(0.342, 1.86)$

D. $(1.39, 1.86)$

E. $(0.342, -1.86)$

33. The area of the region bounded by the graph of $f(x) = |x^2 - 5x + 6|$ and the x-axis on the interval $[1,3]$ is

A. $\dfrac{1}{3}$

B. $\dfrac{2}{3}$

C. 1

D. $\dfrac{4}{3}$

E. $\dfrac{5}{3}$

GO ON TO THE NEXT PAGE.

34. $\lim\limits_{h\to 0}\dfrac{\sin^{-1}\left(\dfrac{1}{2}+h\right)-\dfrac{\pi}{6}}{h}=$

A. 0

B. $\dfrac{1}{2}$

C. $\dfrac{\sqrt{3}}{2}$

D. 1

E. $\dfrac{2}{\sqrt{3}}$

35. Find the trapezoidal approximation of $\displaystyle\int_{\pi/6}^{7\pi/6}\dfrac{\cos(x)}{x}\,dx$ using a uniform partition of the interval $\left[\dfrac{\pi}{6},\dfrac{7\pi}{6}\right]$ into 6 subintervals.

A. –0.339

B. 0.020

C. 0.064

D. 0.156

E. 0.651

36. If $x^2-3xy+y^2=16$, then calculate y' at the point in the first quadrant for which $x=2$.

A. 2.05

B. 1.05

C. 1

D. 0.954

E. 0.489

37. $\dfrac{d}{dx}\left(\displaystyle\int_{x^2}^{x^3}\ln(t)\,dt\right)=$

A. $\dfrac{3}{2}$

B. $(9x^2-4x)\ln(x)$

C. $\dfrac{3}{2}\ln(x)$

D. $\dfrac{1}{x^3}-\dfrac{1}{x^2}$

E. $\dfrac{3}{x^3}-\dfrac{2}{x^2}$

GO ON TO THE NEXT PAGE.

38. Let R be the region bounded by the y-axis and the graphs of $f(x) = \cos(x)$ and $g(x) = \sin(x)$ for $0 \leq x \leq \dfrac{\pi}{4}$. When R is revolved about the x-axis, the volume of the resulting solid is

A. 0.785

B. 1.571

C. 2.356

D. 3.142

E. 3.927

39. $\dfrac{dy}{dx} = 3x + 2y$ and $y(1) = 3$. Estimate $y(2)$ using Euler's method with a step size of 0.25.

A. 0

B. 5.25

C. 8.81

D. 14.34

E. 22.83

40. $\int x\sqrt{x-1}\,dx =$

A. $\dfrac{1}{3}x^2(x-1)^{3/2} + C$

B. $\dfrac{1}{2}x^2\sqrt{x-1} + \dfrac{2}{3}x(x-1)^{3/2} + C$

C. $\dfrac{1}{2}x^2\sqrt{\dfrac{1}{2}x^2 - 1} + C$

D. $\dfrac{2}{15}(3x+2)(x-1)^{3/2} + C$

E. $\dfrac{2}{15}(3x-2)(x-1)^{3/2} + C$

41. Find the real number b such that $\displaystyle\int_1^b \dfrac{1}{t}\,dt = -2$.

A. e^2

B. $e^{1/2}$

C. e^{-2}

D. $\dfrac{1}{\sqrt{3}}$

E. $-\dfrac{1}{\sqrt{3}}$

GO ON TO THE NEXT PAGE.

42. Which of the following functions is strictly monotonic on its entire domain?

A. $f(x) = x^3 - 2x$

B. $f(x) = x^3 - x$

C. $f(x) = x^3 + x$

D. $f(x) = x^2 - x$

E. $f(x) = x^2 + x$

43. Let $A = \int_{-1/2}^{1/2} f(x)\,dx$, $B = \int_{-1/2}^{1/2} g(x)\,dx$, and $C = \int_{-1/2}^{1/2} h(x)\,dx$, where $f(x) = \dfrac{1}{2(x^2 + 1)}$, $g(x) = \dfrac{x^2}{x^2 + 1}$, and $h(x) = e^{-x^2}$. Which of the following inequalities is true?

A. $A < B < C$

B. $A < C < B$

C. $B < A < C$

D. $B < C < A$

E. $C < A < B$

44. Find the area enclosed by the inner loop of the limaçon whose equation is $r = 1 - 2\sin(\theta)$.

A. $\pi - \dfrac{3\sqrt{3}}{2}$

B. $\pi + \dfrac{3\sqrt{3}}{2}$

C. $\pi + 3\sqrt{3}$

D. 2π

E. $2\pi + \dfrac{3\sqrt{3}}{2}$

GO ON TO THE NEXT PAGE.

45. The function $f(x) = 3x^5 - 4x^3 - 3x$ is both decreasing and concave up for all x in which of the following intervals?

A. $\left(-\infty, -\sqrt{\dfrac{2}{5}}\right)$

B. $\left(-\sqrt{\dfrac{2}{5}}, 0\right)$

C. $(-1, 0)$

D. $(0, 1)$

E. $(1, \infty)$

CALCULUS BC – SECTION II

Time – 1 hour and 30 minutes
Number of problems – 6
Percent of total grade – 50

GENERAL INSTRUCTIONS

You may wish to look over the problems before starting to work on them, since it is not expected that everyone will be able to complete all parts of all problems. All problems are given equal weight, but the parts of a particular problem are not necessarily given equal weight.

A GRAPHING CALCULATOR IS REQUIRED FOR SOME PROBLEMS OR PARTS OF PROBLEMS ON THIS SECTION OF THE EXAMINATION.

- You should write all work for each part of each problem in the space provided for that part in the booklet. Be sure to write clearly and legibly. If you make an error, you may save time by crossing it out rather than trying to erase it. Erased or crossed-out work will not be graded.

- Show all of your work. You will be graded on the correctness and completeness of your methods as well as your answers. Correct answers without supporting work may not receive credit.

- Justifications require that you give mathematical (noncalculator) reasons and that you clearly identify functions, graphs, tables, or other objects you use.

- You are permitted to use your calculator to solve an equation, find the derivative of a function at a point, or calculate the value of a definite integral. However, you must clearly indicate the setup of your problem, namely the equation, function, or integral you are using. If you use other built-in features or programs, you must show the mathematical steps necessary to produce your results.

- Your work must be expressed in standard mathematical notation rather than calculator syntax. For example, $\int_{1}^{5} x^2 dx$ may not be written as $fnInt(X^2,X,1,5)$.

- Unless otherwise specified, answers (numeric or algebraic) need not be simplified. If your answer is given as a decimal approximation, it should be correct to three places after the decimal point.

- Unless otherwise specified, the domain of a function f is assumed to be the set of all real numbers x for which $f(x)$ is a real number.

A graphing calculator is required for some problems or parts of problems.

During the timed portion for Part A, you may work only on the problems in Part A.

On Part A, you are permitted to use your calculator to solve an equation, find the derivative of a function at a point, or calculate the value of a definite integral. However, you must clearly indicate the setup of your problem, namely the equation, function, or integral you are using. If you use other built-in features or programs, you must show the mathematical steps necessary to produce your results.

1. Let R be the region bounded by the graphs of $f(x) = x$ and $g(x) = x\ln(x)$. Round all answers to the nearest thousandth.

   A. Find the area of R.

   B. Calculate the distance around R.

   C. Calculate the volume of the solid obtained by revolving R about the line $y = 3$.

GO ON TO THE NEXT PAGE.

2. Consider the infinite series $\displaystyle\sum_{n=1}^{\infty} \frac{(-1)^{n+1}}{n^3}$.

A. Show that the series converges absolutely.

B. Determine the number N of terms needed to approximate the sum of the series with an error of less than 0.001.

C. Compute the partial sum $\displaystyle\sum_{n=1}^{N} \frac{(-1)^{n+1}}{n^3}$. Round your answer to the nearest thousandth.

NO CALCULATOR IS ALLOWED FOR THESE PROBLEMS.

During the timed portion for Part B, you may continue to work on the problems in Part A without the use of any calculator.

3. The graphs of $f(x) = \sqrt{1 + \sin(x)}$ and $g(x) = \sin\left(\dfrac{x}{2}\right) + \cos\left(\dfrac{x}{2}\right)$ are shown below.

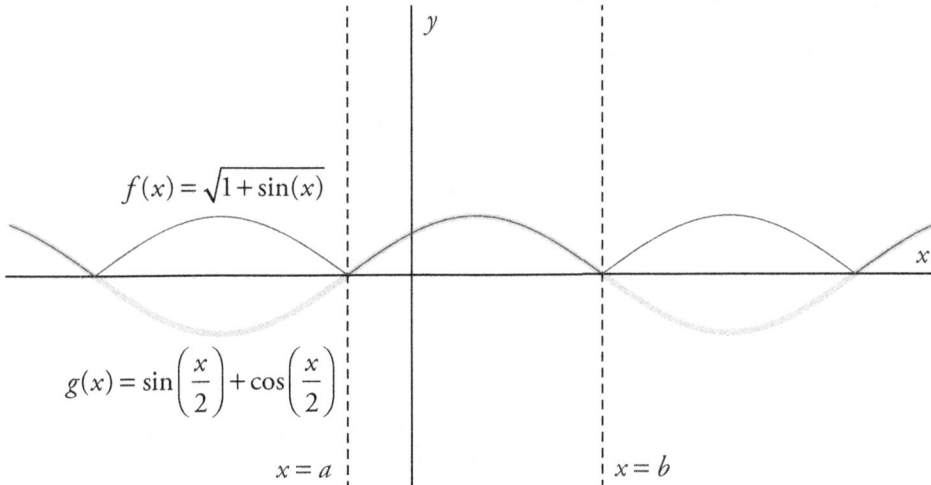

$$f(x) = \sqrt{1 + \sin(x)}$$

$$g(x) = \sin\left(\frac{x}{2}\right) + \cos\left(\frac{x}{2}\right)$$

$$x = a \qquad x = b$$

The two graphs overlap on the interval $[a,b]$.

A. Find a and b.

B. The fact that the two graphs overlap on $[a,b]$ implies that the two functions are *equal* on $[a,b]$. Show that $\sqrt{1 + \sin(x)} = \sin\left(\dfrac{x}{2}\right) + \cos\left(\dfrac{x}{2}\right)$ on $[a,b]$.

C. Use the identity in part (b) to find a fifth degree Maclaurin approximation for $f(x) = \sqrt{1 + \sin(x)}$ on $[a,b]$.

GO ON TO THE NEXT PAGE.

4. A particle moves in the *xy*-plane according to the equations $x(t) = 4\cos\left(\dfrac{2\pi t}{3}\right)$ and $y(t) = -3\sin\left(\dfrac{2\pi t}{3}\right)$ for time t in the interval $[0,6]$.

A. Eliminate the parameter to find the Cartesian equation of the path of the particle.

B. The path of the particle is closed. How many times does the particle go around this path? Explain.

C. Find $\dfrac{dy}{dx}$ for the path in part (a).

D. Find the velocity vector at time $t = 2$.

GO ON TO THE NEXT PAGE.

5. The graph of the **first derivative** of a function f is shown below.

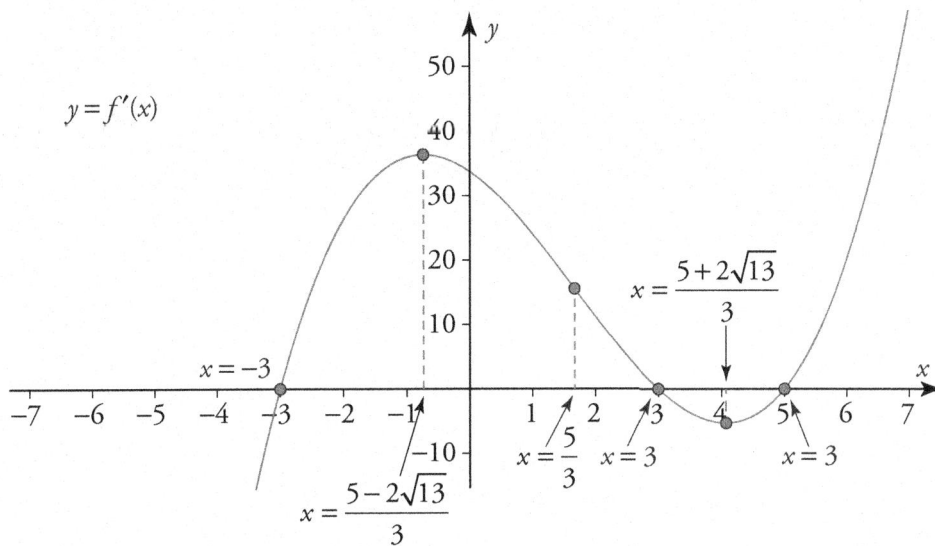

$y = f'(x)$

$x = -3$

$x = \dfrac{5 + 2\sqrt{13}}{3}$

$x = \dfrac{5 - 2\sqrt{13}}{3}$

$x = \dfrac{5}{3}$

$x = 3$

$x = 3$

A. For what value(s) of x does f have a relative maximum? Explain.

B. For what value(s) of x does f have a relative minimum? Explain.

C. For what value(s) of x does f have a point of inflection? Explain.

GO ON TO THE NEXT PAGE.

6. Consider the differential equation $\dfrac{dy}{dx} = -2xy\,\sin\left(x^2\right)$.

A. Solve the equation for y.

B. Find the particular solution for which $y\left(\sqrt{\dfrac{\pi}{2}}\right) = 4$.

C. Find and classify all relative extrema of y in the interval $\left[0, \sqrt{2\pi}\right]$. Justify your answers using either the First or Second Derivative Test.

Answer Key – Section I

1. D	10. E	19. A	28. D	37. B
2. B	11. D	20. E	29. D	38. B
3. D	12. A	21. C	30. E	39. E
4. C	13. C	22. D	31. B	40. D
5. E	14. B	23. C	32. A	41. C
6. A	15. B	24. A	33. C	42. C
7. B	16. A	25. B	34. E	43. C
8. D	17. B	26. E	35. B	44. A
9. C	18. D	27. C	36. A	45. B

Solutions – Section I, Part A

1. Answer: D

Use the Quotient Rule.

$$f'(x) = \frac{\frac{d}{dx}(5x-3)\cdot(2x+7) - (5x-3)\cdot\frac{d}{dx}(2x+7)}{(2x+7)^2}$$

$$f'(x) = \frac{5(2x+7) - (5x-3)2}{(2x+7)^2}$$

$$f'(x) = \frac{41}{(2x+7)^2}$$

2. Answer: B

Integrate the velocity to find the displacement $x(t)$.

$$x(t) = \int v(t)\,dt = \int \frac{2t}{t^2+1}\,dt$$

Let $u = t^2 + 1$. Then $du = 2t\,dt$.

$$x(t) = \int \frac{du}{u} = \ln|u| + C = \ln|t^2+1| + C$$

Use the fact that $x(0) = 5$ to find C.

$$x(0) = \ln|0^2+1| + C = 5 \Rightarrow C = 5$$

$$x(t) = \ln|t^2+1| + 5$$

Now evaluate the displacement at $t = \sqrt{e^4 - 1}$.

$$x\left(\sqrt{e^4-1}\right) = \ln\left|\sqrt{e^4-1}^2 + 1\right| + 5 = \ln\left|e^4 - 1 + 1\right| + 5 = \ln\left|e^4\right| + 5 = 4 + 5 = 9$$

3. Answer: D

Factor a 2 out of the numerator of the summand.

$$\sum_{n=0}^{\infty} \frac{2^{n+1}}{3^n} = \sum_{n=0}^{\infty} 2\frac{2^n}{3^n} = \sum_{n=0}^{\infty} 2\left(\frac{2}{3}\right)^n$$

This is a geometric series with $a = 2$ and $r = \frac{2}{3}$. Its sum is $S = \dfrac{2}{1 - \dfrac{2}{3}} = 6$.

4. Answer: C

By the definition of a derivative, this limit is equal to $\dfrac{d}{dx}\big(\sec(x)\big) = \sec(x)\tan(x)$.

5. Answer: E

Let $u = \dfrac{5}{x}$. Then $du = -\dfrac{5}{x^2}\,dx \Rightarrow \dfrac{1}{x^2}\,dx = -\dfrac{1}{5}\,du$. Furthermore $x = 1 \Rightarrow u = 5$, and $x = 5 \Rightarrow u = 1$.

$$\int_1^5 \frac{e^{5/x}}{x^2}\,dx = -\frac{1}{5}\int_5^1 e^u\,du = -\frac{1}{5}e^u\Big|_5^1 = -\frac{1}{5}\big(e^1 - e^5\big) = \frac{1}{5}\big(e^5 - e\big)$$

6. Answer: A

$$s = \int_0^2 \sqrt{\left(\frac{d}{dt}\big(4t^3\big)\right)^2 + \left(\frac{d}{dt}\big(3t^2\big)\right)^2}\,dt$$

$$s = \int_0^2 \sqrt{\big(12t^2\big)^2 + \big(6t\big)^2}\,dt = \int_0^2 \sqrt{144t^4 + 36t^2}\,dt = \int_0^2 \sqrt{36t^2\big(4t^2+1\big)}\,dt$$

$$s = \int_0^2 6t\sqrt{4t^2+1}\,dt$$

7. Answer: B

From the direction field, we see that $\displaystyle\lim_{x\to 0^-}\frac{dy}{dx} = -\infty$, and $\displaystyle\lim_{x\to 0^+}\frac{dy}{dx} = \infty$. The only equation that satisfies both of these properties is $\dfrac{dy}{dx} = \dfrac{1}{x}$.

8. Answer: D

We first note that for this partition $\Delta x_i = \dfrac{3-0}{3} = 1$. The partition points are $x_0 = 0$, $x_1 = 1, x_2 = 2$, and $x_3 = 3$.

$$\sum_{i=1}^3 f\big(x_{i-1}\big)\Delta x_i = \big(f\big(x_0\big) + f\big(x_1\big) + f\big(x_2\big)\big)\Delta x_i = \big(f\big(0\big) + f\big(1\big) + f\big(2\big)\big)\Delta x_i$$

$$\sum_{i=1}^3 f\big(x_{i-1}\big)\Delta x_i = \big(1 + 4 + k\big)\cdot 1 = 14 \Rightarrow 5 + k = 14 \Rightarrow k = 9$$

9. Answer: C

Even if $f(x)$ and $g(x)$ are both continuous for all x in $[a,b]$, $\dfrac{f(x)}{g(x)}$ will not be continuous for any x in $[a,b]$ for which $g(x) = 0$.

10. Answer: E

The answer is clearly not A or B, as this Maclaurin series does not match those of either $\sin(x)$ or $\cos(x)$. Also, this Maclaurin series has only odd powers of x, which means that it represents an odd function. The functions in C and D are both even. That leaves E.

$$e^x = 1 + x + \frac{x^2}{2!} + \frac{x^3}{3!} + \frac{x^4}{4!} + \frac{x^5}{5!} + \cdots \text{ and } e^{-x} = 1 - x + \frac{x^2}{2!} - \frac{x^3}{3!} + \frac{x^4}{4!} - \frac{x^5}{5!} + \cdots.$$

Subtracting these two series yields the following:

$$e^x - e^{-x} = \left(1 + x + \frac{x^2}{2!} + \frac{x^3}{3!} + \frac{x^4}{4!} + \frac{x^5}{5!} + \cdots\right) - \left(1 - x + \frac{x^2}{2!} - \frac{x^3}{3!} + \frac{x^4}{4!} - \frac{x^5}{5!} + \cdots\right)$$

$$e^x - e^{-x} = 2x + 2\frac{x^3}{3!} + 2\frac{x^5}{5!} + \cdots = 2\left(x + \frac{x^3}{3!} + \frac{x^5}{5!} + \cdots\right)$$

Dividing by 2, we obtain the final result.

$$\frac{1}{2}\left(e^x - e^{-x}\right) = x + \frac{x^3}{3!} + \frac{x^5}{5!} + \cdots$$

11. Answer: D

$x, y,$ and h are related by the Pythagorean Theorem as follows:

$$\left(\frac{y}{2}\right)^2 + h^2 = x^2$$

We use this equation to find h at the instant when $x = 13$ and $y = 10$.

$$\left(\frac{10}{2}\right)^2 + h^2 = 13^2 \Rightarrow h = 12$$

Now we differentiate the equation with respect to t.

$$\frac{d}{dt}\left(\left(\frac{y}{2}\right)^2 + h^2\right) = \frac{d}{dt}\left(x^2\right)$$

$$2\left(\frac{y}{2}\right)\frac{d}{dt}\left(\frac{y}{2}\right) + 2h\frac{dh}{dt} = 2x\frac{dx}{dt}$$

$$2\left(\frac{y}{2}\right) \cdot \frac{1}{2}\frac{dy}{dt} + 2h\frac{dh}{dt} = 2x\frac{dx}{dt}$$

$$\frac{y}{2}\frac{dy}{dt} + 2h\frac{dh}{dt} = 2x\frac{dx}{dt}$$

Now substitute $x = 13, \ y = 10, h = 12, \frac{dy}{dt} = 4\frac{dx}{dt},$ and $\frac{dh}{dt} = a\frac{dx}{dt}$.

$$\frac{10}{2} \cdot 4\frac{dx}{dt} + 24a\frac{dx}{dt} = 26\frac{dx}{dt}$$

$$\left(20+24a\right)\frac{dx}{dt}=26\frac{dx}{dt}$$

Since the length x is increasing, $\dfrac{dx}{dt}\neq0$, and so we can divide both sides by $\dfrac{dx}{dt}$.

$$20+24a=26\Rightarrow a=\frac{1}{4}$$

12. Answer: A

Separate variables by dividing both sides of the differential equation by $\sqrt{y+1}$ and multiplying by dx.

$$\frac{1}{\sqrt{y+1}}\frac{dy}{dx}dx=\frac{4x\sqrt{y+1}}{\sqrt{y+1}}dx$$

$$\frac{1}{\sqrt{y+1}}dy=4x\,dx$$

$$\int\frac{1}{\sqrt{y+1}}dy=4\int x\,dx$$

On the left side, let $u=y+1$, so $du=dy$. Also, write the radicand with a negative fractional exponent.

$$\int u^{-1/2}du=4\int x\,dx$$

$$2u^{1/2}=2x^{2}+C$$

$$2\sqrt{y+1}=2x^{2}+C$$

Plug in $y=8$ and $x=1$ to find C.

$$2\sqrt{8+1}=2\left(1\right)^{2}+C\Rightarrow C=4$$

So $2\sqrt{y+1}=2x^{2}+4$. Now solve this for y.

$$\frac{1}{2}\cdot2\sqrt{y+1}=\frac{1}{2}\cdot\left(2x^{2}+4\right)$$

$$\sqrt{y+1}=x^{2}+2$$

$$\sqrt{y+1}^{\,2}=\left(x^{2}+2\right)^{2}$$

$$y+1=\left(x^{2}+2\right)^{2}$$

$$y=\left(x^{2}+2\right)^{2}-1$$

13. Answer: C

$$\frac{dx}{dt}=-3e^{-3t}\text{ and }\frac{dy}{dt}=\frac{1}{t}.\quad\frac{dy}{dx}=\frac{dy/dt}{dx/dt}=\frac{1/t}{-3e^{-3t}}=-\frac{e^{3t}}{3t}$$

14. Answer: B

$$\int \left(\sin(x)+\cos(x)\right)^2 dx = \int \left(\sin^2(x)+2\sin(x)\cos(x)+\cos^2(x)\right)dx$$

On the right side, use the identities $\sin^2(x)+\cos^2(x)=1$ and $2\sin(x)\cos(x)=\sin(2x)$.

$$\int \left(\sin(x)+\cos(x)\right)^2 dx = \int \left(1+\sin(2x)\right)dx$$

Let $u = 2x$, so $du = 2dx \Rightarrow dx = \frac{1}{2}du$.

$$\int \left(\sin(x)+\cos(x)\right)^2 dx = \frac{1}{2}\int \left(1+\sin(u)\right)du = \frac{1}{2}\left(u-\cos(u)\right)+C$$

$$\int \left(\sin(x)+\cos(x)\right)^2 dx = \frac{1}{2}\left(2x-\cos(2x)\right)+C = x-\frac{1}{2}\cos(2x)+C$$

15. Answer: B

Since the graph of $f(x)=4-x^2$ is easy to sketch without a calculator, we will solve this using graphical reasoning.

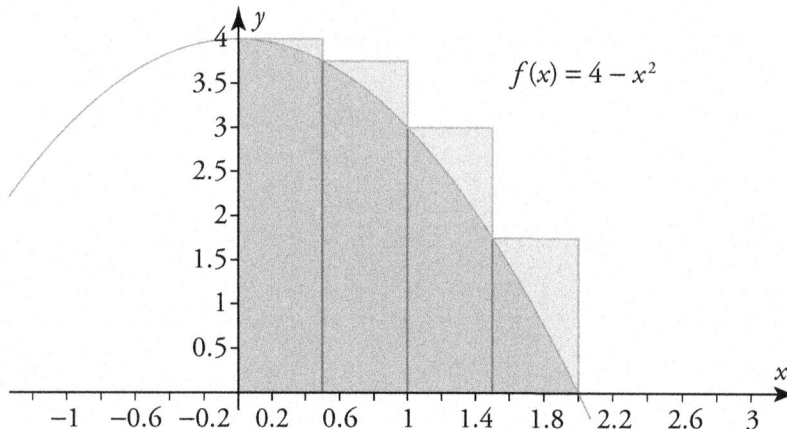

The left Riemann sum is greater than the integral. That is: $\int_0^2 \left(4-x^2\right)dx < L$. Eliminate A, D, and E.

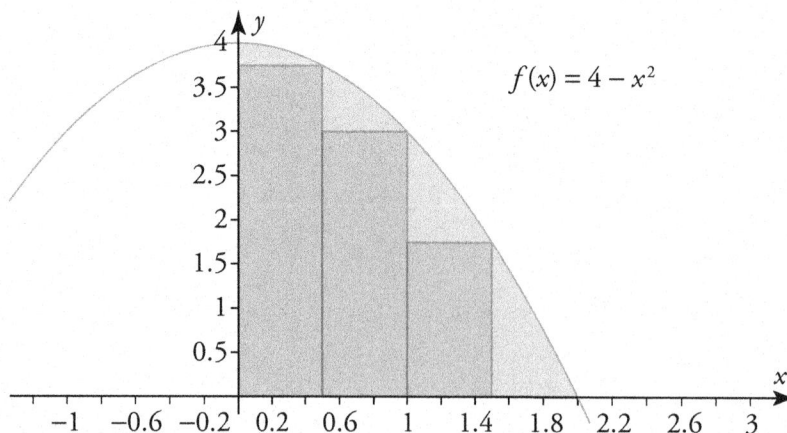

The right Riemann sum is less than the integral, which means that $R < \int_0^2 \left(4-x^2\right)dx$. This does not allow us to eliminate another answer choice.

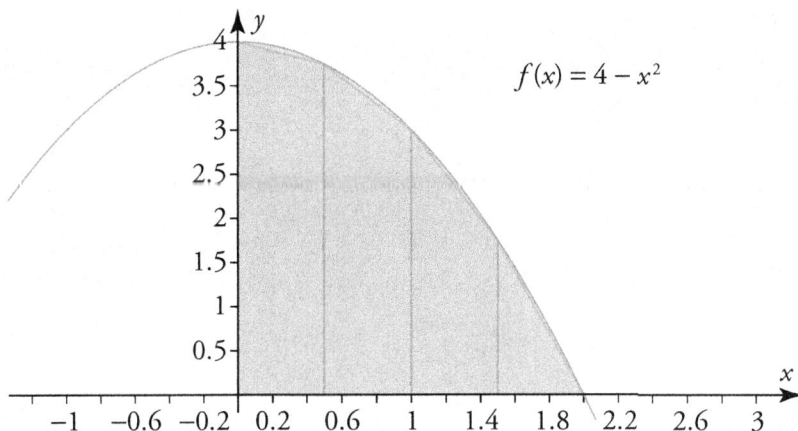

The trapezoidal sum is less than the integral, so $T < \int_0^2 \left(4 - x^2\right) dx$. Eliminate C, and choose B.

16. Answer: A

f is an even function, so f' must be odd. Eliminate C, D, and E, all of which are even. f is increasing on $(-\infty, -1)$, so $f'(x)$ must be positive for all x in this interval. Eliminate B, and choose A.

17. Answer: B

Since $\dfrac{d}{dx}\left(\displaystyle\sum_{n=1}^{\infty} \dfrac{x^n}{n(n+1)}\right) = \displaystyle\sum_{n=1}^{\infty} \dfrac{x^{n-1}}{n+1}$, the two power series have the same *radius* of convergence. We need only to test the second power series at the endpoints of the interval of convergence of the first series.

$x = -1$:

$$\sum_{n=1}^{\infty} \frac{(-1)^{n-1}}{n+1}$$

This is the alternating harmonic series, which converges.

$x = 1$:

$$\sum_{n=1}^{\infty} \frac{(1)^{n-1}}{n+1} = \sum_{n=1}^{\infty} \frac{1}{n+1}$$

This is the harmonic series, which diverges.

The interval of convergence of the second series is $[-1, 1)$.

18. Answer: D

Use the Chain Rule twice, as f is a doubly composite function.

$$f'(x) = \frac{d}{dx}\left(\sin^3\left(x^2\right)\right) = 3\sin^2\left(x^2\right) \cdot \frac{d}{dx}\left(\sin\left(x^2\right)\right)$$

$$f'(x) = 3\sin^2\left(x^2\right)\cos\left(x^2\right) \cdot \frac{d}{dx}\left(x^2\right)$$

$$f'(x) = 3\sin^2\left(x^2\right)\cos\left(x^2\right) \cdot 2x = 6x\sin^2\left(x^2\right)\cos\left(x^2\right)$$

19. Answer: A

$\lim\limits_{n \to \infty} \dfrac{(-1)^n}{n} = 0$, so eliminate B, C, and E.

$\lim\limits_{n \to \infty} (-1)^n$ does not exist, so eliminate D, and choose A.

20. Answer: E

Use Implicit Differentiation to find y'.

$$\frac{d}{dx}\left(x^2 - y^2 - x\right) = \frac{d}{dx}(1)$$

$$2x - 2yy' - 1 = 0$$

$$y' = \frac{2x - 1}{2y}$$

Evaluate y' at the point $(2,1)$ to find the slope m.

$$m = \frac{2x-1}{2y}\bigg|_{(2,1)} = \frac{2(2)-1}{2(1)} = \frac{3}{2}$$

21. Answer: C

Integrate by parts. Let $u = x^2$, so $dv = f''(x)dx$. Then $du = 2x\,dx$, and $v = \int f''(x)\,dx = f'(x)$. We needn't worry about the constant of integration, as this is a definite integral.

$$\int_0^1 x^2 f''(x)\,dx = \left(x^2 f'(x)\right)\bigg|_0^1 - 2\int_0^1 x f'(x)\,dx$$

Integrate by parts again. Let $u = x$, so $dv = f'(x)dx$. Then $du = dx$, and $v = \int f'(x)\,dx = f(x)$. Again, we will not worry about the integration constant.

$$\int_0^1 x^2 f''(x)\,dx = \left(x^2 f'(x)\right)\bigg|_0^1 - 2\left(\left(x f(x)\right)\bigg|_0^1 - \int_0^1 f(x)\,dx \right)$$

$$\int_0^1 x^2 f''(x)\,dx = \left(x^2 f'(x)\right)\bigg|_0^1 - 2\left(x f(x)\right)\bigg|_0^1 + 2\int_0^1 f(x)\,dx$$

$$\int_0^1 x^2 f''(x)\,dx = \left(1^2 f'(1) - 0^2 f'(0)\right) - 2\left(1 f(1) - 0 f(0)\right) + 2\int_0^1 f(x)\,dx$$

$$\int_0^1 x^2 f''(x)\,dx = \left(1^2(-3) - 0^2 f(0)\right) - 2\left(1(2) - 0 f(0)\right) + 2(4) = 1$$

22. Answer: D

Use the Chain Rule.

$$h'(x) = f'(g(x))g'(x)$$

$$h'(1) = f'(g(1))g'(1) - f'(-2)(3) - (2)(3) - 6$$

23. Answer: C

First we find $\dfrac{dy}{d\theta}$.

$$\frac{dy}{d\theta} = \frac{d}{d\theta}(r\sin(\theta)) = \frac{d}{d\theta}\left((1-\cos(\theta))\sin(\theta)\right)$$

$$\frac{dy}{d\theta} = \frac{d}{d\theta}(1-\cos(\theta)) \cdot \sin(\theta) + (1-\cos(\theta)) \cdot \frac{d}{d\theta}(\sin(\theta))$$

$$\frac{dy}{d\theta} = \sin(\theta) \cdot \sin(\theta) + (1-\cos(\theta)) \cdot \cos(\theta) = \sin^2(\theta) + \cos(\theta) - \cos^2(\theta)$$

$$\frac{dy}{d\theta} = \cos(\theta) - \cos(2\theta)$$

Next we find $\dfrac{dx}{d\theta}$.

$$\frac{dx}{d\theta} = \frac{d}{d\theta}(r\cos(\theta)) = \frac{d}{d\theta}\left((1-\cos(\theta))\cos(\theta)\right)$$

$$\frac{dx}{d\theta} = \frac{d}{d\theta}(1-\cos(\theta)) \cdot \cos(\theta) + (1-\cos(\theta)) \cdot \frac{d}{d\theta}(\cos(\theta))$$

$$\frac{dx}{d\theta} = \sin(\theta) \cdot \cos(\theta) + (1-\cos(\theta)) \cdot (-\sin(\theta)) = 2\sin(\theta)\cos(\theta) - \sin(\theta)$$

$$\frac{dx}{d\theta} = \sin(2\theta) - \sin(\theta)$$

Now we find $\dfrac{dy}{dx}$.

$$\frac{dy}{dx} = \frac{dy/d\theta}{dx/d\theta} = \frac{\cos(\theta) - \cos(2\theta)}{\sin(2\theta) - \sin(\theta)}$$

Finally, evaluate $\dfrac{dy}{dx}$ at $\theta = \dfrac{\pi}{6}$ to find the slope m.

$$m = \left.\frac{\cos(\theta) - \cos(2\theta)}{\sin(2\theta) - \sin(\theta)}\right|_{\theta=\frac{\pi}{6}} = \frac{\cos\left(\dfrac{\pi}{6}\right) - \cos\left(\dfrac{\pi}{3}\right)}{\sin\left(\dfrac{\pi}{3}\right) - \sin\left(\dfrac{\pi}{6}\right)} = \frac{\dfrac{\sqrt{3}}{2} - \dfrac{1}{2}}{\dfrac{\sqrt{3}}{2} - \dfrac{1}{2}} = 1$$

24. Answer: A

Decompose the integrand into partial fractions.

$$\frac{1}{x^2-1} = \frac{1}{(x-1)(x+1)} = \frac{A}{x-1} + \frac{B}{x+1}$$

$$1 = A(x+1) + B(x-1)$$

Plug in $x=1$ and $x=-1$ to find A and B.

$$x=1 \Rightarrow 1 = 2A \Rightarrow A = \frac{1}{2}$$

$$x=-1 \Rightarrow 1 = -2B \Rightarrow B = -\frac{1}{2}$$

$$\int \frac{1}{x^2-1}\,dx = \frac{1}{2}\int\left(\frac{1}{x-1} - \frac{1}{x+1}\right)dx = \frac{1}{2}\left(\ln|x-1| - \ln|x+1|\right) + C$$

$$\int \frac{1}{x^2-1}\,dx = \frac{1}{2}\ln\left|\frac{x-1}{x+1}\right| + C$$

25. Answer: B

This limit yields the indeterminate form $\frac{\infty}{\infty}$. Apply L'Hôpital's Rule.

$$\lim_{x\to-\infty} \frac{e^{(x+1)^2}}{x^2+2x} = \lim_{x\to-\infty} \frac{\dfrac{d}{dx}\left(e^{(x+1)^2}\right)}{\dfrac{d}{dx}\left(x^2+2x\right)} = \lim_{x\to-\infty} \frac{e^{(x+1)^2}\cdot\dfrac{d}{dx}\left((x+1)^2\right)}{2x+2}$$

$$\lim_{x\to-\infty} \frac{e^{(x+1)^2}}{x^2+2x} = \lim_{x\to-\infty} \frac{(2x+2)e^{(x+1)^2}}{2x+2} = \lim_{x\to-\infty} e^{(x+1)^2} = \infty$$

26. Answer: E

This integral is improper because the integrand is unbounded at $x=0$, which is in the interval of integration.

$$\int_{-1}^{1} \frac{3}{x^4}\,dx = \int_{-1}^{0} \frac{3}{x^4}\,dx + \int_{0}^{1} \frac{3}{x^4}\,dx$$

We work with the first integral on the right side.

$$\int_{-1}^{0} \frac{3}{x^4}\,dx = \lim_{t\to 0^-}\int_{-1}^{t} \frac{3}{x^4}\,dx = -\lim_{t\to 0^-} \frac{1}{x^3}\Big|_{-1}^{t} = -\lim_{t\to 0^-}\left(\frac{1}{t^3}+1\right) = \infty$$

Since this part diverges, the entire improper integral diverges.

27. Answer: C

We see from the upper limit of summation that $n = 360$. From the rightmost factor in the summand we see that $\Delta x_i = \dfrac{b-a}{n} = \dfrac{\pi}{360}$, so the length of the interval of integration is π. Eliminate A and B. The argument of the sine function in the summand is $x_i = i\Delta x_i$, which indicates that the integral of $\sin(x)$ is being approximated by a right Riemann sum. Since the integrand is $\sin(x)$, eliminate D and E, and choose C.

28. Answer: D

Compute $g''(x)$.

$$g'(x) = \frac{d}{dx}\left(\int_a^x f(t)\,dt\right) = f(x)$$

$$g''(x) = f'(x)$$

The graph of g is concave up whenever $g''(x) > 0$. From the above equation we see that $g''(x) > 0 \Rightarrow f'(x) > 0 \Rightarrow f$ is increasing. Since f is increasing on (a,b) and (d,e), the graph of g is concave up on those intervals as well.

29. Answer: D

The error bound is as follows:

$$\left|R_5\left(0.1\right)\right| \leq \max_{[0,0.1]}\left|\frac{f^{(5)}\left(z\right)}{5!}\left(0.1\right)^5\right| = \frac{\left(0.1\right)^5}{5!}\max_{[0,0.1]}\left|f^{(5)}\left(z\right)\right| = \frac{\left(0.1\right)^5}{5!}\max_{[0,0.1]}\left|-e^{-z}\right|$$

Since the maximum value of $\left|f^{(5)}\left(z\right)\right| = \left|-e^{-z}\right|$ for $0 \leq z \leq 0.1$ is 1, the error bound is given as follows:

$$\left|R_5\left(0.1\right)\right| \leq \frac{\left(0.1\right)^5}{5!} \approx 8.33\times10^{-8}$$

30. Answer: E

Use Logarithmic Differentiation.

$$\ln\left(y\right) = \ln\left(\left(\cos\left(x\right)\right)^{x^2+1}\right)$$

$$\ln\left(y\right) = \left(x^2+1\right)\ln\left(\cos\left(x\right)\right)$$

$$\frac{d}{dx}\left(\ln\left(y\right)\right) = \frac{d}{dx}\left(\left(x^2+1\right)\ln\left(\cos\left(x\right)\right)\right)$$

$$\frac{y'}{y} = \frac{d}{dx}\left(x^2+1\right)\cdot\ln\left(\cos\left(x\right)\right) + \left(x^2+1\right)\cdot\frac{d}{dx}\left(\ln\left(\cos\left(x\right)\right)\right)$$

$$\frac{y'}{y} = 2x\ln\left(\cos\left(x\right)\right) + \left(x^2+1\right)\frac{1}{\cos\left(x\right)}\cdot\frac{d}{dx}\left(\cos\left(x\right)\right)$$

$$\frac{y'}{y} = 2x\ln\left(\cos\left(x\right)\right) + \left(x^2+1\right)\frac{1}{\cos\left(x\right)}\cdot\left(-\sin\left(x\right)\right)$$

$$\frac{y'}{y} = 2x\ln\left(\cos\left(x\right)\right) - \left(x^2+1\right)\tan\left(x\right)$$

$$y' = \left(2x\ln\left(\cos\left(x\right)\right) - \left(x^2+1\right)\tan\left(x\right)\right)y$$

$$y' = \left(2x\ln\left(\cos\left(x\right)\right) - \left(x^2+1\right)\tan\left(x\right)\right)\left(\cos\left(x\right)\right)^{x^2+1}$$

31. Answer: B

In order for f to be continuous at $x = 0$, then f must satisfy the condition $\lim\limits_{x\to0} f\left(x\right) = f\left(0\right)$. On any open interval containing $x = 0$, $f\left(x\right) = \dfrac{\sin\left(2x\right)}{5xe^{3x}}$. So the condition can be restated as follows:

$$f\left(0\right) = \lim_{x\to0}\frac{\sin\left(2x\right)}{5xe^{3x}}$$

This limit yields the indeterminate form $\dfrac{0}{0}$. Apply L'Hôpital's Rule.

$$f(0) = \lim_{x \to 0} \frac{\dfrac{d}{dx}\left(\sin(2x)\right)}{\dfrac{d}{dx}\left(5xe^{3x}\right)} = \lim_{x \to 0} \frac{2\cos(2x)}{\dfrac{d}{dx}(5x)\cdot e^{3x} + 5x \cdot \dfrac{d}{dx}\left(e^{3x}\right)}$$

$$f(0) = \lim_{x \to 0} \frac{2\cos(2x)}{5e^{3x} + 15xe^{3x}} = \frac{2}{5}$$

32. Answer: A

First compute $f'(t)$. We will need to use the Product Rule on the first component and the Chain Rule on the second.

$$f'(t) = \left(\frac{d}{dt}\left(t\sin(t)\right), \frac{d}{dt}\left(\cos\left(t^2\right)\right) \right)$$

$$f'(t) = \left(\frac{d}{dt}(t) \cdot \sin(t) + t \cdot \frac{d}{dt}\left(\sin(t)\right), -\sin\left(t^2\right) \cdot \frac{d}{dt}\left(t^2\right) \right)$$

$$f'(t) = \left(\sin(t) + t\cos(t), -2t\sin\left(t^2\right) \right)$$

Now evaluate f' at $t = \dfrac{\pi}{3}$.

$$f'(t) = \left(\sin\left(\frac{\pi}{3}\right) + \frac{\pi}{3}\cos\left(\frac{\pi}{3}\right), -\frac{2\pi}{3}\sin\left(\left(\frac{\pi}{3}\right)^2\right) \right) \approx (1.39, -1.86)$$

33. Answer: C

We sketch the region below.

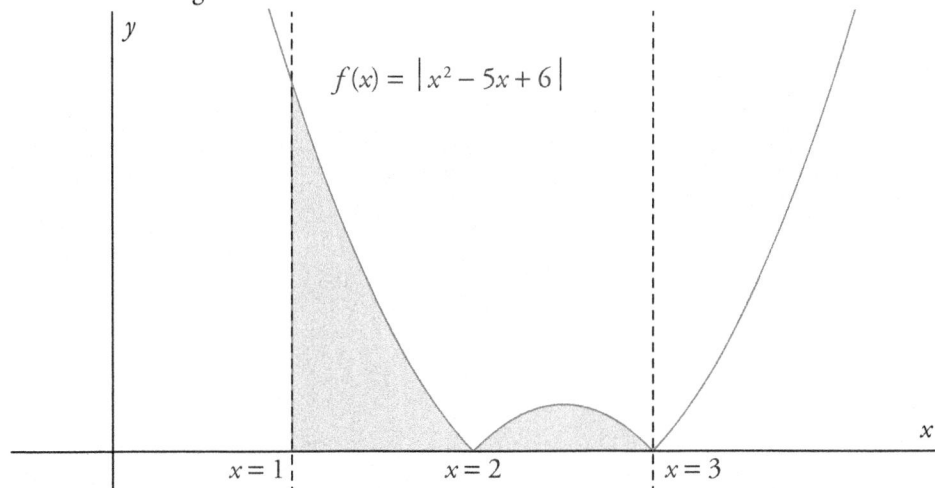

$$f(x) = \left| x^2 - 5x + 6 \right|$$

On $(1,2)$ $f(x) = x^2 - 5x + 6$, and on $(2,3)$ $f(x) = -\left(x^2 - 5x + 6\right) = -x^2 + 5x - 6$. We compute the area A as follows:

$$A = \int_1^3 \left| x^2 - 5x + 6 \right| dx = \int_1^2 \left(x^2 - 5x + 6 \right) dx + \int_2^3 \left(-x^2 + 5x - 6 \right) dx$$

$$A = \frac{5}{6} + \frac{1}{6} = 1$$

34. Answer: E

By definition, this is the derivative of $\sin^{-1}(x)$ evaluated at $x = \dfrac{1}{2}$.

$$\lim_{h \to 0} \frac{\sin^{-1}\left(\dfrac{1}{2} + h \right) - \dfrac{\pi}{6}}{h} = \lim_{h \to 0} \frac{\sin^{-1}\left(\dfrac{1}{2} + h \right) - \sin^{-1}\left(\dfrac{1}{2} \right)}{h} = \frac{d}{dx} \left(\sin^{-1}(x) \right) \Big|_{\frac{1}{2}}$$

$$\lim_{h \to 0} \frac{\sin^{-1}\left(\dfrac{1}{2} + h \right) - \dfrac{\pi}{6}}{h} = \frac{1}{\sqrt{1 - x^2}} \Big|_{\frac{1}{2}} = \frac{1}{\sqrt{1 - \left(\dfrac{1}{2} \right)^2}} = \frac{2}{\sqrt{3}}$$

35. Answer: B

For this partition $x_0 = \dfrac{\pi}{6}, x_1 = \dfrac{2\pi}{6} = \dfrac{\pi}{3}, x_2 = \dfrac{3\pi}{6} = \dfrac{\pi}{2}, x_3 = \dfrac{4\pi}{6} = \dfrac{2\pi}{3}, x_4 = \dfrac{5\pi}{6}, x_5 = \dfrac{6\pi}{6} = \pi,$ and

$x_6 = \dfrac{7\pi}{6}$. Also, we have $\dfrac{b - a}{2n} = \dfrac{\dfrac{7\pi}{6} - \dfrac{\pi}{6}}{2(6)} = \dfrac{\pi}{12}$.

$$\int_{\pi/6}^{7\pi/6} \frac{\cos(x)}{x} dx \approx \frac{\pi}{12} \left[f(x_0) + 2f(x_1) + 2f(x_2) + 2f(x_3) + 2f(x_4) + 2f(x_5) + f(x_6) \right]$$

$$\int_{\pi/6}^{7\pi/6} \frac{\cos(x)}{x} dx \approx \frac{\pi}{12} \left[f\left(\frac{\pi}{6} \right) + 2f\left(\frac{\pi}{3} \right) + 2f\left(\frac{\pi}{2} \right) + 2f\left(\frac{2\pi}{3} \right) + 2f\left(\frac{5\pi}{6} \right) + 2f(\pi) + f\left(\frac{7\pi}{6} \right) \right]$$

$$\int_{\pi/6}^{7\pi/6} \frac{\cos(x)}{x} dx \approx 0.156$$

36. Answer: A

First we will find the y-coordinate of the point in question.

$$2^2 - 3(2)y + y^2 = 16$$

$$4 - 6y + y^2 = 16$$

$$y^2 - 6y - 12 = 0$$

$$y = 3 \pm \sqrt{21}$$

The points on the graph corresponding to $x = 2$ are $\left(2, 3 + \sqrt{21}\right)$ and $\left(2, 3 - \sqrt{21}\right)$. The first of these two points is in the first quadrant, and so that is the one that we'll use. Now find y' using implicit differentiation.

$$\frac{d}{dx}\left(x^2 - 3xy + y^2\right) = \frac{d}{dx}(16)$$

$$2x - \left(\frac{d}{dx}(3x) \cdot y + 3x \cdot \frac{d}{dx}(y)\right) + 2yy' = 0$$

$$2x - 3y - 3xy' + 2yy' = 0$$

$$(2y - 3x)y' = 3y - 2x$$

$$y' = \frac{3y - 2x}{2y - 3x}$$

Finally, evaluate y' at the point $\left(2, 3 + \sqrt{21}\right)$.

$$y'\big|_{\left(2, 3+\sqrt{21}\right)} = \frac{3\left(3 + \sqrt{21}\right) - 2(2)}{2\left(3 + \sqrt{21}\right) - 3(2)} \approx 2.05$$

37. Answer: B

Use Leibniz' Rule.

$$\frac{d}{dx}\left(\int_{x^2}^{x^3} \ln(t)\,dt\right) = \ln\left(x^3\right) \cdot \frac{d}{dx}\left(x^3\right) - \ln\left(x^2\right) \cdot \frac{d}{dx}\left(x^2\right)$$

$$\frac{d}{dx}\left(\int_{x^2}^{x^3} \ln(t)\,dt\right) = 3\ln(x) \cdot 3x^2 - 2\ln(x) \cdot 2x = \left(9x^2 - 4x\right)\ln(x)$$

38. Answer: B

We sketch the region below.

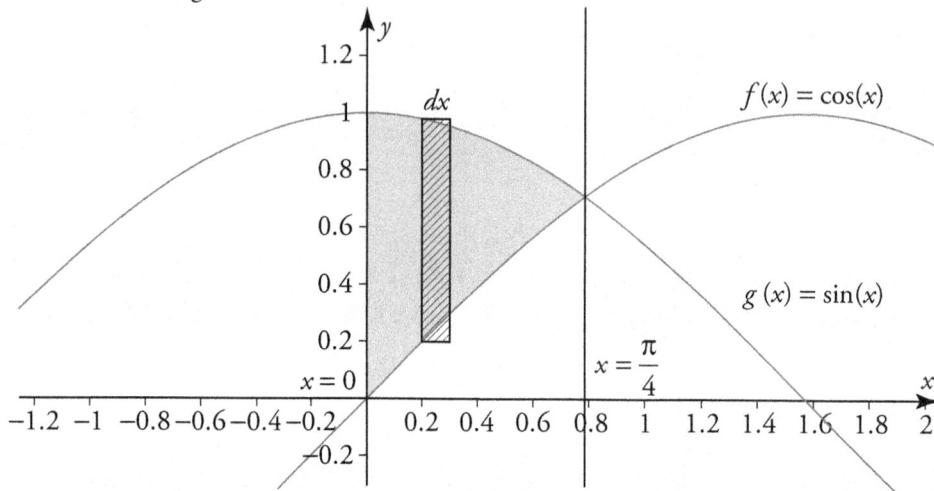

The differential volume element is as follows:

$$dV = \pi\left(\cos^2(x) - \sin^2(x)\right)dx = \pi\cos(2x)dx$$

We find the volume V of the solid of revolution by integrating this on $\left[0, \dfrac{\pi}{4}\right]$.

$$V = \pi \int_0^{\pi/4} \cos(2x)\, dx \approx 1.571$$

39. Answer: E

We use the formulas $x_n = x_{n-1} + h$ and $y_n = y_{n-1} + hF\left(x_{n-1}, y_{n-1}\right)$, where $\left(x_0, y_0\right) = (1, 3)$ and $F(x, y) = 3x + 2y$. With a step size of $h = 0.25$, we will need to take 4 steps to get from $x_0 = 1$ to $x = 2$. The entries in the following table are rounded to the nearest hundredth:

n	0	1	2	3	4
x_n	1	1.25	1.5	1.75	2
y_n	3	5.25	8.81	14.34	22.83

40. Answer: D

Let $u = x - 1$. Then $du = dx$, and $x = u + 1$.

$$\int x\sqrt{x-1}\,dx = \int (u+1)\sqrt{u}\,du = \int\left(u^{3/2} + u^{1/2}\right)du = \frac{2}{5}u^{5/2} + \frac{2}{3}u^{3/2} + C$$

Factor out $\dfrac{2}{15}u^{3/2}$.

$$\int x\sqrt{x-1}\,dx = \frac{2}{15}u^{3/2}\left(3u + 5\right) + C$$

Now undo the substitution.

$$\int x\sqrt{x-1}\,dx = \frac{2}{15}(x-1)^{3/2}\big(3(x-1)+5\big)+C$$

$$\int x\sqrt{x-1}\,dx = \frac{2}{15}(3x+2)(x-1)^{3/2}$$

41. Answer: C

By definition, $\int_{1}^{b}\frac{1}{t}\,dt = \ln(b)$. So the equation in the problem statement can be rewritten as follows:

$$\ln(b) = -2 \Rightarrow b = e^{-2}$$

42. Answer: C

In order for a function to be strictly monotonic on its entire domain, its derivative must not change sign anywhere in its domain. This is only true of answer choice C, as $f'(x)=3x^2+1>0$ for all real numbers x.

43. Answer: C

All three functions are graphed below.

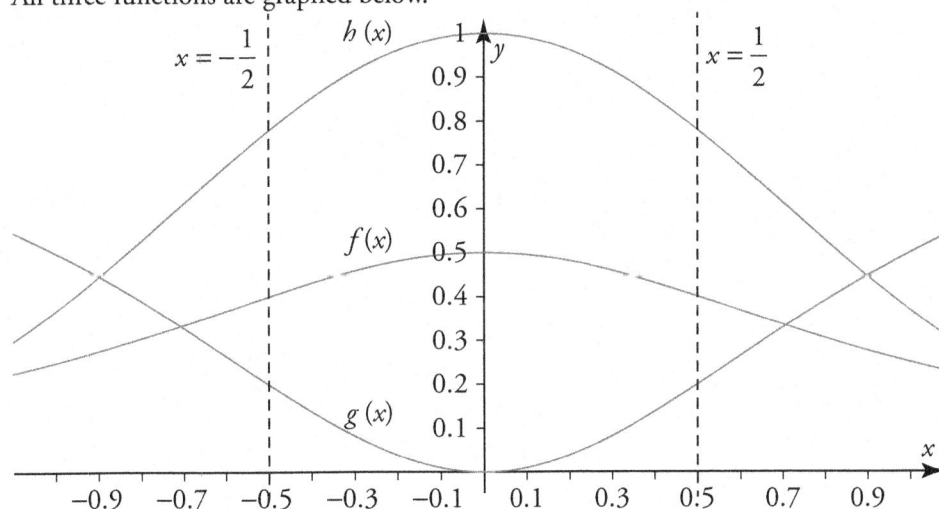

Since, for all x in the interval $\left[-\frac{1}{2},\frac{1}{2}\right]$, $g(x)<f(x)<h(x)$, it follows that

$$\int_{-1/2}^{1/2} g(x)\,dx < \int_{-1/2}^{1/2} f(x)\,dx < \int_{-1/2}^{1/2} h(x)\,dx,$$

or

$B < A < C.$

We sketch the curve below.

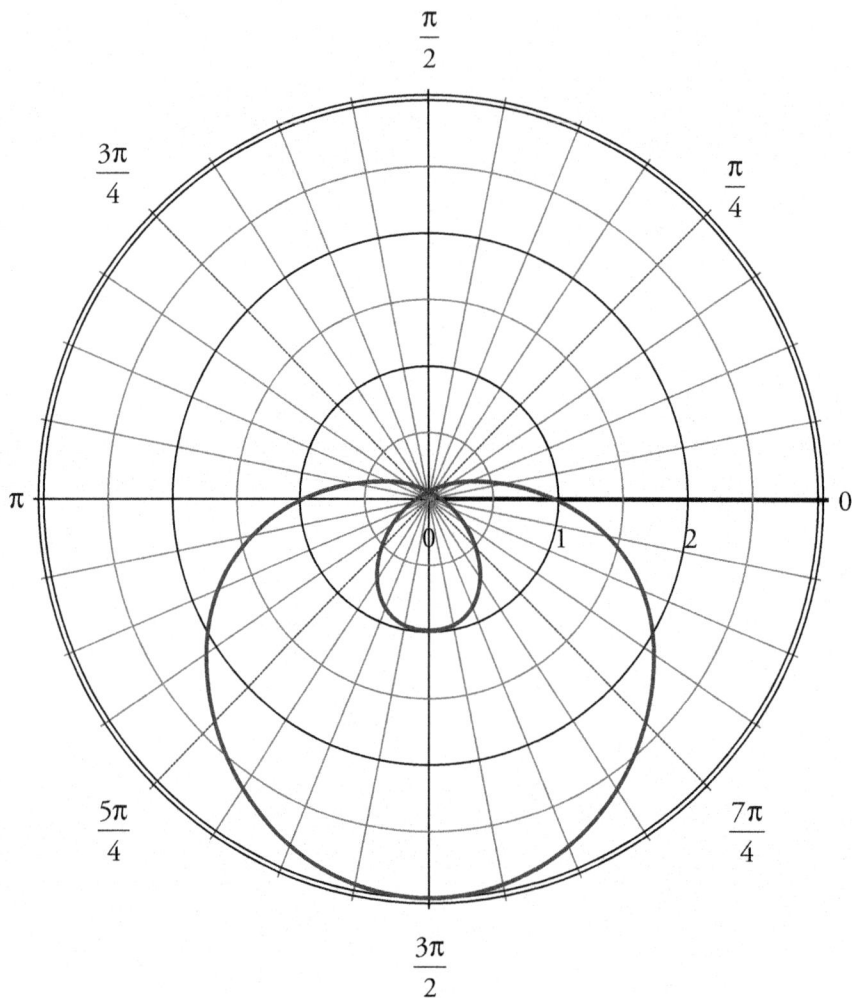

The angular bounds of the inner loop are the first two zeros of $r = 1 - 2\sin(\theta)$ in the interval $[0, 2\pi]$. These zeros are $\theta = \dfrac{\pi}{6}$ and $\theta = \dfrac{5\pi}{6}$. The area A of the inner loop is found as follows:

$$A = \frac{1}{2} \int_{\pi/6}^{5\pi/6} \left(1 - 2\sin(\theta)\right)^2 d\theta = \pi - \frac{3\sqrt{3}}{2}$$

45. Answer: B

We graph the curve below.

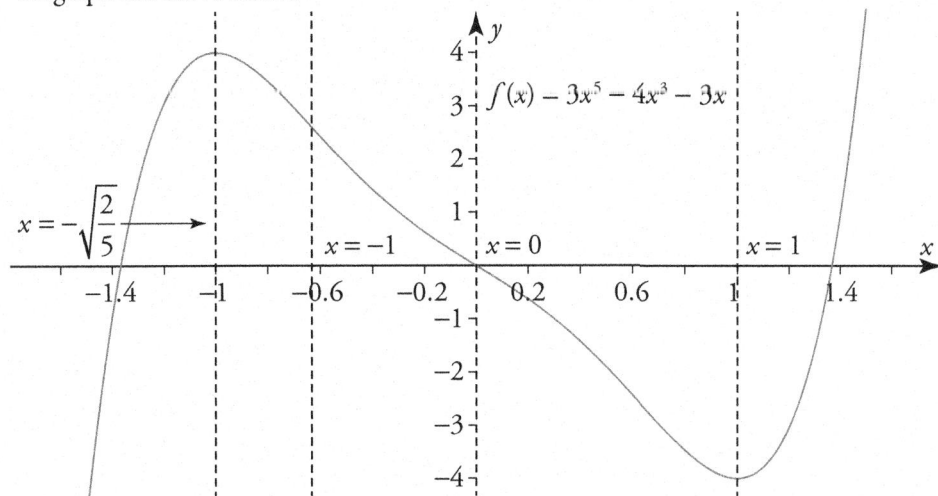

As you can see from the graph, the only interval among the answer choices on which f is both decreasing and concave up is $\left(-\sqrt{\dfrac{2}{5}}, 0\right)$.

1. First find the points of intersection of $f(x)$ and $g(x)$.

$$f(x) = g(x)$$

$$x = x\ln(x)$$

$$x - x\ln(x) = 0$$

$$x(1 - \ln(x)) = 0$$

Setting each factor equal to zero, we get $x = 0$ and $1 - \ln(x) = 0 \Rightarrow x = e$. $g(x)$ is not defined at $x = 0$. However, consider the following limit:

$$\lim_{x \to 0^+} x\ln(x) = \lim_{x \to 0^+} \frac{\ln(x)}{\dfrac{1}{x}}$$

This limit yields the indeterminate form $\dfrac{-\infty}{\infty}$. Apply L'Hôpital's Rule.

$$\lim_{x \to 0^+} x\ln(x) = \lim_{x \to 0^+} \frac{\dfrac{d}{dx}(\ln(x))}{\dfrac{d}{dx}\left(\dfrac{1}{x}\right)} = \lim_{x \to 0^+} \frac{\dfrac{1}{x}}{-\dfrac{1}{x^2}} = -\lim_{x \to 0^+} x = 0$$

The discontinuity of g at $x = 0$ is removable if we define $g(0)$ to be 0. So the interval of integration is $[0, e]$.

So the two points of intersection are $(0,0)$ and (e,e).

A. We sketch the region below.

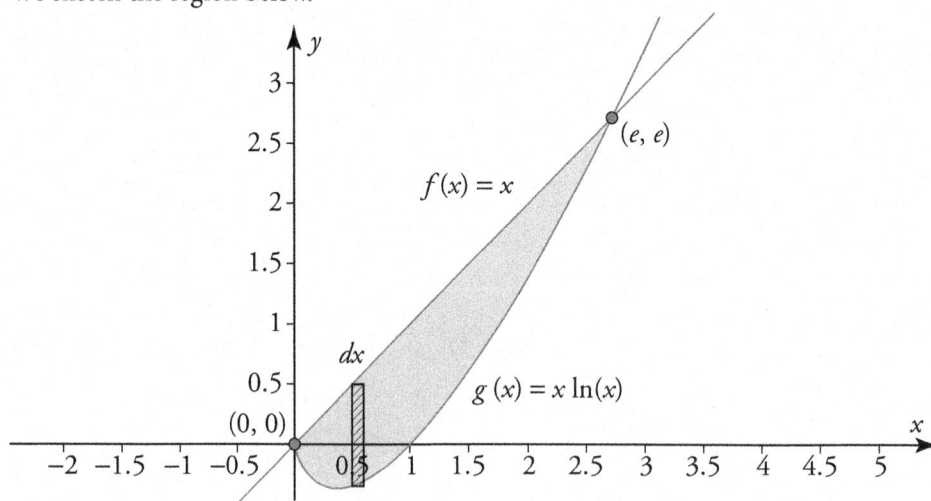

The area A of the region is given by the following integral:

$$A = \int_0^e \left(x - x \ln(x) \right) dx \approx 1.847$$

B. The arc length s_1 of $f(x)$ on $[0,e]$ is given by the following integral:

$$s_1 = \int_0^e \sqrt{1 + \left(f'(x) \right)^2}\, dx = \int_0^e \sqrt{1 + 1^2}\, dx = \int_0^e \sqrt{2}\ dx$$

The arc length s_2 of $g(x)$ on $[0,e]$ is given by the following integral:

$$s_2 = \int_0^e \sqrt{1 + \left(g'(x) \right)^2}\, dx = \int_0^e \sqrt{1 + \left(\ln(x) + 1 \right)^2}\, dx$$

The distance s around the region is then given as follows:

$$s = s_1 + s_2 = \int_0^e \sqrt{2}\ dx + \int_0^e \sqrt{1 + \left(\ln(x) + 1 \right)^2}\, dx$$

$$s = \int_0^e \left(\sqrt{2} + \sqrt{1 + \left(\ln(x) + 1 \right)^2} \right) dx \approx 8.386$$

C. The region is the same as in parts (a) and (b). The differential volume element is as follows:

$$dV = \pi \left(\left(3 - g(x) \right)^2 - \left(3 - f(x) \right)^2 \right) dx$$

$$dV = \pi \left(\left(3 - x \ln(x) \right)^2 - \left(3 - x \right)^2 \right) dx$$

The volume V of the solid of revolution is found by integrating this on $[0,e]$.

$$V = \pi \int_0^e \left(\left(3 - x \ln(x) \right)^2 - \left(3 - x \right)^2 \right) dx \approx 25.472$$

2.

A. The series is an alternating series with $a_n = \dfrac{1}{n^3}$. We apply the Alternating Series Test, which has two conditions to be checked.

- $\lim\limits_{n \to \infty} \dfrac{1}{n^3} = 0$

- $\dfrac{1}{(n+1)^3} < \dfrac{1}{n^3}$ for all integers $n \geq 1$, so $a_{n+1} < a_n$ for all integers $n \geq 1$.

The series converges by the Alternating Series Test. We now test for absolute convergence.

$$\sum_{n=1}^{\infty} \left| \frac{(-1)^{n+1}}{n^3} \right| = \sum_{n=1}^{\infty} \frac{1}{n^3}$$

This series converges by the p-Series Test. Since $\displaystyle\sum_{n=1}^{\infty}\frac{(-1)^{n+1}}{n^3}$ and $\displaystyle\sum_{n=1}^{\infty}\left|\frac{(-1)^{n+1}}{n^3}\right|$ both converge, it follows that $\displaystyle\sum_{n=1}^{\infty}\frac{(-1)^{n+1}}{n^3}$ converges absolutely.

B. An upper bound for the error $\left|R_N\right|$ is $a_{N+1}=\dfrac{1}{(N+1)^3}$, so we solve the following inequality to determine N:

$$\frac{1}{(N+1)^3}<0.001$$

Taking the reciprocal of both sides, we obtain the following:

$(N+1)^3>1000$

$N+1>10$

$N>9$

The smallest integer N that is greater than 9 is $N=10$, so 10 terms are required to approximate the sum to the desired accuracy.

C.

$$\sum_{n=1}^{10}\frac{(-1)^{n+1}}{n^3}=\frac{1}{1^3}-\frac{1}{2^3}+\frac{1}{3^3}-\frac{1}{4^3}+\frac{1}{5^3}-\frac{1}{6^3}+\frac{1}{7^3}-\frac{1}{8^3}+\frac{1}{9^3}-\frac{1}{10^3}\approx0.901$$

3.

A. a and b are zeros of f.

$$f(x) - \sqrt{1 + \sin(x)} = 0$$

$$\sin(x) = -1 \Rightarrow x = \frac{3\pi}{2} + 2\pi n, n = 0, \pm 1, \pm 2, \ldots$$

The two zeros closest to the origin are the ones we are looking for: $a = -\dfrac{\pi}{2}$ and $b = \dfrac{3\pi}{2}$.

B. On the interval $[a, b] = \left[-\dfrac{\pi}{2}, \dfrac{3\pi}{2} \right]$, $\sin\left(\dfrac{x}{2}\right) + \cos\left(\dfrac{x}{2}\right) \geq 0$. This means that we will not introduce any extraneous solutions by squaring both sides of the equation.

$$\sqrt{1 + \sin(x)}^2 = \left(\sin\left(\frac{x}{2}\right) + \cos\left(\frac{x}{2}\right) \right)^2$$

$$1 + \sin(x) = \sin^2\left(\frac{x}{2}\right) + 2\sin\left(\frac{x}{2}\right)\cos\left(\frac{x}{2}\right) + \cos^2\left(\frac{x}{2}\right)$$

Using the Pythagorean Identity $\sin^2\left(\dfrac{x}{2}\right) + \cos^2\left(\dfrac{x}{2}\right) = 1$ and the Double Angle Identity $2\sin\left(\dfrac{x}{2}\right)\cos\left(\dfrac{x}{2}\right) = \sin(x)$, we obtain the following:

$$1 + \sin(x) = 1 + \sin(x)$$

Thus, the identity is proved.

C. Plug $\dfrac{x}{2}$ into the Maclaurin series for $\sin(x)$.

$$\sin\left(\frac{x}{2}\right) = \frac{x}{2} - \frac{(x/2)^3}{3!} + \frac{(x/2)^5}{5!} - \cdots \approx \frac{x}{2} - \frac{x^3}{48} + \frac{x^5}{3840}$$

Now plug $\dfrac{x}{2}$ into the Maclaurin series for $\cos(x)$.

$$\cos\left(\frac{x}{2}\right) = 1 - \frac{(x/2)^2}{2!} + \frac{(x/2)^4}{4!} - \cdots \approx 1 - \frac{x^2}{8} + \frac{x^4}{384}$$

Now find the desired approximation by adding the two previous results together.

$$\sqrt{1 + \sin(x)} = \sin\left(\frac{x}{2}\right) + \cos\left(\frac{x}{2}\right) \approx \left(\frac{x}{2} - \frac{x^3}{48} + \frac{x^5}{3840} \right) + \left(1 - \frac{x^2}{8} + \frac{x^4}{384} \right)$$

$$\sqrt{1 + \sin(x)} \approx 1 + \frac{x}{2} - \frac{x^2}{8} - \frac{x^3}{48} + \frac{x^4}{384} + \frac{x^5}{3840}$$

4.

A. We will use a Pythagorean Identity to eliminate the parameter.

$$x = 4\cos\left(\frac{2\pi t}{3}\right) \Rightarrow \frac{x}{4} = \cos\left(\frac{2\pi t}{3}\right) \Rightarrow \left(\frac{x}{4}\right)^2 = \cos^2\left(\frac{2\pi t}{3}\right)$$

$$y = -3\sin\left(\frac{2\pi t}{3}\right) \Rightarrow \frac{y}{-3} = \sin\left(\frac{2\pi t}{3}\right) \Rightarrow \left(\frac{y}{-3}\right)^2 = \sin^2\left(\frac{2\pi t}{3}\right)$$

Adding these results together, we obtain the following:

$$\left(\frac{x}{4}\right)^2 + \left(\frac{y}{-3}\right)^2 = \cos^2\left(\frac{2\pi t}{3}\right) + \sin^2\left(\frac{2\pi t}{3}\right)$$

$$\frac{x^2}{16} + \frac{y^2}{9} = 1$$

B. The equation that we obtained in part (a) is that of an ellipse. The period of each of the parametric equations is $T = \frac{2\pi}{2\pi/3} = 3$. This is how long it takes to go around the ellipse once. Therefore, the particle goes around the ellipse **twice** for $0 \le t \le 6$.

C. Use Implicit Differentiation on the equation obtained in part (a).

$$\frac{d}{dx}\left(\frac{x^2}{16} + \frac{y^2}{9}\right) = \frac{d}{dx}(1)$$

$$\frac{1}{8}x + \frac{2}{9}y\frac{dy}{dx} = 0$$

$$\frac{dy}{dx} = -\frac{9x}{16y}$$

D. We compute the velocity vector $\vec{v}(t)$ as follows:

$$\vec{v}(t) = \left(x'(t), y'(t)\right)$$

$$\vec{v}(t) = \left(-\frac{8\pi}{3}\sin\left(\frac{2\pi t}{3}\right), -2\pi\cos\left(\frac{2\pi t}{3}\right)\right)$$

Now evaluate this at $t = 2$.

$$\vec{v}(2) = \left(-\frac{8\pi}{3}\sin\left(\frac{4\pi}{3}\right), -2\pi\cos\left(\frac{4\pi}{3}\right)\right) = \left(-\frac{8\pi}{3}\left(-\frac{\sqrt{3}}{2}\right), -2\pi\left(-\frac{1}{2}\right)\right)$$

$$\vec{v}(2) = \left(\frac{4\sqrt{3}\pi}{3}, \pi\right)$$

E. f has a relative maximum whenever $f'(x) = 0$ *and* f' changes sign from positive to negative. This happens only at $x = 3$.

F. f has a relative minimum whenever $f'(x) = 0$ *and* f' changes sign from negative to positive. This happens at $x = -3$ and at $x = 5$.

G. f has a point of inflection whenever f' changes from increasing to decreasing *or* from decreasing to increasing. This happens at $x = \dfrac{5 - 2\sqrt{13}}{3}$ and at $x = \dfrac{5 + 2\sqrt{13}}{3}$.

5.

A. Separate variables by dividing by y and multiplying by dx, and then integrate.

$$\frac{1}{y}\frac{dy}{dx}dx = \frac{-2xy\sin\left(x^2\right)}{y}dx$$

$$\frac{1}{y}dy = -2x\sin\left(x^2\right)dx$$

$$\int\frac{1}{y}dy = -\int 2x\sin\left(x^2\right)dx$$

On the right side, let $u = x^2$. Then $du = 2x\,dx$.

$$\int\frac{1}{y}dy = -\int\sin(u)\,du$$

$$\ln|y| = \cos(u) + C = \cos\left(x^2\right) + C$$

Now convert the equation to exponential form.

$$|y| = e^{\cos\left(x^2\right)+C} = e^C e^{\cos\left(x^2\right)}$$

$$y = \pm e^C e^{\cos\left(x^2\right)}$$

Let $\pm e^C = A$.

$$y = Ae^{\cos\left(x^2\right)}$$

B. Use the initial condition to find A.

$$4 = Ae^{\cos\left(\sqrt{\pi/2}^2\right)} = Ae^0 \Rightarrow A = 4$$

So the particular solution is $y = 4e^{\cos\left(x^2\right)}$, or $y = 4e^{\cos\left(x^2\right)}$.

C. Find the critical numbers of y.

$$\frac{dy}{dx} = \frac{d}{dx}\left(4e^{\cos\left(x^2\right)}\right) = 4e^{\cos\left(x^2\right)} \cdot \frac{d}{dx}\left(\cos\left(x^2\right)\right)$$

$$\frac{dy}{dx} = 4e^{\cos\left(x^2\right)} \cdot \left(-\sin\left(x^2\right)\right) \cdot \frac{d}{dx}\left(x^2\right) = 4e^{\cos\left(x^2\right)} \cdot \left(-\sin\left(x^2\right)\right) \cdot 2x$$

$$\frac{dy}{dx} = -8x\sin\left(x^2\right)e^{\cos\left(x^2\right)} = 0$$

The factor $e^{\cos(x^2)}$ does not equal zero for any x. Setting the factor $-8x = 0$ yields the critical number $x = 0$. Setting the factor $\sin(x^2) = 0$ yields the following:

$$\sin(x^2) = 0$$

$$x^2 = 0, \pi, 2\pi, 3\pi, 4\pi, \ldots$$

$$x = 0, \pm\sqrt{\pi}, \pm\sqrt{2\pi}, \pm\sqrt{3\pi}, \pm 2\sqrt{\pi}, \ldots$$

Of these zeros, the only ones that are in the interval $\left[0, \sqrt{2\pi}\right]$ are $x = 0, \sqrt{\pi}$, and $\sqrt{2\pi}$. The y-coordinates that correspond to these zeros are as follows:

$$y(0) = 4e^{\cos(0^2)} = 4e$$

$$y\left(\sqrt{\pi}\right) = 4e^{\cos\left(\sqrt{\pi}^2\right)} = 4e^{-1}$$

$$y\left(\sqrt{2\pi}\right) = 4e^{\cos\left(\sqrt{2\pi}^2\right)} = 4e$$

We will use the First Derivative Test to classify the critical numbers, evaluating $\dfrac{dy}{dx}$ at

$$x = -\sqrt{\frac{\pi}{2}}, \sqrt{\frac{\pi}{2}}, \sqrt{\frac{3\pi}{2}}, \text{ and } \sqrt{\frac{5\pi}{2}}.$$

$$\left.\frac{dy}{dx}\right|_{x=-\sqrt{\frac{\pi}{2}}} = -8\left(-\sqrt{\frac{\pi}{2}}\right)\sin\left(\left(-\sqrt{\frac{\pi}{2}}\right)^2\right)e^{\cos\left(\left(-\sqrt{\frac{\pi}{2}}\right)^2\right)} = 8\sqrt{\frac{\pi}{2}}\sin\left(\frac{\pi}{2}\right)e^{\cos\left(\frac{\pi}{2}\right)}$$

$$\left.\frac{dy}{dx}\right|_{x=-\sqrt{\frac{\pi}{2}}} = 8\sqrt{\frac{\pi}{2}} > 0$$

$$\left.\frac{dy}{dx}\right|_{x=\sqrt{\frac{\pi}{2}}} = -8\sqrt{\frac{\pi}{2}}\sin\left(\sqrt{\frac{\pi}{2}}^2\right)e^{\cos\left(\sqrt{\frac{\pi}{2}}^2\right)} = -8\sqrt{\frac{\pi}{2}}\sin\left(\frac{\pi}{2}\right)e^{\cos\left(\frac{\pi}{2}\right)} = -8\sqrt{\frac{\pi}{2}} < 0$$

Since the sign of $\dfrac{dy}{dx}$ changes from positive to negative at $x = 0$, y has a relative maximum at $(0, 4e)$.

$$\left.\frac{dy}{dx}\right|_{x=\sqrt{\frac{3\pi}{2}}} = -8\sqrt{\frac{3\pi}{2}}\sin\left(\sqrt{\frac{3\pi}{2}}^2\right)e^{\cos\left(\sqrt{\frac{3\pi}{2}}^2\right)} = -8\sqrt{\frac{3\pi}{2}}\sin\left(\frac{3\pi}{2}\right)e^{\cos\left(\frac{3\pi}{2}\right)}$$

$$\left.\frac{dy}{dx}\right|_{x=\sqrt{\frac{3\pi}{2}}} = 8\sqrt{\frac{3\pi}{2}} > 0$$

Since the sign of $\dfrac{dy}{dx}$ changes from negative to positive at $x = \sqrt{\pi}$, y has a relative minimum at $\left(\sqrt{\pi}, 4e^{-1}\right)$.

$$\frac{dy}{dx}\bigg|_{x=\sqrt{\frac{5\pi}{2}}} = -8\sqrt{\frac{5\pi}{2}}\sin\left(\sqrt{\frac{5\pi}{2}}^2\right)e^{\cos\left(\sqrt{\frac{5\pi}{2}}^2\right)} = -8\sqrt{\frac{5\pi}{2}}\sin\left(\frac{5\pi}{2}\right)e^{\cos\left(\frac{5\pi}{2}\right)}$$

$$\frac{dy}{dx}\bigg|_{x=\sqrt{\frac{5\pi}{2}}} = -8\sqrt{\frac{5\pi}{2}} < 0$$

Since the sign of $\frac{dy}{dx}$ changes from positive to negative at $x = \sqrt{2\pi}$, y has a relative maximum at $\left(\sqrt{2\pi}, 4e\right)$.

AP Calculus BC Practice Examination 2

CALCULUS BC

A CALCULATOR **CANNOT BE USED ON PART A OF SECTION I OR ON PART B OF SECTION II**. A GRAPHING CALCULATOR FROM THE APPROVED LIST **IS REQUIRED ON PART B OF SECTION I AND ON PART A OF SECTION II** OF THE EXAMINATION. CALCULATOR MEMORIES NEED NOT BE CLEARED. COMPUTERS, NONGRAPHING SCIENTIFIC CALCULATORS, CALCULATORS WITH QWERTY KEYBOARDS, AND ELECTRONIC WRITING PADS ARE NOT ALLOWED. CALCULATORS MAY NOT BE SHARED AND COMMUNICATION BETWEEN CALCULATORS IS PROHIBITED DURING THE EXAMINATION. ATTEMPTS TO REMOVE TEST MATERIALS FROM THE ROOM BY ANY METHOD WILL RESULT IN INVALIDATION OF TEST SCORES.

CALCULUS BC – SECTION I

Time – 1 hour and 45 minutes
All questions are given equal weight.
Percent of total grade – 50

Part A: 55 minutes, 28 multiple-choice questions
A calculator is NOT allowed.

Part B: 50 minutes, 17 multiple-choice questions
A graphing calculator is required.

Parts A and B of Section I are printed in this examination booklet. Section II, which consists of longer problems, is printed in a separate booklet.

<u>GENERAL INSTRUCTIONS</u>

DO NOT OPEN THIS BOOKLET UNTIL YOU ARE TOLD TO DO SO.

INDICATE YOUR ANSWERS TO QUESTIONS IN PART A ON PAGE 2 OF THE SEPARATE ANSWER SHEET. THE ANSWERS TO QUESTIONS IN PART B SHOULD BE INDICATED ON PAGE 3 OF THE ANSWER SHEET. No credit will be given for anything written in this examination booklet, but you may use the booklet for notes or scratchwork. After you have decided which of the suggested answers is best, COMPLETELY fill in the corresponding oval on the answer sheet. Give only one answer to each question. If you change an answer, be sure that the previous mark is erased completely.

<u>Example:</u>   <u>Sample Answer:</u> Ⓐ Ⓑ Ⓒ ● Ⓔ

1. What is the value of the numerical expression $4 - 2(3 - 5)$?

   A. -12

   B. -8

   C. 0

   D. 8

   E. 12

Many candidates wonder whether or not to guess the answers to questions about which they are not certain. It is improbable that mere guessing will improve your score significantly; it may even lower your score, and it does take time. If, however, you are not sure of the best answer but have some knowledge of the question and are able to eliminate one or more of the answer choices as wrong, your chance of answering correctly is improved, and it may be to your advantage to answer such a question.

Use your time effectively, working as rapidly as you can without losing accuracy. Do not spend too much time on questions that are too difficult. Go on to other questions and come back to the difficult ones later if you have time. It is not expected that you will be able to answer all of the multiple choice questions.

CALCULUS BC

SECTION I, Part A

Time – 55 Minutes

Number of questions – 28

A CALCULATOR MAY NOT BE USED ON THIS PART OF THE EXAMINATION.

Directions: Solve each of the following problems, using the available space for scratchwork. After examining the form of the choices, decide which is the best of the choices given and fill in the corresponding oval on the answer sheet. No credit will be given for anything written in the test book. Do not spend too much time on any one problem.

In this test: Unless otherwise specified, the domain of a function *f* is assumed to be the set of all real numbers *x* for which $f(x)$ is a real number.

1. If $\sin(xy) + x^2 y = -2$, then $y' =$

    A. $-\dfrac{\cos(xy)}{x^2}$

    B. $-\dfrac{y\cos(xy) + 2xy}{x\cos(xy) + x^2}$

    C. $-\dfrac{\cos(xy) + 2xy}{x^2}$

    D. $\dfrac{y\cos(xy) - 2xy}{x\cos(xy) + x^2}$

    E. $-\dfrac{y\cos(xy) + 2xy + 2}{x\cos(xy) + x^2}$

2. If $y = \dfrac{x}{x+1}$ and $x = \sin(t)$, then $\dfrac{dy}{dt} =$

    A. $\dfrac{1}{\left(\sin(t) + 1\right)^2}$

    B. $\cos(t)$

    C. $-\dfrac{\cos(t)}{\left(\sin(t) + 1\right)^2}$

    D. $\dfrac{\cos(t)}{\left(\sin(t) + 1\right)^2}$

    E. $\dfrac{\cos(t)}{\sin(t) + 1}$

GO ON TO THE NEXT PAGE.

3. $\int \dfrac{dx}{2+25x^2} =$

A. $\tan^{-1}\left(\dfrac{5x}{\sqrt{2}}\right)+C$

B. $\dfrac{1}{\sqrt{2}}\tan^{-1}\left(\dfrac{5x}{\sqrt{x}}\right)+C$

C. $\dfrac{1}{5\sqrt{2}}\tan^{-1}\left(\dfrac{5x}{\sqrt{2}}\right)+C$

D. $\ln\left|2+25x^2\right|+C$

E. $\dfrac{1}{25}\ln\left|2+25x^2\right|+C$

4. If the region bounded by the y-axis, the x-axis, and the graph of $y=-x+2$ is revolved about the x-axis, the volume of the resulting solid is

A. $\dfrac{8\pi}{3}$

B. $\dfrac{4\pi}{3}$

C. $\dfrac{8}{3}$

D. $\dfrac{4}{3}$

E. $\dfrac{2}{3}$

5. $\lim\limits_{x\to\infty} -2x\tan\left(\dfrac{1}{x}\right) =$

A. 2

B. 1

C. –1

D. –2

E. The limit does not exist.

GO ON TO THE NEXT PAGE.

6. Find the interval of convergence of the following power series:

$$\sum_{n=1}^{\infty} \frac{(-1)^n (x-3)^n}{n}$$

A. $2 < x < 4$

B. $2 \le x < 4$

C. $2 < x \le 4$

D. $2 \le x \le 4$

E. $-\infty < x < \infty$

7. $\displaystyle\int \frac{x}{x^2 - 5x + 4}\,dx =$

A. $\dfrac{1}{2}\ln\left|x^2 - 5x + 4\right| + C$

B. $\dfrac{1}{3}\ln\left|(x-4)(x-1)\right| + C$

C. $\dfrac{1}{3}\ln\left|(x-4)^4(x-1)\right| + C$

D. $\dfrac{1}{3}\ln\left|\dfrac{x-4}{x-1}\right| + C$

E. $\dfrac{1}{3}\ln\left|\dfrac{(x-4)^4}{x-1}\right| + C$

8. $\displaystyle\int_0^{\pi/3} \sec^3(x)\tan(x)\,dx =$

A. $\dfrac{7}{3}$

B. $\dfrac{8}{3}$

C. 3

D. $\dfrac{\pi^3}{27}$

E. $\dfrac{\pi^3}{81}$

GO ON TO THE NEXT PAGE.

9. The average value of $f(x) = xe^x$ on the interval $\left[0, \ln(2)\right]$ is

A. 4

B. 2

C. $\dfrac{\ln(4) - 1}{\ln(2)}$

D. $\dfrac{\ln(2) - 1}{\ln(2)}$

E. $\ln(4) - 1$

10. The series $\displaystyle\sum_{n=2}^{\infty} \dfrac{1}{\sqrt{n} - 1}$

A. converges by direct comparison with $\displaystyle\sum_{n=2}^{\infty} \dfrac{1}{n^2}$.

B. converges by direct comparison with $\displaystyle\sum_{n=2}^{\infty} \dfrac{1}{\sqrt{n}}$.

C. diverges by direct comparison with $\displaystyle\sum_{n=2}^{\infty} \dfrac{1}{n^2}$.

D. diverges by direct comparison with $\displaystyle\sum_{n=2}^{\infty} \dfrac{1}{\sqrt{n}}$.

E. diverges by direct comparison with $\displaystyle\sum_{n=2}^{\infty} \dfrac{1}{n^3}$.

GO ON TO THE NEXT PAGE.

11. The graph of the curve described by $x = 4\cos(t)$ and $y = 3\sin(t)$ from $t = 0$ to $t = 2\pi$ is

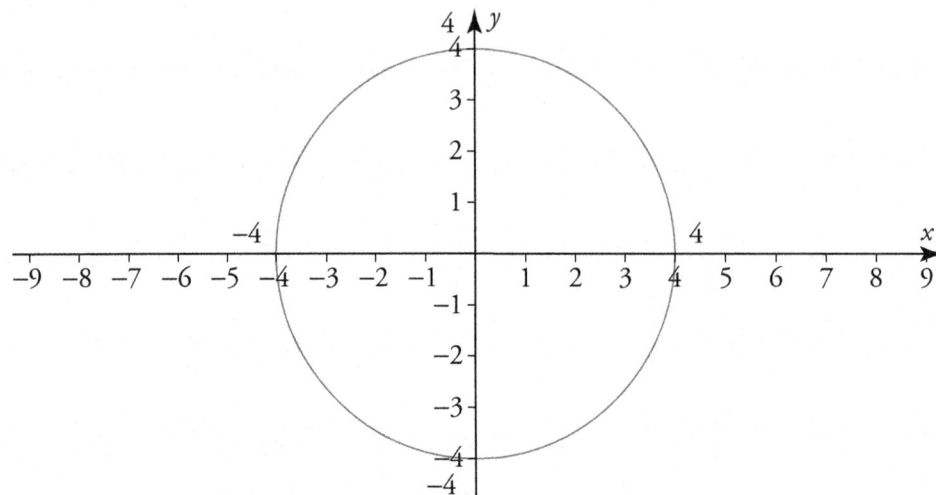

A.

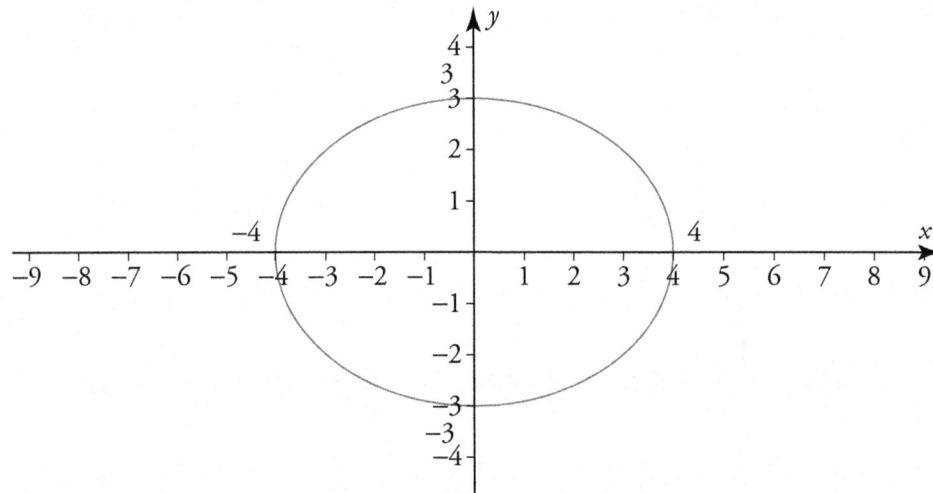

B.

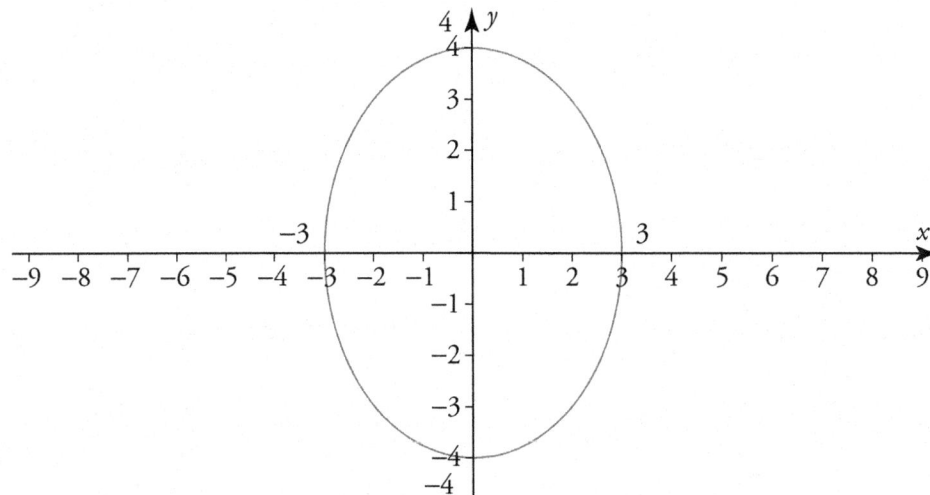

C.

GO ON TO THE NEXT PAGE.

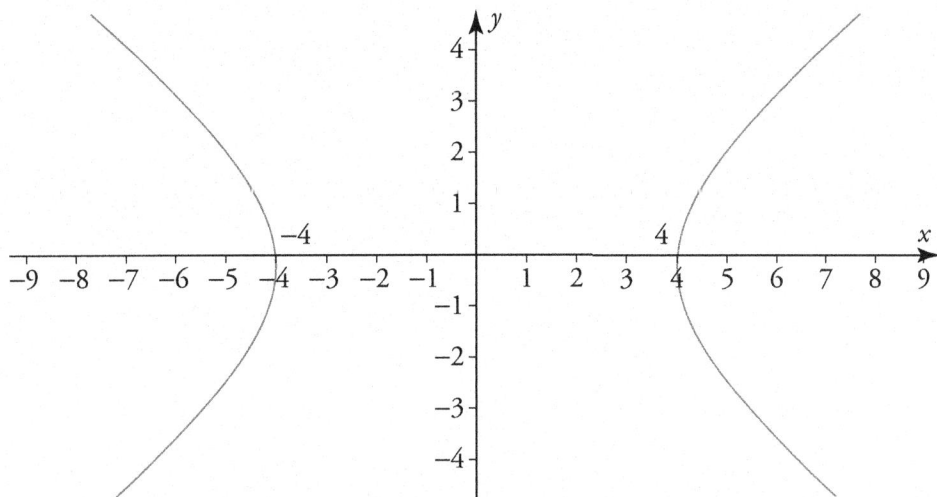

D.

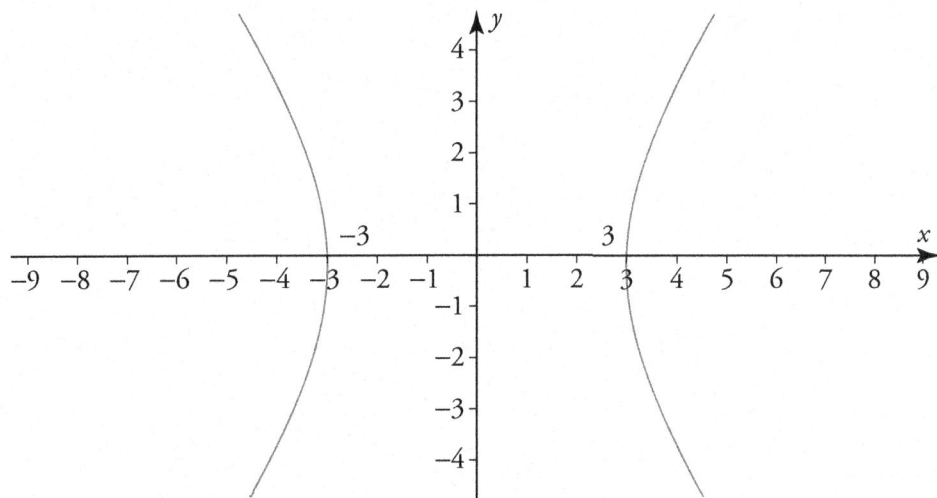

E.

12. The normal line to the graph of $f(x) = x^3 + 2x^2 - 4x + 1$ at the point $(1,0)$ has a y-intercept at

A. $(0,-3)$

B. $\left(0, -\dfrac{1}{3}\right)$

C. $\left(0, \dfrac{1}{3}\right)$

D. $(0,1)$

E. $(0,3)$

GO ON TO THE NEXT PAGE.

13. $\lim\limits_{h \to 0} \dfrac{e^{\ln(2)+h} - e^{\ln(2)}}{h} =$

    A. -2

    B. $-\ln(2)$

    C. $\ln(2)$

    D. 2

    E. The limit does not exist.

14. The length of the curve described by $x = 6t^2$ and $y = 2t^3$ from $t = 1$ to $t = 4$ is

    A. 70

    B. 140

    C. $70\sqrt{5}$

    D. $140\sqrt{5}$

    E. $144\sqrt{5}$

15. If c satisfies the conclusion of the Mean Value Theorem for derivatives for $f(x) = x - \cos(x)$ on the interval $\left[-\dfrac{\pi}{2}, \dfrac{\pi}{2}\right]$, then c could be

    A. $-\pi$

    B. $-\dfrac{\pi}{2}$

    C. 0

    D. $\dfrac{\pi}{2}$

    E. π

GO ON TO THE NEXT PAGE.

16. The graph of $f(x) = \dfrac{4}{x^2+1}$ is concave downward on the interval

A. $(-\infty, 0)$

B. $(0, \infty)$

C. $\left(-\infty, -\sqrt{\dfrac{1}{3}}\right)$

D. $\left(\sqrt{\dfrac{1}{3}}, \infty\right)$

E. $\left(-\sqrt{\dfrac{1}{3}}, \sqrt{\dfrac{1}{3}}\right)$

17. $1 + \ln(3) + \dfrac{(\ln(3))^2}{2!} + \dfrac{(\ln(3))^3}{3!} + \dfrac{(\ln(3))^4}{4!} + \dfrac{(\ln(3))^5}{5!} + \cdots =$

A. e

B. 3

C. π

D. 6

E. The sum is not defined.

18. A particle moves in the xy-plane so that its position vector is $\left(te^{-t}, e^{-t^2}\right)$ for all time $t > 0$. The velocity vector of the particle at $t = 1$ is

A. $\left(\dfrac{1}{e}, \dfrac{1}{e}\right)$

B. $(0, 1)$

C. $\left(\dfrac{1}{e}, -\dfrac{2}{e}\right)$

D. $\left(\dfrac{1}{e}, 1\right)$

E. $\left(0, -\dfrac{2}{e}\right)$

GO ON TO THE NEXT PAGE.

19. $\displaystyle\int_2^\infty \frac{1}{x\left(\ln(x)\right)^2}\,dx$

A. $\dfrac{1}{\ln(2)}$

B. $\dfrac{1}{\left(\ln(2)\right)^2}$

C. 2

D. $\left(\ln(2)\right)^2$

E. $\ln(2)$

20. $\int \sin^3(2x)\,dx$

A. $\dfrac{1}{8}\sin^4(2x)+C$

B. $\dfrac{1}{4}\sin^4(2x)+C$

C. $\dfrac{1}{2}\sin^4(2x)+C$

D. $\dfrac{1}{3}\cos^3(2x)-\cos(2x)+C$

E. $\dfrac{1}{6}\cos^3(2x)-\dfrac{1}{2}\cos(2x)+C$

21. The point on the graph of $f(x)=x^2$ that is closest to the point $\left(2,\dfrac{1}{2}\right)$ is

A. $(0,0)$

B. $(1,1)$

C. $(2,4)$

D. $(3,9)$

E. $(4,16)$

GO ON TO THE NEXT PAGE.

22. If $\left(x^2+16\right)\dfrac{dy}{dx}=xy$, and $y=10$ when $x=3$, then $y=$

A. $x+7$

B. x^2+1

C. $\sqrt{x^2+16}$

D. $2\sqrt{x^2+16}$

E. $-2\sqrt{x^2+16}$

23. The position of a particle moving along the x-axis at time t is given by $x(t)=\tan^{-1}(\sin(t))$, $0\le t\le 2\pi$. At which of the following times is the particle moving to the right?

$$\text{I.}\quad t=\frac{\pi}{4}\qquad \text{II.}\quad t=\frac{3\pi}{4}\qquad \text{III.}\quad t=\frac{7\pi}{4}$$

A. I only

B. II only

C. III only

D. I and III only

E. II and III only

24. $\displaystyle\int\dfrac{e^{2x}-2e^{x}}{e^{x}}\,dx=$

A. e^x-2x+C

B. e^x-2xe^x+C

C. e^x-2+C

D. $\dfrac{1}{2}e^x-2+C$

E. $\dfrac{1}{2}e^{2x}-2e^x+C$

GO ON TO THE NEXT PAGE.

25. The equation of the line tangent to the graph of $f(x) = \dfrac{x}{\sqrt{x^2+5}}$ at the point $\left(2, \dfrac{2}{3}\right)$ is

A. $5x - 3y - 8 = 0$

B. $5x + 27y - 18 = 0$

C. $5x - 27y + 8 = 0$

D. $5x + 27y - 8 = 0$

E. $5x - 27y - 8 = 0$

26. A 25 *foot* ladder is leaning against the side of a building when it begins to slip, as shown in the figure below. The top of the ladder slides down the wall at $-4\dfrac{ft}{s}$. How fast (in $\dfrac{ft}{s}$) is the bottom of the ladder moving away from the base of the building when the top is 7 *feet* from the ground?

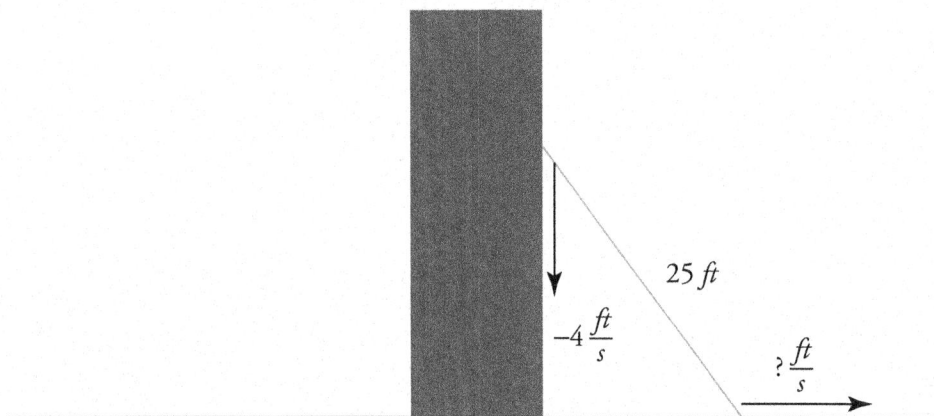

A. $\dfrac{7}{6}$

B. $\dfrac{7}{3}$

C. 7

D. 24

E. 25

27. $\displaystyle\sum_0^{\sqrt{3}}\tan^{-1}(x)\,dx$

A. $-\dfrac{\pi\sqrt{3}}{3}-\ln(2)$

B. $-\dfrac{\pi\sqrt{3}}{3}+\ln(2)$

C. $\dfrac{\pi\sqrt{3}}{3}-\ln(2)$

D. $\pi\sqrt{3}-\ln(2)$

E. $\pi\sqrt{3}+\ln(2)$

28. The sum of the first three terms of the Maclaurin series of $f(x)=e^{2x}$ is

A. $1+x+\dfrac{x^2}{2}$

B. $1-x+\dfrac{x^2}{2}$

C. $1-2x+2x^2$

D. $1+2x+2x^2$

E. $1-2x-2x^2$

STOP

END OF PART A, SECTION I
IF YOU FINISH BEFORE TIME IS CALLED, YOU MAY CHECK YOUR WORK ON PART A ONLY.

DO NOT GO ON TO PART B UNTIL YOU ARE TOLD TO DO SO.

CALCULUS BC

SECTION I, Part B

Time – 50 Minutes

Number of questions – 17

A GRAPHING CALCULATOR IS REQUIRED FOR SOME
QUESTIONS ON THIS PART OF THE EXAMINATION.

Directions: Solve each of the following problems, using the available space for scratchwork. After examining the form of the choices, decide which is the best of the choices given and fill in the corresponding oval on the answer sheet. No credit will be given for anything written in the test book. Do not spend too much time on any one problem.

In this test:

1. The **exact** numerical value of the correct answer does not always appear among the choices given. When this happens, select from among the choices the number that best approximates the exact numerical value.

2. Unless otherwise specified, the domain of a function *f* is assumed to be the set of all real numbers *x* for which $f(x)$ is a real number.

29. If the region bounded by the *y*-axis and the graph of $y = e^{-x^2}$ between $x = 0$ and $x = 1$ is revolved about the *y*-axis, the volume of the resulting solid is

    A. 1.986

    B. 3.972

    C. 3.142

    D. 5.398

    E. 6.283

30. The radius of a ball bearing is measured to be 1.8 *cm*, and the measuring instrument is accurate to within 0.1 *cm*. Use differentials to estimate the propagated error (in *cm*³) when using this measured radius is used to calculate the ball bearing.

    A. 2.26

    B. 2.44

    C. 4.07

    D. 22.6

    E. 40.7

GO ON TO THE NEXT PAGE.

31. The sequence $a_n = \dfrac{2n}{\sqrt{n^2 + 3}}$ converges to

A. $\dfrac{1}{3}$

B. $\dfrac{2}{3}$

C. 1

D. 2

E. 3

32. Use a trapezoidal sum with $n = 4$ to approximate the area of the region bounded by the graph of $f(x) = \sin(x^2)$ and the x-axis for $0 \le x \le \sqrt{\dfrac{\pi}{2}}$.

A. 0.2748

B. 0.5495

C. 1

D. 1.099

E. 1.253

33. $\int x \sec(x) \tan(x)\, dx =$

A. $\sec(x) + C$

B. $x \sec(x) + C$

C. $\dfrac{1}{2} x^2 \sec(x) + C$

D. $x \sec(x) - \ln\left|\sec(x) + \tan(x)\right| + C$

E. $\dfrac{1}{2} x^2 \sec(x) - \ln\left|\sec(x) + \tan(x)\right| + C$

34. If $f(x) = (\sin(x))^x$, then $f'(x) =$

A. $x(\sin(x))^{x-1}$

B. $x(\sin(x))^{x-1} \cos(x)$

C. $(\cos(x))^x$

D. $\ln(\sin(x))(\sin(x))^x$

E. $(\ln(\sin(x)) + x \cot(x))(\sin(x))^x$

GO ON TO THE NEXT PAGE.

35. $\displaystyle\sum_{n=0}^{\infty} e^{-2n} =$

A. 0

B. $\dfrac{1}{e}$

C. $\dfrac{1}{e^2}$

D. $\dfrac{e^2}{e^2-1}$

E. $\dfrac{e^2}{e^2+1}$

36. $\displaystyle\int \frac{2x^3 - 3x^2 + 2x + 2}{x^2 + 1}\,dx$

A. $x^2 - 3x + 5\tan^{-1}(x) + C$

B. $x^2 + 3x + 5\tan^{-1}(x) + C$

C. $x^2 - 3x - 5\tan^{-1}(x) + C$

D. $\dfrac{1}{2}x^4 - x^3 + x^2 + 2x + C$

E. $\dfrac{3x^3 - 6x^2 + 6x + 12}{2x^2 + 6} + C$

37. When the region bounded by the graphs of $f(x) = x^2 + 1$, $x = 0$, $x = 1$, and $y = 0$ is revolved about the y-axis, the volume of the resulting solid is

A. 0.5

B. 1.5

C. 1.571

D. 3.142

E. 4.712

GO ON TO THE NEXT PAGE.

38. The function $f(x) = 2\sin(x) - \cos(2x)$ attains its maximum value in the interval $[0, 2\pi]$ at the point

A. $\left(\dfrac{\pi}{2}, 4\right)$

B. $\left(\dfrac{\pi}{2}, 3\right)$

C. $\left(\dfrac{7\pi}{6}, -\dfrac{3}{2}\right)$

D. $\left(\dfrac{3\pi}{2}, -1\right)$

E. $\left(\dfrac{11\pi}{6}, -\dfrac{3}{2}\right)$

39. The area of the region lying inside one petal of the graph of $r = 3\cos(2\theta)$ is

A. $\dfrac{5\pi}{8}$

B. $\dfrac{7\pi}{8}$

C. $\dfrac{9\pi}{8}$

D. $\dfrac{11\pi}{8}$

E. $\dfrac{13\pi}{8}$

40. Use Euler's Method with $h = 0.2$ to approximate $y(1)$, where y satisfies $y' = x - y$ and passes through $(0,1)$.

A. 0.619

B. 0.624

C. 0.655

D. 0.724

E. 0.819

41. $\displaystyle\int_0^1 x^2 e^x \, dx$

A. 0.718

B. 1.718

C. 2.718

D. 3.718

E. 4.718

GO ON TO THE NEXT PAGE.

42. The horizontal asymptote(s) of the function $f(x) = 2\tan^{-1}\left(\dfrac{x}{2}\right)$ is (are)

A. $y = -\dfrac{\pi}{2}$ and $y = \dfrac{\pi}{2}$

B. $y = -\pi$ and $y = \pi$

C. $y = -\dfrac{3\pi}{2}$ and $y = \dfrac{3\pi}{2}$

D. $y = \dfrac{\pi}{2}$ only

E. $y = \pi$ only

43. The rate of change of y is proportional to y. When $x = 0$, $y = 6$, and when $x = 4$, $y = 15$. When $x = 8$, $y =$

A. 12.5

B. 25.0

C. 37.5

D. 50.0

E. 62.5

44. $\displaystyle\int_1^{3/2} \dfrac{1}{(4x-5)^2}\,dx =$

A. $-\dfrac{1}{2}$

B. $-\dfrac{1}{8}$

C. $\dfrac{1}{4}\ln(6)$

D. $\ln(6)$

E. The integral diverges.

45. Consider the following polar graph:

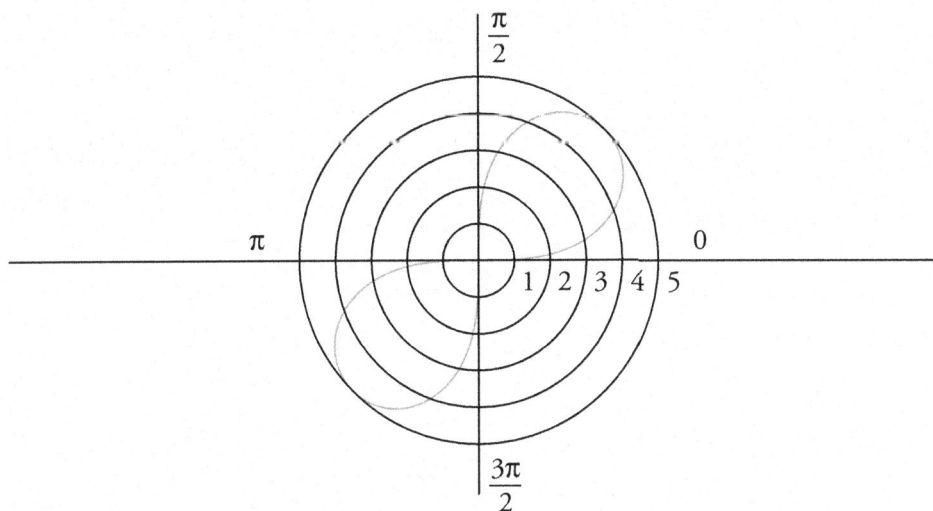

The polar equation of this graph is

A. $r^2 = 25 \cos(2\theta)$

B. $r^2 = 25 \sin(2\theta)$

C. $r = 5 \cos(2\theta)$

D. $r = 5 \sin(2\theta)$

E. $r = 5 \sin(\theta)$

STOP

END OF PART B, SECTION I
IF YOU FINISH BEFORE TIME IS CALLED, YOU MAY CHECK YOUR WORK ON PART B ONLY.

DO NOT GO ON TO SECTION II UNTIL YOU ARE TOLD TO DO SO.

CALCULUS BC – SECTION II

Time – 1 hour and 30 minutes
Number of problems – 6
Percent of total grade – 50

GENERAL INSTRUCTIONS

You may wish to look over the problems before starting to work on them, since it is not expected that everyone will be able to complete all parts of all problems. All problems are given equal weight, but the parts of a particular problem are not necessarily given equal weight.

A GRAPHING CALCULATOR IS REQUIRED FOR SOME PROBLEMS OR PARTS OF PROBLEMS ON THIS SECTION OF THE EXAMINATION.

- You should write all work for each part of each problem in the space provided for that part in the booklet. Be sure to write clearly and legibly. If you make an error, you may save time by crossing it out rather than trying to erase it. Erased or crossed-out work will not be graded.

- Show all your work. You will be graded on the correctness and completeness of your methods as well as your answers. Correct answers without supporting work may not receive credit.

- Justifications require that you give mathematical (noncalculator) reasons and that you clearly identify functions, graphs, tables, or other objects you use.

- You are permitted to use your calculator to solve an equation, find the derivative of a function at a point, or calculate the value of a definite integral. However, you must clearly indicate the setup of your problem, namely the equation, function, or integral you are using. If you use other built-in features or programs, you must show the mathematical steps necessary to produce your results.

- Your work must be expressed in standard mathematical notation rather than calculator syntax. For example, $\int_1^5 x^2 dx$ may not be written as $fnInt(X^2,X,1,5)$.

- Unless otherwise specified, answers (numeric or algebraic) need not be simplified. If your answer is given as a decimal approximation, it should be correct to three places after the decimal point.

- Unless otherwise specified, the domain of a function f is assumed to be the set of all real numbers x for which $f(x)$ is a real number.

A graphing calculator is required for some problems or parts of problems.

During the timed portion for Part A, you may work only on the problems in Part A.

On Part A, you are permitted to use your calculator to solve an equation, find the derivative of a function at a point, or calculate the value of a definite integral. However, you must clearly indicate the setup of your problem, namely the equation, function, or integral you are using. If you use other built-in features or programs, you must show the mathematical steps necessary to produce your results.

1. Consider the function $f(x)$ given by the following graph:

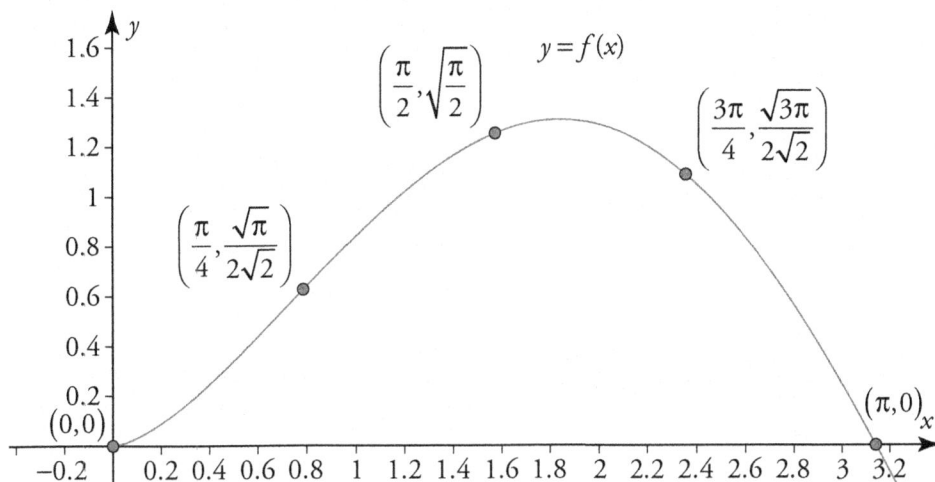

In this problem you are going to approximate $I = \int_0^\pi f(x)\,dx$ with a uniform partition of $[0,\pi]$ into four subintervals. Round all of your answers to the nearest thousandth.

A. Approximate I using a left Riemann sum.

B. Approximate I using a right Riemann sum.

C. Approximate I using a trapezoidal sum.

2. Consider the polar function $f(\theta) = 2 - 2\sin(\theta)$. Express your answers in terms of π whenever possible. When this is not possible, round your answers to the nearest thousandth.

   A. Find all points of horizontal tangency.

   B. Find all points of vertical tangency.

   C. Find all tangents at the pole.

NO CALCULATOR IS ALLOWED FOR THESE PROBLEMS.

During the timed portion for Part B, you may continue to work on the problems in Part A without the use of any calculator.

3. Consider the equation $x^2 y^2 - 2x = 3$.

A. Find $\dfrac{dy}{dx}$.

B. Find the general form of the equation of the line tangent to the graph of the equation at the point $(3,1)$.

C. Find $\dfrac{d^2 y}{dx^2}$ in terms of x and y.

GO ON TO THE NEXT PAGE.

4. Consider the function given by $f(x) = \sum_{n=1}^{\infty} \dfrac{x^n}{n}$.

   A. Find the interval of convergence for $f(x)$.

   B. Find the interval of convergence for $f'(x)$.

   C. Find the interval of convergence for $\int f(x)\,dx$.

GO ON TO THE NEXT PAGE.

5. Let $F(x) = \int_0^x t \sin(t)\,dt$ for $0 \le x \le 2\pi$.

   A. Find $F(\pi)$ and $F(2\pi)$.

   B. Find the critical numbers of F for $0 \le x \le 2\pi$.

   C. Find the coordinates of all relative maxima and minima of F for $0 \le x \le 2\pi$.

GO ON TO THE NEXT PAGE.

6. Let $f(x) = (x - a)(x - b)(x - c)$.

A. Find $f'(x)$ in terms of a, b, and c.

B. Find $f''(x)$ in terms of a, b, and c.

C. Show that the graph of f has a point of inflection when x is equal to the average of a, b, and c.

STOP

END OF EXAM

Answer Key – Section I

1. C	10. D	19. A	28. D	37. E
2. D	11. B	20. E	29. A	38. B
3. C	12. C	21. B	30. C	39. C
4. A	13. D	22. D	31. D	40. C
5. D	14. C	23. D	32. B	41. A
6. C	15. C	24. A	33. D	42. B
7. E	16. E	25. C	34. E	43. C
8. A	17. B	26. A	35. D	44. E
9. C	18. E	27. C	36. A	45. B

Solutions

Solutions – Section I, Part A

1. Answer: C

Use Implicit Differentiation. On the left side of the equation, we will need to use the Product Rule on the second term, and we will need to use both the Chain Rule *and* the Product Rule on the first term.

$$\frac{d}{dx}\left(\sin(xy)+x^2y\right)=\frac{d}{dx}(-2)$$

$$\cos(xy)\cdot\frac{d}{dx}(xy)+\left(2xy+x^2y'\right)=0$$

$$\cos(xy)\cdot\left(y+xy'\right)+\left(2xy+x^2y'\right)=0$$

$$y\cos(xy)+xy'\cos(xy)+2xy+x^2y'=0$$

$$\left(x\cos(xy)+x^2\right)y'=-y\cos(xy)-2xy$$

$$y'=-\frac{y\cos(xy)+2xy}{x\cos(xy)+x^2}$$

2. Answer: D

Use the Chain Rule.

$$\frac{dy}{dt}=\frac{dy}{dx}\cdot\frac{dx}{dt}$$

$$\frac{dy}{dt}=\frac{d}{dx}\left(\frac{x}{x+1}\right)\cdot\frac{d}{dt}\left(\sin(t)\right)$$

Use the Quotient Rule on the first factor on the right side.

$$\frac{dy}{dx}=\frac{(1)(x+1)-x(1)}{(x+1)^2}\cdot\cos(t)$$

$$\frac{dy}{dx}=\frac{1}{(x+1)^2}\cdot\cos(t)$$

Now make the substitution $x=\sin(t)$.

$$\frac{dy}{dx}=\frac{\cos(t)}{(\sin(t)+1)^2}$$

3. Answer: C

Let $u = 5x$. Then $du = 5dx \Rightarrow dx = \dfrac{1}{5}du$.

$$\int \frac{dx}{2+25x^2} = \frac{1}{5}\int \frac{du}{2+u^2} = \frac{1}{5} \cdot \frac{1}{\sqrt{2}} \tan^{-1}\left(\frac{u}{\sqrt{2}}\right) + C = \frac{1}{5\sqrt{2}} \tan^{-1}\left(\frac{5x}{\sqrt{2}}\right) + C$$

4. Answer: A

We sketch the region below. We will use the method of discs.

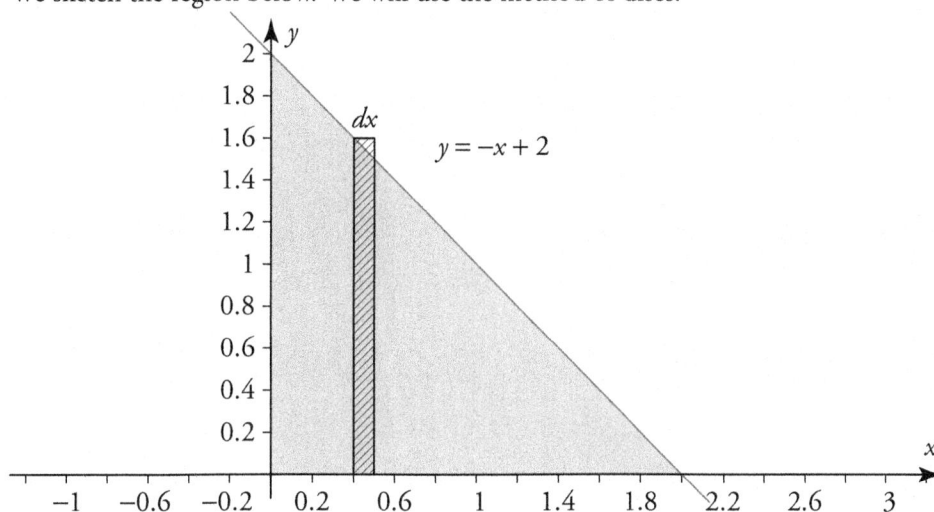

The differential volume element is given as follows:

$$dV = \pi(-x+2)^2\, dx$$

$$dV = \pi(x^2 - 4x + 4)\, dx$$

The total volume V of the solid is given by the following integral:

$$V = \pi\int_0^2 (x^2 - 4x + 4)\, dx = \pi\left(\frac{1}{3}x^3 - 2x^2 + 4x\right)\Bigg|_0^2 = \pi\left(\left(\frac{8}{3} - 8 + 8\right) - 0\right) = \frac{8\pi}{3}$$

5. Answer: D

Direct substitution yields the indeterminate form $-\infty \cdot 0$. Work it into the indeterminate form $\dfrac{0}{0}$.

$$\lim_{x\to\infty} -2x\tan\left(\frac{1}{x}\right) = -2\lim_{x\to\infty} \frac{\tan\left(\dfrac{1}{x}\right)}{\dfrac{1}{x}}$$

Now apply L'Hôpital's rule.

$$\lim_{x \to \infty} -2x \tan\left(\frac{1}{x}\right) = -2 \lim_{x \to \infty} \frac{\frac{d}{dx}\left(\tan\left(\frac{1}{x}\right)\right)}{\frac{d}{dx}\left(\frac{1}{x}\right)} = -2 \lim_{x \to} \frac{-\frac{1}{x^2}\sec\left(\frac{1}{x}\right)}{-\frac{1}{x^2}} = -2 \lim_{x \to \infty} \sec\left(\frac{1}{x}\right)$$

$$\lim_{x \to \infty} -2x \tan\left(\frac{1}{x}\right) = -2\sec(0) = -2$$

6. **Answer: C**

Apply the Ratio Test. $u_n = \frac{(-1)^n (x-3)^n}{n}$ and $u_{n+1} = \frac{(-1)^{n+1}(x-3)^{n+1}}{n+1}$.

$$\lim_{n \to \infty}\left|\frac{u_{n+1}}{u_n}\right| < 1$$

$$\lim_{n \to \infty}\left|\frac{(-1)^{n+1}(x-3)^{n+1}}{n+1} \cdot \frac{n}{(-1)^n (x-3)^{n+1}}\right| < 1$$

$$|x-3|\lim_{n \to \infty}\frac{n}{n+1} < 1$$

$$|x-3| < 1$$

$$-1 < x - 3 < 1$$

$$2 < x < 4$$

We must now test the endpoints of the interval.

$x = 2$:

$$\sum_{n=1}^{\infty}\frac{(-1)^n (2-3)^n}{n} = \sum_{n=1}^{\infty}\frac{1}{n}$$

This is the harmonic series, which diverges.

$x = 4$:

$$\sum_{n=1}^{\infty}\frac{(-1)^n (4-3)^n}{n} = \sum_{n=1}^{\infty}\frac{(-1)^n}{n}$$

This is the alternating harmonic series, which converges conditionally.

The power series diverges at $x = 2$ and converges at $x = 4$, so the interval of convergence is given by $2 < x \le 4$.

7. **Answer: E**

Decompose the integrand into partial fractions.

$$\frac{x}{x^2 - 5x + 4} = \frac{x}{(x-1)(x-4)} = \frac{A}{x-1} + \frac{B}{x-4}$$

$$x = A(x-4) + B(x-1)$$

Plug in $x = 1$ and $x = 4$ to determine A and B, respectively.

$$x = 1 \Rightarrow 1 = -3A \Rightarrow A = -\frac{1}{3}$$

$$x = 4 \Rightarrow 4 = 3B \Rightarrow B = \frac{4}{3}$$

We rewrite the integral as follows:

$$\int \frac{x}{x^2 - 5x + 4} \, dx = -\frac{1}{3} \int \frac{1}{x-1} \, dx + \frac{4}{3} \int \frac{1}{x-4} \, dx$$

We will do one integral at a time. For the first integral, let $u = x - 1$. Then $du = dx$.

$$-\frac{1}{3} \int \frac{1}{x-1} \, dx = -\frac{1}{3} \int \frac{1}{u} \, du = -\frac{1}{3} \ln|u| = -\frac{1}{3} \ln|x-1|$$

We needn't worry about the constant of integration until we are finished. For the second integral, let $u = x - 4$. Then $du = dx$.

$$\frac{4}{3} \int \frac{1}{x-4} \, dx = \frac{4}{3} \int \frac{1}{u} \, du = \frac{4}{3} \ln|u| = \frac{4}{3} \ln|x-4|$$

Combining these results, we obtain the following:

$$\int \frac{x}{x^2 - 5x + 4} \, dx = -\frac{1}{3} \ln|x-1| + \frac{4}{3} \ln|x-4| + C$$

$$\int \frac{x}{x^2 - 5x + 4} \, dx = \frac{1}{3} \left(-\ln|x-1| + 4\ln|x-4| \right) + C$$

Applying the rules of logarithms, we obtain our final result.

$$\int \frac{x}{x^2 - 5x + 4} \, dx = \frac{1}{3} \ln \left| \frac{(x-4)^4}{x-1} \right| + C$$

8. **Answer: A**

Rewrite the integral as follows:

$$\int_0^{\pi/3} \sec^3(x) \tan(x) \, dx = \int_0^{\pi/3} \sec^2(x) \sec(x) \tan(x) \, dx$$

Let $u = \sec(x)$. Then $du = \sec(x)\tan(x)\,dx$. Furthermore, $x = 0 \Rightarrow u = \sec(0) = 1$, and

$$x = \frac{\pi}{3} \Rightarrow u = \sec\left(\frac{\pi}{3}\right) = 2 \, .$$

$$\int_0^{\pi/3} \sec^3(x)\tan(x)\,dx = \int_1^2 u^2\,du = \frac{1}{3}u^3 \Big|_1^2 = \frac{1}{3}(2^3 - 1^3) = \frac{7}{3}$$

9. Answer: C

Let $\overline{f}$ be the average value of $f(x)$ on $[0, \ln(2)]$.

$$\overline{f} = \frac{1}{\ln(2) - 0} \int_0^{\ln(2)} xe^x \, dx = \frac{1}{\ln(2)} \int_0^{\ln(2)} xe^x \, dx$$

Integrate by parts. Let $u = x$, so $dv = e^x dx$. Calculate du and v.

$u = x \Rightarrow du = dx$

$dv = e^x dx \Rightarrow v = \int e^x \, dx = e^x$

We needn't concern ourselves with the constant of integration, since this is a definite integral.

$$\overline{f} = \frac{1}{\ln(2)} \left(xe^x \Big|_0^{\ln(2)} - \int_0^{\ln(2)} e^x \, dx \right) = \frac{1}{\ln(2)} \left(xe^x \Big|_0^{\ln(2)} - e^x \Big|_0^{\ln(2)} \right)$$

$$\overline{f} = \frac{1}{\ln(2)} \left(\left(\ln(2) e^{\ln(2)} - 0 \right) - \left(e^{\ln(2)} - 1 \right) \right) = \frac{1}{\ln(2)} \left(2\ln(2) - (2 - 1) \right)$$

$$\overline{f} = \frac{1}{\ln(2)} \left(\ln(2^2) - 1 \right) = \frac{\ln(4) - 1}{\ln(2)}$$

10. Answer: D

The series $\sum_{n=2}^{\infty} \frac{1}{n^2}$ is a converging p-series, so it cannot be used to show divergence in a direct comparison. Eliminate C. For all $n \geq 2$, $n^2 > \sqrt{n} - 1 \Rightarrow \frac{1}{\sqrt{n} - 1} > \frac{1}{n^2}$, so direct comparison of $\sum_{n=2}^{\infty} \frac{1}{\sqrt{n} - 1}$ with the converging p-series $\sum_{n=2}^{\infty} \frac{1}{n^2}$ couldn't possibly tell us anything anyway. Eliminate A.

The series $\sum_{n=2}^{\infty} \frac{1}{n^3}$ is a converging p-series, so it cannot be used to show divergence in a direct comparison. Eliminate E.

The series $\sum_{n=2}^{\infty} \frac{1}{\sqrt{n}} = \sum_{n=2}^{\infty} \frac{1}{n^{1/2}}$ is a diverging p-series, so it cannot be used to show convergence in a direct comparison. Eliminate B.

For all $n \geq 2, \sqrt{n} > \sqrt{n} - 1 \Rightarrow \frac{1}{\sqrt{n} - 1} > \frac{1}{\sqrt{n}}$. Since $\sum_{n=2}^{\infty} \frac{1}{\sqrt{n}}$, $\sum_{n=2}^{\infty} \frac{1}{\sqrt{n} - 1}$ also diverges by direct comparison.

11. Answer: B

Evaluate the parametric equations at $t = 0$.

$$x(0) = 4\cos(0) = 4$$

$$y(0) = 3\sin(0) = 0$$

The curve passes through the point $(4,0)$. Since the graphs in answer choices C and E do not do this, eliminate them.

Next evaluate the parametric equations at $t = \dfrac{\pi}{2}$.

$$x(0) = 4\cos\left(\frac{\pi}{2}\right) = 0$$

$$y(0) = 3\sin\left(\frac{\pi}{2}\right) = 3$$

The curve passes through the point $(0,3)$. Since the graphs in answer choices A and D do not do this, eliminate them.

To see why B is the correct answer, we will eliminate the parameter and obtain the rectangular equation for this curve.

$$x = 4\cos(t) \Rightarrow \frac{x}{4} = \cos(t) \Rightarrow \frac{x^2}{16} = \cos^2(t)$$

$$y = 3\sin(t) \Rightarrow \frac{y}{3} = \sin(t) \Rightarrow \frac{y^2}{9} = \sin^2(t)$$

Adding these results together, we obtain the following:

$$\frac{x^2}{16} + \frac{y^2}{9} = \cos^2(t) + \sin^2(t)$$

$$\frac{x^2}{16} + \frac{y^2}{9} = 1$$

This is the standard form of the equation of an ellipse that is centered at the origin with a horizontal semi-major axis of length 4 and a vertical semi-minor axis of length 3. Only the graph in answer choice B matches this description.

12. Answer: C

First find the slope m_t of the tangent line.

$$f'(x) = 3x^2 + 4x - 4$$

$$m_t = f'(1) = 3$$

The slope m_n of the normal line is found as follows:

$$m_n = -\frac{1}{m_t} = -\frac{1}{3}$$

Next, find the equation of the normal line in slope-intercept form.

$$y = -\frac{1}{3}(x-1) + 0$$

$$y = -\frac{1}{3}x + \frac{1}{3}$$

The y-intercept of this line is $\left(0, \frac{1}{3}\right)$.

13. Answer: D

According to the definition of the derivative at a point, this limit is the derivative of e^x evaluated at $x = \ln(2)$.

$$\lim_{h \to 0} \frac{e^{\ln(2)+h} - e^{\ln(2)}}{h} = \left(\frac{d}{dx}(e^x)\right)\Bigg|_{\ln(2)} = e^x\Big|_{\ln(2)} = e^{\ln(2)} = 2$$

14. Answer: C

The differential arc length element is given by the following:

$$ds = \sqrt{\left(\frac{d}{dt}(6t^2)\right)^2 + \left(\frac{d}{dt}(2t^3)\right)^2} \, dt = \sqrt{(12t)^2 + (6t^2)^2} \, dt = \sqrt{144t^2 + 36t^4} \, dt$$

$$ds = \sqrt{36t^2(4+t^2)} \, dt = 6t\sqrt{4+t^2} \, dt$$

The total arc length is found by integrating this element on $[1,4]$.

$$s = 6\int_1^4 t\sqrt{4+t^2} \, dt$$

Let $u = 4 + t^2$. Then $du = 2t\, dt \Rightarrow t\, dt = \frac{1}{2}du$. Furthermore, $x = 1 \Rightarrow u = 4 + 1^2 = 5$, and $x = 4 \Rightarrow u = 4 + 4^2 = 20$.

$$s = 6 \cdot \frac{1}{2}\int_5^{20} \sqrt{u}\, du = 3\int_5^{20} u^{1/2}\, du = 3 \cdot \frac{2}{3} u^{3/2}\Big|_5^{20} = 2\left(20^{3/2} - 5^{3/2}\right) = 2\left(\sqrt{20}^3 - \sqrt{5}^3\right)$$

$$s = 2\left(20\sqrt{20} - 5\sqrt{5}\right) = 2\left(20 \cdot 2\sqrt{5} - 5\sqrt{5}\right) = 70\sqrt{5}$$

15. Answer: C

We find the mean rate of change of f on $\left[-\dfrac{\pi}{2}, \dfrac{\pi}{2} \right]$ below.

$$\frac{f\left(\dfrac{\pi}{2}\right) - f\left(-\dfrac{\pi}{2}\right)}{\dfrac{\pi}{2} - \left(-\dfrac{\pi}{2}\right)} = \frac{\left(\dfrac{\pi}{2} - \cos\left(\dfrac{\pi}{2}\right)\right) - \left(-\dfrac{\pi}{2} - \cos\left(-\dfrac{\pi}{2}\right)\right)}{\dfrac{\pi}{2} - \left(-\dfrac{\pi}{2}\right)} = 1$$

The instantaneous rate of change of f at $x = c$ is given as follows:

$$f'(c) = 1 + \sin(c)$$

We now equate these two results and solve for c.

$$1 + \sin(c) = 1$$

$$\sin(c) = 0$$

$$c = 0 \, .$$

16. Answer: E

First find $f'(x)$. Note that we will need the Chain Rule to do this.

$$f'(x) = \frac{d}{dx}\left(\frac{4}{x^2+1}\right) = 4\frac{d}{dx}\left(\left(x^2+1\right)^{-1}\right) = -4\left(x^2+1\right)^{-2} \cdot \frac{d}{dx}\left(x^2+1\right)$$

$$f'(x) = -\frac{4}{\left(x^2+1\right)^2} \cdot 2x = -\frac{8x}{\left(x^2+1\right)^2}$$

Now find $f''(x)$ using the Quotient Rule and the Chain Rule.

$$f''(x) = -\frac{\left(x^2+1\right)^2 \cdot \dfrac{d}{dx}(8x) - 8x \cdot \dfrac{d}{dx}\left(x^2+1\right)^2}{\left(\left(x^2+1\right)^2\right)^2}$$

$$f''(x) = -\frac{8\left(x^2+1\right)^2 - 8x \cdot 2\left(x^2+1\right) \cdot \dfrac{d}{dx}\left(x^2+1\right)}{\left(x^2+1\right)^4}$$

$$f''(x) = -\frac{8\left(x^2+1\right)^2 - 16x\left(x^2+1\right) \cdot 2x}{\left(x^2+1\right)^4}$$

$$f''(x) = -\frac{8\left(x^2+1\right)^2 - 32x^2\left(x^2+1\right)}{\left(x^2+1\right)^4}$$

$$f''(x) = -\frac{8(x^2+1)(x^2+1-4x^2)}{(x^2+1)^4}$$

$$f''(x) = -\frac{8(1-3x^2)}{(x^2+1)^3}$$

We are interested in the interval on which the graph of $f(x)$ is concave downwards, so $f''(x) < 0$ on this interval.

$$f''(x) = -\frac{8(1-3x^2)}{(x^2+1)^3} < 0$$

$$-8(1-3x^2) < 0$$

$$1-3x^2 > 0$$

$$x^2 < \frac{1}{3}$$

$$-\sqrt{\frac{1}{3}} < x < \sqrt{\frac{1}{3}}$$

So the interval we seek is $\left(-\sqrt{\frac{1}{3}}, \sqrt{\frac{1}{3}}\right)$.

17. Answer: B

Rewrite the sum using summation notation.

$$1 + \ln(3) + \frac{(\ln(2))^2}{2!} + \frac{(\ln(3))^3}{3!} + \frac{(\ln(3))^4}{4!} + \frac{(\ln(3))^5}{5!} + \cdots = \sum_{n=0}^{\infty} \frac{(\ln(3))^n}{n!}$$

Now recognize that the sum on the right-hand side is equal to the Maclaurin series for e^x, evaluated at $x = \ln(3)$.

$$1 + \ln(3) + \frac{(\ln(2))^2}{2!} + \frac{(\ln(3))^3}{3!} + \frac{(\ln(3))^4}{4!} + \frac{(\ln(3))^5}{5!} + \cdots = e^{\ln(3)} = 3$$

18. Answer: E

Let $\vec{r}(t) = \left(te^{-t}, e^{-t^2}\right)$. Then the velocity vector $\vec{v}(t)$ is found as follows:

$$\vec{v}(t) = \frac{d\vec{r}(t)}{dt} = \frac{d}{dt}\left(te^{-t}, e^{-t^2}\right) = \frac{d}{dt}\left(1 \cdot e^{-t} + t \cdot e^{-t} \cdot \frac{d}{dt}(-t), e^{-t^2} \cdot \frac{d}{dt}(-t^2)\right)$$

$$\vec{v}(t) = \left(e^{-t} + te^{-t} \cdot (-1), e^{-t^2} \cdot (-2t)\right)$$

$$\vec{v}(t) = \left((1-t)e^{-t}, -2te^{-t^2}\right)$$

Note that when differentiating the second vector component we used the Chain Rule, and when differentiating the first vector component we used the Product Rule *and* the Chain Rule. Now evaluate the velocity vector at $t = 1$.

$$\vec{v}(1) = \left((1-1)e^{-1}, -2(1)e^{-1^2} \right) = \left(0, -\frac{2}{e} \right)$$

19. Answer: A

This integral is improper because its upper limit of integration is infinite.

$$\int_{2}^{\infty} \frac{1}{x(\ln(x))^2} \, dx = \lim_{b \to \infty} \int_{2}^{b} \frac{1}{x(\ln(x))^2} \, dx$$

Let $u = \ln(x)$, so $du = \frac{1}{x} dx$. Furthermore $x = 2 \Rightarrow u = \ln(2)$, and $x = b \Rightarrow u = \ln(b)$.

$$\int_{2}^{\infty} \frac{1}{x(\ln(x))^2} \, dx = \lim_{b \to \infty} \int_{\ln(2)}^{\ln(b)} \frac{1}{u^2} \, du = \lim_{b \to \infty} \int_{\ln(2)}^{\ln(b)} u^{-2} \, du = -\lim_{b \to \infty} u^{-1} \Big|_{\ln(2)}^{\ln(b)}$$

$$\int_{2}^{\infty} \frac{1}{x(\ln(x))^2} \, dx = -\lim_{b \to \infty} \left(\frac{1}{\ln(b)} - \frac{1}{\ln(2)} \right) = -\left(0 - \frac{1}{\ln(2)} \right) = \frac{1}{\ln(2)}$$

20. Answer: E

Rewrite the integral as follows:

$$\int \sin^3(2x) \, dx = \int \sin^2(2x)\sin(2x) \, dx$$

Use the Pythagorean identity $\sin^2(\theta) + \cos^2(\theta) = 1$ to rewrite $\sin^2(2x)$ as follows:

$$\int \sin^3(2x) \, dx = \int \left(1 - \cos^2(2x) \right) \sin(2x) \, dx$$

Now let $u = \cos(2x)$. Then $du = -2\sin(2x) \, dx \Rightarrow \sin(2x) \, dx = -\frac{1}{2} du$.

$$\int \sin^3(2x) \, dx = -\frac{1}{2} \int \left(1 - u^2 \right) du = -\frac{1}{2} \left(u - \frac{1}{3} u^3 \right) + C$$

$$\int \sin^3(2x) \, dx = -\frac{1}{2} \cos(2x) + \frac{1}{6} \cos^3(2x) + C$$

21. Answer: B

Let $(x, y^2) = (x, x^2)$ be the point on the graph of f that we are to determine, and let r be the distance between (x, x^2) and the point $\left(2, \dfrac{1}{2}\right)$. See the following figure:

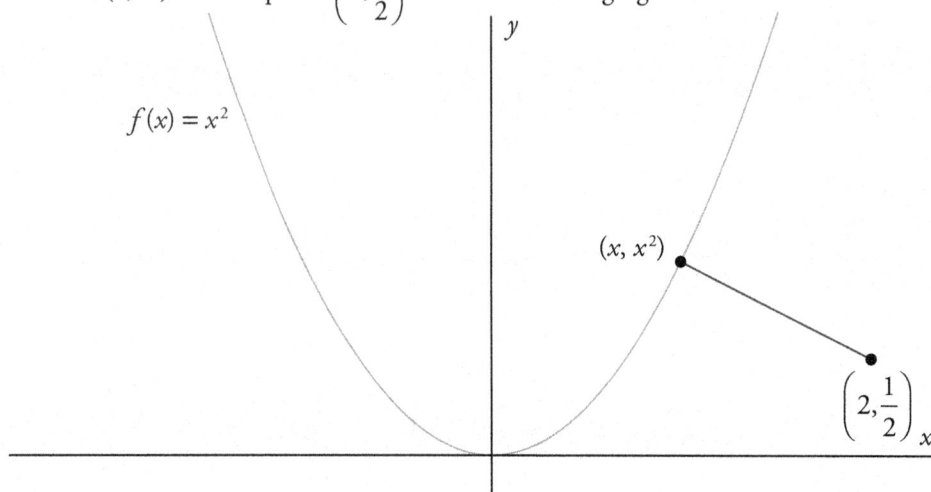

Using the distance formula, we get $r = \left(\left(x-2\right)^2 + \left(x^2 - \dfrac{1}{2}\right)^2\right)^{1/2}$. We wish to minimize r. To that end, we find the critical numbers of r by setting $\dfrac{dr}{dt} = 0$ and solving for x. We will need to use the Chain Rule twice, as you will see below.

$$\frac{dr}{dt} = \frac{d}{dt}\left(\left(x-2\right)^2 + \left(x^2 - \frac{1}{2}\right)^2\right)^{1/2}$$

$$\frac{dr}{dt} = \frac{1}{2}\left(\left(x-2\right)^2 + \left(x^2 - \frac{1}{2}\right)^2\right)^{-1/2} \cdot \frac{d}{dt}\left(\left(x-2\right)^2 + \left(x^2 - \frac{1}{2}\right)^2\right)$$

$$\frac{dr}{dt} = \frac{1}{2}\left(\left(x-2\right)^2 + \left(x^2 - \frac{1}{2}\right)^2\right)^{-1/2}\left(2(x-2) + 2\left(x^2 - \frac{1}{2}\right) \cdot \frac{d}{dt}\left(x^2\right)\right)$$

$$\frac{dr}{dt} = \frac{1}{2}\left(\left(x-2\right)^2 + \left(x^2 - \frac{1}{2}\right)^2\right)^{-1/2}\left(2(x-2) + 2\left(x^2 - \frac{1}{2}\right) \cdot 2x\right)$$

$$\frac{dr}{dt} = \frac{-2 + 2x^3}{\left(\left(x-2\right)^2 + \left(x^2 - \frac{1}{2}\right)^2\right)^{1/2}} = \frac{2\left(x^3 - 1\right)}{r}$$

$$\frac{dr}{dt} = 0 \Rightarrow x^3 - 1 = 0$$

The equation $x^3 - 1 = 0$ has only one real solution, namely $x = 1$. To verify that we have found a minimum (as opposed to a maximum), we apply the First Derivative Test. To the left of $x = 1$, at $x = 0$,

$\frac{dr}{dt} = -\frac{2}{r} < 0$. So r is decreasing at $x = 0$. To the right of $x = 1$, at $x = 2$, $\frac{dr}{dt} = \frac{14}{r} > 0$. So r is increasing at $x = 2$. Since r changes from decreasing to increasing at $x = 1$, we have indeed found a minimum. So the point on the graph of $f(x) = x^2$ that is closest to the point $\left(2, \frac{1}{2}\right)$ is $(1,1)$.

22. Answer: D

Multiply both sides of the differential equation by $\frac{dx}{(x^2+16)y}$ to separate the variables.

$$\frac{dx}{(x^2+16)y} \cdot (x^2+16)\frac{dy}{dx} = \frac{dx}{(x^2+16)y} \cdot xy$$

$$\frac{dy}{y} = \frac{x}{x^2+16}dx$$

Now integrate both sides.

$$\int \frac{dy}{y} = \int \frac{x}{x^2+16}dx$$

On the right side, let $u = x^2 + 16$. Then $du = 2x\,dx \Rightarrow x\,dx = \frac{1}{2}du$.

$$\int \frac{dy}{y} = \frac{1}{2}\int \frac{1}{u}du$$

$$\ln|y| = \frac{1}{2}\ln|u| + C = \frac{1}{2}\ln(x^2+16) + C = \ln\sqrt{x^2+16} + C$$

Exponentiating both sides with base e, we obtain the following:

$$e^{\ln|y|} = e^{\ln\sqrt{x^2+16}+C} = e^C e^{\ln\sqrt{x^2+16}}$$

$$|y| = e^C\sqrt{x^2+16}$$

$$y = \pm e^C\sqrt{x^2+16}$$

Let $K = \pm e^C$.

$$y = K\sqrt{x^2+16}$$

Now use the fact that $y = 10$ when $x = 3$ to find K.

$$10 = K\sqrt{3^2+16} \Rightarrow K = 2$$

So the solution is $y = 2\sqrt{x^2+16}$.

23. Answer: D

The particle is moving to the right whenever the velocity $v(t)$ is positive. The particle is moving to the left whenever $v(t)$ is negative. First compute $v(t)$. We will need to use the Chain Rule for this.

$$v(t) = x'(t) = \frac{d}{dt}\left(\tan^{-1}\left(\sin(t)\right)\right)$$

$$v(t) = \frac{1}{1+\sin^2(t)} \cdot \frac{d}{dt}\left(\sin(t)\right) = \frac{\cos(t)}{1+\sin^2(t)}$$

Now plug in the three values of t given in the problem.

$$v\left(\frac{\pi}{4}\right) = \frac{\cos\left(\frac{\pi}{4}\right)}{1+\sin^2\left(\frac{\pi}{4}\right)} = \frac{\frac{\sqrt{2}}{2}}{1+\left(\frac{\sqrt{2}}{2}\right)^2} = \frac{\sqrt{2}}{3} > 0$$

$$v\left(\frac{3\pi}{4}\right) = \frac{\cos\left(\frac{3\pi}{4}\right)}{1+\sin^2\left(\frac{3\pi}{4}\right)} = \frac{-\frac{\sqrt{2}}{2}}{1+\left(\frac{\sqrt{2}}{2}\right)^2} = -\frac{\sqrt{2}}{3} < 0$$

$$v\left(\frac{7\pi}{4}\right) = \frac{\cos\left(\frac{7\pi}{4}\right)}{1+\sin^2\left(\frac{7\pi}{4}\right)} = \frac{\frac{\sqrt{2}}{2}}{1+\left(\frac{\sqrt{2}}{2}\right)^2} = \frac{\sqrt{2}}{3} > 0$$

So the particle is moving to the left at $t = \frac{\pi}{4}$ and $t = \frac{7\pi}{4}$ only.

24. Answer: A

Simplify the integrand by dividing each term in the numerator by the single term in the denominator, and then integrate.

$$\int \frac{e^{2x} - 2e^x}{e^x}\,dx = \int \left(e^x - 2\right)dx = e^x - 2x + C$$

25. Answer: C

First compute $f'(x)$. We will need to use both the Quotient Rule and the Chain Rule.

$$f'(x) = \frac{d}{dx}\left(\frac{x}{\sqrt{x^2+5}}\right) = \frac{d}{dx}\left(\frac{x}{\left(x^2+5\right)^{1/2}}\right)$$

$$f'(x) = \frac{\left(x^2+5\right)^{1/2} \cdot \frac{d}{dx}(x) - x \cdot \frac{d}{dx}\left(x^2+5\right)^{1/2}}{\left(\left(x^2+5\right)^{1/2}\right)^2}$$

$$f'(x) = \frac{(x^2+5)^{1/2} \cdot 1 - x \cdot \frac{1}{2}(x^2+5)^{-1/2} \frac{d}{dx}(x^2+5)}{x^2+5}$$

$$f'(x) = \frac{(x^2+5)^{1/2} - x \cdot \frac{1}{2}(x^2+5)^{-1/2}(2x)}{x^2+5}$$

$$f'(x) = \frac{\sqrt{x^2+5} - \dfrac{x^2}{\sqrt{x^2+5}}}{x^2+5}$$

Multiply the numerator and denominator by $\sqrt{x^2+5}$.

$$f'(x) = \frac{\sqrt{x^2+5} - \dfrac{x^2}{\sqrt{x^2+5}}}{x^2+5} \cdot \frac{\sqrt{x^2+5}}{\sqrt{x^2+5}} = \frac{x^2+5-x^2}{(x^2+5)^{3/2}}$$

$$f'(x) = \frac{5}{(x^2+5)^{3/2}}$$

Now find the slope m of tangent line.

$$m = f'(2) = \frac{5}{(2^2+5)^{3/2}} = \frac{5}{27}$$

Finally, find the equation of the tangent line.

$$y - \frac{2}{3} = \frac{5}{27}(x-2)$$

$$y - \frac{2}{3} = \frac{5}{27}x - \frac{10}{27}$$

$$-\frac{5}{27}x + y - \frac{8}{27} = 0$$

$$-27\left(-\frac{5}{27}x + y - \frac{8}{27}\right) = -27(0)$$

$$5x - 27y + 8 = 0$$

26. Answer: A

Use the coordinate system shown in the figure below.

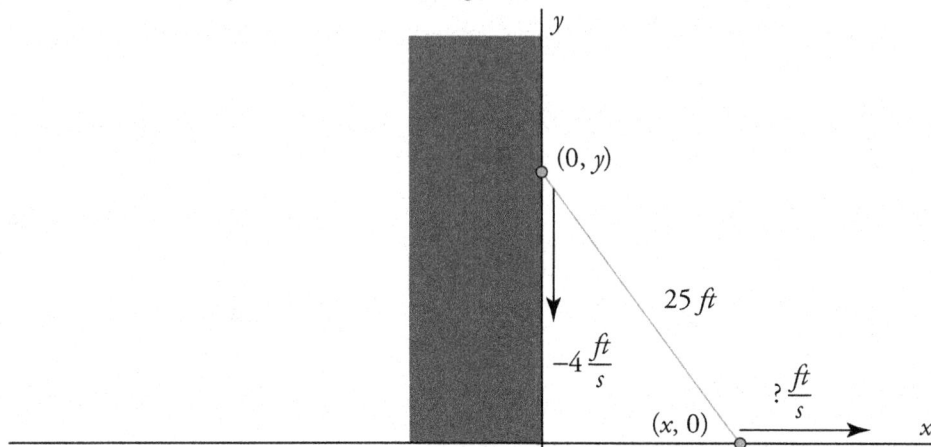

As shown in the figure, the point at which the top of the ladder touches the building is $(0, y)$, and the point at which the bottom of the ladder touches the ground is $(x, 0)$. Using the Pythagorean Theorem, we obtain the following relation between x and y:

$x^2 + y^2 = 625$

Differentiate both sides with respect to t.

$\dfrac{d}{dt}\left(x^2 + y^2\right) = \dfrac{d}{dt}(25)$

$2x\dfrac{dx}{dt} + 2y\dfrac{dy}{dt} = 0$

From the figure, we see that $\dfrac{dy}{dt} = -4$. $\dfrac{dx}{dt}$ is the unknown rate whose value we seek. When the top is 7 feet from the ground, $y = 7$ and $x = \sqrt{625 - 7^2} = 24$. Plug these numbers into the last equation, and solve for $\dfrac{dx}{dt}$.

$2(24)\dfrac{dx}{dt} + 2(7)(-4) = 0$

$48\dfrac{dx}{dt} - 56 = 0$

$\dfrac{dx}{dt} = \dfrac{7}{6}$

27. Answer: C

Integrate by parts. Let $u = \tan^{-1}(x)$, so $dv = dx$. Calculate du and v as follows:

$$u = \tan^{-1}(x) \Rightarrow du = \frac{1}{x^2 + 1}\, dx$$

$$dv = dx \Rightarrow v = \int dx = x$$

We needn't worry about the constant of integration, because this is a definite integral.

$$\int_0^{\sqrt{3}} \tan^{-1}(x)\, dx = x\tan^{-1}(x)\Big|_0^{\sqrt{3}} - \int_0^{\sqrt{3}} \frac{x}{x^2 + 1}\, dx$$

We will evaluate the integral on the right side using a substitution of variables. Let $u = x^2 + 1$. Then $du = 2x\, dx \Rightarrow x\, dx = \frac{1}{2}\, du$. Furthermore, $x = 0 \Rightarrow u = 1$, and $x = \sqrt{3} \Rightarrow u = 4$.

$$\int_0^{\sqrt{3}} \tan^{-1}(x)\, dx = x\tan^{-1}(x)\Big|_0^{\sqrt{3}} - \frac{1}{2}\int_1^4 \frac{1}{u}\, du = x\tan^{-1}(x)\Big|_0^{\sqrt{3}} - \frac{1}{2}\ln|u|\Big\|_1^4$$

$$\int_0^{\sqrt{3}} \tan^{-1}(x)\, dx = \left(\sqrt{3}\tan^{-1}\left(\sqrt{3}\right) - 0\right) - \frac{1}{2}\left(\ln(4) - \ln(1)\right) = \left(\sqrt{3}\cdot\frac{\pi}{3} - 0\right) - \frac{1}{2}\left(\ln(4) - 0\right)$$

$$\int_0^{\sqrt{3}} \tan^{-1}(x)\, dx = \frac{\pi\sqrt{3}}{3} - \ln\left(4^{\frac{1}{2}}\right) = \frac{\pi\sqrt{3}}{3} - \ln(2)$$

28. Answer: D

The Maclaurin series for $g(x) = e^x = \sum_{n=0}^{\infty} \frac{x^n}{n!}$. The Maclaurin series for $f(x) = e^{2x}$ is found by evaluating $g(2x)$.

$$f(x) = g(2x) = \sum_{n=0}^{\infty} \frac{(2x)^n}{n!} = \frac{(2x)^0}{0!} + \frac{(2x)^1}{1!} + \frac{(2x)^2}{2!} + \cdots = 1 + 2x + 2x^2 + \cdots$$

29. Answer: A

The method of shells is preferable here, as the method of disks would require two integrations. We sketch the region below.

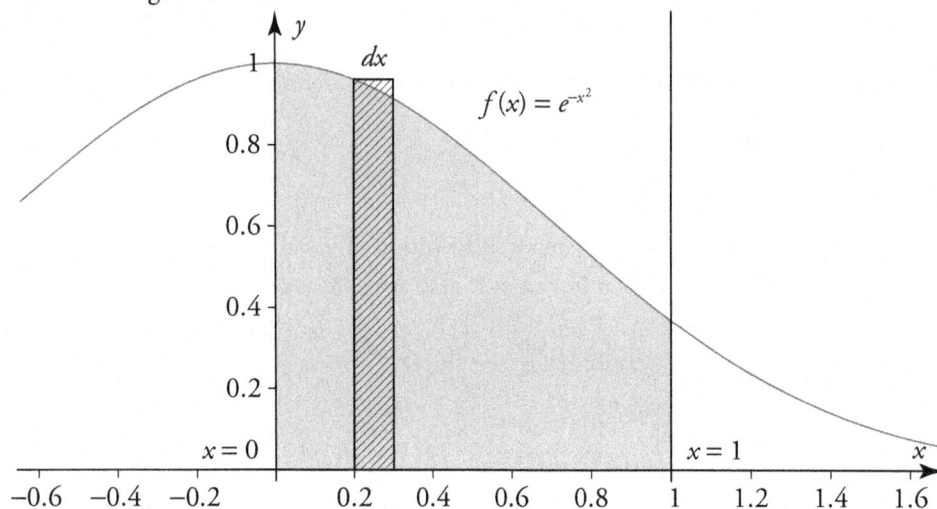

The differential volume element is given as follows:

$$dV = 2\pi x e^{-x^2}\, dx$$

The total volume V of the solid is given by the following integral, which you can do on your calculator:

$$V = 2\pi \int_0^1 x e^{-x^2}\, dx \approx 1.986$$

30. Answer: C

The volume V of the (spherical) ball bearing is given in terms of the radius r by the following relation:

$$V = \frac{4}{3}\pi r^3$$

The propagated error ΔV is approximated by the differential dV.

$$\Delta V \approx dV = 4\pi r^2\, dr$$

The measured value of the radius is $r = 1.8$ *cm*, and the accuracy of the measuring instrument is $dr = 0.1$ *cm*. Evaluating the propagated error using these numbers gives us our result.

$$\Delta V \approx 4\pi(1.8)^2(0.1) \approx 4.07\, cm^3$$

31. Answer: D

$$\lim_{n \to \infty} a_n = \lim_{n \to \infty} \frac{2n}{\sqrt{n^2 + 3}}$$

Divide the numerator by n and the denominator by $\sqrt{n^2}$ (which also equals n).

$$\lim_{n \to \infty} a_n = \lim_{n \to \infty} \frac{\dfrac{2n}{n}}{\dfrac{\sqrt{n^2 + 3}}{\sqrt{n^2}}} = \lim_{n \to \infty} \frac{2}{\sqrt{1 + \dfrac{3}{n^2}}} = \frac{2}{\sqrt{1 + 0}} = 2$$

32. Answer: B

The area A of the region is equal to $\int_0^{\sqrt{\pi/2}} \sin(x^2)\, dx$. We will approximate this integral with a trapezoidal sum.

We have $\Delta x_i = \dfrac{\sqrt{\dfrac{\pi}{2}} - 0}{4} = \dfrac{1}{4}\sqrt{\dfrac{\pi}{2}}$, and the partition points are $x_0 = 0$, $x_1 = \dfrac{1}{4}\sqrt{\dfrac{\pi}{2}}$, $x_2 = \dfrac{1}{2}\sqrt{\dfrac{\pi}{2}}$, $x_3 = \dfrac{3}{4}\sqrt{\dfrac{\pi}{2}}$, and $x_4 = \sqrt{\dfrac{\pi}{2}}$.

$$A \approx \frac{\sqrt{\dfrac{\pi}{2}} - 0}{2(4)} \left[f(0) + 2f\left(\frac{1}{4}\sqrt{\frac{\pi}{2}}\right) + 2f\left(\frac{1}{2}\sqrt{\frac{\pi}{2}}\right) + 2f\left(\frac{3}{4}\sqrt{\frac{\pi}{2}}\right) + f\left(\sqrt{\frac{\pi}{2}}\right) \right]$$

$$A \approx \frac{1}{8}\sqrt{\frac{\pi}{2}} \left[\sin(0) + 2\sin\left(\frac{\pi}{32}\right) + 2\sin\left(\frac{\pi}{8}\right) + 2\sin\left(\frac{9\pi}{32}\right) + \sin\left(\frac{\pi}{2}\right) \right] \approx 0.5495$$

33. Answer: D

Integrate by parts. Let $u = x$, and so $dv = \sec(x)\tan(x)\,dx$. Calculate du and v.
$u = x \implies du = dx$

$$dv = \sec(x)\tan(x)\,dx \implies v = \int \sec(x)\tan(x)\,dx = \sec(x)$$

We needn't worry about the constant of integration until the end.

$$\int x \sec(x)\tan(x)\,dx = x\sec(x) - \int \sec(x)\,dx = x\sec(x) - \ln\left|\sec(x) + \tan(x)\right| + C$$

34. Answer: E

Use logarithmic differentiation.

$$\ln(f(x)) = \ln\left((\sin(x))^x\right)$$

$$\ln(f(x)) = x\ln(\sin(x))$$

$$\frac{d}{dx}(\ln(f(x))) = \frac{d}{dx}(x\ln(\sin(x)))$$

$$\frac{f'(x)}{f(x)} = \frac{d}{dx}(x)\cdot\ln(\sin(x)) + x\cdot\frac{d}{dx}(\ln(\sin(x)))$$

$$\frac{f'(x)}{f(x)} = 1\cdot\ln(\sin(x)) + x\cdot\frac{1}{\sin(x)}\cdot\frac{d}{dx}(\sin(x))$$

$$\frac{f'(x)}{f(x)} = \ln(\sin(x)) + x\cdot\frac{1}{\sin(x)}\cdot\cos(x) = \ln(\sin(x)) + x\cdot\frac{\cos(x)}{\sin(x)}$$

$$\frac{f'(x)}{f(x)} = \ln(\sin(x)) + x\cot(x)$$

$$f'(x) = \left(\ln(\sin(x)) + x\cot(x)\right)f(x) = \left(\ln(\sin(x)) + x\cot(x)\right)(\sin(x))^x$$

35. Answer: D

$$\sum_{n=0}^{\infty} e^{-2n} = \sum_{n=0}^{\infty}\left(\frac{1}{e^2}\right)^n$$

This is a geometric series with a common ratio of $\dfrac{1}{e^2}$ between successive terms. Since $\dfrac{1}{e^2} < 1$, the series converges. Its sum is given below.

$$\sum_{n=0}^{\infty} e^{-2n} = \frac{1}{1 - \dfrac{1}{e^2}} = \frac{e^2}{e^2 - 1}$$

36. Answer: A

Dividing $2x^3 - 3x^2 + 2x + 2$ by $x^2 + 1$ gives $2x - 3 + \dfrac{5}{x^2 + 1}$.

$$\int\frac{2x^3 - 3x^2 + 2x + 2}{x^2 + 1}dx = \int\left(2x - 3 + \frac{5}{x^2 + 1}\right)dx = x^2 - 3x + 5\tan^{-1}(x) + C$$

37. Answer: E

We will use the method of shells. The region is sketched below.

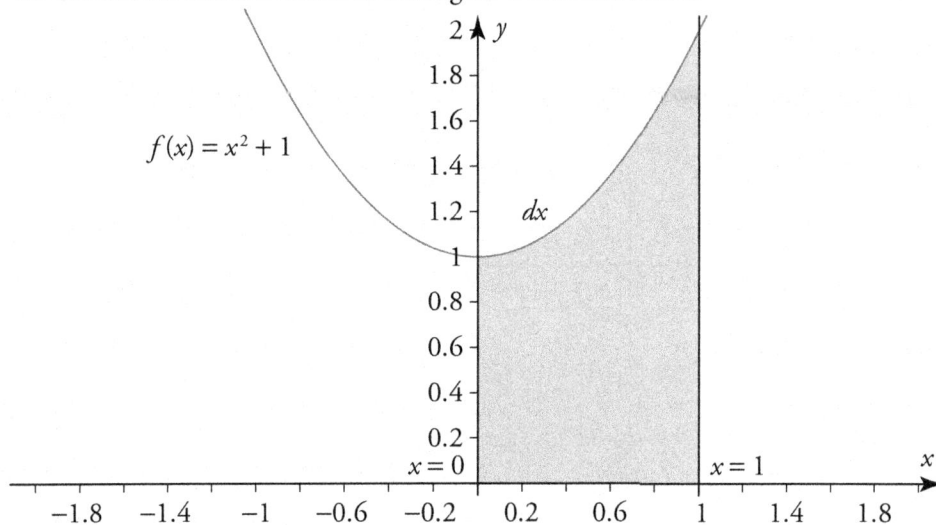

$f(x) = x^2 + 1$

The differential volume element is given below.

$$dV = 2\pi x\left(x^2 + 1\right)dx = 2\pi\left(x^3 + x\right)dx$$

The total volume is given by the following integral:

$$V = 2\pi\int_0^1\left(x^3 + x\right)dx \approx 4.712$$

38. Answer: B

Use the calculator to graph the function and find the maximum. The point at which f attains its maximum has been labeled in the figure below.

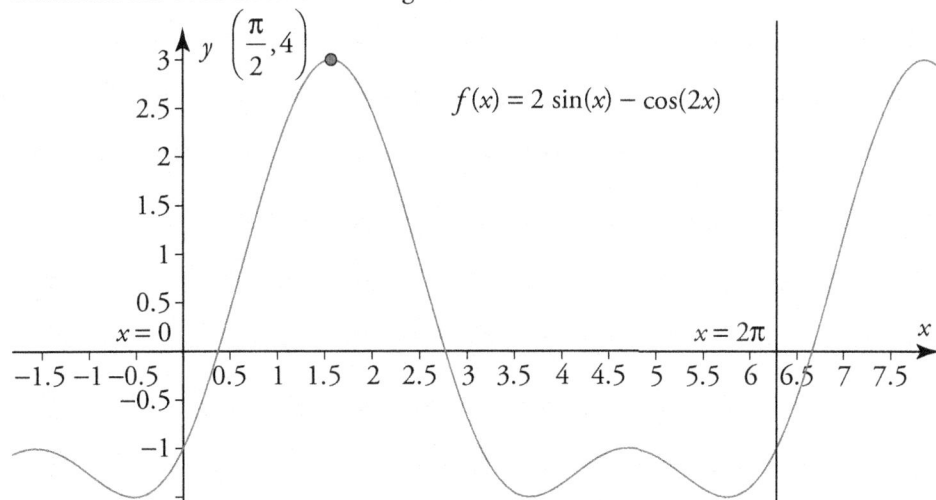

$\left(\dfrac{\pi}{2}, 4\right)$

$f(x) = 2\sin(x) - \cos(2x)$

39. Answer: C

We sketch the region below.

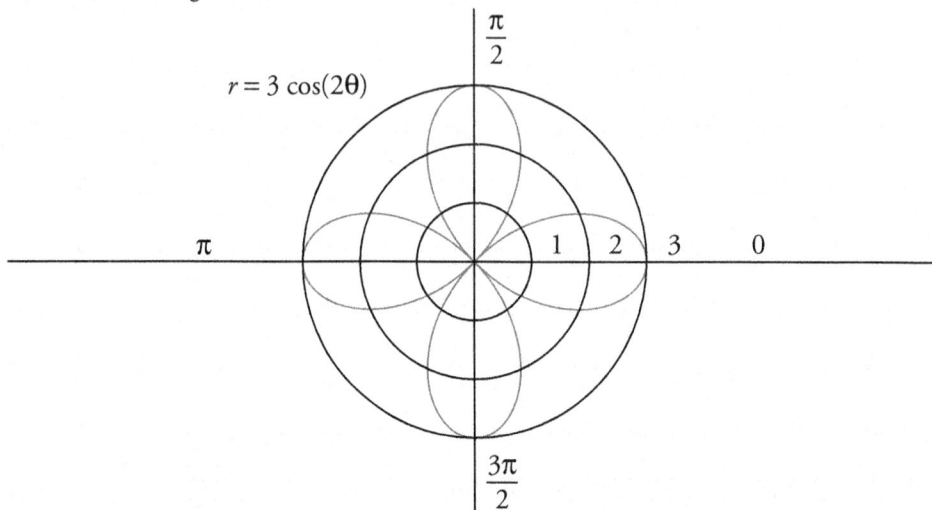

$r = 3\cos(2\theta)$

To find the angular limits of integration, we solve the equation $3\cos(2\theta) = 0$.

$$3\cos(2\theta) = 0 \Rightarrow 2\theta = \pm\frac{\pi}{2}, \pm\frac{3\pi}{2}, \pm\frac{5\pi}{2}, \ldots \Rightarrow \theta = \pm\frac{\pi}{4}, \pm\frac{3\pi}{4}, \pm\frac{5\pi}{4}, \ldots$$

The solutions $\theta = \pm\dfrac{\pi}{4}$ are the limits that determine the right petal. The area A of the petal is determined by the following integral, which you can do on your calculator:

$$A = \frac{1}{2}\int_{-\pi/4}^{\pi/4} \left(3\cos(2\theta)\right)^2 d\theta = \frac{9\pi}{8}$$

40. Answer: C

We use the formulas $x_n = x_{n-1} + h$ and $y_n = y_{n-1} + hF(x_{n-1}, y_{n-1})$, where $(x_0, y_0) = (0,1)$ and $F(x,y) = x - y$. With a step size of $h = 0.2$, we will need to take 5 steps to get from $x_0 = 0$ to $x = 1$.

n	0	1	2	3	4	5
x_n	0	0.2	0.4	0.6	0.8	1
y_n	1	0.8	0,68	0.624	0.6192	0.65536

41. Answer: A

Do the integral on your calculator.

$$\int_0^1 x^2 e^x \, dx \approx 0.718$$

42. Answer: B

Use your calculator to find the asymptotes. They are labeled in the figure below.

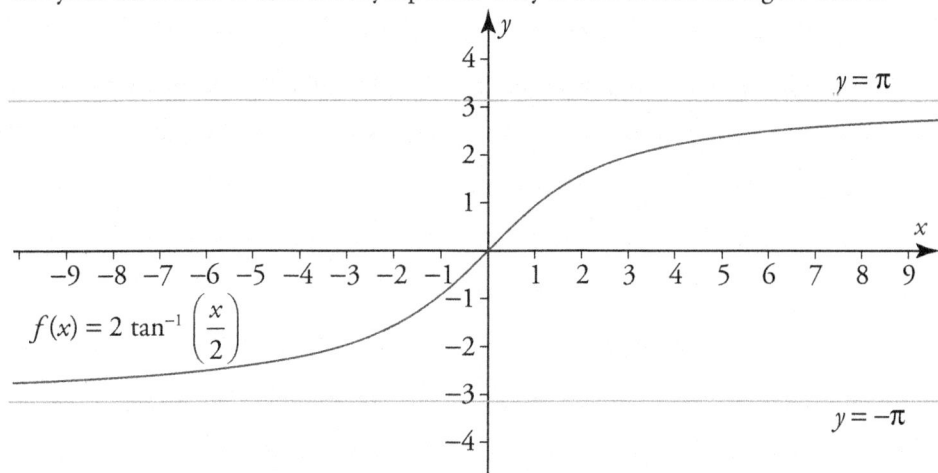

$$f(x) = 2 \tan^{-1}\left(\frac{x}{2}\right)$$

43. Answer: C

The differential equation is $\dfrac{dy}{dx} = cy$, and the general solution is $y(x) = Ae^{cx}$. Use the fact that $y = 6$ when $x = 0$ to find A.

$$6 = Ae^{0c}$$
$$6 = A$$

So thus far, we have $y(x) = 6e^{cx}$. Now use the fact that $y = 15$ when $x = 4$ to find c.

$$15 = 6e^{4c}$$

$$e^{4c} = \frac{5}{2} \Rightarrow c = \frac{1}{4}\ln\left(\frac{5}{2}\right)$$

So our particular solution is $y(x) = 6e^{\frac{1}{4}\ln\left(\frac{5}{2}\right)x}$. Evaluate this at $x = 8$ to get the answer.

$$y(8) = 6e^{\frac{1}{4}\ln\left(\frac{5}{2}\right)\cdot 8} = 37.5$$

44. Answer: E

The integral is improper because the integrand is unbounded at $x = \dfrac{5}{4}$, which is in the interval of integration.

$$\int_{1}^{3/2} \frac{1}{(4x-5)^2}\,dx = \int_{1}^{5/4} \frac{1}{(4x-5)^2}\,dx + \int_{5/4}^{3/2} \frac{1}{(4x-5)^2}\,dx$$

We start by evaluating the first integral on the right side.

$$\int_{1}^{5/4} \frac{1}{(4x-5)^2}\,dx = \lim_{t\to\frac{5}{4}^-} \int_{1}^{t} \frac{1}{(4x-5)^2}\,dx$$

Let $u = 4x - 5$. Then $du = 4dx \Rightarrow dx = \dfrac{1}{4}du$. Furthermore $x = 1 \Rightarrow u = -1$, and $x = t \Rightarrow u = 4t - 5$.

$$\int_{1}^{5/4} \frac{1}{(4x-5)^2}\,dx = \frac{1}{4}\lim_{t\to\frac{5}{4}^-}\int_{-1}^{4t-5}\frac{1}{u^2}\,du = \frac{1}{4}\lim_{t\to\frac{5}{4}^-}\int_{-1}^{4t-5}u^{-2}\,du = -\frac{1}{4}\lim_{t\to\frac{5}{4}^-}u^{-1}\Big|_{-1}^{4t-5}$$

$$\int_{1}^{5/4} \frac{1}{(4x-5)^2}\,dx = -\frac{1}{4}\lim_{t\to\frac{5}{4}^-}\frac{1}{u}\Big|_{-1}^{4t-5} = -\frac{1}{4}\lim_{t\to\frac{5}{4}^-}\left(\frac{1}{4t-5}+1\right) = \infty$$

Since this piece of the improper integral diverges, the entire improper integral diverges.

45. Answer: B

Graph this equation on your calculator and verify that it matches the graph given in the problem statement.

1. For each part the length of each subinterval is $\Delta x_i = \dfrac{\pi - 0}{4} = \dfrac{\pi}{4}$, and the partition points are
    $x_0 = 0$, $x_1 = \dfrac{\pi}{4}$, $x_2 = \dfrac{\pi}{2}$, $x_3 = \dfrac{3\pi}{4}$, and $x_4 = \pi$.

A. Left Riemann sum:

$$\int_0^\pi f(x)\,dx \approx \sum_{i=1}^4 f(x_{i-1})\Delta x_i = \left[f(x_0) + f(x_1) + f(x_2) + f(x_3) \right]\Delta x_i$$

$$\int_0^\pi f(x)\,dx \approx \left[0 + \frac{\sqrt{\pi}}{2\sqrt{2}} + \sqrt{\frac{\pi}{2}} + \frac{\sqrt{3\pi}}{2\sqrt{2}} \right]\frac{\pi}{4} \approx 2.329$$

B. Right Riemann sum:

$$\int_0^\pi f(x)\,dx \approx \sum_{i=1}^4 f(x_i)\Delta x_i = \left[f(x_1) + f(x_2) + f(x_3) + f(x_4) \right]\Delta x_i$$

$$\int_0^\pi f(x)\,dx \approx \left[\frac{\sqrt{\pi}}{2\sqrt{2}} + \sqrt{\frac{\pi}{2}} + \frac{\sqrt{3\pi}}{2\sqrt{2}} + 0 \right]\frac{\pi}{4} \approx 2.329$$

C. Trapezoidal sum:

$$\int_0^\pi f(x)\,dx \approx \frac{\pi - 0}{2(4)}\left[f(x_0) + 2f(x_1) + 2f(x_2) + 2f(x_3) + 2f(x_4) \right]$$

$$\int_0^\pi f(x)\,dx \approx \frac{\pi}{8}\left[0 + 2\cdot\frac{\sqrt{\pi}}{2\sqrt{2}} + 2\cdot\sqrt{\frac{\pi}{2}} + 2\cdot\frac{\sqrt{3\pi}}{2\sqrt{2}} + 0 \right] \approx 2.329$$

Note that all three approximations are equal. That is due to the fact that $f(x_0) = f(x_4) = 0$.

2. If you sketch the graph of the function, you will see that the entire curve is traced out if we restrict the domain to $0 \le \theta < 2\pi$. We can also see in the figure below that there are five horizontal tangents (one of which is at the pole), which are drawn green, and five vertical tangents (again, one of which is at the pole), which are drawn in red.

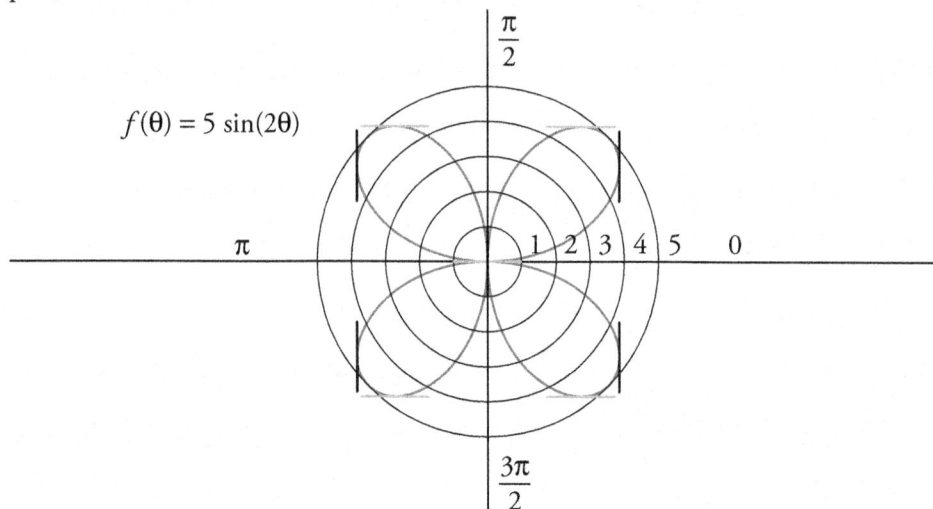

$f(\theta) = 5\sin(2\theta)$

For the first two parts, we will need to know $\dfrac{dx}{d\theta}, \dfrac{dy}{d\theta}$, and their zeros in the interval $[0, 2\pi]$.

$$\frac{dx}{d\theta} = \frac{d}{d\theta}\left(f(\theta)\cos(\theta)\right) = -f(\theta)\sin(\theta) + f'(\theta)\cos(\theta)$$

$$\frac{dx}{d\theta} = -\left[5\sin(2\theta)\right] \cdot \sin(\theta) + \frac{d}{d\theta}\left[5\sin(2\theta)\right] \cdot \cos(\theta)$$

$$\frac{dx}{d\theta} = -5\sin(2\theta)\sin(\theta) + 10\cos(2\theta)\cos(\theta)$$

Use the trigonometric identities $\sin(2\theta) = 2\sin(\theta)\cos(\theta)$ and $\cos(2\theta) = 1 - 2\sin^2(\theta)$ to simplify the result.

$$\frac{dx}{d\theta} = -5\left(2\sin(\theta)\cos(\theta)\right)\sin(\theta) + 10\left(1 - 2\sin^2(\theta)\right)\cos(\theta)$$

$$\frac{dx}{d\theta} = -30\sin^2(\theta)\cos(\theta) + 10\cos(\theta)$$

$$\frac{dx}{d\theta} = -10\cos(\theta)\left(3\sin^2(\theta) - 1\right)$$

The zeros of $\dfrac{dx}{d\theta}$ in $[0, 2\pi)$ are $\theta = \dfrac{\pi}{2}, \dfrac{3\pi}{2}$, $\alpha, \pi - \alpha, \pi + \alpha$, and $2\pi - \alpha$, where $\alpha = \sin^{-1}\left(\dfrac{1}{\sqrt{3}}\right) \approx 0.615$.

$$\frac{dy}{d\theta} = \frac{d}{d\theta}\left(f(\theta)\sin(\theta)\right) = f(\theta)\cos(\theta) + f'(\theta)\sin(\theta)$$

$$\frac{dy}{d\theta} = \left[5\sin(2\theta)\right] \cdot \cos(\theta) + \frac{d}{d\theta}\left[5\sin(2\theta)\right] \cdot \sin(\theta)$$

$$\frac{dy}{d\theta} = 5\sin(2\theta)\cos(\theta) + 10\cos(2\theta)\sin(\theta)$$

Use the trigonometric identities $\sin(2\theta) = 2\sin(\theta)\cos(\theta)$ and $\cos(2\theta) = 2\cos^2(\theta) - 1$ to simplify the result.

$$\frac{dy}{d\theta} = 5\big(2\sin(\theta)\cos(\theta)\big)\cos(\theta) + 10\big(2\cos^2(\theta) - 1\big)\sin(\theta)$$

$$\frac{dy}{d\theta} = 30\sin(\theta)\cos^2(\theta) - 10\sin(\theta)$$

$$\frac{dy}{d\theta} = 10\sin(\theta)\big(3\cos^2(\theta) - 1\big)$$

The zeros of $\dfrac{dy}{d\theta}$ in $[0, 2\pi)$ are $\theta = 0$, π, β, $\pi - \beta$, $\pi + \beta$, and $2\pi - \beta$, where $\beta = \cos^{-1}\left(\dfrac{1}{\sqrt{3}}\right) \approx 0.955$.

A. The points of horizontal tangency are the points at which $\dfrac{dy}{d\theta} = 0$ and $\dfrac{dx}{d\theta} \neq 0$. The polar coordinates (r, θ) of these points are $(0, 0)$, $(4.714, 0.955$, $(-4.714, 4.097)$, $(4.714, 4.097)$, and $(-4.714, 5.328)$. The point $(0, \pi)$ can be discarded because it is at the same location as the point $(0, 0)$. Note that $(0, 0)$ is also a point of tangency at the pole.

B. The points of vertical tangency are the points at which $\dfrac{dx}{d\theta} = 0$ and $\dfrac{dy}{d\theta} \neq 0$. The polar coordinates (r, θ) of these points are $\left(0, \dfrac{\pi}{2}\right)$, $(4.714, 0.615)$, $(-4.714, 2.526)$, $(4.714, 3.757)$, and $(-4.714, 5.668)$. The point $\left(0, \dfrac{3\pi}{2}\right)$ can be discarded because it is at the same location as the point $\left(0, \dfrac{\pi}{2}\right)$. Note that $\left(0, \dfrac{\pi}{2}\right)$ is also a point of tangency at the pole.

C. The tangents at the pole are all lines whose equations are of the form $\theta = \gamma$, where $f(\gamma) = 0$ and $f'(\gamma) \neq 0$. The zeros of $f(\theta) = 5\sin(2\theta)$ in $[0, 2\pi)$ are $\theta = 0, \dfrac{\pi}{2}, \pi$, and $\dfrac{3\pi}{2}$. The zeros of $f'(\theta) = 10\cos(2\theta)$ in $[0, 2\pi)$ are $\theta = \dfrac{\pi}{4}, \dfrac{3\pi}{4}, \dfrac{5\pi}{4}$, and $\dfrac{7\pi}{4}$. Since none of the zeros of f are also zeros of f', all of the zeros of f are tangents at the pole. However, since $\theta = \pi$ is the same line as $\theta = 0$, we may discard $\theta = \pi$. Likewise, since $\theta = \dfrac{3\pi}{2}$ is the same line as $\theta = \dfrac{\pi}{2}$, we may discard $\theta = \dfrac{3\pi}{2}$. So there are two tangents at the pole: one horizontal ($\theta = 0$) and one vertical $\left(\theta = \dfrac{\pi}{2}\right)$.

3.

A. Use Implicit Differentiation. We will need to use the Product Rule on the first term on the left side.

$$\frac{d}{dx}\left(x^2 y^2 - 2x\right) = \frac{d}{dx}(3)$$

$$\frac{d}{dx}\left(x^2\right) \cdot y^2 + x^2 \cdot \frac{d}{dx}\left(y^2\right) - 2 = 0$$

$$2xy^2 + 2x^2 yy' - 2 = 0$$

$$2x^2 yy' = 2\left(1 - xy^2\right)$$

$$y' = \frac{2\left(1 - xy^2\right)}{2x^2 y}$$

$$y' = \frac{1 - xy^2}{x^2 y}$$

B. The slope m of the tangent line is found by evaluating y' at the point of tangency.

$$m = y'\big|_{(x,y)=(3,1)} = \frac{1 - xy^2}{x^2 y}\Bigg|_{(x,y)=(3,1)} = \frac{1 - (3)(1)^2}{(3)^2 (1)} = -\frac{2}{9}$$

We now find the equation of the tangent line, starting from the point-slope form.

$$y = -\frac{2}{9}(x - 3) + 1$$

$$y = -\frac{2}{9}x + \frac{5}{3}$$

$$\frac{2}{9}x + y - \frac{5}{3} = 0$$

$$2x + 9y - 15 = 0$$

C. Use Implicit Differentiation. We will need to use both the Quotient Rule and the Product Rule.

$$y'' = \frac{dy'}{dx} = \frac{d}{dx}\left(\frac{1 - xy^2}{x^2 y}\right)$$

$$y'' = \frac{\frac{d}{dx}\left(1 - xy^2\right) \cdot x^2 y - \left(1 - xy^2\right) \cdot \frac{d}{dx}\left(x^2 y\right)}{\left(x^2 y\right)^2}$$

$$y'' = \frac{-\left(\frac{d}{dx}(x) \cdot y^2 + x^2 \cdot \frac{d}{dx}(y^2)\right)x^2 y - (1 - xy^2) \cdot \left(\frac{d}{dx}(x^2) \cdot y + x^2 y'\right)}{x^4 y^2}$$

$$y'' = \frac{-\left(y^2 + 2x^2 yy'\right)x^2 y - (1 - xy^2)(2xy + x^2 y')}{x^4 y^2}$$

$$y'' = \frac{x^2 y^3 - 2xy - x^2 y' - x^3 y^2 y'}{x^4 y^2}$$

The instructions indicate that we are to express y'' in terms of x and y, so we must make the substitution $y' = \dfrac{1 - xy^2}{x^2 y}$ and simplify.

$$y'' = \frac{x^2 y^3 - 2xy - x^2 \left(\dfrac{1 - xy^2}{x^2 y}\right) - x^3 y^2 \left(\dfrac{1 - xy^2}{x^2 y}\right)}{x^4 y^2}$$

$$y'' = \frac{x^2 y^3 - 2xy - \dfrac{1 - xy^2}{y} - xy(1 - xy^2)}{x^4 y^2}$$

$$y'' = \frac{2x^2 y^3 - 3xy - \dfrac{1 - xy^2}{y}}{x^4 y^2} \cdot \frac{y}{y}$$

$$y'' = \frac{2x^2 y^4 - 2xy^2 - 1}{x^4 y^3}$$

4.

A. Use the Ratio Test. $u_n = \dfrac{x^n}{n}$ and $u_{n+1} = \dfrac{x^{n+1}}{n+1}$.

$$\lim_{n \to \infty} \left| \frac{u_{n+1}}{u_n} \right| < 1$$

$$\lim_{n \to \infty} \left| \frac{x^{n+1}}{n+1} \cdot \frac{n}{x^n} \right| < 1$$

$$|x| \lim_{n \to \infty} \frac{n}{n+1} < 1$$

$$|x| < 1$$

$$-1 < x < 1$$

We must test the endpoints.

$x = -1$:

$$\sum_{n=1}^{\infty} \frac{1^n}{n} = \sum_{n=1}^{\infty} \frac{1}{n}$$

This is the harmonic series, which diverges.

$x = 1$:

$$\sum_{n=1}^{\infty} \frac{(-1)^n}{n}$$

This is the alternating harmonic series, which converges.

The interval of convergence is given by $-1 < x \le 1$ or $(-1, 1]$.

B. First we compute $f'(x)$.

$$f'(x) = \frac{d}{dx} \sum_{n=1}^{\infty} \frac{x^n}{n} = \sum_{n=1}^{\infty} \frac{nx^{n-1}}{n} = \sum_{n=1}^{\infty} x^{n-1} = \sum_{n=0}^{\infty} x^n$$

$f'(x)$ has the same *radius* of convergence as $f(x)$, but it may have different behavior at the endpoints. We must test them.

$x = -1$:

$$\sum_{n=0}^{\infty} (-1)^n$$

Since $\lim_{n \to \infty} (-1)^n \ne 0$, this series diverges by the nth Term Test.

$x = 1$:

$$\sum_{n=0}^{\infty} 1^n = \sum_{n=0}^{1} 1$$

Since $\lim_{n \to \infty} 1 \ne 0$, this series diverges by the nth Term Test.

The interval of convergence is given by $-1 < x < 1$ or $(-1, 1)$.

C. First we compute $\int f(x)\, dx$.

$$\int f(x)\, dx = \int \sum_{n=1}^{\infty} \frac{x^n}{n}\, dx = \sum_{n=1}^{\infty} \frac{x^{n+1}}{n(n+1)} + C$$

$\int f(x)\, dx$ has the same *radius* of convergence as $f(x)$, but it may have different behavior at the endpoints. We must test them. The constant of integration has no effect on the interval of convergence, so we needn't consider it.

$x = -1$:

$$\sum_{n=1}^{\infty} \frac{(-1)^{n+1}}{n(n+1)}$$

This series converges by the Alternating Series Test.

$x = 1$:

$$\sum_{n=1}^{\infty} \frac{1^{n+1}}{n(n+1)} = \sum_{n=1}^{\infty} \frac{1}{n(n+1)}$$

This series converges by direct comparison with the converging p-series $\sum_{n=1}^{\infty} \frac{1}{n^2}$.

The interval of convergence is then given by $-1 \le x \le 1$ or $[-1, 1]$.

5.

A. Integrate by parts. Let $u = t$, and so $dv = \sin(t)\,dt$. Use these to calculate du and v.

$u = t \Rightarrow du = dt$

$dv = \sin(t)\,dt \Rightarrow v = \int \sin(t)\,dt = -\cos(t)$

Since this is a definite integral, we needn't worry about the constant of integration.

$$F(\pi) = -t\cos(t)\Big|_0^\pi - \int_0^\pi \left(-\cos(t)\right)dt = -t\cos(t)\Big|_0^\pi + \int_0^\pi \cos(t)\,dt$$

$$F(\pi) = -t\cos(t)\Big|_0^\pi + \sin(t)\Big|_0^\pi = \left(-\pi\cos(\pi) - 0\right) + \left(\sin(\pi) - 0\right)$$

$$F(\pi) = \pi$$

$$F(2\pi) = -t\cos(t)\Big|_0^{2\pi} - \int_0^{2\pi} \left(-\cos(t)\right)dt = -t\cos(t)\Big|_0^{2\pi} + \int_0^{2\pi} \cos(t)\,dt$$

$$F(2\pi) = -t\cos(t)\Big|_0^{2\pi} + \sin(t)\Big|_0^{2\pi} = \left(-2\pi\cos(2\pi) - 0\right) + \left(\sin(2\pi) - 0\right)$$

$$F(2\pi) = -2\pi$$

B. We compute $F'(x)$ using the Second Fundamental Theorem of Calculus.

$$F'(x) = \frac{d}{dx}\int_0^x t\sin(t)\,dt = x\sin(x)$$

We then find the critical numbers of F by solving $F'(x) = 0$ for x.

$$F'(x) = x\sin(x) = 0 \Rightarrow x = 0, \pi, 2\pi$$

C. Compute $F''(x)$ in order to apply the Second Derivative Test. We will need to use the Product Rule.

$$F''(x) = \frac{d}{dx}\left(x\sin(x)\right) = \frac{d}{dx}(x)\cdot\sin(x) + x\cdot\frac{d}{dx}\left(\sin(x)\right)$$

$$F''(x) = \sin(x) + x\cos(x)$$

Now test each critical number.

$x = 0$:

$F''(0) = \sin(0) + 0\cos(0) = 0$, so the Second Derivative Test fails. Apply the First Derivative Test using $x = -\dfrac{\pi}{2}$ and $x = \dfrac{\pi}{2}$ as test points.

$x = -\dfrac{\pi}{2}$:

$$F'\left(-\frac{\pi}{2}\right) = -\frac{\pi}{2}\sin\left(-\frac{\pi}{2}\right) = \frac{\pi}{2} > 0$$

$x = \dfrac{\pi}{2}$:

$$F'\left(\frac{\pi}{2}\right) = \frac{\pi}{2}\sin\left(\frac{\pi}{2}\right) = \frac{\pi}{2} > 0$$

Since F' does not change sign at $x = 0$, this critical number does not correspond to a relative extremum.

$x = \pi$:

$F''(\pi) = \cos(\pi) + \pi\sin(\pi) = -1 < 0 \Rightarrow x = \pi$ corresponds to a relative maximum. We found in part (a) that $F(\pi) = \pi$, so F has a relative maximum at (π, π).

$x = 2\pi$:

$F''(2\pi) = \cos(2\pi) + 2\pi\sin(2\pi) = -1 > 0 \Rightarrow x = 2\pi$ corresponds to a relative minimum. We found in part (a) that $F(2\pi) = -2\pi$, so F has a relative maximum at $(2\pi, -2\pi)$.

6.

A. We will need to use the Product Rule for *three* factors.

$$f'(x) = \frac{d}{dx}\big((x-a)(x-b)(x-c)\big)$$

$$f'(x) = \frac{d}{dx}(x-a)\cdot(x-b)(x-c) + (x-a)\cdot\frac{d}{dx}(x-b)\cdot(x-c) + (x-a)(x-b)\cdot\frac{d}{dx}(x-c)$$

$$f'(x) = (x-b)(x-c) + (x-a)(x-c) + (x-a)(x-b)$$

B. Again, we will need the Product Rule.

$$f''(x) = \frac{d}{dx}\big((x-b)(x-c) + (x-a)(x-c) + (x-a)(x-b)\big)$$

$$f''(x) = \left(\frac{d}{dx}(x-b)\cdot(x-c) + (x-b)\cdot\frac{d}{dx}(x-c)\right)$$
$$+ \left(\frac{d}{dx}(x-a)\cdot(x-c) + (x-a)\cdot\frac{d}{dx}(x-c)\right)$$
$$+ \left(\frac{d}{dx}(x-a)\cdot(x-b) + (x-a)\cdot\frac{d}{dx}(x-b)\right)$$

$$f''(x) = 2(x-a) + 2(x-b) + 2(x-c)$$

$$f''(x) = 6x - 2(a+b+c)$$

C. We solve the equation $f''(x) = 0$ for x.

$$f''(x) = 6x - 2(a+b+c) = 0$$

$$6x = 2(a+b+c)$$

$$x = \frac{a+b+c}{3}$$

The solution is indeed the average of a, b, and c. However, in order to show that this corresponds to an inflection point, we must show that f'' changes sign at this value of x. We will look at f'' on either side of $x = \frac{a+b+c}{3}$.

$x = \dfrac{a+b+c}{3} - 1:$

$$f''\left(\frac{a+b+c}{3} - 1\right) = 6\left(\frac{a+b+c}{3} - 1\right) - 2(a+b+c)$$

$$f''\left(\frac{a+b+c}{3} - 1\right) = 2(a+b+c) - 6 + 2(a+b+c) = -6 < 0$$

$x = \dfrac{a+b+c}{3} + 1:$

$$f''\left(\frac{a+b+c}{3} + 1\right) = 6\left(\frac{a+b+c}{3} + 1\right) - 2(a+b+c)$$

$$f''\left(\frac{a+b+c}{3} + 1\right) = 2(a+b+c) + 6 + 2(a+b+c) = 6 > 0$$

Since the sign of f'' changes at $x = \dfrac{a+b+c}{3}$, the graph of f does indeed have a point of inflection corresponding to that value of x.

CPSIA information can be obtained
at www.ICGtesting.com
Printed in the USA
BVOW04s0445120817

491837BV00022B/343/P